D0151231

ROBOTICS

INTRODUCTION, PROGRAMMING, AND PROJECTS

ROBOTICS

INTRODUCTION, PROGRAMMING, AND PROJECTS

Second Edition

JAMES L. FULLER

Prentice Hall
Upper Saddle River, New Jersey   Columbus, Ohio

Library of Congress Cataloging-in-Publication Data
Fuller, James L.
Robotics : introduction, programming, and projects / James L.
Fuller. — 2nd ed.
p. cm.
Includes bibliographical references and index.
ISBN 0-13-095543-4 (case)
1. Robotics. I. Title.
TJ211.F85 1999
670.42'72—dc21 98-4165
CIP

Cover art: © Uniphoto
Editor: Stephen Helba
Production Editor: Louise N. Sette
Design Coordinator: Karrie M. Converse
Cover Designer: Tom Mack
Production Manager: Deidra M. Schwartz
Marketing Manager: Frank Mortimer, Jr.

This book was set in Sabon by The Clarinda Company and was printed and bound by R. R. Donnelley & Sons Company. The cover was printed by Phoenix Color Corp.

Printed in the United States of America

10 9 8 7 6 5 4 3 2 1

ISBN: 0-13-095543-4

Prentice-Hall International (UK) Limited, *London*
Prentice-Hall of Australia Pty. Limited, *Sydney*
Prentice-Hall Canada, Inc., *Toronto*
Prentice-Hall Hispanoamericana, S.A., *Mexico*
Prentice-Hall of India Private Limited, *New Delhi*
Prentice-Hall of Japan, Inc., *Tokyo*
Simon & Schuster Asia Pte. Ltd., *Singapore*
Editora Prentice-Hall do Brasil, Ltda., *Rio de Janeiro*

PREFACE

This book is designed for use in an introductory robotics course for community or four-year colleges. While the text is aimed at engineering and technology students, it can be understood by all college students. To gain maximum benefit from the book, the student should already have taken courses in algebra, trigonometry, introductory computer programming, general physics and mechanics, and elementary electronics. However, sufficient background information in most subject areas is provided to enable anyone with an interest in robotics to learn about this subject. Part Two, the section on detailed programming of robots, may be skimmed or skipped if close coverage of this area is not needed. Engineers or company managers planning to install robots can also profit from reading this book.

Robots have begun the transition from manufacturing to office work. Many robotic applications for the business office are on the horizon.

The main commercial use of robots today is in the manufacturing industry. Several personal hobbyist-type robots are now available for home use, but they lack any practical everyday use. Still, it should not be too long before we see domestic robots that will do simple practical tasks. At present, many jobs in this country directly or indirectly depend on computers. In the near future, many jobs may similarly depend on robots.

This book is divided into three parts. Part One, Chapters 1 through 13, provides an introduction to robotics. Since an engineer dealing with robotics must be a generalist, this section should be read by everyone. This material may be all that can be covered in a one-quarter or one-semester course.

Chapter 1 answers the elementary questions, What is robotics? What is a robot? and What is an industrial robot? It explores the interrelations of automation and robotics and concludes with an inquiry into several areas of nonindustrial robots. Chapter 2 discusses the history of robotics. Chapter 3 provides detailed definitions of industrial robot and manipulator configurations. Chapter 4 provides detailed information on the control unit, the power supply, and vehicles. Chapter 5 discusses grippers and other end-of-arm tooling for robots. Chapter 6 deals with robot sensors and their value in expanding the potential applications for robots.

The many applications for the industrial robot—both present and future—are the subject of Chapter 7. These applications are expanded to include some nonindustrial robots, as well as some industrial robots in nonindustrial situations.

Chapter 8 discusses robot maintenance—an important area if the robot is to continue to work at its best. Chapter 9 deals with robots and human safety. Chapter 10 introduces the field of artificial intelligence and explains why improvements in robotic intelligence depend on advances in artificial intelligence. Chapter 11 reviews several more ways of classifying the industrial robot. Chapter 12 discusses how to justify the use of robots. After all, the engineer or manager generally must prove on paper that a robot will be successful before the robot gets the chance to prove itself on the factory floor. Some educated guesses about the future of the robot are offered in Chapter 13.

Part Two of the book, Chapters 14 through 21, investigates the intelligence of the robot and reviews robot sensors in increased detail. While this section focuses on software, it also devotes considerable attention to the electronics (computer hardware) of the robot. The connection of the two subjects is logical, since what is done in software or programs one year might be done in the electronic circuits or hardware the next.

Chapter 14 deals with the robot's operating system, where all the robot's abilities are stored. As the robot becomes more intelligent, the operating system necessarily becomes more complex. Chapter 15 looks at applications programming for the robot. Such programming instructs the robot about the tasks it must perform. Chapter 16 discusses some of the myriad details involved in carrying out a successful robotics project.

The remaining chapters of Part Two inquire into sensory systems used by robots. Chapters 17, 18, and 19 look at interfacing and programming vision systems, interfacing tactile systems, and interfacing proximity sensors, respectively. Chapter 20 looks at speech output (speech synthesis) for robots, and deals with the difficult area of speech understanding.

Part Three, Chapter 21, deals with some simple robotics projects. Many senior projects in robotics have bogged down because their complexity was underestimated. It is best to start with a very simple project and add to it later—especially when dealing with unfamiliar areas of technology.

Appendixes offer background on mechanisms and on fundamentals of electricity, electronics, and computers. The book also has a glossary, a bibliography, and lists of robot manufacturers and robotics organizations.

Each chapter of the book contains the following elements:

1. A list of objectives
2. An overview of the chapter
3. Key terms in the chapter
4. A summary
5. A list of formulas (if any) that have appeared in the chapter
6. Review questions

Acknowledgments

I thank Robert Armstrong and Professor Lew Jones for their help in preparing my manuscript, as well as all my robotics class students for their help. Finally, I appreciate the suggestions of the following reviewers: Philip N. Miller, Community College of Rhode Island; Dr. Lee Rosenthal, Fairleigh Dickinson University; and Pascal Zanzano, Niagara County Community College.

CONTENTS

PART 1

AN INTRODUCTION TO ROBOTICS

This section provides a broad overview of robotics and introduces robotics terminology. It covers the following areas: definition of a robot, the history of robotics, parts that make up a robot, robot grippers, robot sensors, applications of robots, robot maintenance, robot safety, robotics and artificial intelligence, classification of robots, justifying the use of robots, and the future of robots.

1

WHAT IS A ROBOT?

OVERVIEW

Everyone has heard of robots, but most people are not really sure what a robot is. Are there other types of robots besides industrial robots? If so, what are they? How are automation and robots related? How are computers used in connection with robots? What is robotics? These are a few of the questions that are answered in this chapter.

OBJECTIVES

When you have completed this chapter, you should be familiar with:

- The meaning of the term *robotics*
- The meaning of the term *robot*
- Industrial automation and industrial robotics
- Nonindustrial robots
- How the hobbyist contributes to robotics
- The criteria used to decide whether a job should be done by humans, by robots, or by automation

KEY TERMS

AGV (automated guided vehicle)
ASRS (automated storage and retrieval system)
Automation
CAD (computer-aided design)
CAE (computer-aided engineering)
CAM (computer-aided manufacturing)
CAR (computer-aided robotics)
Domestic robot
Educational robot
Flexible automation
Four Ds of Robotics
Hard automation
Hobbyist robot
Industrial robot
LAN (local area network)
Laws of robotics

MAP (manufacturing automation protocol)
Military robot
Numerical control
PBX (private-branch exchange)
PC (personal computer)
PC (programmable controller)
PLC (programmable logic controller)
Remote control
Robot
Robotics
Show robot
Teleoperator
TOP (technical and office protocol)
WAN (wide-area network)

ROBOTS AND ROBOTICS

The terms *robot* and *robotics* are of recent origin, but the ideas underlying them predate recorded history. The word **robot** was first used by a Czechoslovakian dramatist, Karel Capek, in his 1921 play "Rossum's Universal Robots." Capek's robots were designed to be perfect and tireless workers who performed manual labor for human beings. Other meanings for the word *robot* will be covered later in this chapter, but the idea of a perfect and tireless worker is a good starting point.

Isaac Asimov in his science fiction stories about robots in the 1940s coined the term **robotics** as the science or study of robots. The *Merriam Webster Dictionary,* 1998, defines *robotics* as "technology dealing with the design, construction, and operation of robots." In Europe, *robotics* is defined as "the science of robotology," and *robotology* is defined as "the means by which robot machines are put together and made to work."

Even though the word is relatively new, work in robotics research has been going on for hundreds of years.

Many people think of robotics as a single area of technology, but in fact robotics encompasses such diverse areas of technology as mechanical, electrical, and electronic systems; computer hardware; and computer software.

INDUSTRIAL ROBOTS

Like Karel Capek's robots, the **industrial robot** is designed to be a perfect and tireless worker; it is intended to help human workers, not replace them.

Many people misunderstand what an industrial robot is. They confuse the terms *remote-controlled, automation,* and *numerical-controlled* with the term *industrial robot.* This is due in part to false impressions created by science fiction and in part to the way robots actually developed. We will take up these other terms shortly, but first we need to settle on a definition of *industrial robot.*

Even within the industrial world, this term has several definitions. For instance, the Japanese have defined a robot as an all-purpose machine equipped with a memory device and a terminal, and capable of rotation and of replacing human labor by automatic performance of movement. This definition could include some hard-wired automation devices such as bottling machines or even a modern microprocessor-controlled washing machine, but it does exclude human beings. The Robot Institute of America defines a robot as a reprogrammable, multifunctional manipulator designed to move materials, parts, tools, or specialized devices, through variable programmed motions, for the performance of a variety of tasks. This definition does not exclude human beings. This book will use the Robot Institute of America's definition.

Most people today would not recognize an industrial robot if they saw one. They tend to mistake industrial robots for just another machine tool. In fact, however, an industrial robot is a specialized machine tool—one that is consistent, unchanging, unthinking, and untiring. Such robots are best suited for repetitive, unskilled or semiskilled, monotonous, and burdensome tasks. They are also useful for jobs that would be hazardous for a person to perform. Robots are just now starting to do some jobs that require a level of precision that a person could not match.

Remote control refers to a form of human operation in which the human is not physically present at the site of operations. Still, the human remains the controller or brain of the operation. **Automation** involves using specialized machines to do a specific operation. A much more detailed description of automation appears later in this chapter. **Numerical control** is one type of specialized machine operation used in automation. It relies on punched paper or plastic tapes, tape cassettes, or floppy disks to control the manufacturing of some part.

THE STUDY OF ROBOTS

This book describes seven distinct types of robots: industrial robots, military robots, show or promotional robots, educational robots, medical robots, domestic or personal robots, and hobbyist robots. The emphasis, however, will be placed on industrial robots.

Isaac Asimov envisioned the robot as a helper of humankind. Toward this end, Asmiov set forth three basic "laws" for robots. These "laws" would better be defined as "rules of robotics," much like the "Golden Rule" for people.

1. A robot must not harm a human being, nor through inaction allow one to come to harm.
2. A robot must always obey human beings, unless that is in conflict with the first "law."
3. A robot must protect itself from harm, unless that is in conflict with the first two "laws."

Gerald Norman, an Oregon Institute of Technology professor of manufacturing engineering technology, has suggested that a law formulated by Skoles be adopted as the fourth "law" or "rule of robotics":

4. A robot may take a human being's job. But it may not leave that person jobless!

Attempts are being made to adhere to these laws of robotics, but there is no automatic way to implement them. For instance, the military robot, by its very nature, is likely to be designed with the intention of breaking these laws.

Most industrial robots of today are designed to replace human workers in boring, unpleasant, hazardous, or too precise jobs. The robot designs used are often patterned after human functions. But robots do not have to imitate human ways of doing and thinking in every way, any more than airplanes have to copy a bird's method of flying right down to the flapping of its wings. Sometimes a machine can improve on human performance by using methods that are not available to humans. For example, robots can make use of infrared and ultrasonic sensors that human workers do not possess. A robot might also have its hearing or vision located down on its wrist or fingers—something a person cannot replicate. Robot hands can be equipped with built-in devices that use electromagnetic force, vacuum cups, or even inflatable bladders to pick up objects, whereas the human hand does not have these capabilities. Moreover, a robot's hands can be designed to handle very hot or very cold objects that the human hand cannot handle safely.

Isaac Asimov's books on robots inspired Joseph F. Engelberger in the 1950s to try to design a working robot.[1] Engelberger, the founder and first president of the Unimation Robotics Company, is now a leading authority on robots and wrote the book *Robots in Practice.*[2]

Another technique for designing robots is to study human beings. Analyzing how humans use their arms, hands, legs, and other moving parts is helpful in developing moving parts for robots, not to mention in designing replacement parts for human body parts that have been lost or rendered useless by injury or disease. The study of human hearing, sight, touch, balance, and other senses is helpful in making sensors for robots and in teaching the robot how to interpret sensory information. The study of how humans remember and interpret information is helpful in adapting artificial intelligence to use in robots.

INDUSTRIAL AUTOMATION AND ROBOTS

From the moment people started doing work, they began trying to find methods of automating the work. Progress in such methods can be seen in the use of automated machines, computer-aided design, computer-aided manufacturing, computer-aided robotics, and industrial robots. While industrial robots and automated machines are usually treated as two separate topics, most industrial robots work in cooperation with other automated machines. That is, industrial robots usually do not work alone but are used with other machines.

Persons working with industrial robots must be familiar with techniques for communicating with other machines and devices on the production line. These may include the use of *LAN, MAP,* and *TOP.*

LAN, short for "local area network," is used to interconnect computers and terminals. The LAN system has also been called *data highways,* since it is used to move electronic data or information between the various devices on the network. Many offices and school computer laboratories make use of LAN systems.

[1]See *I, Robot* (Greenwich, Conn.: Fawcett Crest, 1950), a collection of novels by Asimov about robots.
[2]American Management Association, 1980.

MAP stands for "manufacturing automation protocol"; it is a communications standard developed for General Motors. MAP is a fault-tolerant industrial protocol developed for multivendor data communication in a factory environment; it continues to be studied and used worldwide.

TOP, an acronym for "technical and office protocol," was developed for use in office automation by Boeing Computer Services. It exchanges information using electronic mail and allows the interconnection of multiple offices, each having a LAN, and connects them to wide-area networks (**WAN**) and digital private-branch exchanges (**PBX**), for long-distance exchange of information.

Industrial robots may also communicate with special-purpose computers (once called *PC,* for "programmable controller," but more recently named *PLC,* for "programmable logic controller," to avoid confusion with another kind of **PC,** the personal computer) that are used to control many modern production lines.

Internet, the world's largest WAN, is now connecting together many LAN, TOP, and MAP sites.

Automated Machines

Automated machines can be subdivided into two classes: *hard automation* machines and *flexible automation* machines. **Hard automation** deals with specialized machines designed for a specific operation or a narrow range of operations. **Flexible automation** deals with relatively general-purpose machines, such as the industrial robot.

Automated and numerical-controlled machines are generally considered hard automation machines, because their functions are built-in and cannot be changed to any great extent. Thus, for example, an automated canning machine can be adjusted to accommodate different can sizes and different labels, but it cannot be changed into a photographic film development and printing machine. Automated machines may also be known as numerical-controlled (NC) machine tools, dedicated computer numerical-controlled (CNC) machine tools, or direct computer numerical-controlled (DNC) machine tools. All of these types of automated machines are seen on production lines. Aircraft frame tube-bending machines, canning machines, bottling machines, photographic film development and printing machines, newspaper press machines, and even photocopying machines are very efficient at doing their particular jobs.

Early automated machines had all their functions built into their mechanical gears, cams, and levers. A wind-up clock is thus an old-fashioned automated machine.

One of the first automated programmable industrial machines was the automatic loom, invented by Joseph Marie Jacquard in 1801. It used cards with holes punched in them to control the pattern being woven on the loom. Jacquard gave a vivid demonstration of how powerful his invention was by using 10,000 punched cards to program the loom to weave a portrait of himself in black and white silk.

The advent of hard-wired, electronically controlled circuits and devices had a tremendous impact on automated machines. Electronic automated machines relied on the same digital computer circuits used by general-purpose electronic computers, but their programs were built into the electronic circuits at the factory and could not be changed. These machines often received their setup instructions from paper tape or computer punch cards, allowing them to change automatically from one size of product to another. Automated aircraft frame tube-bending machines could turn out 100 parts of one shape and then switch automatically to turn out parts of another shape, without human intervention. Similarly, an automatic wire-processing machine can measure, cut, strip, lug, stamp, label, and package 1,000 identical wires and then switch to another size and quantity of wire. These automated machines work much faster and more accurately than human beings can. Still, the tube-bending machine cannot be made to process wire, nor the wire-processing machine to bend tubing.

Computer-Aided Design (CAD)

Computer-aided design (CAD) and **computer-aided engineering** (CAE) are involved in the product design and engineering phases of manufacturing. These areas include computer-aided drafting, computer-aided analytical testing, group technology, and computer-aided process planning.

Computer-aided drafting enables an engineer to enter information into a computer by using a keyboard and a mouse and then to have the computer produce the engineering drawings. The engineer can then add text of different sizes to the drawing. This increases the engineer's control over the drawing process and yields faster results. In addition, the engineer can use the mouse later to alter a drawing quickly and easily. The drawings needed in designing integrated circuits have become so complex that they can no longer be done by hand and have any chance of getting into production before the competition's circuits do.

Computer-aided analytical testing lets an engineer give a computer information on how to draw a model of a part and then test that part. In many systems, the drawing is produced in three dimensions and can be rotated on the screen by the engineer. The engineer can then propose changes to the part and have the computer do testing on the new configuration. Some of the newest CAD workstations have a digitizer that resembles a robot arm and can be used to trace the shape of an existing three-dimensional part and thus enter its shape into the computer. This saves much time.

CAD supports group technology for a family of related parts. The engineer enters the parameters of the basic part into the computer and then uses the computer to create the rest of the family of parts by specifying to the computer the ways in which each related part differs from the basic part.

Computer-aided process planning helps an engineer determine the appropriate production process for a product. It helps show the difference in cost of different production processes that could be used to make a part. This aids the engineer in selecting the most desirable process from among a group of alternative processes. If the CAD system is connected to or includes computer-aided manufacturing, it will be easier for the engineer to determine the order and cost of different processing techniques for producing the needed part.

Where CAD workstations are used, engineers' productivity shows a threefold to sixfold increase, because repeated shapes do not have to be redrawn, savings on modifications can be made quickly and inexpensively, and the engineer can work alone rather than having to turn the sketch over to a draftsperson. Computer modeling, or simulation, allows more parts to be tested before the final choice must be made, and there is no need for model-building. Thus CAD has enabled companies to reduce the size of their engineering staffs without impairing productivity.

Computer-Aided Manufacturing (CAM)

Computer-aided manufacturing (CAM) employs computers in many different aspects of the manufacturing or part-machining process. One recent growth in CAM has been the marriage of numerical-control (NC) machinery and on-line computers. One CAM application is the automated bill of materials, a means of tracking a product from its highest inventory level to its lowest inventory level. Other applications include automatic storage and retrieval systems with computer-operated part pickers and stockers, automatic reorder point inventory control, routing or process sheets, master production scheduling, and material requirements planning. The computerization of these reports and processes allows the information to be kept up to date and makes cost accounting virtually automatic. Computer-controlled inventory methods can greatly reduce the stock of parts that must be kept on hand, thus lowering inventory costs.

CAM applications can also enlist the services of computer-controlled automated machines of all types, including robots.

If the scheduling computer can communicate with the computer-controlled automated machine tools, the automated machines can be run on the basis of the automated schedule. This begins to look like a fully automated factory.

Computer-Aided Robotics (CAR)

The field of **computer-aided robotics** (CAR) uses electronic computers in the design, manufacture, installation, and programming of robots. Computer-aided robotics is a new entry in the arena of computer-aided whatevers. In fact, the idea of CAR first occurred to me in April 1985, while I was reading an article called "Anatomy of an Off-line Programming System" by John J. Craig.[3] CAR is an offshoot of computer-aided design (CAD) and computer-aided manufacturing (CAM). The field of CAR will come into its own as the number of robots being manufactured and used increases. That is, CAR will emerge from CAD and CAM just as the field of computers emerged from the field of electronics and just as the field of electronics emerged from the field of electricity.

In their present state, CAD and CAM allow us to generate simulations of robots, robot parts, and even complete robot systems. Such simulations help in the design of robots, in the training of robot operators, and in the actual programming

[3]*Robotics Today* (February 1985): 45–47.

of robots off-line. Since off-line programming is done without the physical use of a robot, this technique can be used to determine the feasibility of having a robot do a specific task—before purchasing the robot.

A CAR system already exists under the tradename ROBOCAM.[4] It can be run on the Apollo family of CAD/CAM workstations. Directions are entered into the CAR system either through a high-level computer language or through a computer mouse. This system allows the user to construct simulations of robots and tasks, and then it carries out the simulated task with an interactive arm-independent language. The system can determine whether a proposed placement of a robot in its workspace will allow it to reach all desired points and whether the robot has enough axes of movement to orient a workpiece or tool properly.

The SILMA Corporation has developed a CAR system called CimStation. It can be run on an Apollo computer or on an IBM PC AT computer. CimStation can perform the following functions: workcell layout, programming workcell devices, multiple process simulation, workcell optimization, output generation, program translation and generation, and adaptation to changes.

Industrial Robots

The industrial robot is intended to serve as a general-purpose unskilled or semi-skilled laborer. Typically it does not resemble a human worker physically, and it may not do the job the same way a human worker would. Nor does the industrial robot look like most hobbyist robots. For one thing, most industrial robots are stationary, while most hobbyist robots move about. The industrial robot generally has a single manipulator somewhat similar to a human arm and hand. Figure 1–1(a) shows a rectangular-coordinates industrial robot, and Figure 1–1(b) shows a polar-coordinates industrial robot. (These names will be explained in Chapter 3.) These robots look more like machine tools than like human beings. In fact, the general public would not think of these machines as robots.

In most applications, robots do not work as fast as humans, but they are more reliable than humans in some applications. They are neither as fast nor as efficient as special-purpose automated machine tools. However, industrial robots are easily retrained or reprogrammed to perform an array of different tasks, whereas an automated special-purpose machine tool can work on only a very limited class of tasks.

The first machine tools to resemble the modern industrial robot were probably the automatic spray-painting machines designed in the late 1930s by Pollard and Roseland. The first industrial robots showed up in American manufacturing plants in the late 1950s. In 1970 only about 200 industrial robots existed in the United States. By 1995 the number had grown to 66,000, and by 2000 the number of industrial robots is expected to reach 105,000.

The industrial robot is intended to take over work currently done by humans in areas that are dull, dirty, dangerous, or difficult. A hazardous atmosphere in a

[4]Ibid.

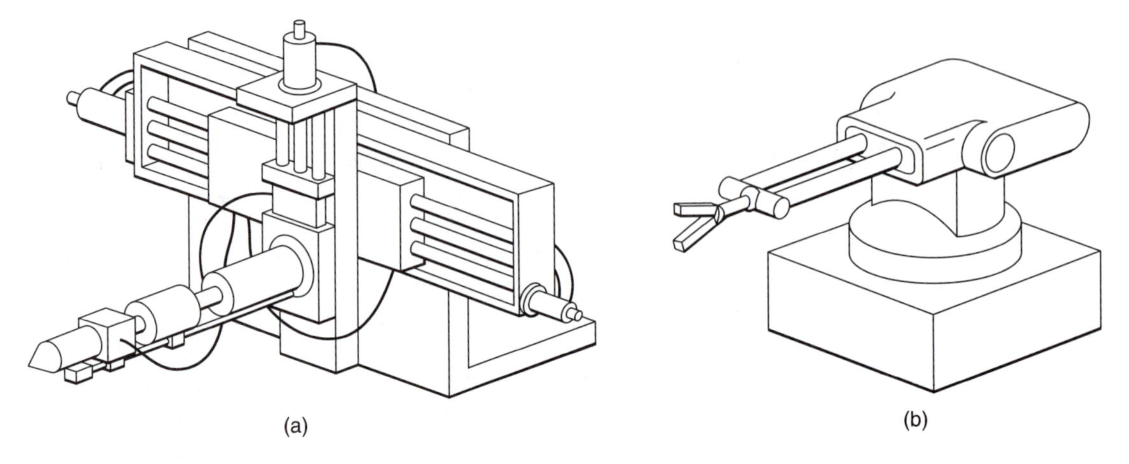

Figure 1-1 Industrial Robots.
(a) A rectangular-coordinates robot moves along three linear axes. It looks very much like a numerical-controlled machine tool. (b) A polar-coordinates robot rotates along two of its three axes. It looks less like a machine tool but very unlike what the average person would call a robot.

workplace requires expensive protective devices for human workers, and even these may not completely protect the human from harm. The use of robots to work with radioactive materials, toxic chemicals, cotton lint, coal dust, and asbestos fibers could eliminate human exposure and save lives. Other dangerous areas include work in outer space, undersea, or in deep mines. Humans usually do not perform well in jobs they find dull or tiresome; industrial robots do not become tired or bored, so their efficiency does not diminish.

Another type of robot used in industry is the *automated guided vehicle (AGV)*. **AGVs** (which may also be called Robovans) are unmanned, computer-controlled material transport vehicles. Although they are designed to follow painted lines on the floor or buried cables, they require control or guidance from a master computer or programmable controller telling them where to go. The AGV is a special-purpose robot. That is, it is designed exclusively to transport parts, and it cannot be reprogrammed to perform unrelated tasks.

Most *automated storage and retrieval systems (ASRS)* use AGVs. **ASRSs** are automated warehouses that use computers and robots to store and retrieve parts. ASRSs can be several stories high.

CHOOSING AMONG HUMANS, ROBOTS, AND AUTOMATION

When should a person be used to do a task? When should an industrial robot be used? When should automation be used? The answers to these questions are not simple and straightforward. However, some rules of thumb can help suggest significant factors to keep in mind.

The first rule to consider is known as the **Four Ds of Robotics.** Is the task *d*irty, *d*ull, *d*angerous, or *d*ifficult? If so, a human will probably not be able to do the job efficiently for hours on end. Therefore, the job may be appropriate for automation or robotic labor.

The second rule recalls the fourth law of robotics: A robot may not leave a human jobless. Robotics and automation must serve to make our lives more enjoyable, not to eliminate jobs for people.

A third rule involves asking whether you can find people who are willing to do the job. If not, the job is a candidate for automation or robotics. Indeed, this should be a primary reason for the growth of automation and robotics.

A fourth rule of thumb is that the use of robots or automation must make short-term and long-term economic sense. As a general starting point, consider the following. A task that has to be done only once or a few times and is not dangerous probably is best done by a human. After all, the human is the most flexible of all machines. A task that has to be done a few hundred to a few hundred thousand times, however, is probably best done by a flexible automated machine such as an industrial robot. And a task that has to be done 1 million times or more is probably best handled by building a special-purpose hard automated machine to do it.

NONINDUSTRIAL ROBOTS

While the focus of this book is the industrial robot, you should be aware that there are other types of robots: military robots, show or promotional robots, educational robots, medical robots, domestic or personal robots, and hobbyist robots. Each of these nonindustrial types is described in one of the subsections that follow.

Military Robots

Military engineers consider any machine that can be operated without a person being present a robot. This definition encompasses most remote-controlled devices, since almost any such device can be operated by a computer in place of a human controller. Remote-control tanks, radio-controlled airplanes, and remote-control devices for detonating bombs over enemy lines are examples of **military robots.**

Robot bombs or "smart bombs" that can be steered by remote television control after they leave an airplane represent another type of military robot. They increase the accuracy of the bombing while lowering the risk to the airplane and its crew. The French Exocet missile, for example, can be launched from a plane 25 miles away from a target ship. A low-flying plane at that distance is unlikely even to be detected by the ship. The Exocet and similar missiles are military robots by anyone's definition; however, these robots are extremely special-purpose.

The military also uses remote-controlled robots to defuse or detonate bombs and shells. Bomb disposal teams have always suffered a high fatality rate. Thus fire and police departments are eager to use remote-controlled robots in certain dangerous situations.

Many functions in a modern military aircraft have become too complicated for the pilot to handle. Although the pilot still initiates most operations, computers and specialized robots carry out the details. One of the earliest robot devices used on airplanes was the autopilot (or robot pilot, as it is known in Europe). Today, some pilotless aircraft known as drones are used for target practice and reconnaissance.

Pilots and other military personnel use flight simulators for training. These simulators are in effect robots that control people. Modern flight simulators have "working" instruments and realistic visual and motion capabilities. One of the first robotic flight simulators in use was the F-4C weapons system training set (see Figure 1–2). Built in 1965 by Link Trainer, it had three axes of roll and five axes of pitch, and used 45 gallons of hydraulic fluid. A more modern trainer requires a clearance of 38 feet just to handle its vertical movement.

Military interest in atomic energy led to the development of mechanical remote-control devices known as **teleoperators** to handle dangerous radioactive materials. These manipulators lacked any sensory feedback other than that provided by the operator's vision. It soon became obvious that some type of pressure or feeling feedback was needed for the device to pick up and put down objects reliably. As a result, attempts were made to build mechanical feeling feedback into the device.

A more modern remote-controlled device known as the Mobot is made by Hughes Aircraft Company for the Atomic Energy Commission. This successor to the teleoperator has two arms, remote vision (through the use of television), remote hearing (through the use of microphones at the wrist), and "soft" hands. Because the soft hands give some pressure or feeling feedback information to the operator, they overcome one of the problems of the early teleoperators.

A direct predecessor of the robot manipulator was the prosthetic (artificial) arm. The earliest prostheses were probably the peg leg and the hook hand. The first artificial hand that looked like a human hand may have been the iron hand made in 1509 for Goetz von Berlichinger, a German knight. The development of artificial manipulators for humans has provided valuable insights into designing robot manipulators.

The military has developed other sensor devices that are useful to both humans and robots. Radar, radio detection, and range finders are types of artificial vision useful for missiles and spy satellites. Infrared vision, digital television vision, sonar, and ultrasonic hearing are other sensor areas pioneered by the military.

The development of the electronic computer—the brain for the sophisticated robots—received much of its early funding and research from military and space programs. Without the development of the integrated circuit, which allows thousands of transistor circuits to be placed inside a tiny package, the microcomputer would not be possible and the intelligent robot would not have a brain.

The space program has developed some of the most sophisticated special-purpose robots ever made. The unmanned lunar lander vehicle, and the *Viking* and the *Voyager* vehicles have performed fantastic feats of exploration.

The *Viking* landers on Mars were equipped with two cameras with stereoscopic vision capabilities. They had a low-frequency seismometer for an ear and a

Figure 1–2 F-4C Weapons System Trainer.
(a) Exterior view of the cockpit. (b) Partial view of trainer control stations. Built in 1965 by Link Trainer, the F-4C Weapons System Trainer was one of the first robotic flight simulators. It has three axes of roll and five axes of pitch.

gas chromatograph/mass spectrometer for a nose and taste buds. Each *Viking* probe could detect organic molecules at a concentration of a few parts per million. Its single arm, which could be withdrawn completely into its body, was mainly used to gather soil and rock samples. Finally, each *Viking* had biological sensors for measuring microbial metabolism.

On July 4, 1997, the United States landed the *Pathfinder* spacecraft on Mars. The *Pathfinder* carried its own six-wheeled robot rover called *Sojourner.* The *Sojourner* was used to take pictures and study the rocks and soil of Mars. Information gathered by *Sojourner* was then relayed to Earth, through the *Pathfinder.*

The military is faced with many unpleasant jobs it would prefer to have robots do. Repairing reactors or the outside of ships at sea would be a good job for robots. Using robots to repair space vehicles will also be desirable in the near future. Even the boiler room onboard ships would be a good place to use robots. And of course the military would like to replace foot soldiers with robots.

Unfortunately, the robot is not yet advanced enough to handle most of these tasks. But the military remains very interested in robots and will continue to contribute to their development.

Show (Promotional) Robots

Some robots used by industrial companies more closely resemble personal or hobbyist robots than industrial robots. Thus **show robots** (also known as play or promotional robots) are nonindustrial robots that might be better described as remote-control devices. Such robots are seen at electronic shows and conventions, where they move about in the aisles, mingle with the many visitors, and appear to converse with them.

Figure 1–3 shows Woodi, a promotional robot that acts as mascot of the Oregon Institute of Technology rally squad. Figure 1–4 shows Harry, one of the Remarkable robots made by Robots Northwest, making a promotional appearance at a store opening.

Max, from Nationwide Robots, is radio-controlled, stands 4½ feet tall, talks, shakes hands, carries a tray, blows up balloons, makes robotlike noises, blushes when kissed, has flashing lights, and is battery-powered. Essentially though, Max is just a remote-control device that requires a human to furnish all controls, including the fancy conversation.

The Robot Factory offers a line of radio-controlled robots, ranging from the 40-inch-tall Hot Tot, which rides a bicycle or tricycle, to the 8-foot-tall Ralph Roger robot. The company's most popular and sophisticated model is the in-between size SIX-T robot. Among other things, SIX-T can tip its hat, hand out business cards, talk, and blink its eyes.

Making a remote-control toy tank move about and making a show robot move about are very similar undertakings. The swiveling of the tank's turret is analogous to the turning of the show robot's head. The raising and lowering of the tank's gun is much like the raising and lowering of the robot's arm(s). The firing of

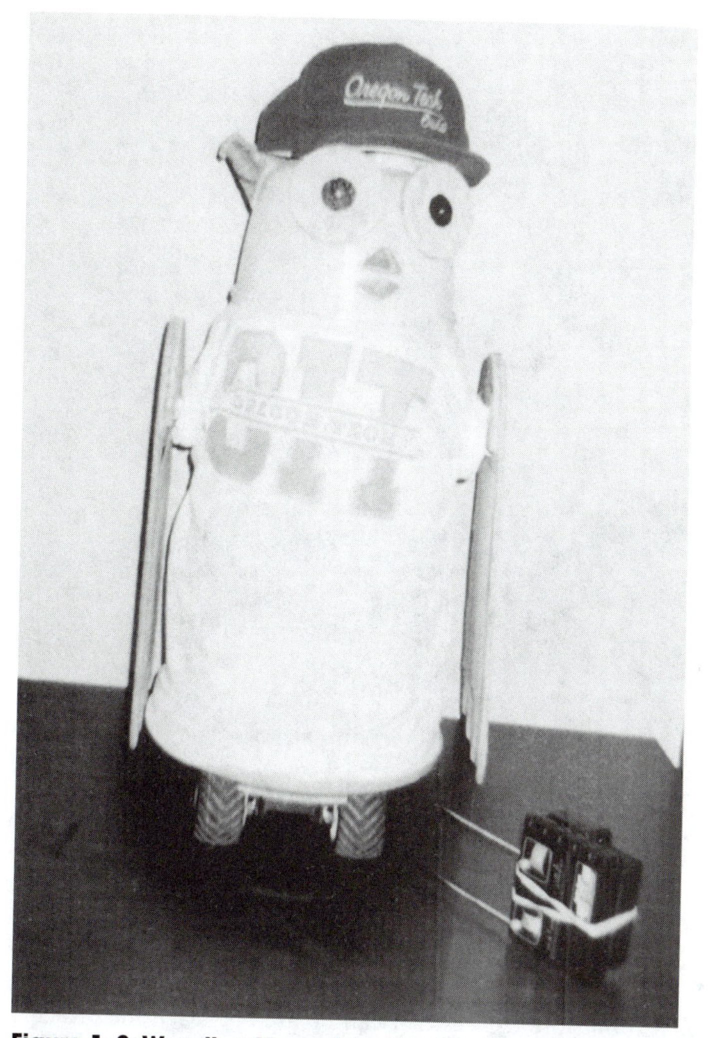

Figure 1-3 Woodi, a Mascot for the Oregon Institute of Technology Rally Squad.
Woodi is a remote-control device that entertains at football and basketball games.

the tank's gun could be replaced by the opening and closing of the robot's hand. Add a remote-controlled speaker, a microphone, and some flashing lights, and you have yourself a show robot. The construction of the show robot GARCAN (short for "garbage can"), a homemade show robot, is explained in the *Handbook of Advanced Robotics.*[5]

[5]Blue Ridge Summit, Penn.: TAB Books, 1982.

Figure 1–4 Harry, a Promotional Robot.
Harry is a remote-control device that functions as a professional entertainer. Harry works with a handler (or roboteer) and a clown, and earns $75 per hour.

Educational Robots

Educational robots are devices that can be used to teach the principles of robotics. They may be manipulators, computers, sensor kits, or any part related to robotics. All of the robots mentioned in this chapter could be classified as educational robots.

Some companies specialize in this area of robotics. The robots from Eshed Company come with the manipulator, controller, and power supply. Figure 1–5 shows the SCORBOT-ER V plus from Eshed. Figure 1–6 shows the SCORBOT ER VII robot. The robots from Feedback Inc., for example, are manipulators that require the user to supply the controller (computer). Figure 1–7 shows Feedback's ARMSORT PPR1030 manipulator with the AIM65 computer. Figure 1–8 shows Feedback's ARMDRAULIC EHA1052A manipulator with the IBM PC computer.

Fischertechnik also supplies educational manipulator kits. The Rhino educational robot and the HeathKit Hero I robot come with a built-in controller.

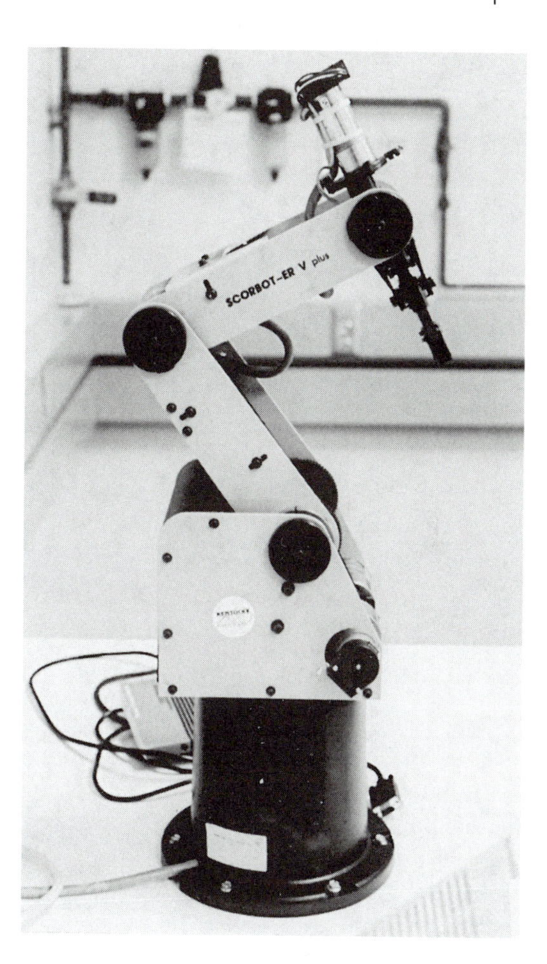

Figure 1-5
Eshed Company SCORBOT-ER V plus electric-powered educational robot comes with manipulator, controller, and power supply.

Medical Robots

Medical robots include all robotlike devices that either give medical aid or substitute for or restore functions that a disabled person lacks. Show robots are used in medical therapy work with abused children. Industrial robots are used as lab assistants to handle dangerous fluids and as aids to disabled persons to do tasks the person can no longer do. Bionic arms, hands, legs, and feet are just now reaching the useful stage. Artificial hearing and vision are under development.

AGVs are being used as hospital orderlies to deliver food and medication.

Domestic (Personal) Robots

The **domestic** (or personal) **robot** has yet to get off the ground. When it does, however, interest in robotics can be expected to grow phenomenally. Progress in the

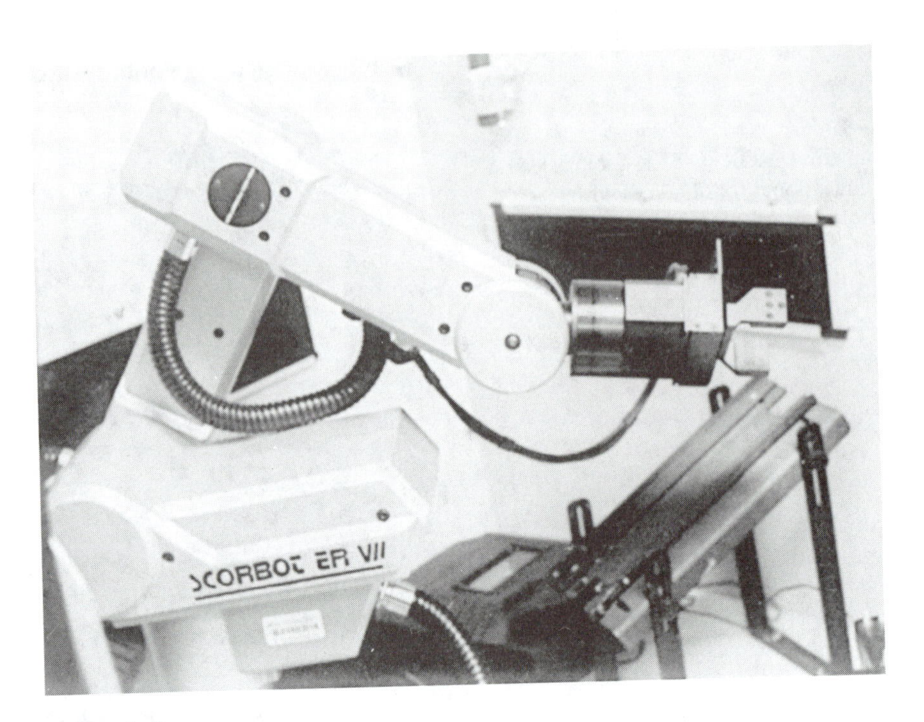

Figure 1-6
Eshed Company SCORBOT ER VII electric-powered educational robot comes with the above manipulator, controller, and power supply. The upper arm has power outlets for powering either an electric- or pneumatic-operated gripper.

Figure 1-7 ARMSORT PPR1030 with AIM65.
The ARMSORT PPR1030 manipulator and interface are connected to an AIM65 computer. ARMSORT is an electrically powered cylindrical-coordinates manipulator. (Courtesy of Feedback Inc., Berkeley Heights, New Jersey)

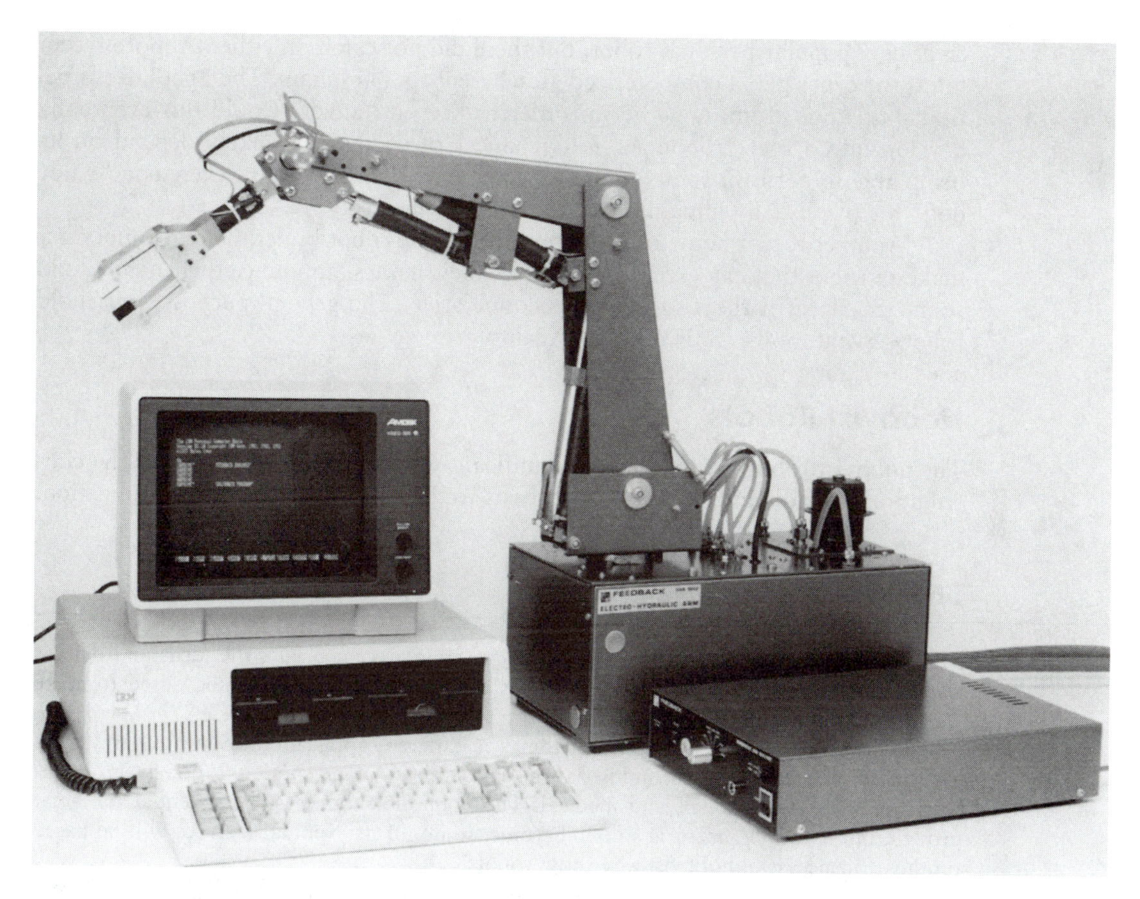

Figure 1-8 ARMDRAULIC EHA1052A with IBM PC.
The ARMDRAULIC EHA1052A manipulator and interface are connected to an IBM PC computer. It is a hydraulic-powered jointed-arm manipulator. (Courtesy of Feedback Inc., Berkeley Heights, New Jersey)

areas of hobbyist and show robotics will provide the basis for making personal robots. One U.S. manufacturer was hoping to have a personal robot pet on the market by 1986. Although this did not happen, the personal robot may well appear at any time now.

The first generation of personal robots will serve primarily as mechanical pets. Later versions will be capable of entertaining, baby-sitting, vacuuming the floor, bringing in the mail and the newspaper, mowing the lawn, and taking out the trash. They will also be useful for home protection, security, health care for the aged or infirm, fire detection, and (perhaps) repair of robot-built appliances.

A simpler electronic pet—the Petster series of toys from Axlon—was marketed for under $100 in 1986. Robotics hobbyists liked it, but the general public did not. HeathKit marketed the Hero I, Hero Jr, and Hero 2000 robots in hopes of creating a popular personal robot, but these did not catch on either. A more recent fad is the cyber pet, a computerized attachment to a keychain. The "pet" needs periodic attention from its owner in order to keep it "alive." Could this eventually develop into a real "robot pet"? Its chances of success will likely depend on its being able to perform some useful service for its owner, perhaps even as a "guide dog" for the visually impaired or physically challenged.

Another recent home robot application is the robot butler. One manufacturer makes a robot that can greet visitors, take their names, announce their names, and sound an alarm if the visitors are uncooperative. The general-purpose household robot servant, however, is not yet a reality.

Hobbyist Robots

The **hobbyist robot** will contribute significantly to the field of robotics, if the contributions of radio and computer hobbyists to their technologies are any indication.

SUMMARY

While the words *robot, robotics,* and *computer* are of recent origin, the ideas and devices they signify have been around for thousands of years. Only the technology used to make these devices is new.

The word *robot* means different things in different applications. Seven general areas of robotics can be identified: industrial, hobbyist, show or promotional, domestic or personal, military, educational, and medical. The domestic robot has the potential for great growth, but its development is still embryonic. Many of its eventual features will be based on those of successful hobbyist and show robots.

Automated industrial tools are special-purpose machines designed to do one task very efficiently. Industrial robots are general-purpose tools that can be programmed to do many different tasks. Consequently, robots are used in areas that require flexible automation. Robots generally work in concert with hard-automation devices.

In simplest terms, a robot is a computer whose primary purpose is to produce motion.

REVIEW QUESTIONS

1. What is robotics?
2. What is an industrial robot?
3. Name several different areas of robotics.
4. Differentiate among hard automation, flexible automation, and robotics.
5. How does work on hobbyist robots contribute to industrial robots?
6. How has the military affected the development of robots?
7. What tasks might a robot be able to do better than a human?
8. List some tasks you presently do that you would like a robot to do for you.
9. What is meant by the statement, "An industrial robot does not work by itself"?
10. Suppose that you have been given the job of setting up a production line for a new product. List the criteria you would use in deciding what jobs

should be done by humans, what jobs should be done by robots, and what jobs should be done by automation.

11. Differentiate between a teleoperator and an industrial robot.
12. What is an AGV and how might it be used in an ASRS?
13. What is CAD?
14. What is CAE?
15. What is CAM?
16. What is CAR?
17. Differentiate between PC and PLC.
18. Differentiate among LAN, MAP, and TOP.
19. When might LAN, MAP, and TOP be connected together?
20. Why hasn't the personal robot become a reality yet?

2

THE HISTORY OF ROBOTS

OVERVIEW

What is the history of robots? Why study the history of robots? Where can more information on the history of robots be obtained? These are the major questions addressed in this chapter.

OBJECTIVES

When you have completed this chapter, you should be familiar with:

- The present-day significance of the history of robots and computers in robotics
- The key stepping stones in the development of robots

KEY TERMS

Abacus
Android
Automatons
Auto-pilot
Babbage, Charles
Cray I
Devol, George
Droid
ENIAC
Humanoid
Jacquard's loom
Mark I
Mistakes in history
Mistakes in the future
Pollard and Roseland
Roboteer
Rossum's Universal Robots
Slide rule

THE VALUE OF STUDYING THE HISTORY OF ROBOTS

Studying the history of robots enables students to see the rate and order of the robot's development. It reveals the historical steps by which the robot developed, shows how far ahead of technology people's dreams move, and points out which dreams remain unfulfilled. Unfulfilled dreams suggest the direction the future development of robots will take.

While this chapter doesn't attempt to cover the entire history of robots in detail, a thorough study of robot history might help predict where future failures will occur, based on areas where past failures were concentrated. For example, studying the first four generations of commercial computers would reveal that numerous problems arose as a result of not putting sufficient thought and care into the design of the computer's power supply and air conditioning. Robots are now using computers for their controllers or brains, and robotics companies without previous computer experience are starting to design and make their own computers for the robot. Therefore, we can expect power supplies and air conditioning to be a big problem in these early-generation units. Since these same companies are also designing their own robot languages and writing robot operating systems and applications programs without prior experience in these areas, we can expect the problems that showed up in these areas in the first four generations of computers to recur in the case of robot programming.

Another mistake from history deals with the handling of service. UNIVAC, the leading computer company of the early 1960s, did not provide prompt service; a new computer company, IBM, did and it soon became the leading computer company.

Historically, engineers have often started their own companies in order to make a product of their own design. However, many of these companies failed because of poor marketing efforts. It is highly instructive to observe the business fate of engineers who have good ideas but do not know how and where to sell them.

KEY EVENTS IN THE HISTORY OF ROBOTS

The idea of robots goes back to ancient times, as Table 2–1 indicates. The successful building of robots—or at least of robotlike mechanisms—goes back hundreds of years and would go back farther, except for the lack of adequate technology.

Fictional writings mention robotic devices as far back as 3,000 years ago, in the *Iliad*'s reference to "mobile tripods" (book IX). The Greek myth of Jason and the Argonauts includes an encounter with Talos, a giant bronze sentinel "programmed" by the gods to defend the island of Crete against intruders. India gives us an ancient legend about mechanically moving elephants.

The abacus or bead-adding machine was the first digital computer. Perfected around 1000 B.C., it is still used today. Figure 2–1 shows a modern abacus.

As early as 300 B.C., engineers created water-powered automatons. **Automatons** were scale models of living creatures that moved like those creatures. Figure 2–2 on page 27 is a drawing of an automaton made by Hero of Alexandria in 300 B.C. It depicts Hercules killing a dragon with an arrow.

In 1621 William Oughtred invented both the rectilinear and circular slide rules (see Figure 2–3 on page 28). The slide rule is an analog computer. Many early automated and robotic devices used analog computers. In the United States, slide rules have been replaced by pocket calculators.

Table 2-1
Time Line of the History of Robotics

Early history		–India's legend of mechanical elephants –Witches and magicians causing inanimate objects to become alive
1000 B.C.		–Abacus (bead-adding machine)
800 B.C.	Homer	–Walking tripods in the *Iliad*
300 B.C.		–Water-powered scale models of creatures that move (automatons)
100 B.C.		–Greek ship with analog navigational computer on board
A.D. 1621	William Oughtred	–Rectilinear and circular slide rules
1623	Wilhelm Schickard	–Four-function calculator
1642	Blaise Pascal	–Calculator for adding and subtracting (mass produced)
1730–1780		–Clockwork automatons
1801	Joseph Jacquard	–Automated loom using holes punched in cards for control (first NC machine)
1812	Charles Babbage	–Theory of automated computing
1822		–Difference engine
1830	Christopher Spencer	–Cam-operated lathe
1833		–Analytical engine
1868	Zadoc P. Dederick	–Steam-powered rickshaw man
1880s	Railroads	–Automaton railroad signal that looks like a man and operates on electricity
1890	Herman Hollerith	–Punch card tabulating equipment for U.S. census
1892	Seward Babbitt	–Motorized crane with gripper to remove ingots from a furnace
1898	Nikola Tesla	–Radio-controlled submersible boat
1900	James Power	–1900 census with simultaneous punching
1921	Karel Capek	–First use of the word *robot,* in the play *Rossum's Universal Robots*
1926	Fritz Lang	–The word *robot* is used in the movie *Metropolis*
1938	Pollard and Roseland	–Programmable paint-spraying machine
1939	John V. Atanasoff	–ABC model computer (Atanasoff-Berry computer), first electronic computer
1939–1944	Howard H. Aiken	–Harvard/IBM Mark I electromechanical calculator
1940	Isaac Asimov	–First use of the word *robotics*
1945	Mauchly and Ecker	–ENIAC, first useful electronic computer
1946	George Devol	–Patent for general-purpose playback device for controlling machines (using magnetic process recorder)
1946	John Von Neumann	–Paper on stored-program computer
1951	Mauchly and Ecker	–UNIVAC I, first mass-produced commercial computer
1951	Raymond Goertz	–Teleoperator-equipped articulated arm
1953	Seiko Corp.	–Assembly of watches by miniature robots
1956	George Devol	–First programmable robot designed for what will become Unimation. U.S. patent issued in 1961.

1958	Joseph F. Engelberger and George C. Devol	–Started the company that would become Unimation and named the first robot "Unimate"
1959	Planet Corp.	–Marketing of first commercially available robot
1960	Condec Corp.	–Purchase of Unimation and development of Unimate Robot System
1960	AMF Corp.	–Marketing of Versatran robot
1961	Unimation	–First Unimate robot installed to tend a die casting machine
1964		–Artificial intelligence research laboratories opened at MIT, Stanford, and University of Edinburgh
1968		–Shakey robot with vision capability developed at Stanford Research Institute (SRI)
1968		–Minicomputer
1968	Unimation	–Unimation takes its first multirobot order from General Motors
1970	Stanford University	–Stanford arm
1971	Intel	–The 4004, the first microprocessor
1973	Cincinnati Milacron	–First commercially available minicomputer-controlled industrial robot
1974	Vicarm Inc.	–Company formed to market an industrial version of the Stanford arm (minicomputer controlled)
1975	Cray Inc.	–Cray I array-processor computer
1976		–Robot arms on *Viking 1* and *2* space probes
1976	Vicarm Inc.	–Vicarm design now includes a microcomputer
1976	Tandy Inc.	–TRS-80 home microcomputer
1976	NASA	–The robotic spacecraft *Viking* lands on the Martian surface
1977	ASEA	–Two microcomputer-controlled robots
1977	Unimation	–Purchase of Vicarm
1978	Unimation and GM	–The PUMA (programmable universal machine for assembly) robot
1980		–Rapid growth of robotics industry, with a new robot or company added each month
1984	Joseph F. Engelberger	–The "Father of Robotics," starts new company called Transition Research Corp. concentrating on service robots
1987	James L. Fuller	–Hooter designed and built, becomes first college rally squad robot mascot
1988	CNN News	–SCAMP designed as first robot pet with personality and feelings
1991	Transition Research	–First HelpMate robot becomes operational
1995	Transition Research	–Changes its name to HelpMate Robotics, Inc.
1997	NASA	–*Pathfinder* spacecraft lands on Mars with *Sojourner* rover robot to explore Mars

Figure 2–1 Abacus, or Bead-adding Machine.
The abacus was developed in India around 1000 B.C. The abacus here shows the value 0000987654321.

In the eighteenth century, several elaborate automatons were made—mainly as toys for the very rich. They were also mechanical masterpieces. One was a fake chess-playing robot invented by Wolfgang Von Kempelen around 1769. It actually relied on the presence of a small child hidden inside for control. Von Kempelen also made a doll that demonstrated the mechanism of human speech. Another automaton was a model of a duck composed of over 4,000 parts, with a working digestive system. The duck walked, quacked, ate, drank, and eliminated. If you touched the duck, it would turn its head towards you and protest. Another well-known automaton was an artificial female organ player with simulated breathing, head movements, and hand movements.

Around 1775 Jaquet Droz made a series of doll automatons, including one doll that played a model piano and another doll that could write words. These dolls are still functional.

In 1801 Joseph Marie Jacquard constructed the first numerical-controlled machine—an automatic loom that used cards with holes punched in them to control the pattern being woven. In successive decades, Charles Babbage formulated the theory of automated computing and designed and partly built the analytical machine and the difference engine. These were the forerunners of the modern electronic digital computer. In 1898 Nikola Tesla demonstrated a radio remote-controlled submersible boat. He also tried to develop machines that he hoped would possess intelligence.

Figure 2–2 Automaton Made by Hero of Alexandria.
This automaton is called *Hercules Killing the Dragon.* The dragon squirts water at Hercules, and Hercules kills the dragon with an arrow. The automaton is water-powered and was built by Hero of Alexandria around 300 B.C.

In the 1880s railroad companies found a use for a less expensive type of automaton, an electric railroad signal that looked like a man. In the late 1960s the Queens Devices Corporation of New York began producing a lifelike automaton known as Silent Sam for directing traffic. Sam, which is still being made, costs less than $1,000 and can run on batteries for five to seven days at a time. Sam moves a flag held in one hand and from over 20 feet away is mistaken for a human. Since human flagmen are in short supply in the United States and since being a flagman is a hazardous job, even the unions think Sam is great and would like to see more Sams put in use.

In the 1930s aircraft designers developed the **autopilot** for airplanes. In Europe, autopilots were known as *robot pilots.* Radio remote-controlled aircraft were also under development at this time. During the 1930s the first spray-painting industrial robots were made, although they were called automatic machine tools instead of robots. Two separate patents were issued for these devices, to Pollard in

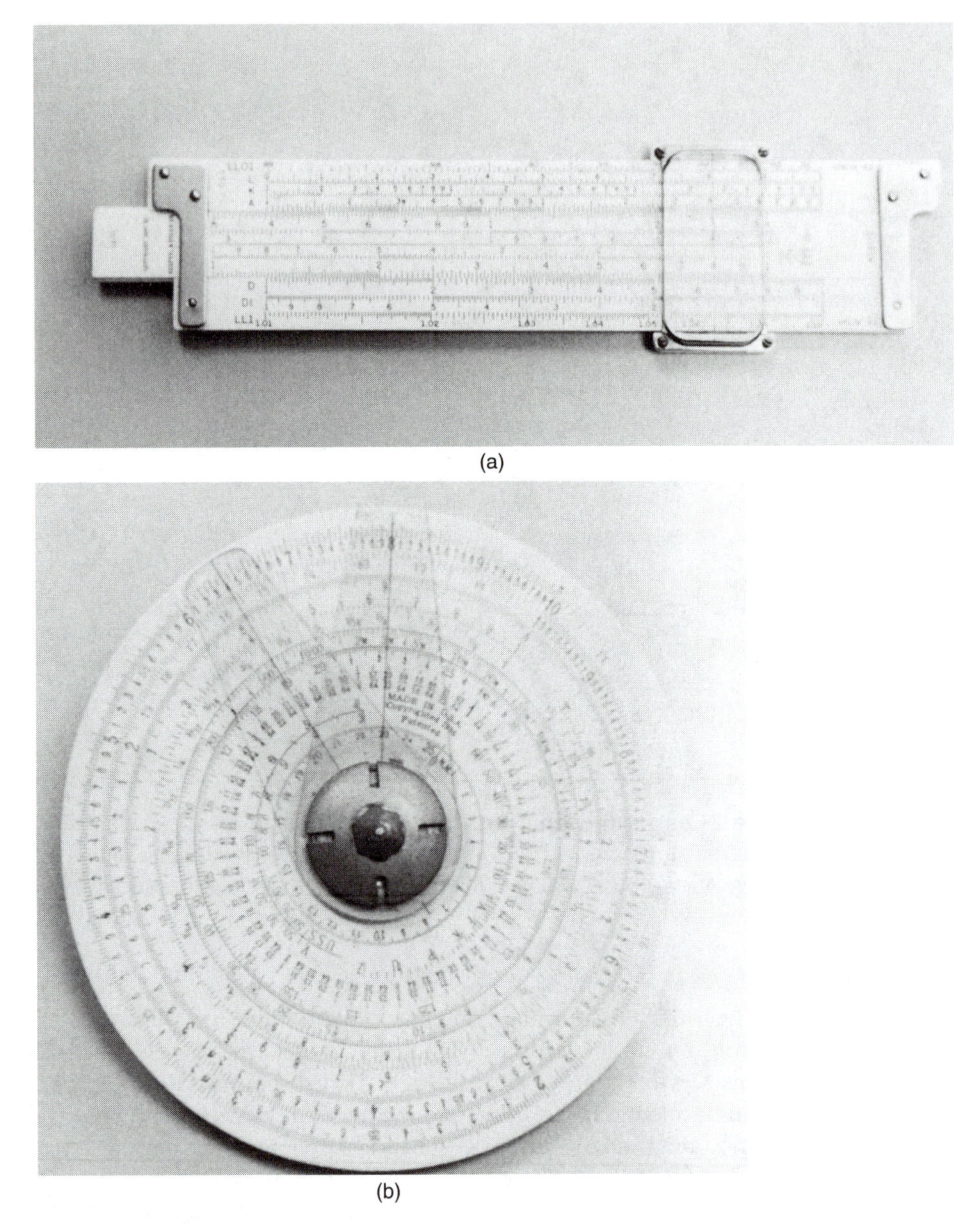

(a)

(b)

Figure 2-3 Rectilinear and Circular Slide Rules.
(a) Rectilinear slide rule. (b) Circular slide rule. Both are based on logarithmic scales and are analog computing devices. Both slide rules have been set to compute 8×8.

1942 and to Roseland in 1944. These machines were trained by being led through a task as they recorded their motion information on a phonograph-like device. The robots then played back the information continuously as they replicated their painting task. The controller thus used an analog computer.

In the 1950s the first generation of commercial digital computers appeared. These used vacuum tubes as their main components. They were large and slow and could not be used as a robot controller.

Around 1953 Seiko of Japan developed what appeared to be a miniature robot system to assemble a wide range of mechanical watch parts. Then in 1956 George C. Devol applied for a patent (issued in 1961 by the U.S. Patent Bureau) for the first practical industrial robot. It was labeled "Programmed Article Transfer."

In the 1960s a second generation of computers was developed. They replaced vacuum tubes with transistors, which made them smaller and faster. Very simple second-generation computers were used as controllers for automated machine tools and robots.

In the 1940s most homemade robots were mechanical units whose programming was fixed in gears and cams. In the 1950s and 1960s the transistor and early integrated circuits allowed homemade robots to use electronic circuits for some built-in programming. Simultaneously, industry was building automation equipment and numerical-controlled equipment, and the military was working on inertial guidance (or navigation) systems for planes and rockets. All of these used built-in electronic programs or instructions. Changing the program thus required redesigning the equipment.

By 1970 a third generation of computers had replaced transistors with integrated circuits. An integrated circuit only slightly larger than a couple of transistors could contain from ten to twenty complete transistor circuits. This made the minicomputer available for use as a programmable logic controller on automated and numerical-controlled machines. Although these controllers could be reprogrammed, the machines they ran were not general purpose enough to benefit from reprogramming. Industrial robots, which had begun appearing in the late 1960s, were getting a firm foothold in the industry, with approximately 2,500 robots in the United States by 1978.

In 1974 Vicarm, Inc., marketed a robot that used a minicomputer for a controller. In 1975 Cray, Inc., introduced the Cray I supercomputer, which used array processing so that it could work on many parts of a mathematical array problem in parallel. In the future, such computing will make vision processing much more practical.

By 1976 at least three companies were selling home personal microcomputers: Apple, Commodore, and Tandy. Figure 2–4 shows the TRS-80 computer from Tandy. This represented the fourth generation of computers, replacing simple integrated circuits with large-scale integrated circuits. A large-scale integrated circuit could contain a complete computer. By this time, Vicarm had designed a robot with a microcomputer for a controller. Microcomputers are the basis for all modern robot controllers. Figure 2–5 shows components from each of the first four generations of computers.

Figure 2-4 TRS-80 Microcomputer.
The Radio Shack TRS-80 Model I computer by Tandy was one of the first ready-made personal microcomputers. It was first offered for sale in the summer of 1977.

In the 1980s robots continued to move into industry. By the end of the 1980s a second generation of industrial robots with more powerful sensor devices was on the brink of emerging.

FICTIONAL ROBOTS

Fictional writings reveal many human fantasies and often predict future events. The human wish to fly led to the invention of the airplane. The human wish for independent means of transportation led to the invention of the automobile. The dream of putting a person on the moon led to the U.S. space program that did put a man on the moon. Throughout history people have dreamed of creating mechanical creatures to do their work for them.

Many folk tales deal with witches and magicians who could give life to inanimate objects and cause them to move about. How nice it would be to train a broom or vacuum cleaner to clean the house automatically or an axe to chop wood by itself. Then there is the story of Pinocchio, a string puppet, who comes to life and becomes his maker's son.

The story of a flying mechanical horse is told in the *Arabian Nights*. And Tik-Tok of Oz (not to mention the Tin Woodman) come to life in the Oz book series. Then we have Dr. Frankenstein's attempts to make a man from pieces of dead men and then bring it to life.

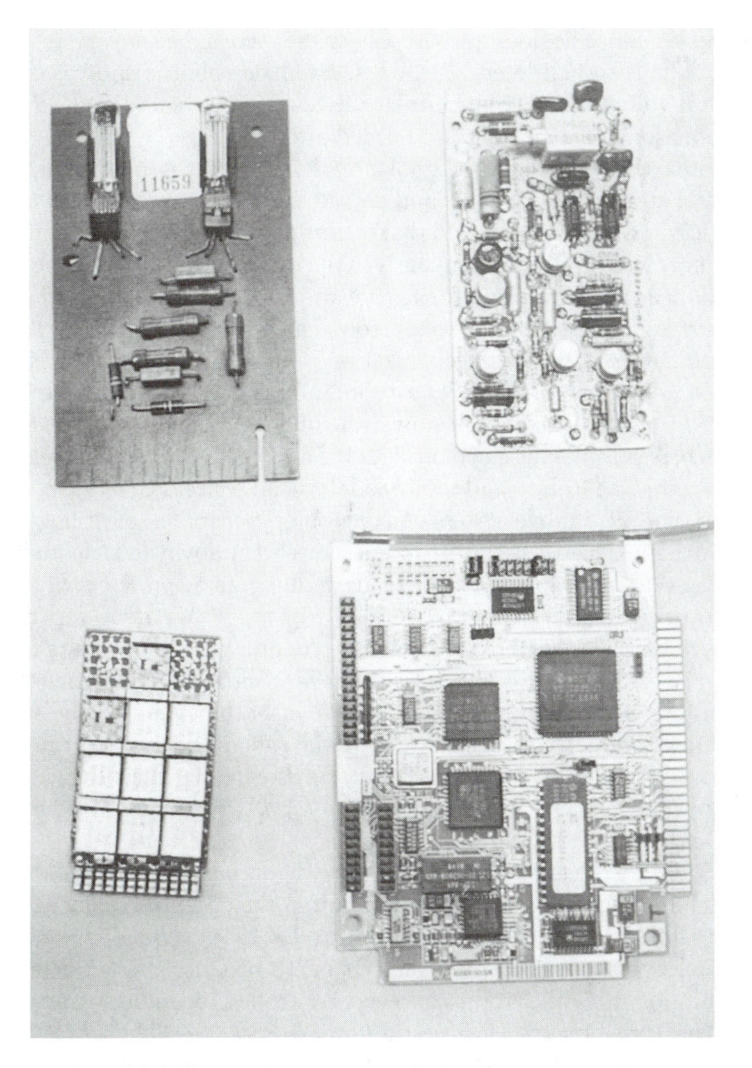

Figure 2–5 Four Generations of Computer Components.
(a) A first-generation computer circuit that uses vacuum tubes as a major component. (b) A second-generation computer circuit that uses transistors as a major component. (c) A third-generation computer circuit in which each rectangle is an integrated circuit (the major component). (d) A fourth-generation computer circuit in which each black square is a large-scale integrated circuit (the major component).

In 1921 Karel Capek first used the word *robot,* in the play *Rossum's Universal Robots*. The inventor Rossum is trying to produce a perfect and tireless worker, and his robot is manufactured wholesale for every purpose. Someone gets the idea that giving the robots feelings and nerves will make them better servants. However, the robots' first emotion is to hate human beings. The robots rebel and wipe out all humans except one skilled mason, whom they order to help them continue their robot species. Since he does not know how they work, he states that he needs to dissect one if he is to help them. But when a female robot is chosen as the subject, a male robot volunteers to take her place. Recognizing the love agleam in the robot's eyes, the mason sees that he is not needed after all.

More recently, we have the book *I, Robot,* by Isaac Asimov. Asimov has written many other stories about robots and coined the word *robotics* in 1940. Many other science fiction writers have found that robots make good characters in their books.

Robots have appeared frequently in the movies. The 1926 movie *Metropolis* introduced the machine woman (or female android), Olympia. The 1934 movie serial *The Vanishing Shadows* has a robot, too. The 1936 movie *The Birth of the Robot* deals with automatons and mechanization in modern times. The 1940 movie serial *The Mysterious Dr. Stan* features a robot, and the 1952 movie *The Day the Earth Stood Still* features Gort, an all-powerful robot. In the 1954 movie *Gog,* the robot Gog goes berserk and kills a person. The 1956 movie *Forbidden Planet* introduces the lovable Robbie the Robot, and the 1965 television series *Lost in Space* features a similar robot. The 1971 movie *2001* has a computer-controlled spaceship (that is, a robot spaceship) named HAL. HAL suffers a nervous breakdown and tries to kill all the men aboard the ship, which it almost succeeds in doing. In *Silent Running,* made in 1972, robot workers are used to tend trees. The 1973 movie *Westworld* is full of robots or androids that are designed to give humans pleasure but later try to take over.

Star Wars, which first appeared in 1976, introduces the friendly and likable robots R2D2 and C3PO, who also appear in both sequels to the film. The Buck Rogers movie and television show feature the robot Tweaky. Most futuristic movies depict robots taking over humans' work or threatening their lives. A more benign vision appears in the 1979 movie *C.H.O.M.P.S.,* in the form of a computerized watchdog designed to ward off burglars.

In 1988 there was Johnny Five (formerly Number Five), of *Short Circuit 2.* In real life, Johnny cost $2 million and required the combined talents of four roboteers (operators in charge of a remote-controlled show robot). One roboteer wore a form of teleoperator called a telemetry suit, to produce the major motions except for locomotion; the other three roboteers controlled locomotion and facial features. The movies *Jurassic Park* (1993) and *Jurassic Park: Lost World* (1997) used robot dinosaurs and computer graphics to portray "live" dinosaurs. *Jurassic Park 3,* due out May 2001, will use even more robots. The first underwater pictures of the real *Titanic* were photographed using a remote-controlled camera.

SOURCES OF ADDITIONAL INFORMATION

More information on the history of robots can be found in *Robots, Robots, Robots* edited by Geduld and Gottesman.[1] Two other areas closely related to the history of robots are the history of automation and the history of computers. Both are covered in *BIT by BIT: An Illustrated History of Computers* by Stan Augarten.[2]

[1]Boston: New York Graphic Society, 1978.
[2]New York: Ticknor & Fields, 1984.

SUMMARY

The human wish to have robots goes back many centuries. Literature and the cinema have shown what robots of the future may look like and act like. After all, human progress receives its first impulse from dreams and imagination.

While a few successful mechanical robots have been made over the last few centuries, major advances in robot technology have had to wait for the electronic age and the computer. With a computer for a brain, the modern robot has an unlimited future.

As technology makes the production of computers and robots easier, new companies will begin making computers and robots. Because these new companies tend not to study past history of computers and robots, they are bound to repeat the many mistakes of previous companies.

REVIEW QUESTIONS

1. How do humanity's dreams affect its future?
2. Why might science fiction and other fictional literature accurately foretell what robots may someday be able to do?
3. What devices from history constitute stepping stones toward the industrial robot?
4. What effect does the portrayal of robots in fiction have on people's acceptance of them?
5. When and by whom were the words *robot* and *robotics* coined?
6. Suppose that you are founding a new robotics company. What lessons from history will help you in your new venture?
7. Why is space exploration a good task to assign to robots?
8. How does using a digital computer for the controller of a robot make it easier to program and reprogram?
9. Why is the military interested in remote-controlled devices (teleoperators) and robots?
10. How are nonindustrial robots gaining use in hospitals and offices?
11. Why are people fascinated with robots in the movies?
12. Computers and robots are just starting to have some features shown in the movies *Star Trek* and *Star Wars*. Do you think robots and computers will ever be as intelligent as those in the movies? If so, when?

3

COMPONENTS OF AN INDUSTRIAL ROBOT: PART I

OVERVIEW

This chapter offers a detailed description of the parts that make up an industrial robot. It also introduces much of the terminology related to parts of the robot and explains their origin. For example, the parts of the manipulator are named after corresponding parts in the human body, and work area shapes are named after their geometric shapes.

What styles of manipulators are used for an industrial robot? What type of power is used? What are its components? These are some of the questions answered in this chapter.

OBJECTIVES

When you have completed this chapter, you should be familiar with:

- The components of an industrial robot
- The many styles of manipulators
- The three types of power used for robots
- The specialized field of robot grippers
- How to begin selecting a robot for a specific task

KEY TERMS

Actuator
Anthropomorphic robot
Arm
Articulate robot
Automatic control
Bang-bang robot
Cell
Closed-loop
Computer interface
Controller
Cylindrical-coordinates robot
Gear motor
Gripper
Hand
Hydraulic
Jointed-arm robot
Manipulator
Manual control
Memory
Nonservo-controlled

- Open-loop
- Pitch
- Pneumatic
- Point-to-point control unit
- Polar-coordinates robot
- Power supply
- Rectangular-coordinates robot
- Repeatability
- Revolute-coordinates robot
- Robot choreography
- Roll
- Safety interlocks
- SCARA
- Servo-controlled
- Shoulder
- Spherical-coordinates robot
- Vane motor
- Vehicle
- Work envelope
- Wrist
- Yaw

GENERAL CHARACTERISTICS OF AN INDUSTRIAL ROBOT

An industrial robot has the following parts and characteristics: hand, wrist, arm, base, lifting power, repeatability, manual control, automatic control, memory, library of programs (programmed by the user), safety interlocks, speed of operation, computer interface, reliability, and easy maintenance.

The **"hand"** of a robot is known as a **gripper,** an *end effector,* an **actuator,** or *end-of-arm tooling.* It consists of the driven mechanical device(s) attached to the end of the **manipulator,** by which objects can be grasped or acted upon. The robot may require a different type and design of hand for each different object it is to grasp or each different tool it is to hold. In some cases, the hand itself acts as the tool. Clearly, designing grippers properly is a key task in robotics.

The **wrist** of the robot is used to aim the hand at any part of the workpiece. The wrist may use any of three motions: a **pitch,** or up-and-down motion; a **yaw,** or side-to-side motion; and a **roll,** or rotating motion.

The **arm** is used to move the hand within reach of a part or workpiece. It can pivot at its elbow and at its shoulder joint. The type of joints used and the motion of the arm determine the work area, or **cell,** of the robot.

The waist, or **base,** of the robot, which serves to support the arm, is called the **shoulder.** In some robots, the arm can rotate about the shoulder; in others, the arm is mounted on a **vehicle** and can move back and forth on a shoulder track. Generally, however, the shoulder's capacity to move is very limited.

The lifting power a robot must have depends on the weight and shape of the object to be moved, along with how fast and in what direction the object is to be moved. The lifting power for small robots may be supplied by vacuum, pneumatic (air-pressure), or even electrical force. Medium-size and large robots run on electricity or hydraulic power (the force of liquid).

To replace a human worker successfully, a robot must be able to replicate the required work motions within some specified precision or tolerance. This replication of motion with precision is known as **repeatability.** The narrower the tolerance, the more expensive the robot. In very close-tolerance work, such as that needed for assembly or machining, the robot is equipped with a remote center compliance (RCC) device, which acts as a multiaxis float to help pull the hand or tool into the required position. RCC devices are discussed in detail in Chapter 5.

A **manual control** device is used to teach the robot how to do a new task. The robot can learn either by having the sequence of required motions fed directly into its memory, or by being given on-the-job training in which a manual control is used to walk the robot through each motion needed for the task.

An **automatic control** system is used to carry out the instructions stored in the robot's memory. Without the automatic control system, the robot would just be a remote-controlled device. Automatic control systems are of two types: open-loop and closed-loop controlled systems.

The **open-loop** nonservo-controlled system assumes that everything is working and does not check the robot's orientation. It is sometimes referred to simply as a **nonservo-controlled** system.

The **closed-loop** servo-controlled system senses where the manipulator is and corrects its position as needed. The closed-loop servo control can function in either a **point-to-point** or a **continuous-path** mode of operation. Closed-loop servo-controlled systems are sometimes referred to simply as **servo-controlled** systems.

The robot's **memory** holds a library of programs to use in executing different tasks. Its memory saves the robot from having to relearn a task every time the task must be resumed. If the memory can hold only a few programs at a time, it is handy to have a way to store the memory (e.g., magnetic tape or disk) and as a result build up a library of program tasks.

Safety interlocks prevent the robot from inserting a hand into a machine, such as a closing press, and causing damage to both the robot and the machine. Most robots are unable to sense the presence of unexpected objects, so interlocks should be placed around the robot's working area to prevent a human from accidentally getting in the robot's way and being injured. Safety for both robot and humans should be a prime consideration in setting up a robot task. Chapter 9 is devoted to this subject.

The robot's **speed of operation** in performing a task should at least equal that of the human worker it is replacing. Equaling human speed allows a robot to be inserted in the midst of a human-operated production line. Of course, in an automated production line, it may be possible for the robot to work much faster than a human could.

The robot may have a **computer interface** to enable the robot to use the computer's larger memory to hold more task programs and to synchronize its actions with a complete production line of robots and other automated machines, all under the control of the computer. A single robot can learn a task and then transfer the program to the computer; then other robots can get the program from the computer without having to go through the learning process. However, because there are slight differences in tolerances of the positioning circuits for each robot, a position translation table may have to be used when the program is transmitted to another robot.

Industrial robots must be reliable and easy to maintain; the average time between failures should be at least 400 hours. When a breakdown does occur, the parts of the robot should be easily accessible and interchangeable. Built-in diagnostic routines or programs should help locate the problem. Spare parts, such as those listed in the manufacturers' spare parts list, must be kept on hand.

GENERAL COMPONENTS OF AN INDUSTRIAL ROBOT

An industrial robot has three types of components: physical parts or anatomy, built-in instructions or instinct (placed there by the manufacturer), and learned behavior or task programs (on-the-job training).

The physical portion of an industrial robot is made up of four or five parts: the mechanical part or manipulator, which peforms the work through motion; the

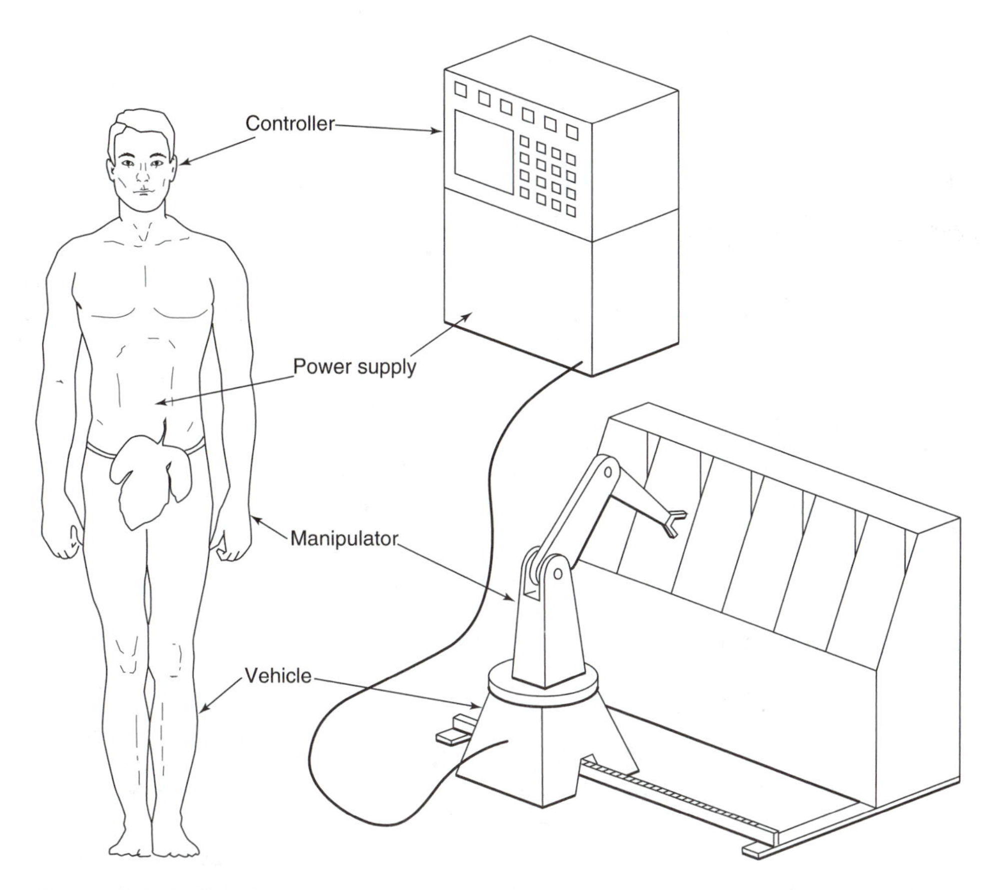

Figure 3-1 Parts of a Robot.
The controller for the human is the brain; the controller for the industrial robot is an electronic computer. The power supply for the human is the digestive system; for the industrial robot, it can be electrical (as in this figure), pneumatic, or hydraulic. The main manipulator for the human is the arm, while for the robot it is the manipulator arm. The vehicle for the human is the legs; for the industrial robot in this figure, it is wheels that move along a fixed rail.

controller, which directs motions; the power supply, which supplies energy to the manipulator; end-of-arm tooling (EOAT), or gripper; and (optionally) the vehicle, which transports the base of the robot to where it is to do its work. Figure 3–1 draws comparisons between the physical parts of the human being and those of the industrial robot.

The *manipulator* is the part of the robot that physically performs the task. Attached to it is the gripper, or hand, which actually contacts the parts or materials being processed.

The **controller** is the brain or computer center of the robot. It may be an adjustable timer and a set of mechanical stops and limit switches, or as complex as a complete minicomputer or microcomputer. The controller must recall all the motions that the robot is to perform, in the proper order and with the correct timing. When two or more robots are designed to work together, their individual controllers may receive directions from a large electronic computer that controls the entire factory. Plotting the movements of several robots working together is known as **robot choreography.**

The **power supply** for the manipulator depends on whether the type of power used by the manipulator is electric, hydraulic, pneumatic, or vacuum.

The robot may be stationary or sit on a vehicle. The vehicle may use wheels on a track, free wheels, treads, or some type of legs and feet. Whether the robot is stationary or mobile, the base of the robot must be stable and accurately situated if the robot's manipulator is to perform accurately.

The manipulator, controller, and power supply for the robot may be housed together as a single unit or they may be housed separately as discrete units. To apply spray paint, for example, only the manipulator need actually be located in the spray-paint booth. The controller and power supply can be kept in another area with a less severe environment. If a robot works in an explosive hazard area, again only the manipulator need be present in the area.

MANIPULATOR CONFIGURATIONS

Without the manipulator, the robot would not be a robot since it would not be able to produce motion. The parts of a robot's manipulator are named after similar parts in a human's chief manipulators, the arm and the hand. The points at which a robot's manipulator bends, slides, or rotates are called *joints,* or axes of motion; they include the shoulder, the elbow, the wrist, and the finger joints. The number of joints an industrial robot has determines its degrees of freedom of motion. Each axis equals one degree of freedom.

Figure 3–2 draws comparisons between human and robot manipulators. The human arm and hand constitute the finest universal manipulator in nature. Thus they serve as a model for the industrial robot to copy. The industrial robot of today generally has only a single manipulator, with one arm and one hand on it. In a few cases, the arm may have more than one hand on it. The method by which the arm positions the hand varies from one robot to another.

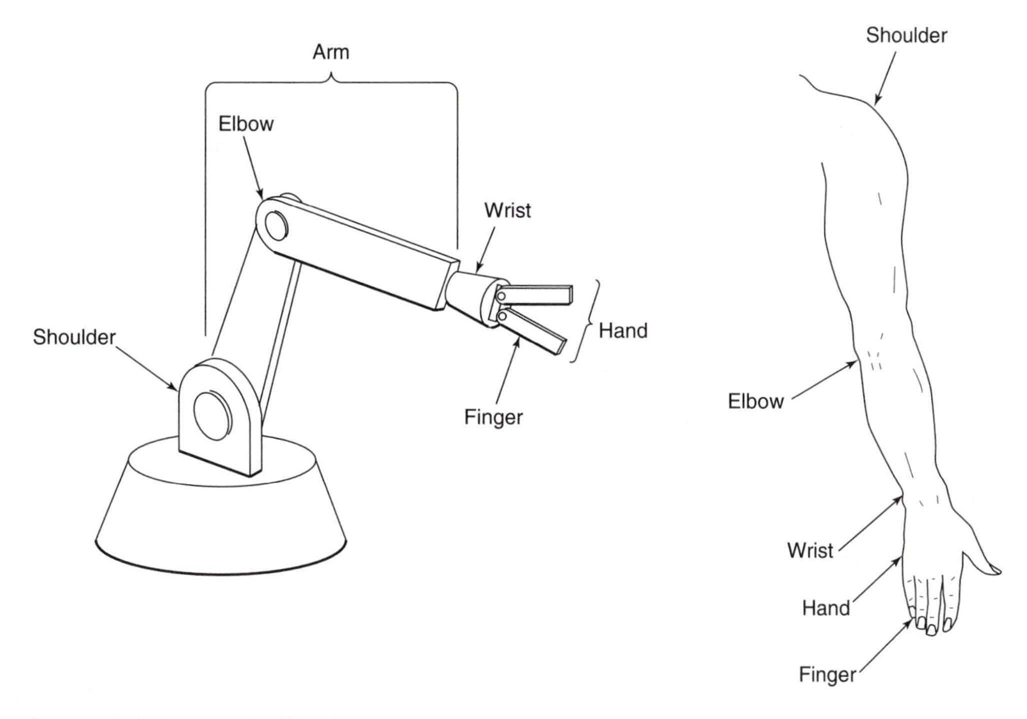

Figure 3-2 Parts of a Manipulator.
The industrial robot manipulator has an arm, a wrist, and a hand with fingers. These names match those of the corresponding human parts. The jointed-arm manipulator shown here also has an upper arm and a forearm.

Robots are often classified by the shape of the space, or **work envelope,** that their manipulator can reach. For example, the manipulator of a **rectangular,** or **Cartesian-coordinates, robot** operates within a cube or box-shaped work envelope.

Cartesian Coordinates

Positioning may be done by straight, or linear, motion along three axes: back and forth, in and out, and up and down. These axes are known, respectively, as the Cartesian axes x, y, and z. Figure 3–3(a) shows a typical manipulator arm for a Cartesian-coordinates robot. The work area, work envelope, or cell reached or serviced by the Cartesian-coordinates robot's arm is a big box-shaped area (see Figure 3–3(b)). This is similar to the type of work area that an overhead crane in a foundry or nuclear power plant would have (see Figure 3–3(c)). If such a crane is controlled by a computer, it acts as the manipulator of a robot. If the crane is controlled by a person, it is simply a remote-control device.

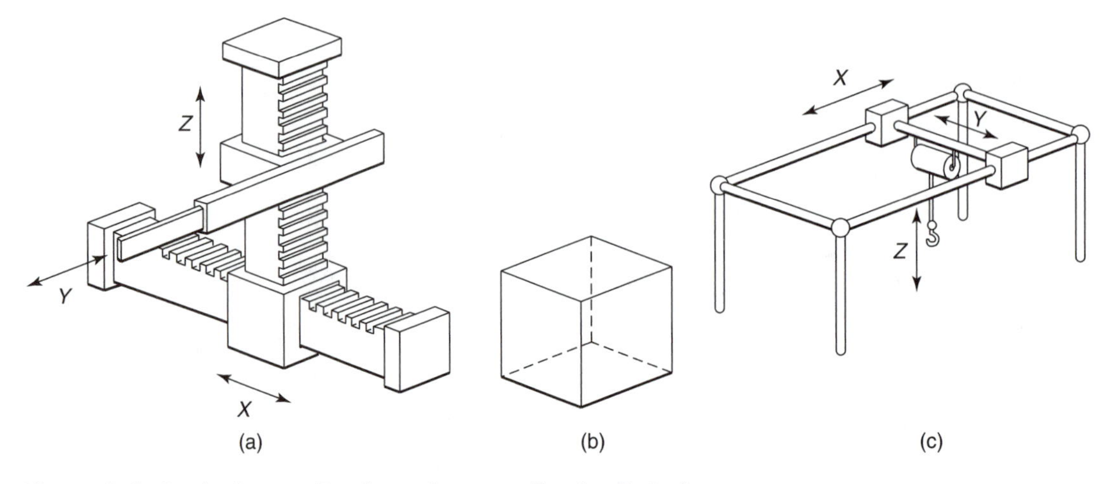

Figure 3–3 Cartesian- or Rectangular-coordinates Robot.
(a) A Cartesian- or rectangular-coordinates arm. It moves in three linear axes. (b) The box-shaped work envelope within which a Cartesian-coordinates manipulator operates. (c) An overhead crane. Its movements are similar to those of a Cartesian-coordinates arm.

The Cartesian-coordinates robot can only reach out in front of itself. The mechanism used by an automatic back-plane wiring machine for the computer industry is similar to the manipulator of a Cartesian-coordinates robot. Programming motion for a Cartesian-coordinates robot consists of specifying to the controller the *x*, *y*, and *z* values of a desired point to be reached. The robot then moves along each axis to the desired point. Figure 3–4 shows a rectangular-coordinates robot from Cincinnati Milacron Corporation.

Cylindrical Coordinates

If the positioning of the manipulator allows the arm to rotate about the base or shoulder, and if the other two axes allow up-and-down and in-and-out motions, the robot is known as a **cylindrical-coordinates robot.** The axes for the cylindrical coordinates are θ (the Greek letter theta), the base rotational axis; *R* (reach), the in-and-out axis; and *z*, the up-and-down axis. The work area, or cell, serviced by a cylindrical-coordinates robot arm is the space between two concentric cylinders of the same height. The inner cylinder represents the reach of the arm with the arm fully retracted, and the outer cylinder represents the reach of the arm with the arm fully extended. Figure 3–5(a) shows a typical manipulator arm for a cylindrical-coordinates robot. Figure 3–5(b) shows its work envelope, and Figure 3–5(c) shows a construction crane used for working on tall buildings that has a cylindrical-coordinates manipulator.

In most robots, the rotational axis does not have a full 360° turning capacity, but the cylindrical-coordinates robot can nonetheless reach things all around it.

Figure 3–4 Rectangular-coordinates Robot from Cincinnati Milacron.
This is the model T^3800 rectangular-coordinates robot. It is also called a gantry robot. It is electrically driven, with a 12-meter *x*-axis of travel, a 5-meter *y*-axis of travel, and a 200-pound maximum payload. The hand is located near the upper-right corner of the robot frame. (Courtesy of Cincinnati Milacron)

Programming a cylindrical-coordinates robot involves translating the *x* and *y* values for a desired point into a base rotation angle and the in/out value. The *z* value of the point is not changed and can be found directly along the *z* axis. Figure 3–6 shows a cylindrical-coordinates robot from Cincinnati Milacron.

SCARA Robot

The **SCARA** (Selective Compliance Assembly Robot Arm) robot has the same work area as a cylindrical-coordinates robot. However, the reach axis includes a rotational joint in a plane parallel to the floor. This feature allows the SCARA robot to reach around some obstacles in its work envelope. The SCARA arm is also more compact than the cylindrical-coordinates arm.

The SCARA robot was developed at Yamanashi University in Japan in 1978 and was intended for parts assembly work. Figure 3–7 shows the SCARA manipulator.

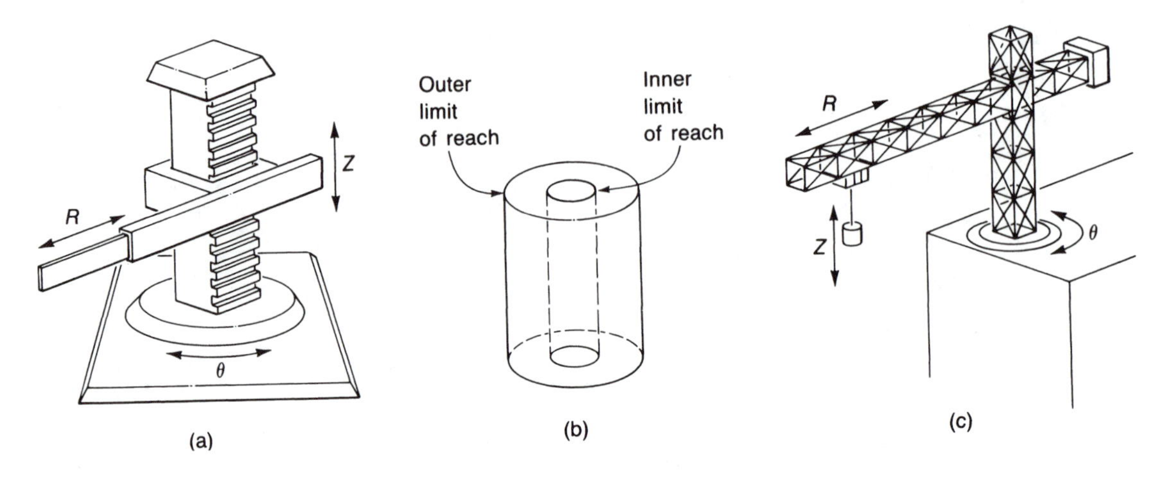

Figure 3-5 Cylindrical-coordinates Robot.
(a) A cylindrical-coordinates arm. It rotates about the base, moves in and out, and moves up and down. (b) The space between the two cylinders shown is the work envelope occupied by a cylindrical-coordinates manipulator. (c) A construction crane on top of a tall building. Its movements are similar to those of a cylindrical-coordinates manipulator.

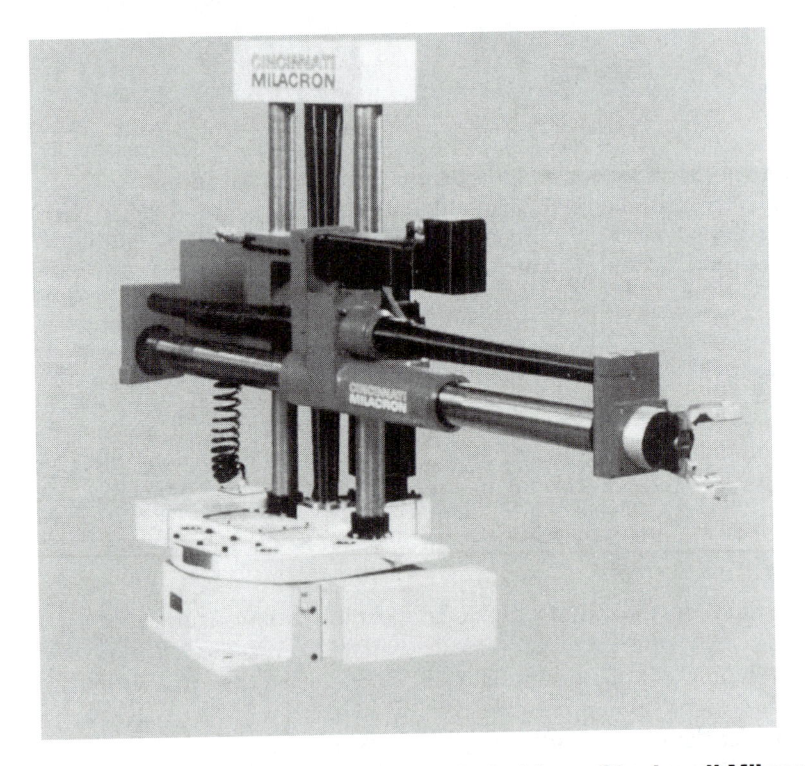

Figure 3-6 Cylindrical-coordinates Robot from Cincinnati Milacron.
This is the model $T^3$363 cylindrical-coordinates robot. It is electrically driven and can handle a 110-pound maximum payload. (Courtesy of Cincinnati Milacron)

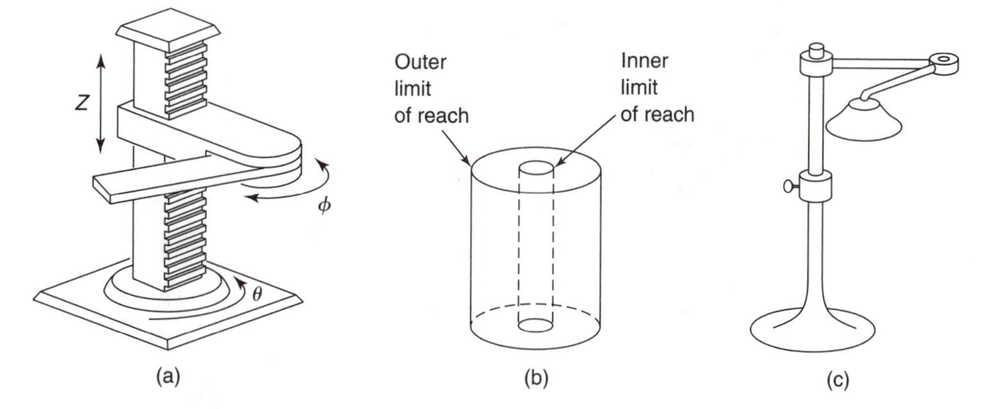

Figure 3–7 SCARA Robot.
(a) A SCARA manipulator. It rotates in two axes in the horizontal plane and moves linearly up and down. (b) The work envelope for the SCARA manipulator is the space between the two cylinders. The SCARA manipulator can reach around obstacles. (c) A folding lamp has movements similar to those of a SCARA manipulator.

Figure 3–8 shows a SCARA robot from Intelledex, Inc. The SCARA robot can also be called a horizontally articulate revolute robot.

Polar Coordinates

If the arm not only rotates about the base, but also rotates about an axis in the vertical plane, the robot is known as a **polar-coordinates,** or **spherical-coordinates, robot.** The axes for the spherical coordinates are θ, the rotational axis; *R*, the reach axis; and β (the Greek letter beta), the bend-up-and-down axis. The work area, or cell, serviced by a polar-coordinates robot is the space between two concentric hemispheres. The inner hemisphere is defined by the reach of the arm when it is fully retracted along the *R* axis. The outer hemisphere is defined by the reach of the arm when it is fully straightened along the *R* axis. Figure 3–9(a) shows a typical manipulator arm for a polar-coordinates robot. Figure 3–9(b) shows its work envelope, and Figure 3–9(c) shows the ladder mechanism of a hook-and-ladder truck that resembles the manipulator of a polar-coordinates robot.

In theory the polar-coordinates robot can reach things almost directly above itself, as well as around itself. In actuality, the rotational axis of most robots is limited to a turning range of less than 360° and the bend-up-and-down axis may be limited to 60° or less. Most polar-coordinates robots use some type of positional feedback information, which makes them closed-loop servo-controlled robots. Programming a polar-coordinates robot involves translating the *x*, *y*, and *z* values for a desired point into two angles of rotation and an in/out value. Many polar-coordinates robots cannot reach directly above themselves.

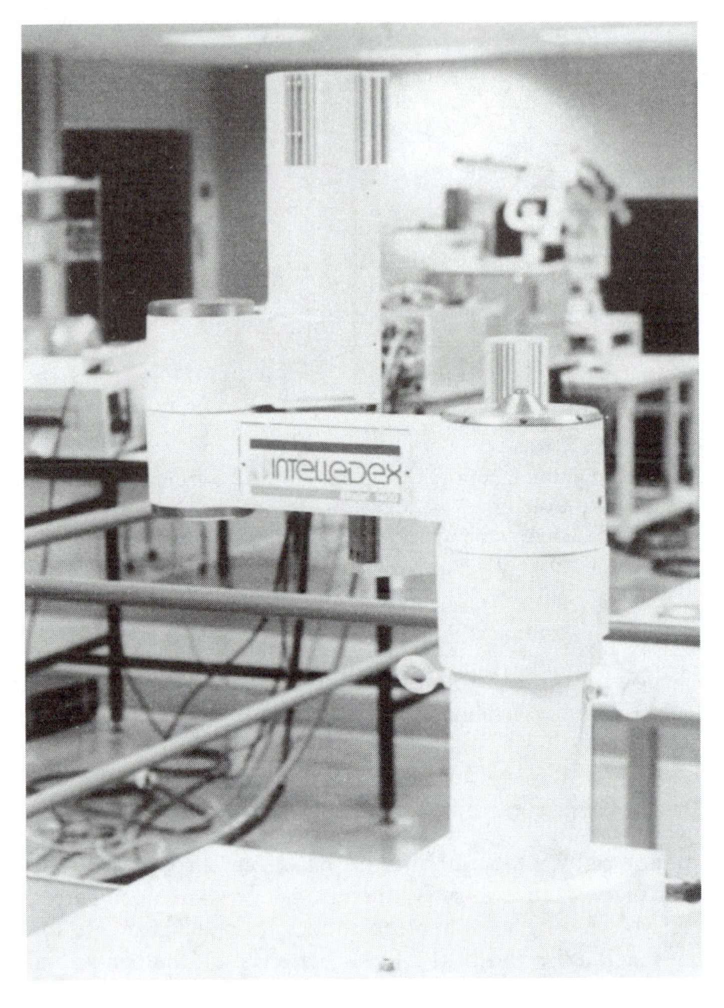

Figure 3–8 SCARA Robot from Intelledex.
This is an electrically powered SCARA manipulator robot made by Intelledex, Inc.

Jointed Arm

If the arm can rotate about all three axes, the robot is called a **revolute-coordinates, articulate, anthropomorphic, vertically articulated,** or **jointed-arm robot.** *Anthropomorphic* means human-shaped and here refers to the fact that this design resembles the human arm. The axes for the revolute coordinates are θ, the base rotational axis; β, the upper arm rotational axis; and α (the Greek letter alpha), the lower arm rotational axis. The inner hemisphere of the work envelope is not necessarily spherical but is the shape of the base that supports the arm. The outer hemisphere is defined by the reach of the arm when it is fully extended. Figure 3–10(a) shows a

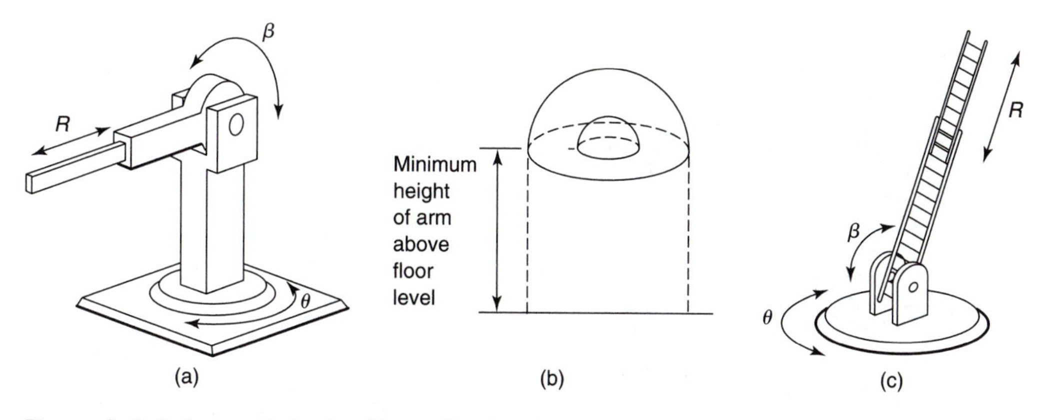

Figure 3-9 Polar- or Spherical-coordinates Robot.
(a) A polar or spherical-coordinates manipulator. It rotates about the base and about the shoulder and moves linearly in and out. (b) The work envelope for a polar-coordinates manipulator is the space between the two hemispheres. (c) A ladder on a hook-and-ladder truck has movements similar to those of a polar-coordinates manipulator.

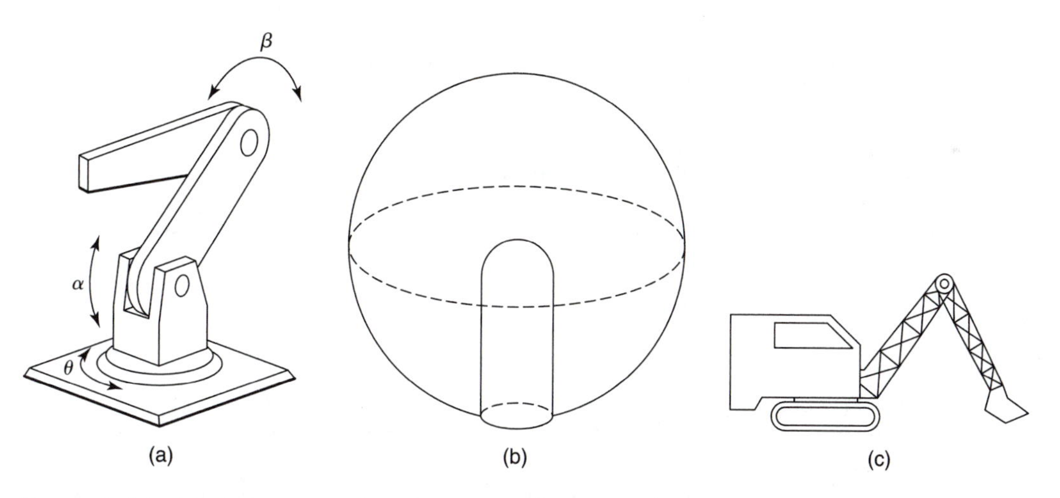

Figure 3-10 Jointed-arm, or Revolute-coordinates, Robot.
(a) A jointed-arm, or revolute-coordinates, manipulator. All three of its axes are rotational. (b) The area between the sphere and the column (representing the base support) is the work envelope for the jointed-arm manipulator. The jointed-arm manipulator can reach above and below an obstacle. (c) A power shovel has movements similar to those of a jointed-arm manipulator.

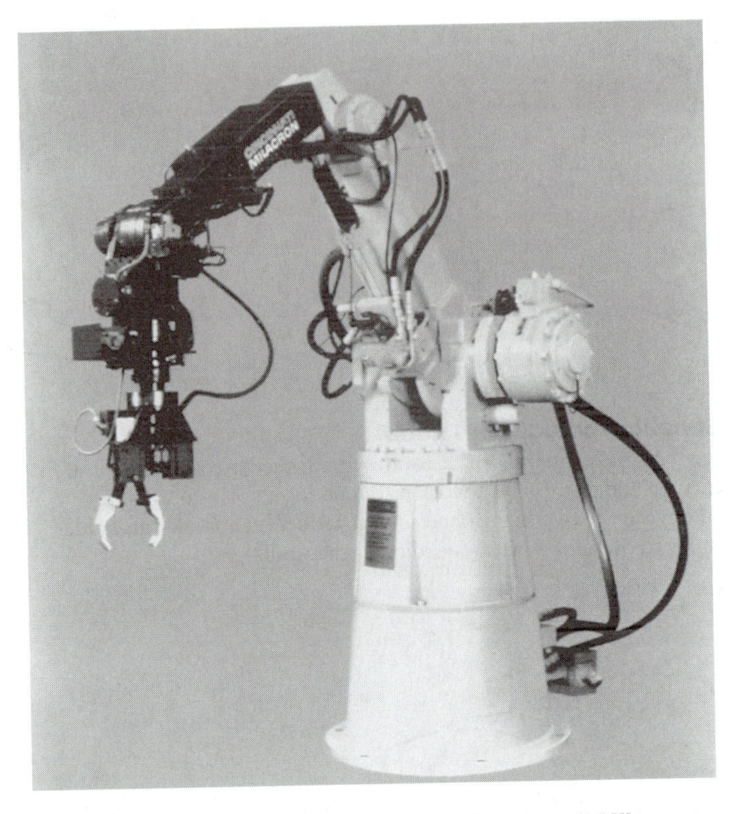

Figure 3–11 Jointed-arm Robot from Cincinnati Milacron.
This is the model $T^3$566 hydraulic-powered jointed-arm robot. It uses an electrohydraulic closed-loop servo system and can handle a maximum 100-pound payload. (Courtesy of Cincinnati Milacron)

typical manipulator arm for a revolute-coordinates robot. Figure 3–10(b) shows its work envelope, and Figure 3–10(c) shows a power shovel whose mechanism resembles a revolute-coordinates arm.

In theory, the revolute-coordinates arm can reach things all around it, including things above and below it. But again, most rotational axes do not rotate a full 360°. In any case, the revolute-coordinates, or jointed-arm, robot can serve the largest work area for the smallest floor space. These robots are almost always closed-loop servo-controlled. Programming a jointed-arm robot to reach a desired point involves translating the *x*, *y*, and *z* values into three angles of rotation and choosing between two possible sets of angles that can reach the same point. Figure 3–11 shows a jointed-arm robot from Cincinnati Milacron.

Table 3–1
Assessment of the Five Basic Robot Manipulator Configurations

Configuration	Advantages	Disadvantages
Cartesian coordinates (x, y, z—base travel, reach, and height); three linear axes	Easy to visualize Rigid structure Easy to program off-line Linear axes make for easy mechanical stops	Can only reach in front of itself Requires large floor space for size of work envelope Axes hard to seal
Cylindrical coordinates (θ, *r, z*—base rotation, reach, and height); two linear axes, one rotating axis	Can reach all around itself Reach and height axes rigid Rotational axis easy to seal	Cannot reach above itself Base rotation axis is less rigid than a linear axis Linear axes hard to seal Won't reach around obstacles Horizontal motion is circular
SCARA coordinates (θ, ϕ, *z*—base rotation, reach angle, height); one linear axis, two rotating axes	Height axis is rigid Large work area for floor space Can reach around obstacles	Two ways to reach a point Difficult to program off-line Highly complex arm
Spherical coordinates (θ, *R*, β—base rotation, reach, elevation angle); one linear axis, two rotating axes	Long horizontal reach	Can't reach around obstacles Generally has short vertical reach
Revolute coordinates (θ, β, α—base rotation, elevation angle, reach angle); three rotating axes	Can reach above or below obstacles Largest work area for least floor space	Difficult to program off-line Two or four ways to reach a point Most complex manipulator

Comparison of Manipulator Configurations

Each manipulator configuration described previously has some advantages and some disadvantages. These are summarized in Table 3–1.

Wrist

The wrist of a manipulator may add one, two, or three axes of motion to the three the manipulator already possesses. These wrist motions are called *pitch, yaw,* and *roll,* after an aircraft's motions of bend, twist, and swivel (see Figure 3–12). They are used to position the hand or end effector to the desired orienta-

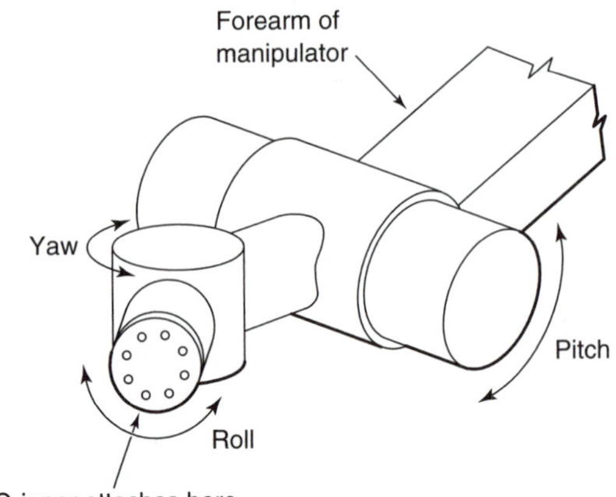

Figure 3-12 Wrist Motions.
This is a three-roll wrist. All three of its motions are rotational. The axes are known as *pitch* (up and down), *roll* (rotation around the axis of forward motion), and *yaw* (back and forth).

tion with the work. Figure 3–13 shows a three-axis wrist from IBM. Figure 3–14 shows three wrists that have only pitch and roll axes. A wrist with only a roll axis is shown in Figure 3–15, and Figure 3–16 shows a differential wrist that effectively has only pitch and roll axes.

Grippers

The choice of grippers can determine whether a robot succeeds or fails in performing a task. A very accurate and expensive robot equipped with the wrong gripper may be useless for a specific task.

The gripper, or hand, of the robot attaches to the wrist end of the arm and serves to handle or process parts. The human hand is one of the most versatile end effectors in existence. Besides having pitch, yaw, and roll control, the hand also has three joints in each finger and two in the thumb. While these joints have only a simple rotational ability, the joints at the base of the fingers can also move sideways, thus allowing the hand to assume many shapes and grip objects of many different sizes and shapes. The hand is also covered with positional feedback indicators, in the form of pressure and heat sensors. These tell what (if anything) is being touched, how hard it is, and how hot or cold it is relative to the hand's temperature. Figure 3–17 on page 53 shows a human gripper and a robot gripper. Grippers are discussed in detail in Chapter 5.

MANIPULATOR POWER SUPPLIES

Three types of power supplies are used to energize robot manipulators: pneumatic, electrical, and hydraulic.

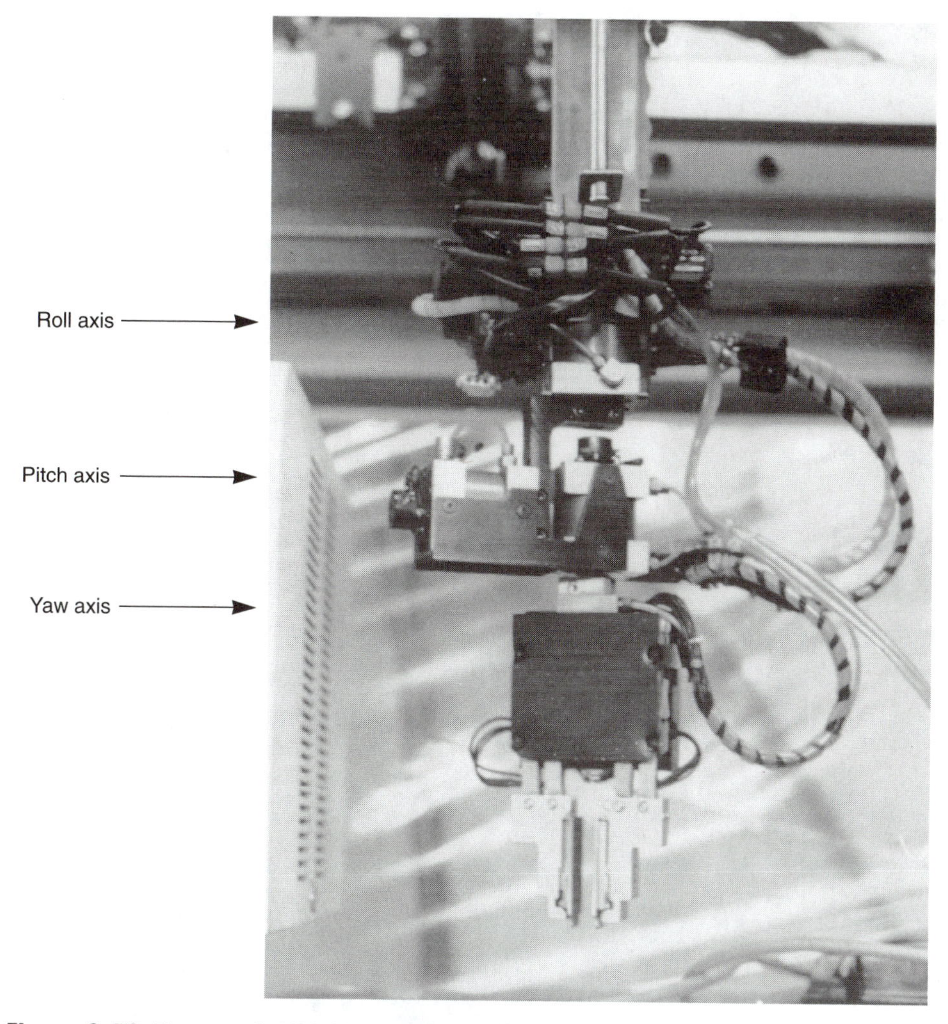

Figure 3–13 Three-axis Wrist from IBM 7576 Assembly Robot.
The wrist on this IBM robot has three rotational axes.

Pneumatic Power

Pneumatic-powered devices use a compressible fluid such as air for their power. The movement of compressed fluid can also be used to create a vacuum in some devices. The pneumatic power supply consists of a compressor, a storage tank for holding the pressurized fluid, and a motor or engine to run the compressor (see Figure 3–18). Figure 3–19 shows actual components of a pneumatic system, including a pneumatic power supply and parts of a regulator for injecting oil into compressed air to lubricate the devices it powers. The regulator also removes

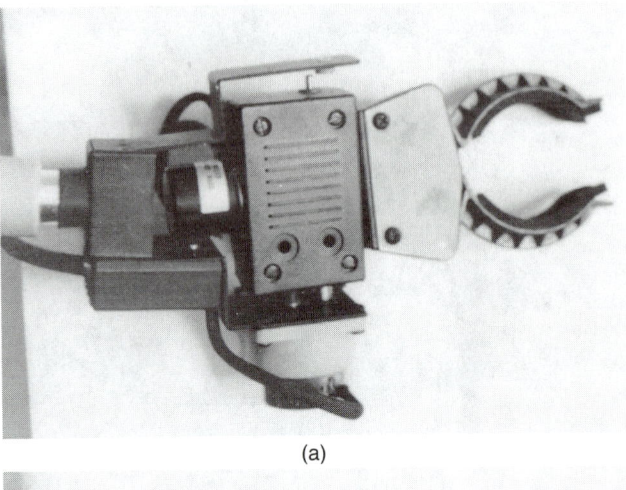

(a)

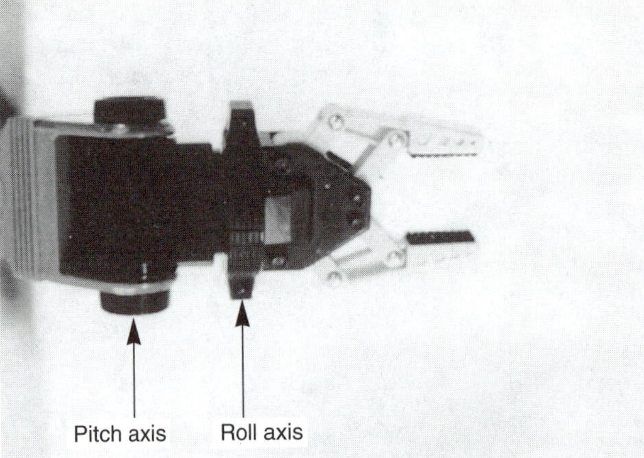

(b)

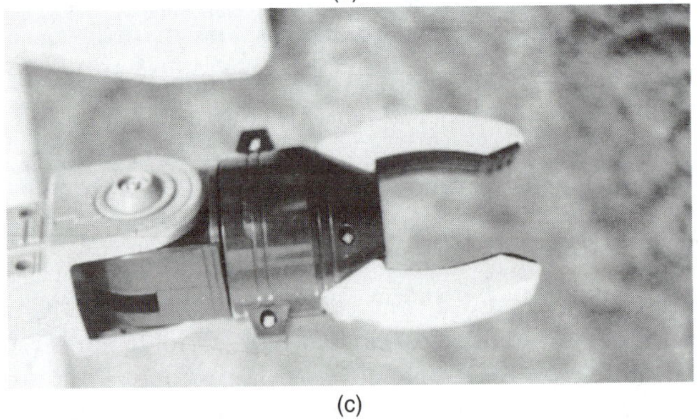

(c)

Figure 3-14 Two-axis Wrists.
(a) The Hero I wrist. (b) The Armatron wrist. (c) The Mobile Armatron wrist. Each of these wrists has two rotational axes.

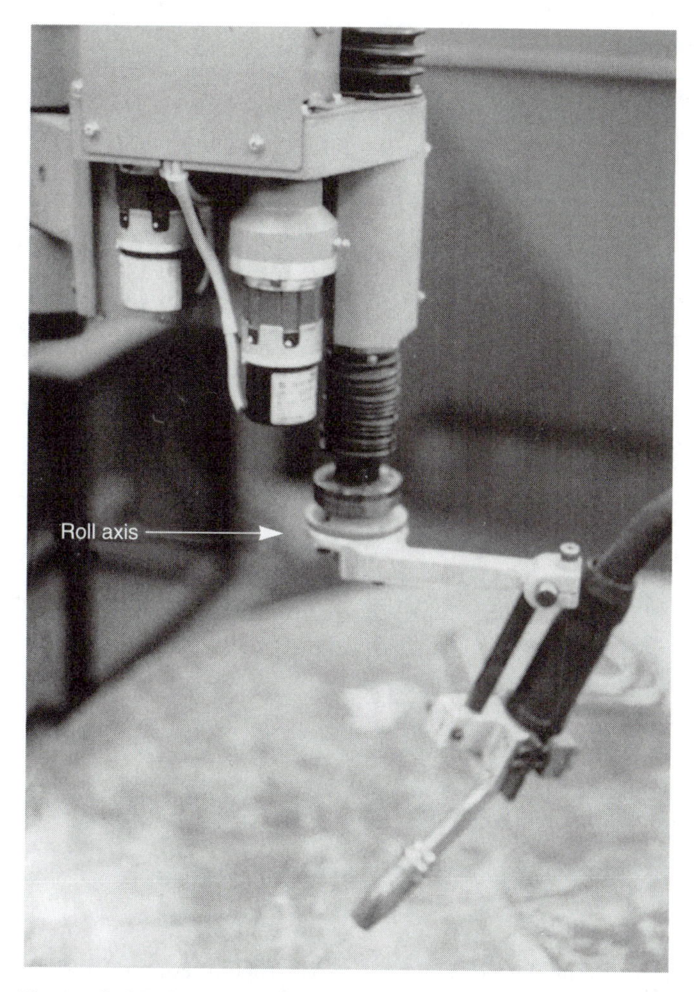

Figure 3-15 Single-axis Wrist.
The wrist on the AR-1 welding robot has only a roll axis.

water from the compressed air. This pressurized fluid can be stored in tanks and/or transported through high-pressure tubing to the place(s) where it is used.

If ambient air is used as the fluid, the used air can be vented back into the atmosphere. Hydraulic fluids, on the other hand, must be returned to the storage tanks. Air that picks up moisture becomes very corrosive. To solve the moisture problem, some pneumatic systems use an inert gas in place of ambient air. The pressures in a pneumatic-powered system may be 100 pounds per square inch (psi) or more.

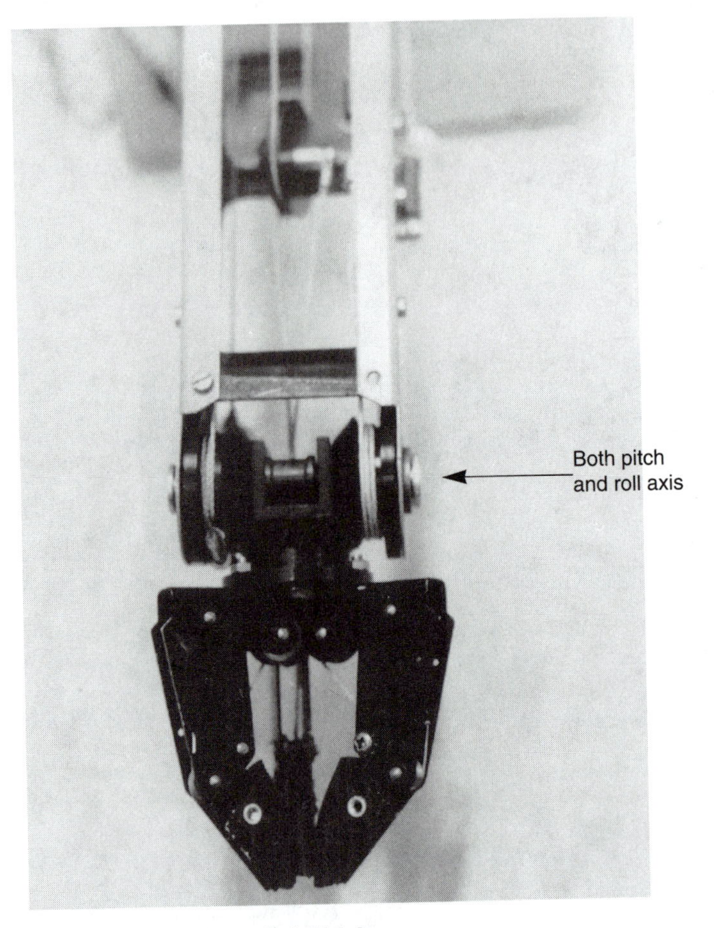

Figure 3–16 Differential Wrist.
The differential wrist on the Armdroid educational robot uses two motors. When these motors move in the same direction, the wrist moves up or down; when they move in opposite directions, the wrist rotates.

The "muscles" for a pneumatic-powered manipulator generally take the form of some type of actuating cylinder. These cylinders produce a linear (straight-line) motion, which may be converted into rotational motion through the use of a linkage similar to that used in steam engines. Figure 3–20 shows a single-action cylinder. The force (F) on the outward stroke of the piston is equal to 0.7854 times the square of the diameter (D) of the piston times the pressure (P) of the fluid entering the cylinder, minus the sum of the return spring pressure (S) and the friction force (Ff) of the piston. The formula is thus

$$F = (0.7854 \times D^2 \times P) - (S + Ff)$$

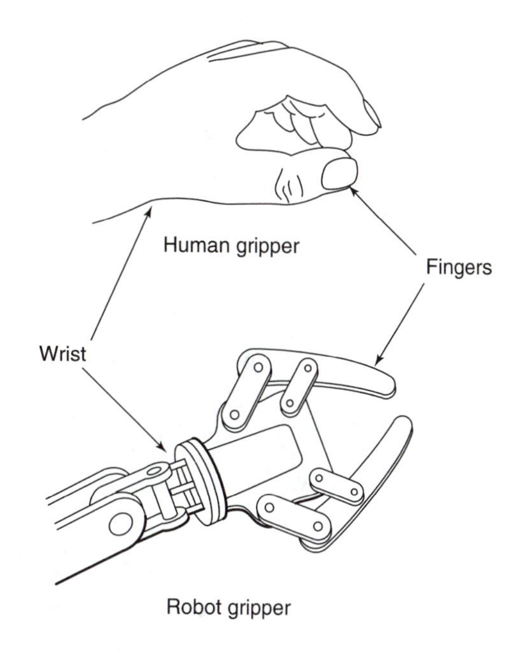

Figure 3–17 Human and Robot Grippers. Human grippers and robot grippers use similar nomenclature. The moving parts that do the gripping are called *fingers*.

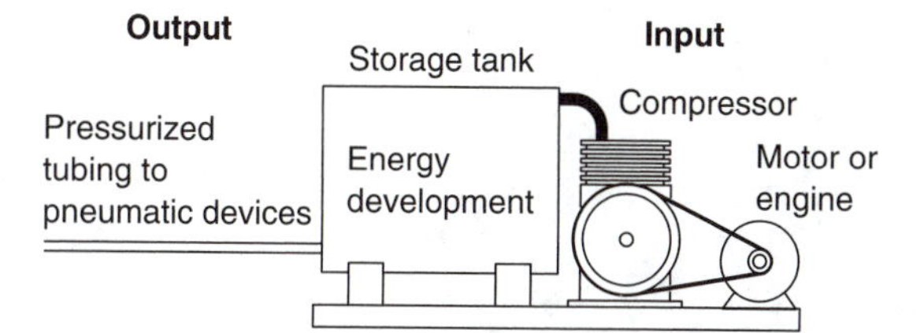

Figure 3–18 Pneumatic Power Supply. A pneumatic power supply always includes a motor or engine, a compressor, and a storage tank.

The force (F) on the return stroke of the piston is equal to the spring pressure (S) minus the friction force (Ff) of the piston. Expressed mathematically this is

$$F = S - Ff$$

If the fluid used is a compressible one (such as air), the pressure applied to the piston will be uneven (as a result of gravity and turbulence) and will be affected by the weight of the load on the piston.

Figure 3–21 shows a double-action cylinder. The force (F) on the outward stroke of the piston is equal to 0.7854 times the square of the diameter (D) of the piston times the pressure (P) of the fluid entering the cylinder, minus the friction force (Ff) of the piston. The formula is thus

$$F = (0.7854 \times D^2 \times P) - Ff$$

(a)

(b)

Figure 3-19 Components of a Pneumatic Power Supply.
(a) A working pneumatic power supply, showing all its components in place. (b) A water condenser and a lubricator. The condenser removes water from the compressed air, rendering the air less corrosive. The lubricator adds oil to the air so that it will lubricate the parts it passes through.

The force (F) on the inward stroke of the piston is equal to 0.7854 times the difference between the square of the diameter (D) of the piston and the square of the diameter of the piston rod (Dr), times the pressure (P) of the fluid entering the cylinder, minus the friction force (Ff) of the piston. Expressed mathematically, this is

$$F = [0.7854 \times (D^2 - Dr^2) \times P] - Ff$$

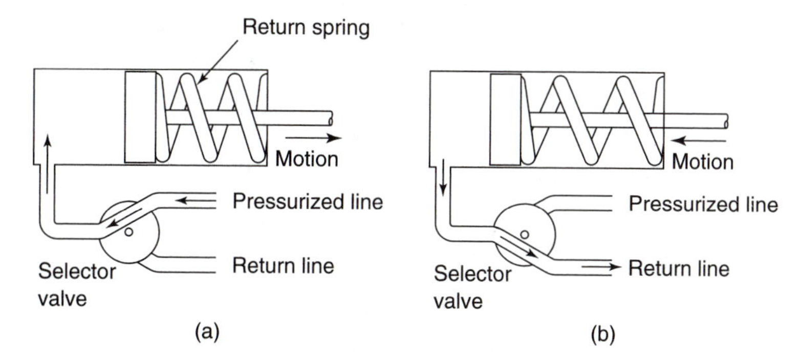

Figure 3-20 Single-action Cylinder.
(a) The power stroke of the piston in a single-action cylinder. Compressed air moves the piston out against the return spring. (b) The return stroke of the piston. The return spring furnishes the power to restore the piston to its original position and push the air out of the piston.

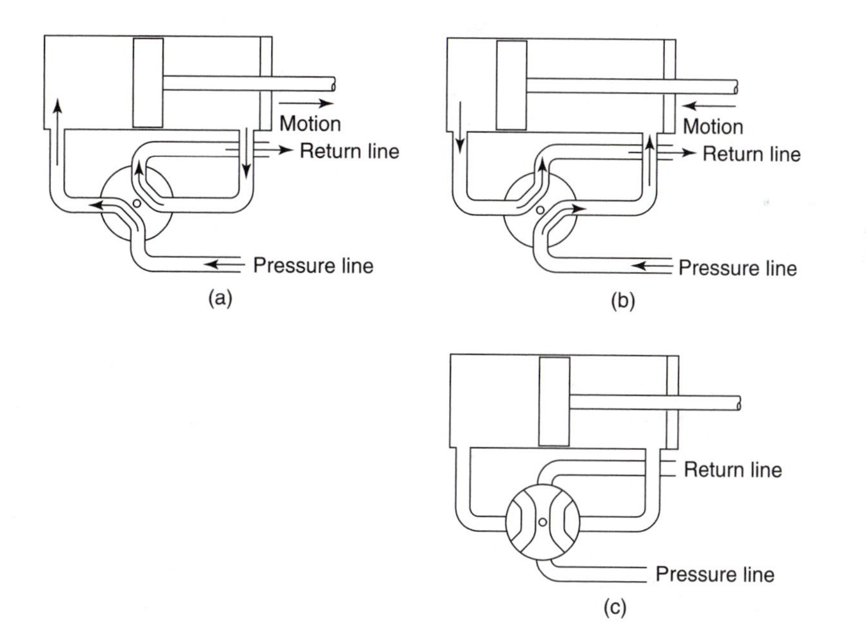

Figure 3-21 Double-action Cylinder.
(a) The outward stroke of the piston in a double-action cylinder. Compressed air moves the piston out and pushes the air out the return side of the piston. (b) The inward or return stroke of the piston. Compressed air moves the piston back in and pushes the air out the other side of the piston. (c) The piston in a holding position, with both sides of the piston sealed. Since air is compressible, the actual holding position depends on the weight of the load.

The double-action cylinder has an advantage over the single-action cylinder: It can be set to hold an in-between position. If a compressible fluid is used, the accuracy of this in-between position will be affected by the weight of the load on the piston.

SAMPLE PROBLEM 1.

A 1-inch-diameter double-action cylinder has an air pressure of 150 psi applied against the piston face, moving it outward. If the friction force of the piston is 20 pounds, what force will the piston rod apply to its load?

Answer

The force is equal to 0.7854 times the piston diameter squared times the air pressure, minus the friction force.

$$\begin{aligned} F &= (0.7854 \times 1^2 \times 150) - 20 \\ &= (0.7854 \times 1 \times 150) - 20 \\ &= 117.81 - 20 \\ &= 97.81 \text{ pounds} \end{aligned}$$

Pneumatic-powered manipulators are normally used only on small, speedy **pick-and-place robots** with open-loop nonservo-controlled systems. These robots handle light loads and move from point to point without any positional feedback information. Since there is no control over their speed of motion, the robots usually can be stopped at only two or three positions in their range of travel along each axis. The stop points are set up by mechanical switches or stops. Because these robots make a noise each time they run into one of their mechanical stops, they are also known as **bang-bang robots.** The pressure of the pneumatics holds the manipulator against the mechanical stop and thus provides very accurate positioning.

Pneumatic power is the least expensive of the power types for robots and is relatively maintenance-free. Small pneumatic-powered robots can achieve motion accuracies of up to 0.001 inch at endpoints. Small open-loop nonservo-controlled pick-and-place robots may have cycle times of 3 to 5 seconds for light loads. The heavier the part, the longer the cycle time and the less accurate the placement of the part. While bang-bang robots are the simplest type of robot, they also represent the largest class of industrial robots in use. Because they handle many of industry's material transfer tasks, they are sometimes classified as material transfer devices rather than as robots.

Electrical Power

Electrical power is clean and can be distributed through wires. The electrical power supplied to small and medium-size robots is usually direct current (DC) electricity for use in controlling DC electric motors. Since most companies use alternating current

(AC) electricity for input power to the robot's power supply, the electrical power supply must convert the AC power to DC power. Usually, more than one voltage of DC power is required. Figure 3–22 shows a sketch of an electrical power supply system.

The "muscles" of an electrically powered manipulator are furnished by some type of electromagnetic device such as relays and motors. DC electric motors, including linear solenoids, pulse motors, rotational solenoids, servomotors, and stepping motors, are used for small and medium-size robots. Most hobbyist robots are electrically powered.

Solenoids have two possible states: on and off. They move an armature to one of two positions, depending on which state they are in. Solenoids are popular for controlling values and for starting and stopping motors and are used, for example, on washing machines and automatic dishwashers. Stepping motors move incrementally each time they receive a pulse. The direction in which they move depends on the phasing of the pulse. By keeping track of the number and phase of the pulses sent to the stepper motor, you can determine where the device driven by the motor is positioned. Stepping motors are popular in robots that lack positional sensors. Servomotors are used to generate feedback about where devices are and to control their movement.

Electric motors become very expensive when they are used to run large and powerful manipulators. Such motors require voltages large enough to constitute a fire hazard in some industrial atmospheres. Electric closed-loop servo-controlled jointed-arm robots have been successfully used for transporting parts, inspecting dimensions of parts, arc welding, assembly, light-duty processing, and for precise placement of needles during brain surgery.

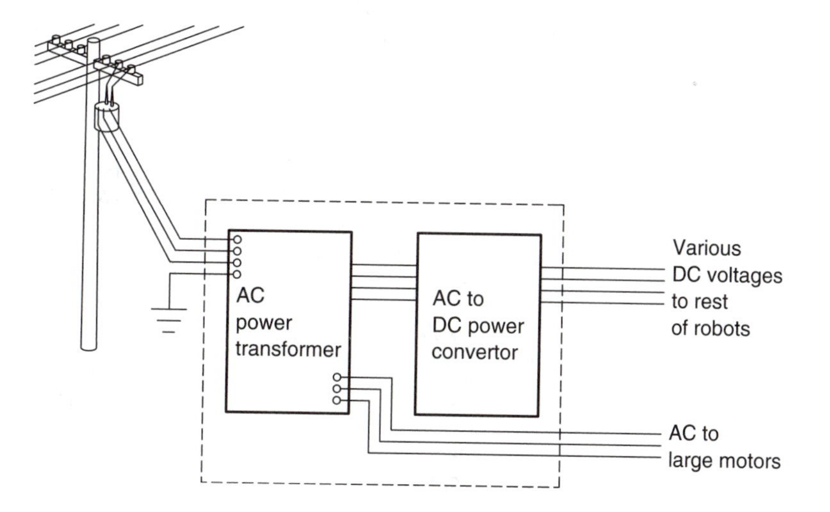

Figure 3–22 Electrical Power Supply System.
Electrical power supply systems usually use 60-cycle AC input, outputting the various DC voltages required by the robot and the controller. A large electric robot would also receive input of up to 460 volts of three-phase AC for its motor.

Very large, electrically powered robots use 220- or 440-VAC three-phase motors and are as big and powerful as the largest hydraulic robots.

If you are accustomed to the 5-volt logic of the modern computer, beware! The voltages of any electric robot are dangerous.

Hydraulic Power

Hydraulic power uses a noncompressible fluid to transmit energy. The hydraulic power supply uses pumps to supply the force needed to move the hydraulic fluid under high pressure. The power transmitted by the hydraulic fluid comes from the pressure the fluid exerts against things, and not from compression of the fluid itself. The fluid is used to activate hydraulic actuators or hydraulic motors. The power from the pump is then transmitted by the fluid through tubing to the hydraulic devices. After the fluid has done its work, it must be returned to the fluid supply tank, because it is too messy and dangerous to release into the environment and because it costs money.

Since hydraulic fluid works under high pressure, very small solid particles in the fluid can damage the hydraulic actuators and motors. Therefore, the hydraulic power supply must filter such particles out of the fluid before placing it into the hydraulic lines. Hydraulic fluid can also be damaged by reaching too high a temperature. When this happens, the fluid changes color and is consid-

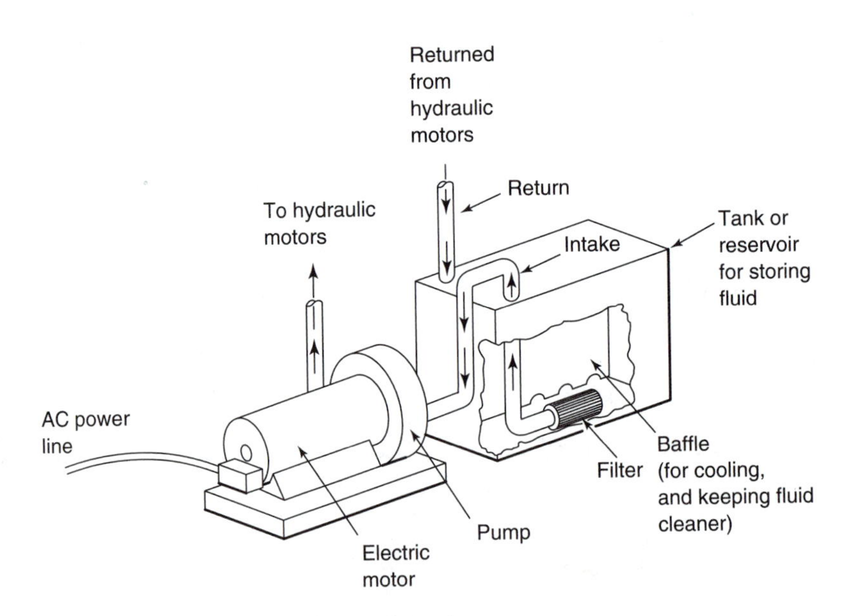

Figure 3-23 Hydraulic Power Supply.
A hydraulic power supply always includes a motor or engine to run a pump, a reservoir to hold a supply of fluid, a filter to remove contaminants from the returning fluid, and baffles to help cool the fluid and separate foreign material from the fluid.

ered burnt. The fluid needs to be inspected periodically and replaced if it is damaged or contaminated.

Figure 3–23 illustrates the main components of a hydraulic power supply. The supply tank has a filter to keep impurities out of the hydraulic system, and baffle plates to prevent turbulence, to trap air from the fluid, to separate foreign material from the fluid and allow it to settle to the bottom of the tank, and to increase heat transfer through the reservoir walls. Figure 3–24 shows actual parts of a hydraulic system.

The "muscles" for a hydraulic-powered manipulator usually take the form of some type of cylinder or motor. Hydraulic systems may use the same single- and double-action linear-motion cylinders that pneumatic power supplies use. When supported by a double-action cylinder and some type of feedback information, hydraulic-powered manipulators can achieve very accurate linear positions.

(a)

(b)

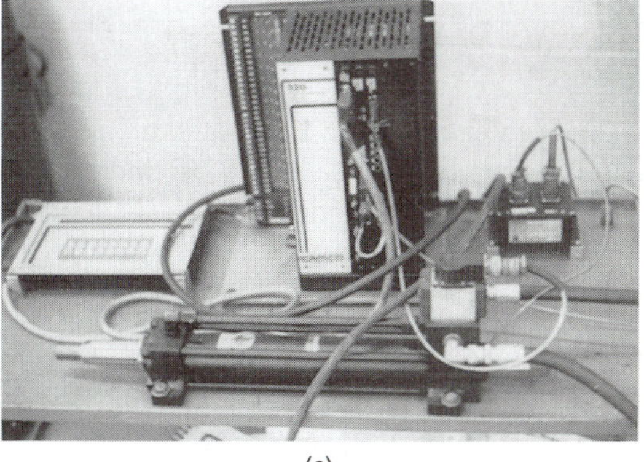

(c)

Figure 3-24 Components of a Hydraulic System.
(a) A working hydraulic power supply. (b) A hydraulic cylinder (above) and a hydraulic motor (below). (c) A hydraulic controller on a trainer.

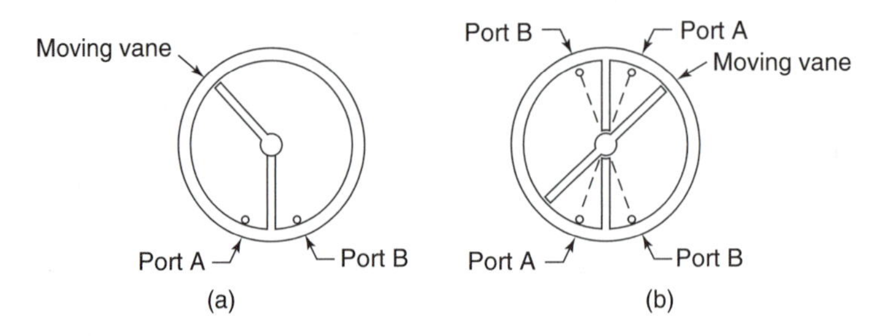

Figure 3-25 Rotary Piston.
(a) A single-action rotary actuator. If fluid enters port A, the vane moves clockwise. If the fluid enters port B, the vane moves counterclockwise. (b) A double-action rotary actuator. It has approximately half the rotational angle of the single-action actuator, but it gives twice the output power.

Hydraulic-powered systems can also use rotary actuators or cylinders. Figure 3–25 shows single- and double-action rotary actuators. The single-action rotary actuator can move through a rotational angle of approximately 280°. The torque (T) developed by the actuator is equal to the fluid pressure (P) times the vane area (A) times the center radius (Rc) of the vane, minus any friction torque (Tf). The formula is thus

$$T = (P \times A \times Rc) - Tf$$

The double-action actuator can move through only half the rotation of the single-action rotary actuator—or approximately 140°—but it can develop twice the torque by applying pressure to twice the vane area. The formula in this case is

$$T = (2 \times P \times A \times Rc) - Tf$$

SAMPLE PROBLEM 2.

A single-action rotary cylinder has a vane center radius of 1 inch and a vane area of 0.75 square inch. If the fluid pressure against the vane is 1,500 psi and the piston has 20 inch-pounds of friction torque, what torque is delivered to the load on the cylinder shaft?

Answer

The torque output of the cylinder is equal to pressure times vane area times vane center radius, minus friction torque.

$$\begin{aligned} T &= (1{,}500 \times 0.75 \times 1) - 20 \\ &= 1{,}125 - 20 \\ &= 1{,}105 \text{ inch-pounds} \\ &= 92.08 \text{ foot-pounds} \end{aligned}$$

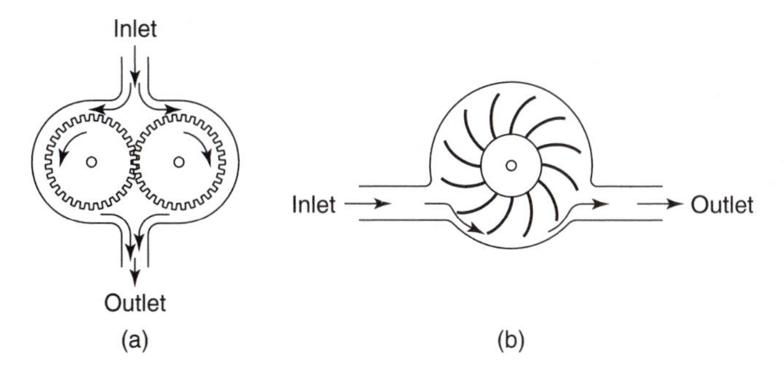

Figure 3–26 Hydraulic Motors.
(a) A gear motor. Fluid is forced between the gear teeth, causing the gears to rotate. One of the two gears is attached to an axle that has a load on it. (b) A vane motor. Fluid pushes against vanes, causing the shaft to rotate.

Several types of motors can be used in hydraulic systems. Figure 3–26 shows two of these types, the gear motor and the vane motor. The latter works much as a waterwheel does.

Hydraulic-powered manipulators are used in medium-size and large robots. The pressure in a hydraulic system may reach 1,000 pounds or more. Relatively small hydraulic motors can move loads weighing thousands of pounds. For powering large robots, hydraulic motors are less expensive than electric motors. Hydraulic actuators require relatively low voltages and can be used in explosive atmospheres without danger of causing a fire or explosion. This makes hydraulic-powered robots very popular in spray-painting applications. Large closed-loop servo-controlled unit robots can attain position accuracies of up to 0.001 inch. Hydraulic servo-controlled jointed-arm robots have been successfully used for spray painting, arc welding, and material transfer. Since hydraulic devices can be moved by hand when the hydraulic pressure is released from the system, these robots are generally trained by the walk-through teaching method.

Other Power Sources

As the robot emerges from the factory into the surrounding world, it and its manipulator may be powered by sources other than pneumatic, electrical, and hydraulic energy. In the past, robots have used water, steam, and metal springs for power. These sources may again be used in the future, along with gasoline, diesel fuel, engines, solar power, wind power, and nuclear power.

SUMMARY

The industrial robot is made up of four or five parts: the manipulator, the controller, the power supply, end-of-arm tooling, and (optionally) the vehicle. Manipulators are named after their geometric shape and the shape of their work envelope, or cell. Common types include rectangular or Cartesian coordinates, cylindrical coordinates, SCARA, spherical or polar coordinates, and revolute coordinates or jointed arm. The manipulator performs the required work motions for the robot. At the end of the manipulator is an end effector—a gripper (hand) or an end-of-arm tool. The power supply furnishes the power needed to move the manipulator. The most common power supplies used are electrical, pneumatic, and hydraulic. The control unit—always some type of computer—acts as the robot's brain.

The vehicle is used to move the robot if it cannot reach everything it needs from one base position.

FORMULAS

Single-action Cylinder

Outward stroke: $F = (0.7854 \times D^2 \times P) - (S + Ff)$

where F = force
D = diameter of piston
P = fluid pressure
Ff = friction force
S = return spring pressure

Return stroke: $F = S - Ff$

Double-action Cylinder

Outward stroke: $F = (0.7854 \times D^2 \times P) - Ff$

Return stroke: $F = [0.7854 \times (D^2 - Dr^2) \times P] - Ff$

where Dr = piston rod diameter

Single-action Rotary Actuator

$$T = (P \times A \times Rc) - Tf$$

where T = torque
A = vane area
Rc = center radius of vane
Tf = friction torque

Double-action Rotary Actuator

$$T = (2 \times P \times A \times Rc) - Tf$$

REVIEW QUESTIONS

1. What are the five styles of manipulators?
2. What are the three types of power used by robots?
3. Which style of manipulator has the longest reach for the amount of floor space it occupies?
4. Which type of robot manipulator most closely resembles the human arm and hand?
5. How does an industrial robot differ from a general-purpose electronic computer that might be used (for example) in the accounting department?
6. Given the following robot manipulators and tasks, find the best type of robot for each task.

Robots

a. Cartesian-coordinates robot with a reach of 4 feet along the x-axis, 2 feet along the y-axis, and 3 feet along the z-axis.
b. Cylindrical-coordinates robot with a vertical movement of 2 feet, from an inner radius of 3 feet, to an outer radius of 5 feet, and with a rotational angle of 190°.
c. SCARA robot with upper and lower arm sizes of 20 inches each, vertical movement of 2 feet, and a rotational angle of 190°.
d. Polar-coordinates robot with a base rotation of 190°, a vertical movement of 20° from the horizontal, and an in-and-out movement of 24 inches.
e. Jointed-arm robot with a combined 24-inch upper and lower arm, vertical movement of 160°, and horizontal movement of 190°.

Tasks

i. Retrieve a 6-inch part from a machine, and drop it into a bin behind the robot. A horizontal bar runs 9 inches in front of the machine, at the same level as the part to be grasped.
ii. Retrieve a 6-inch part from a machine, and drop it into a bin behind the robot. A vertical bar runs 9 inches in front of the machine, at the same left/right position as the part.
iii. Wire a 2-foot back plane that is placed on its side in front of the robot.
iv. Pick up a part from a machine at the robot's side, and place the part in a machine in front of it. Use a second gripper to retrieve the finished part and place it in a bin behind the robot.
v. Reach up above the robot, remove a painted part from a hook on a line, and place it in a bin in front of the robot.

7. A 1¼-inch-diameter double-action cylinder, with ½-inch-diameter piston rod, has a pressure of 1,500 psi applied against the back of the piston to move it inward. If the friction force is 25 pounds, with what force will the piston rod pull against the load?
8. A double-action rotary cylinder has a vane radius of 1½ inches and a vane area of 2 square inches. If the input fluid pressure is 1,250 psi and the friction loss is 25 pounds, what will be the torque delivered to the output shaft?
9. Why are the robots used in hospitals and offices electrically powered instead of using pneumatic or hydraulic power?
10. While processing semiconductors in a clean-room environment using robots, why are electric-powered robots preferred over hydraulic- and pneumatic-powered robots?
11. What are the four parts of a robotics system?
12. Why might a hydraulic-powered robot be preferred for handling heavy loads?

COMPONENTS OF AN INDUSTRIAL ROBOT: PART II

OVERVIEW

This chapter gives a detailed description of the robot's controller and its optional vehicle. It continues from where the last chapter left off in describing the parts of the robot.

The manipulator is the part of the robot that does the work and that people can see working. The power supply furnishes the energy to do the work. However, without a controller for a brain, it isn't a robot—it is just a remote-controlled devise, or teleoperator. Some of the robots in the factory and most of the robots used in the office need a vehicle so they can move about the office and do their task.

How is a robot controlled? Are industrial robots stationary or do they move about? These are some of the questions answered in this chapter.

OBJECTIVES

When you have completed this chapter, you should be familiar with:

- The three types of robot controllers
- The robot vehicle and mounting

KEY TERMS

Closed-loop
Continuous-path control unit
Controller
Control unit
Feedback
High-technology control unit
Low-technology control unit
Medium-technology control unit
Nonservo-controlled
Pick-and-place control unit
PLC
Point-to-point control unit
Servo-controlled
Stepper motor
Vehicle

CONTROL UNITS

The **control unit** is the brain of the robot. It contains the instructions that direct the manipulator to perform various movements in the proper manner and at the proper time to accomplish a particular task.

Feedback is the process of receiving information on how well a task or positioning is progressing. Control units either receive some type of feedback to follow the actual positioning of the manipulator, or have no feedback control and must assume that the manipulator is where it is supposed to be. Systems that use continuous feedback information are known as *closed-loop servo-controlled systems*. Human beings use a closed-loop servo-controlled system known as kinesthesia for knowing where their limbs are. Humans can add to their servo-controlled system any information received from sight and touch inputs relating to the limb's position. Systems without feedback information are called *open-loop nonservo-controlled systems*. A nondigital watch functions without feedback about how well it is positioning its hands when telling time; a music box works without feedback about how well it is playing a tune.

Human beings use a closed-loop servo system for most movements. The original signal sent to a muscle gets the arm or leg to the approximate position desired, although it generally falls a little short of its goal. Feedback messages from the muscles then tell the brain that the position is short of the goal, and the brain sends the muscle a supplementary correction command. If feedback signals are not received from the muscles, positioning the arm or leg is difficult and inexact. Visual feedback can be used to overcome the loss of neurological feedback. But even with visual feedback, a person's grip is very weak without muscle and pressure feedback signals.

Control systems and devices are nothing new to us. Our cars contain control devices in the form of engine timing, automatic transmissions, cruise control, and automatic windows. Our radios and televisions have control devices in the form of automatic gain control (AGC), automatic volume control (AVC), automatic frequency control (AFC), and automatic fine tuning (AFT). Such devices may be open-loop servo-controlled or closed-loop servo-controlled.

Servo Systems

Servo systems are used to control machinery. A system that works without sensors is known as an open-loop nonservo system. A system that works with internal sensors to provide feedback information is called a closed-loop servo system, or simply a servo system.

Open-loop Nonservo Systems Systems controlled entirely by on/off switches or by simple speed adjustments without feedback are open-loop servo systems. Common examples of open-loop nonservo systems include light switches, light dimmers, gas pedals on cars, volume controls on a radio or television set, stepper motors, mechanical stops, water faucets, manual transmissions, and clocks. Figure 4–1 diagrams two simple open-loop nonservo systems.

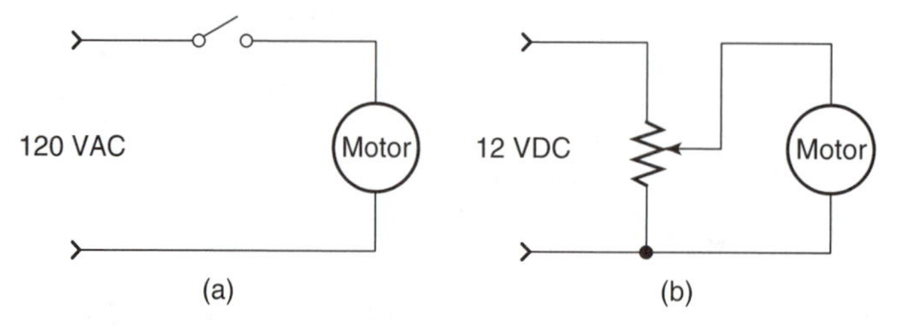

Figure 4-1 Open-loop Nonservo Systems.
(a) A circuit controlled by a switch—perhaps the simplest open-loop nonservo system. (b) A circuit controlled by a potentiometer.

Closed-loop Servo Systems Systems that use internal sensors to see how well they are doing their jobs are called closed-loop servo systems. The sensors may be as simple as the limit switch on an automatic garage door opener or as complicated as an optical incremental encoder on an advanced robot. Common examples of closed-loop servo systems include a heater with a thermostat, an oven with a thermostat, a governor on an engine, a DC motor with a tachometer or encoder, a motor with a limit switch, a stepper motor with an encoder, an automatic gain control, an automatic transmission, and an automatic volume control.

All closed-loop servo systems have at least the following six parts: a command or input signal, a comparer, an amplifier, an output device, a sensor device, and a feedback signal. The command or input signal specifies a desired position, speed, or other end state (depending on what the machine is supposed to do). The feedback signal identifies how well the machine has carried out the command by noting the actual position, speed, or whatever of the machine. The comparer compares the input signal and the feedback signal. If the signals are the same, the comparer issues no signal and the output device stops. If the two signals are different, the comparer

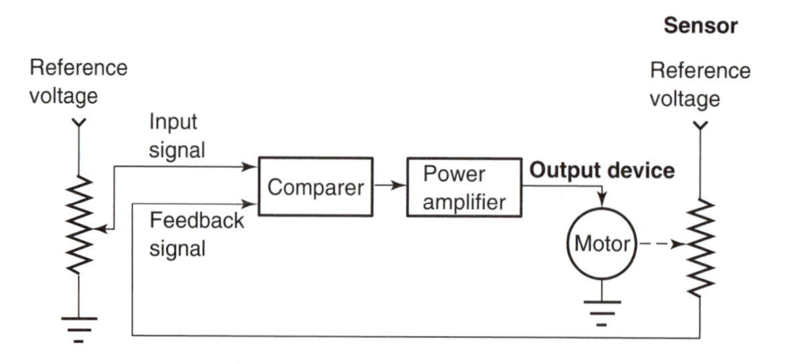

Figure 4-2 Closed-loop Servo System.
The closed-loop servo system has a feedback reference signal from the output telling how well the system is progressing toward carrying out the input command signal.

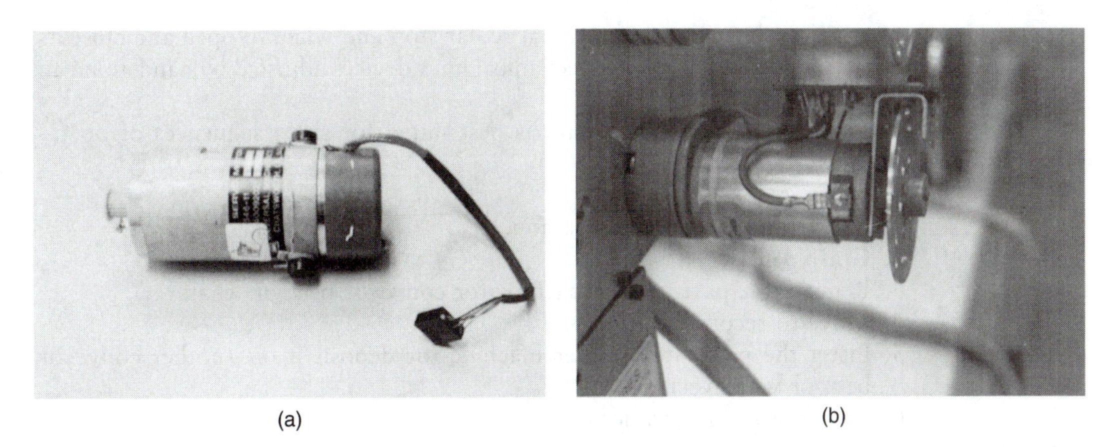

(a) (b)

Figure 4–3 Servomotors.
(a) A servomotor from a Qume printer. (b) A servomotor from a Rhino robot. Both servomotors are DC motors with optical encoders.

transmits some type of difference signal, causing the output device to move. The amplifier is used to raise the comparer signal to a power level capable of activating the output device. The output device is the motor (or whatever) that is to be controlled. Figure 4–2 shows a simple closed-loop servo system.

Motors used in a closed-loop servo system are often called **servomotors.** The motor portion of a servomotor can be a DC motor, an AC motor, a stepper motor, or even a hydraulic motor. What makes the servomotor special is that it includes some type of feedback system for assessing how well it is doing its task. Figure 4–3 shows two servomotors, both of which are DC motors that use optical incremental encoders for feedback. Encoders are explained in Chapter 6.

OPERATING METHODS OF ROBOT CONTROL UNITS

Control units can also be classified according to the operating method used by the robot. These methods include limited-sequence pick-and-place units, point-to-point units, and continuous-path units.

Pick-and-Place Control Units

Limited-sequence **pick-and-place control unit** robots are generally small and pneumatic powered with no position information feedback. That is, they are open-loop nonservo-controlled robots. Pick-and-place control units are sometimes referred to as **low-technology** control units. The simplest type of control unit for a pick-and-place robot uses a rotating control drum—each rotation represents one work cycle for the robot. On the drum's surface are places to insert pegs to determine where and when to position the robot against a particular stop and when to open and close its gripper. The stops themselves are mechanical limit devices adjusted by hand along an axis.

The pick-and-place control unit is best-suited for short sequences of operations, such as:

1. Move robot to starting position.
2. Grasp a part.
3. Remove the part from a machine (or conveyor belt, or whatever).
4. Move to second position.
5. Insert the part into another machine (or deposit it on another conveyor belt, or whatever).
6. Prepare to start another cycle.

Pick-and-place robots are programmed by adjusting the mechanical limit stops on the axes of the manipulator and by placing pegs as desired on the control unit's drum memory. Most open-loop nonservo-controlled pick-and-place robots are Cartesian-coordinates or cylindrical-coordinates robots, since the linear slides work well for the placement of mechanical stops and limit switches.

The main style of control device used on pick-and-place robots is some type of on/off mechanism. The control drum on a washer, called a timer, is comparable. Electrical contacts on the timer start and stop the motor, turn the water on and off, and engage or disengage the spinning of the tube. These electrical contacts are really electrical on/off switches. Typically, they activate a solenoid motor or electric magnet that performs the mechanical turning on and off of some device. Figure 4–4 shows a timer drum and an electrical solenoid motor.

A modern pick-and-place robot may use an electronic controller instead of an electromechanical drum. Either way, the controller activates solenoids that turn on

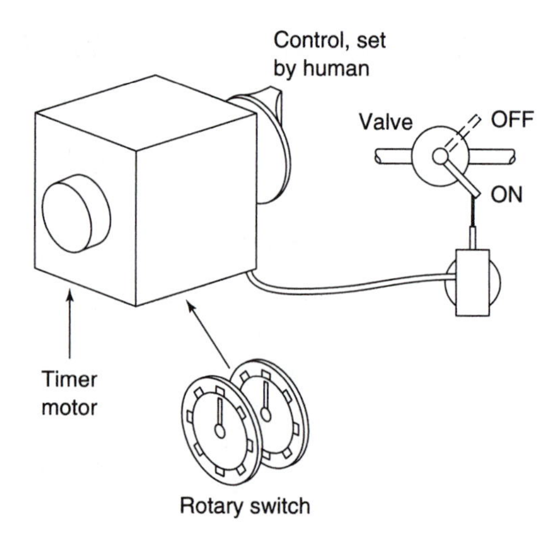

Figure 4–4 Timer Drum and Solenoid Motor. A timing drum uses a timer motor to turn a rotary switch. Contacts on the rotary switch connect to various devices. For example, a solenoid motor that opens and closes a valve could be one device.

or off the air supply in the cylinders of the pneumatic manipulator. This causes that axis of the manipulator to extend or retract.

It is also possible to buy a pick-and-place robot manipulator without a controller. In this case, the manipulator may be controlled by the PLC (programmable logic controller) that controls the rest of the machines on the production line. A **PLC** is a special-purpose electronic computer designed to handle the input and output signals to mechanical devices (see Figure 4–5). PLCs have replaced the old relay controller logic.

Point-to-Point Control Units

The **point-to-point control unit** can reach any point within its work envelope and can have as many points in its work sequence as a particular task may require, limited only by the size of the control unit's memory. A robot with this type of control unit can do much more complicated tasks than can a robot with pick-and-place controls. Point-to-point control units are also known as **medium-technology** control units. Point-to-point robots are programmed through the use of a manual teaching device whereby a person moves the robot through the sequence of points that it will be required to repeat in performing the task. Alternatively, the point's coordinates may be entered directly into the robot's memory.

The manual teaching method requires that, at each desired point of the sequence, the training person press a button telling the robot's control unit to remember that point. This programming phase is done at a slow speed. Later, when the robot is running at normal speed, its movement from point to point in the task sequence will occur in an approximately straight line. The exact path taken by the robot between points is not programmable.

Remote programming, an alternative method of programming the robot, is most often done when the robot is part of a group of computer-controlled robots. The programmer enters the point coordinates into the computer, and the computer calculates the necessary point and slide positions to get the robot to that point. Since the position sensors on different robots (even robots of the same make and model) differ slightly, no two robots will agree about exactly where the manipulator is located. Consequently, the computer may have to develop a correction table for translating a program from one robot to another.

Point-to-point controlled robots are generally more expensive and can handle heavier loads than pick-and-place robots. Most point-to-point robots are hydraulic-powered units, although some electrically powered point-to-point robots are manufactured.

An electrical point-to-point robot not equipped with position feedback could use stepper motors to control its position. Stepper motors have been used for years on electronic computer input and output devices—printers, flexible diskette drives, and hard disk drives. Stepper motors, for example, move the paper in a printer forward or backward, one step at a time, and position a disk drive's read/write head to the proper place for reading or writing a specific circle of information.

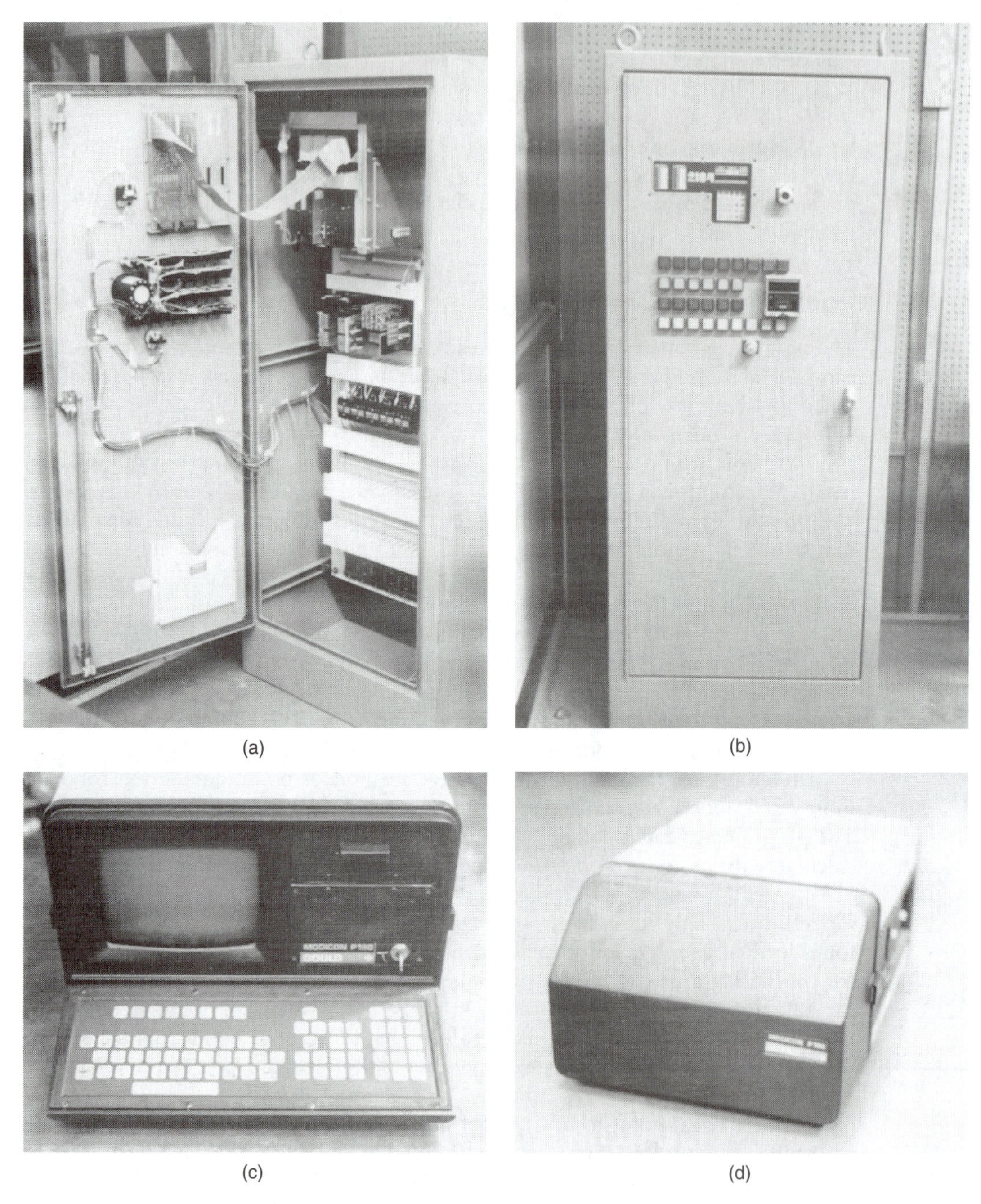

Figure 4-5 Programmable Logic Controller (PLC).
(a) The inside of a small programmable logic controller that handles eight input and eight output lines. (b) The outside of the controller. (c) The removable terminal for the controller. (d) The terminal with its protective cover in place.

Stepper motors are multipole motors that advance their rotor by one pole for each voltage pulse they receive. Figure 4–6 shows a cross-section of a three-phase stepper motor. To function, a stepper motor must have a minimum of two phase windings. The rotor in Figure 4–6 is shown after the phase A winding has been activated. Residual magnetism is usually sufficient to hold the rotor in this position after the phase A winding is deactivated, although the winding can be kept activated to increase the holding force, if necessary.

If the phase B winding is activated next, the rotor will turn counterclockwise to align with the B stator poles. (The counterclockwise direction of movement is due to the fact that the rotor pole between stator poles B and C is closer to stator pole B than is the rotor pole at stator pole A.) If the phase windings are alternately activated in the sequence A, B, C, A, B, C, . . . , the rotor will rotate counterclockwise. If instead the phase windings are alternately activated in the sequence A, C, B, A, C, B, . . . , the rotor will rotate clockwise.

Each activation of a phase winding causes the rotor to move one-half the distance between two stator poles. With six poles, then, the rotor rotates one-twelfth of a circle (30°). Every increase in the number of poles decreases the number of degrees of rotation per phase-winding activation.

Continuous-path Control Units

The **continuous-path control unit** robot can reach any point within its work envelope and can have as many points in its sequence as a particular task may require, just as the point-to-point robot can. But in addition it can map the exact path to take between points. This allows the robot to be used for some semiskilled tasks that require control of the path motion between points, as in automobile spot-welding or spray-painting. In fact, a continuous-path robot is more consistent at replicating an exact path than a human worker can be.

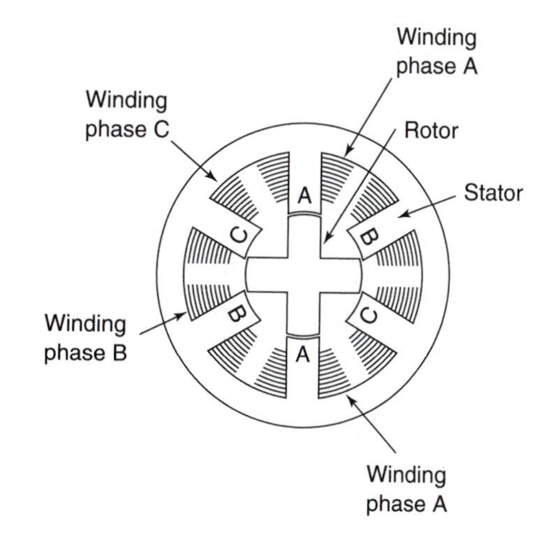

Figure 4–6 Stepping Motor.
This is a three-phase permanent magnet stepper motor.

Continuous-path control units are the most expensive of all control units. They require a large memory capacity to be able to record many times a second the exact position along a path. Continuous-path control units are also known as **high-technology** control units. Manually programming a continuous-path robot consists of leading the robot through its task at full operating speed. Most continuous-path robots need some type of positional feedback system in order to keep their positioning accurate.

THE VEHICLE AND THE ROBOT'S BASE

The vehicle for a robot is the device used to move it into a position in which the manipulator can do its work. The base of the robot provides a stable foundation for the manipulator.

Stationary Robots

Many industrial robots have fixed-position bases and thus do not have a vehicle. Even with a fixed-base robot, a stable mounting is essential. If the mounting of the robot is unstable, the endpoint accuracy of the robot will be impaired. In some instances, it has been necessary to tear out the old factory floor and lay a new, more stable floor to get the robot to work accurately.

Most fixed-base robots use floor mountings, although side or overhead mounts may be used to save floor space; however, getting a stable wall mount is more difficult than getting a stable floor mount.

Overhead mounting involves supporting the arm from above. An overhead pendulum mounting that allows the robot to swivel at approximately its center of gravity can achieve approximately 50 percent higher acceleration than can a conventional floor-mounted jointed-arm robot. While it is possible to mount a floor-mount robot from a ceiling mount, it is difficult to do so. A more common overhead mount for a robot arm is used by the gantry robot. Here, the arm is suspended in much the same way as the manipulator of an overhead crane is.

Figure 4–7 shows some possible robot base mountings.

Mobile Robots

An industrial robot that serves several machines may be mounted on a carriage vehicle that takes the robot along a track from machine to machine. The vehicle must hold the robot base in a prescribed position and must make the base very stable. The robot's center of gravity must be kept within the wheelbase of the vehicle. The vehicle may possess from one to six axes of motion.

If the vehicle uses free-moving wheels for its motion, some method of steering the vehicle must be provided. At least three wheels, placed to form a triangle, are required for stability. The odd wheel can be either the front wheel or the rear

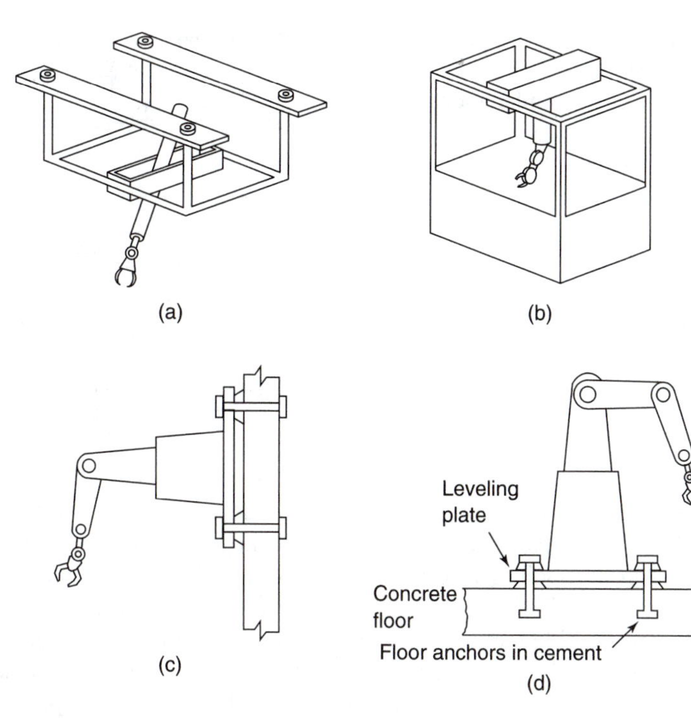

Figure 4–7 Fixed-base Mounted Robots.
(a) A ceiling-mounted robot. This is generally not a practical arrangement. (b) An overhead gantry mount, the most common way to mount a robot from above. (c) A wall-mounted robot. This arrangement saves floor space, but unless the wall is very rigid, the robot will lose some accuracy. (d) A floor mounted robot. This setup requires a very stable floor.

wheel, and the steering can be done by either a single wheel (the simpler method) or a pair of wheels. Likewise, the drive power can be applied to a single drive wheel or to two or more wheels. Using a single drive wheel is the simplest approach. Using two drive wheels works best with some type of differential between the wheels, if the vehicle is to make turns—unless, of course, the steering is to be accomplished by turning the two wheels at two different speeds. Figure 4–8 shows some possible three- and four-wheel vehicle configurations. A vehicle-mounted robot must be equipped with some type of sensors to position the vehicle so that the manipulator can do its work accurately.

A robot's **center of gravity** on a vehicle should be as low as possible. This will make the robot more stable during acceleration, deceleration, turning, and going up and down nonlevel surfaces. If the center of gravity ever falls outside one of the wheels, the vehicle will tip over (see Figure 4–9). Spreading the wheels out as far as possible improves stability. Large wheels allow the robot to move over rougher surfaces than can be negotiated on smaller wheels. Treaded vehicles can move over rougher surfaces than can wheeled vehicles but are more difficult to construct.

Figure 4–8 Configurations for Wheeled Vehicles.
These are some of the combinations that can be used to drive and steer a three- or four-wheel robot vehicle.

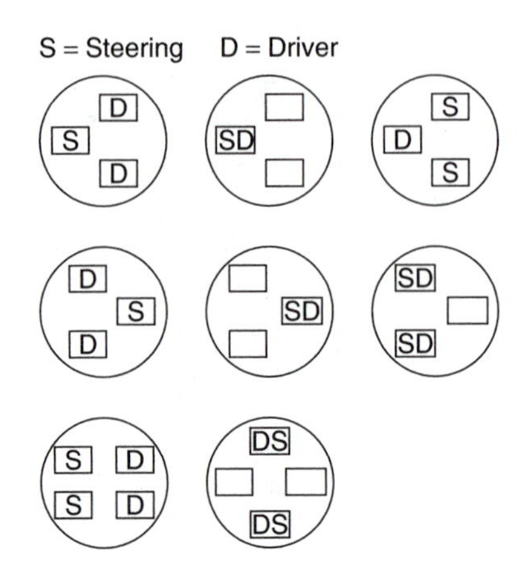

Wheels are the most common locomotive device for a robot vehicle. Some vehicles make use of treaded devices similar to those used on tanks and Caterpillar machinery. A few rough-terrain robots are given multiple-leg devices that move the way a spider's legs do. The Odex I robot, from Odetics, Inc., was the first functional robot to be able to walk on any terrain (see Figures 4–10 and 4–11).

Robots generally do not try to copy the two-legged style of locomotion used by humans because this style is inherently unstable. The human processes of walking and running are both forms of controlled falling. If you find this difficult to imagine, just watch a baby human who is learning to walk. Walking on two legs requires sophisticated stabilization sensors, stabilization recovery or correction devices, and the ability to shift weight from one leg to the other.

Figure 4–9 Vehicles and Center of Gravity.
A vehicle is stable as long as its center of gravity falls within the wheelbase of the vehicle. The lower the center of gravity, the steeper the incline the vehicle can safely ascend or descend.

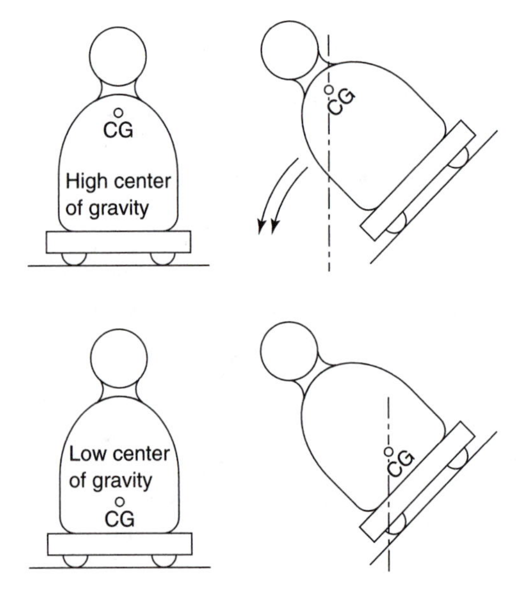

Figure 4–10 Odex I Climbing into a Truck.
The Odex I robot here is climbing into the bed of a pickup truck through remote control.

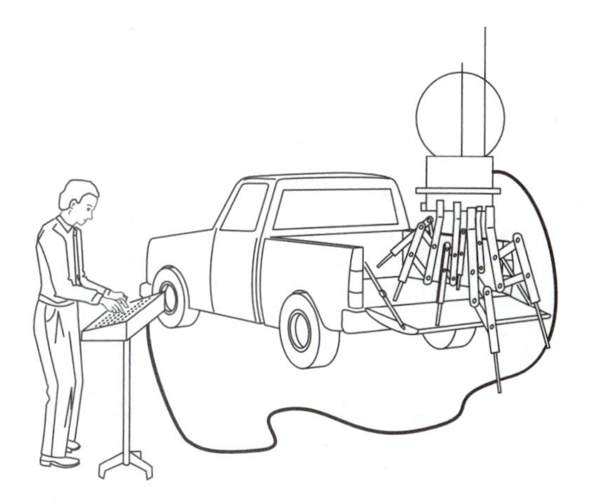

(a)

Figure 4–11 Odex I on Stairs.
(a) The original Odex I in a museum. (b) The Odex I climbing stairs under internal control. (Courtesy of Odetics, Inc., Anaheim, California)

(b)

Figure 4-11 ***(continued)***

SUMMARY

The industrial robot is made up of four or five parts: the manipulator, the controller, the power supply, end-of-arm tooling, and (optionally) the vehicle.

The control unit—always some type of computer—acts as the robot's brain. Control units are named for the type of control they provide: limited-sequence, point-to-point, or continuous-path. All controllers require some type of memory. Limited-sequence controllers have the least amount of memory and may be mechanical in nature. Point-to-point controllers keep track of where they are located and record the points they must pass through while doing a task. Continuous-path controllers require the most memory and record where they are many times per second.

The vehicle is used to move the robot if it cannot reach everything it needs from one base position. This makes the base less stable and may reduce the accuracy of the robot's positioning.

REVIEW QUESTIONS

1. Why do most industrial robots not have a vehicle?
2. What problems may be encountered by a robot that uses a vehicle?
3. Why do industrial robots require a stable base?
4. Why might a spot-welding robot be able to use a point-to-point controller while a spray-painting robot would need a continuous-path controller?

5

END-OF-ARM TOOLING

OVERVIEW

Robot end-of-arm tooling is not limited to various kinds of gripping devices. Such tooling may also take the form of tools for doing such work as drilling, grinding, painting, and welding. In this chapter, the term *gripper* refers both to devices that grip materials and parts, and to devices that do work on parts.

What specific types of end-of-arm tooling are used? How important are these to a robotics project? How much force should a gripper exert? These are a few of the questions answered in this chapter.

OBJECTIVES

When you have completed this chapter, you should be familiar with:

- The great variety of grippers available
- The importance of choosing the right gripper
- How to calculate the force required to do a job

KEY TERMS

Acceleration
Center of gravity
Coefficient of friction
Compliance surface
Deceleration
End effector
End-of-arm tooling
Gripper
Hand
Mandrel lifting device
Pin lifting device
Pneumatic finger
Remote center compliance
Safety factor
Springs
Teleoperator
Torque

GRIPPERS AND HUMAN HANDS

The first gripper, or artificial hand, was probably a hook replacing a lost human hand. An artificial hand that looked like a human hand is first recorded in 1509—an iron hand made for a German knight, Goetz von Berlichinger. Medical robotics continues to improve artificial hands or grippers.

As research in nuclear science became active, grippers became popular for moving things remotely. Remote-controlled grippers are called **teleoperators.** Figure 5–1 shows two short-range teleoperators. Most teleoperators rely on visual feedback to supply status reports on how a task is progressing; more recent teleoperators have been equipped with touch feedback sensors. Astronauts train to manipulate robotic teleoperators in space from within the safety of the spacecraft.

The robot's end-of-arm tooling may also be known as a gripper, a hand, or an end effector. It attaches to the wrist end of the arm and may serve either to handle parts or to process parts. End-of-arm tooling is not always included with the basic robot. If you buy a welding robot or a painting robot, the wrist and gripper may well be included as part of a complete package, but more general-purpose robots, such as for machine loading and unloading, where a gripper specifically suited to the part being moved is needed, are commonly sold without grippers.

Figure 5–2 shows the human hand, with its twenty-seven bones and twenty-two degrees of freedom. No robot gripper even approaches the human hand's versatility. When a robot is substituted for a human in a task, it is natural to think of using a gripper that resembles the human hand, yet this may not prove to be the best gripper for the task.

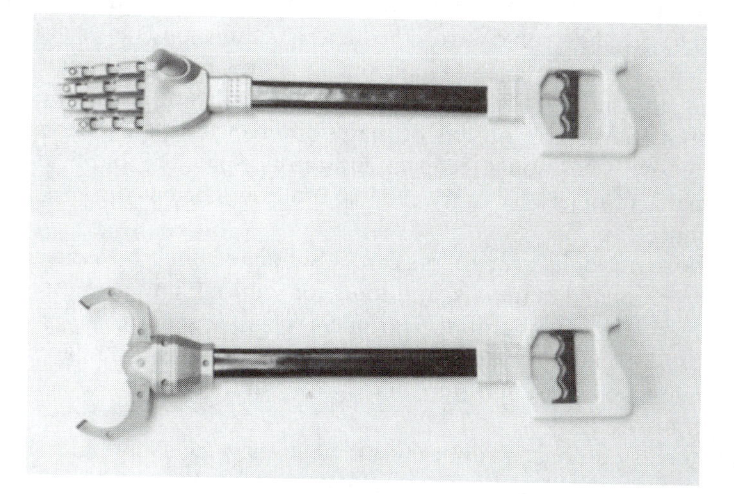

Figure 5–1 Teleoperators.
These teleoperators are from a toy store, but teleoperators are nonetheless a very serious matter when it comes to handling hazardous materials or working in dangerous surroundings.

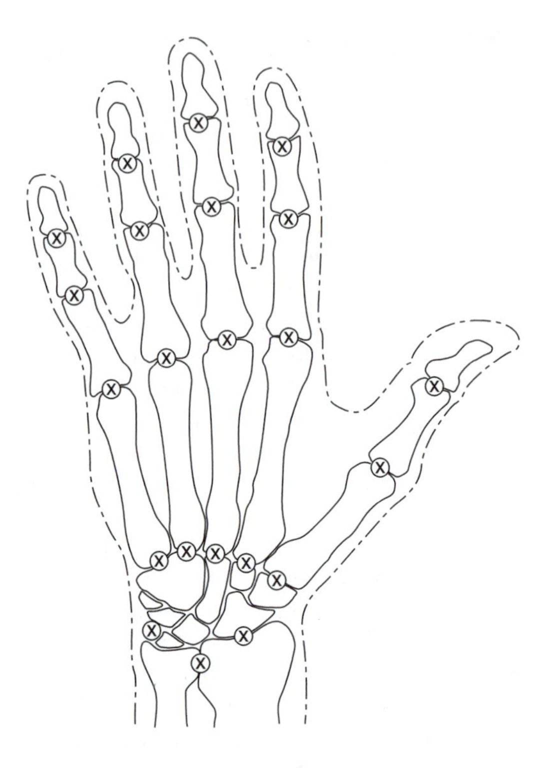

Figure 5–2 Human Hand.
The human hand is the most effective and versatile general-purpose gripper in the world. With twenty-seven bones and twenty-two degrees of freedom, it is also very complex.

The human wrist and hand are extremely complicated. In addition to the wrist's pitch, yaw, and roll control, the hand has three joints in each finger and two in the thumb. While these joints have only a simple rotational ability, the joints at the base of the fingers can also move sideways, thus allowing the hand to take many shapes and to grip objects of many different sizes and shapes. The hand is also covered with positional feedback indicators—pressure and heat sensors—that tell how hard an object being touched is and how hot or cold it is relative to the hand's temperature.

End-of-arm tooling for robots can be subdivided into two categories: *grippers* for moving parts and materials, and *tools* for working on parts and materials. Examples of grippers are multifinger grippers, clamps, scoops, ladles, hooks, forks, vacuum cups, vacuum surfaces, magnetic devices, and sticky fingers. Examples of tools are drills, torches, screwdrivers, spray-painting heads, spinning heads, and riveting heads.

In some situations, a robot must change its gripper during its task. If so, the robot's wrist must be fitted with a quick-disconnect device. The robot then selects the proper gripper at the proper time from a gripper carousel. Figure 5–3 shows an ISI Manufacturing Inc. two-finger gripper mounted on a low-profile pneumatic cylinder with a manual quick-disconnect pin. The ring through the head of the pin is visible at the lower middle portion of the picture.

Figure 5-3 ISI Two-finger Gripper Assembly.
The ISI two-finger gripper assembly includes a low-profile pneumatic cylinder (on the left). In the middle is the quick-disconnect head adapter, with the head of the release pin showing. On the right is the two-finger gripper. (Photo courtesy of ISI Manufacturing Inc., Fraser, Michigan)

Some companies specialize in the designing and manufacture of robot grippers. Two such companies are Barry Wright Corporation and ISI Manufacturing Inc. Figure 5–4 shows some of the grippers offered by the latter company.

CHARACTERISTICS OF END-OF-ARM TOOLING

The industrial robot's hand has not reached the state of development of the human hand, although experiments have been conducted with the aim of designing a mechanical hand with fingers and a thumb that will have the same rotational joints as on a human hand.[1] Even so, an industrial robot's hand can be designed to do a better job at some particular tasks than can the human hand. For instance, with the proper metal hand, a robot can handle red-hot objects; with suction cups, a robot

[1]Caporali and Shahinpoor, "Design and Construction of a Five-fingered Robotic Hand," *Robotics Age* (February 1984): 14–20.

Figure 5-4 ISI Gripper Parts.
The top center item is a low-profile pneumatic cylinder. The upper left item is an MGH5 gripper. The lower left item is an MGH1 gripper. The lower right item is a GHAO gripper head, without any fingers. The upper right item is an MGH2 gripper. At the center is the pin used for connecting and disconnecting grippers. (Photo courtesy of ISI Manufacturing Inc., Fraser, Michigan)

can handle large sheets of glass or other smooth materials without ever dropping them; and with an electromagnet attachment, it can be used to pick up and deposit objects made from magnetic materials such as iron, steel, and nickel.

The robot's hand may also have sensors in it—proximity switches, fiber optics and light sensors, pressure switches, magnetic-field sensors, vibration detectors, or speed-of-motion sensors. The wrist of the robot may be equipped with microphones or television cameras to provide artificial hearing or sight. Experimental grippers are now being made with a plastic skin that contains a matrix of tiny wires to give the gripper a sense of feeling or pressure.

Specialized Utility for a Job

For most industrial applications the robot's hand or end effectors must be designed to perform a specific job. Consequently, when a robot is assigned a new task, it generally needs a different gripper or end effector to do the work. Even a change in the size of the parts being processed by a machine that a robot is tending may require a change in the robot's gripper.

A robot's end effectors are not limited to replicating what a human hand can do. In some cases, two different types of end effectors may be used on the same arm: one to load the unmachined part, and another to unload the machined part. A robot tending more than one machine may need to change end effectors as it goes from one machine to another. Something similar to this occurs when a human puts down a hammer and picks up a saw during a construction task. When a robot needs to switch to a drill or welding gun, that tool becomes the end effector.

Figure 5–5 shows a few of the many available types of grippers. Figure 5–6 shows several two-finger grippers. Figure 5–7 shows two more types of end-of-arm tooling.

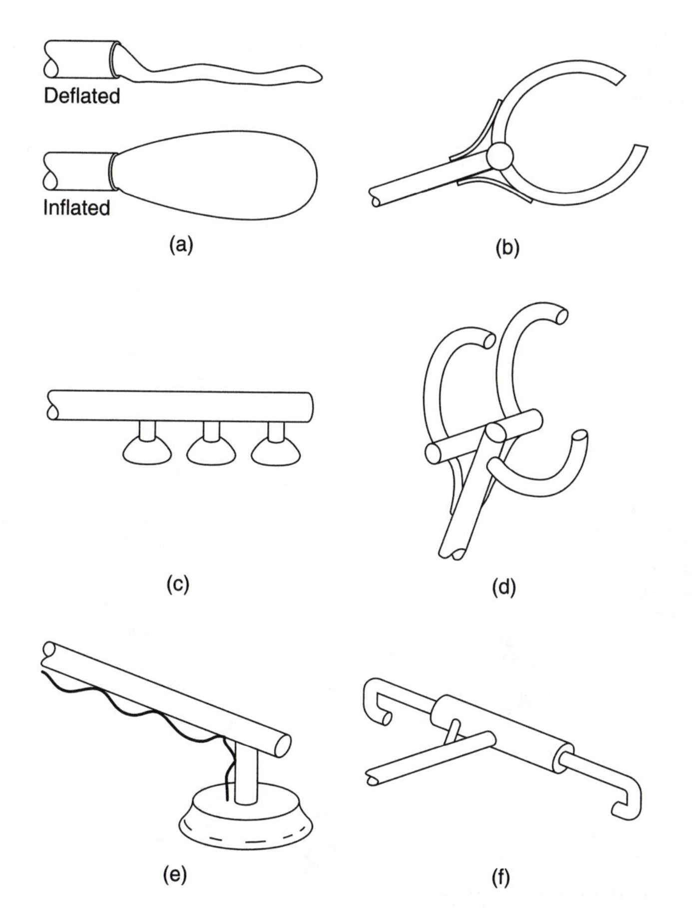

Figure 5–5 Types of Robotic Grippers or Hands.
(a) Inflatable bladder. (b) Two-finger gripper. (c) Vacuum cups. (d) Three-finger gripper. (e) Magnet head. (f) Tubing pickup device.

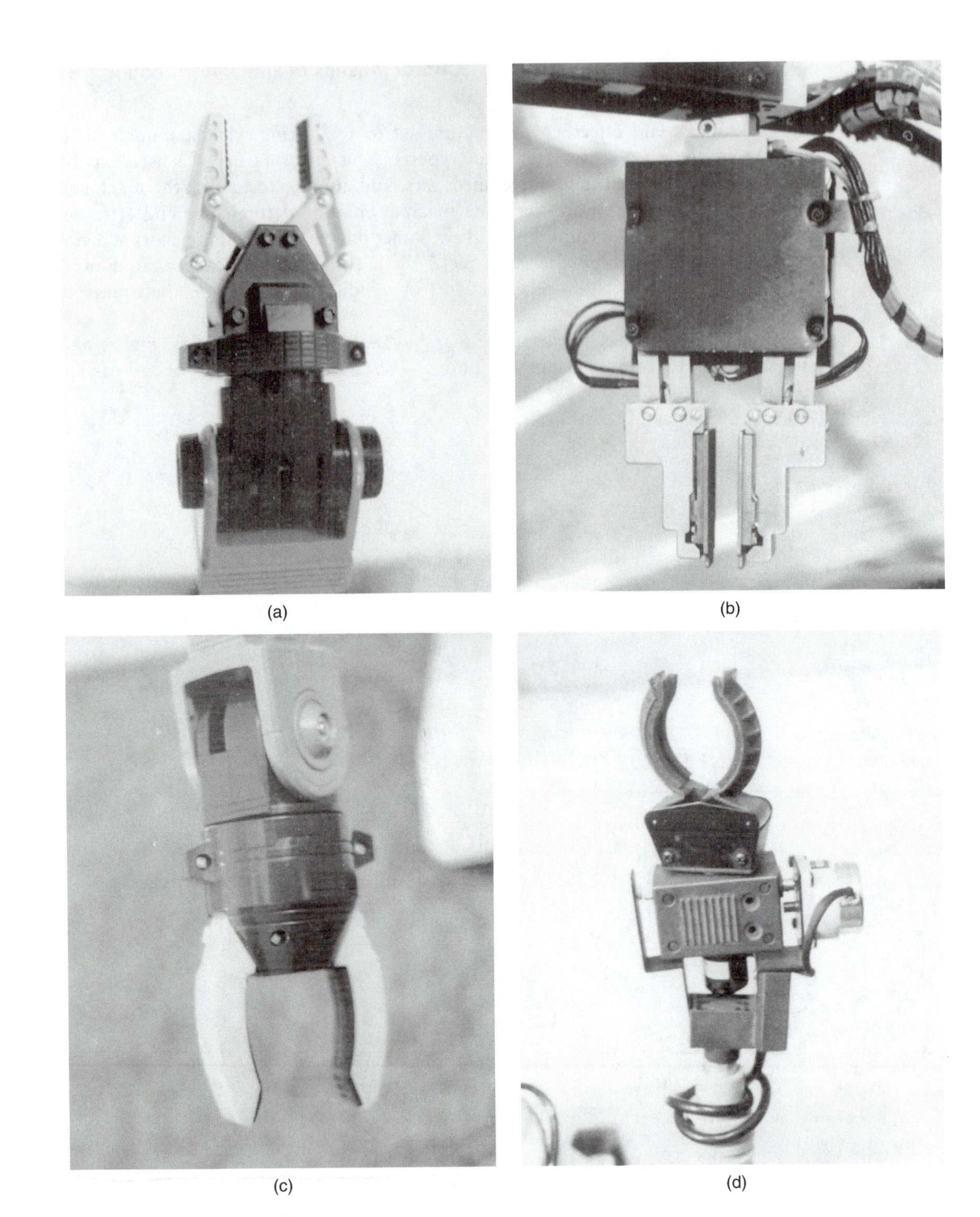

Figure 5-6 Two-finger Grippers.
(a) The two-finger parallel-jaw gripper of the Armatron. (b) The two-finger parallel jaw of an IBM gripper. (c) The nonparallel two-finger gripper of the Mobile Armatron. (d) The nonparallel two-finger gripper of the Hero I robot.

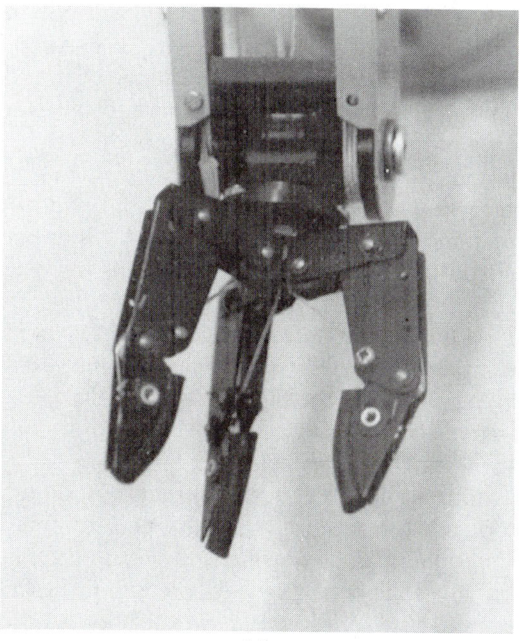

(a)

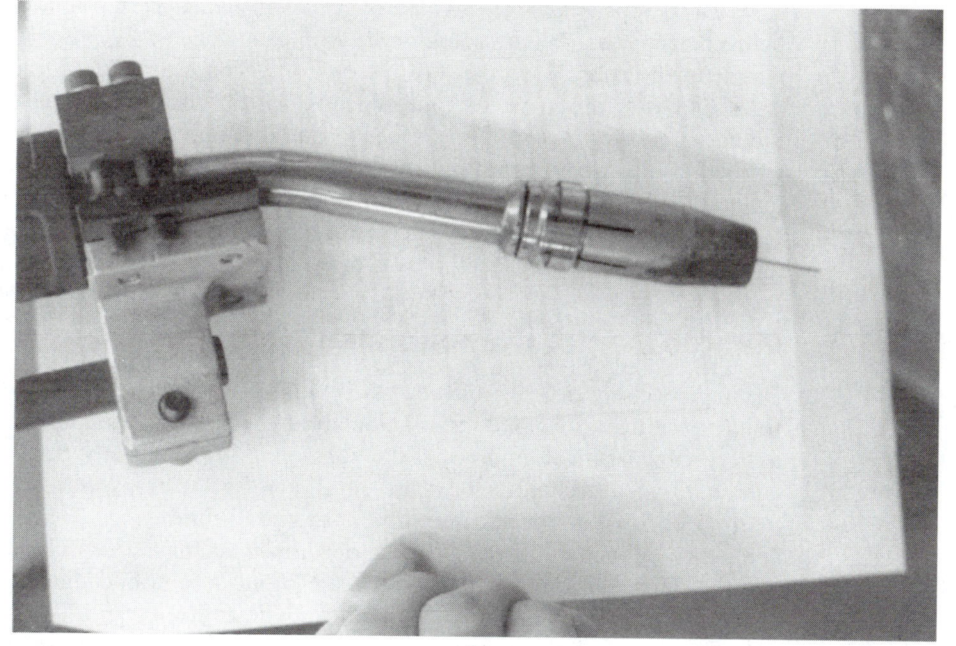

(b)

Figure 5-7 Three-finger Gripper and Arc-welding End Effector.
(a) The three-finger gripper of the Armdroid educational robot. (b) The arc-welding end effector of the AR-1 welding robot.

Designing a two-finger clamplike gripper involves taking many different details into consideration. The size and shape of the part or parts to be handled by the robot during a task greatly influence the size and shape of the required gripper. The weight, hardness, and surface material of the part to be handled determine whether the fingers or jaws of the gripper should be smooth, rough, or padded and how powerfully they should grip the part. The speed at which the part is moved during handling also affects how strong a grip is required. A padded or soft surface can be used to adjust for minor mispositioning of the robot's gripper; in this case, the material used is called a **compliance surface.** Handling very cool or very hot parts requires special gripper materials. Jaws may need to be curved (to handle round parts) or straight (to handle flat-sided parts). Straight jaws may be designed to move in parallel, like parallel-jaw pliers, or at a varying angle, like a regular pair of pliers; they may grasp the part from the outside (as they clamp together) or from inside (as they spread apart).

Other factors influencing the design of the gripper are the gripper's power source, its mode of activation, the degree of accuracy that must be attained in the positioning of the part or tool, and the need to change grippers (or not) during the task. Like the manipulator, the gripper can be powered by electricity, pneumatics, or hydraulics. Hydraulic power gives the best weight-to-strength ratio. Pneumatic power makes the fastest movement possible. Electric power, while clean and reliable, requires larger and heavier grippers, but it does produce better positional accuracy.

The gripper may be activated by a spring-loaded device as it contacts a part, by a proximity sensor, by some other type of sensor, or by preprogrammed commands in the robot's sequence of instructions. The ultimate accuracy of positioning achieved with a part or tool depends on the positional accuracy of the manipulator as well as on the gripper's construction. If a robot changes its own grippers while performing a task, positional accuracy will be more difficult to maintain. Generally, the cost of a manipulator goes up exponentially as its capacity for positional accuracy increases. The gripper can counteract some of the manipulator's inaccuracies by means of a **remote center compliance** (RCC) device.

Remote Center Compliance (RCC)

Drilling precisely drilled holes or assembling small parts requires very tight positional tolerances, and making manipulators capable of these positional tolerances is very expensive. Human workers position things correctly by using eye–hand coodination—a technique few present-day robots can use effectively; however, many robots use a gripper with a remote center compliance (RCC) device, which contains a built-in multiaxis floating joint to adjust for these misalignments—much like the universal joint in the power train of an automobile compensates for misalignment between the axle and the transmission.

Figure 5–8 presents a sketch of one type of RCC. As a small part is being inserted into a hole in an assembly, the RCC device centers the part in the hole. To accomplish this result the hole must have an enlarged tapered edge. As the part contacts the tapered edge, it applies pressure on the RCC device mechanism to

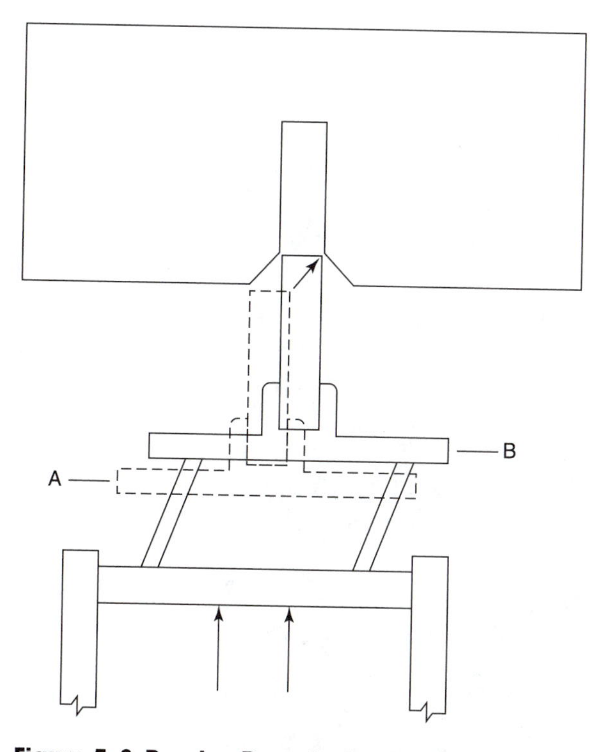

Figure 5-8 Passive Remote Center Compliance Device.
A passive RCC device uses an adjustment mechanism to move a part toward the center of a hole. Here the RCC device shifts the piece to be inserted from its initial position A to centered position B.

adjust its position toward the center of the hole, thereby moving the part into correct alignment.

A drilling jig or template, fastened to the surface of a part to be drilled, helps guide a human worker in positioning the drill correctly. A device similar to a drilling jig can be used to help a robot equipped with an RCC device in its gripper to position a drill correctly.

The RCC device discussed to this point is a passive compliance device. It is also possible to use force sensors and computer circuits to make an active RCC device. To do this, sensor information is organized into a history of the offsets used when positioning the parts. The history can then be used to make adjustments for the wearing or aging of a robot and to detect wear and tear on the machines that make the parts used in the robot's task.

Power for Grippers

Four types of power are used for grippers: pneumatic, electrical, hydraulic, and springs. For robotic manipulators sold without the gripper, the user must buy a gripper separately and find a way to route power to it. A few manipulators have

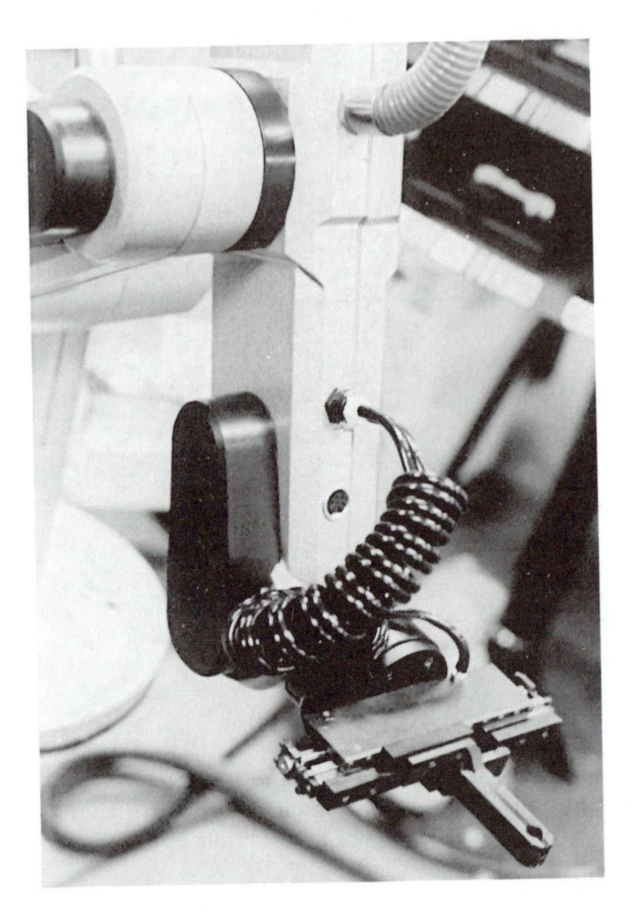

Figure 5–9
The end of the manipulator arm of the Eshed SCORBOT ER VII robot. The top connector, which is being used, supplies the pneumatic-powered gripper; the unused bottom connector can supply electric power and control signals to a gripper.

provisions for gripper power built into them. Figure 5–9 shows the end of the manipulator for an Eshed SCORBOT ER VII robot. The top connector, which is being used, supplies the pneumatic-powered gripper; the unused bottom connector can supply electric power and control signals to a gripper.

Pneumatic Power Pneumatic power for a gripper can come from an external source of compressed air or a vacuum, as well as from a pneumatic-powered manipulator. While an electrically powered or hydraulic-powered manipulator could use a pneumatic-powered gripper, such a manipulator generally uses a more sophisticated type of gripper. Vacuum cups and plate grippers require pneumatic power, as do pneumatic gripper devices such as **pneumatic fingers, mandrel lifting devices,** and **pin lifting devices.** A pneumatic finger grips an

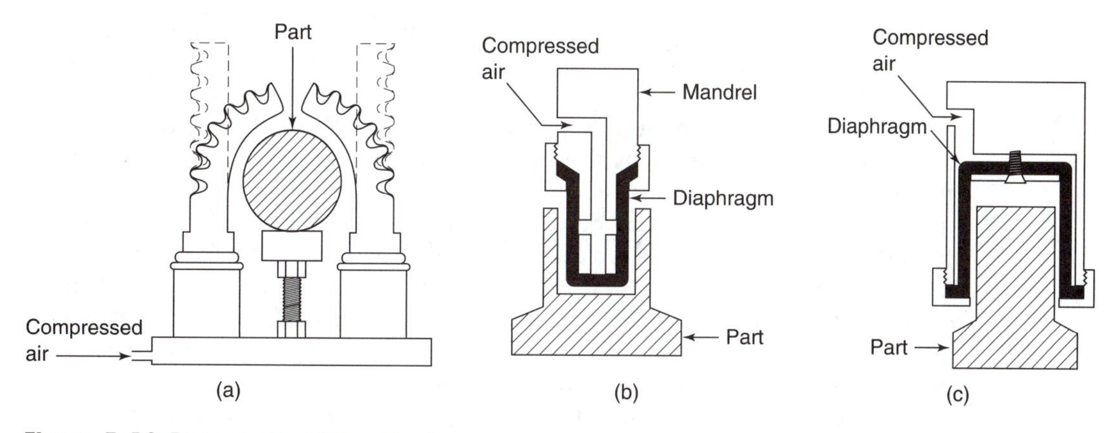

Figure 5–10 Pneumatic Lifting Devices.
(a) A pneumatic finger. When compressed air enters the finger, the folded edge expands and the finger bends toward the smooth edge. (b) A mandrel lifting device. It picks up the part by expanding to fill the inside of the part. (c) A pin lifting device. It lifts a part by contracting around the outside of the part.

object by bending it, a mandrel lifting device by expanding inside it, and a pin lifting device by contracting around the outside of it. Figure 5–10 presents sketches of these devices.

Electric Power Electric power may be used by itself or in conjunction with some other type of gripper power. Using an electromagnetic pickup device on a gripper necessitates using electric power, as does a gripper equipped with sensing devices. Most electrically powered manipulators have electrically powered grippers. If the end effector is a tool rather than a gripper, it is likely to be electrically powered.

Hydraulic Power Hydraulic power can be used to open and close grippers and to power some rotating tools. Hydraulic-powered grippers are the strongest grippers available and possess the best power-to-weight ratios. Because of the specialized requirements of hydraulic systems, however, hydraulic grippers are normally used only on hydraulic-powered manipulators.

Springs are generally used to open a gripper; some other source of power must then be used to close the gripper. That way, a power failure will not cause a gripper to close suddenly on an unsuspecting human. Any type of robot can use spring power on a gripper.

How Grippers Work

At least seven methods are used by grippers to grip a part: grasp it, hook it, scoop it, inflate around it, attract it magnetically, attract it by a vacuum, or stick to it. Grasping a part involves applying mechanical force to it at at least two points—for example, by pinching it between two or more fingers, or by spreading two fingers apart against the inside of the part. Hooking a part consists of placing a hook through a hole in the part and lifting it. Scooping involves using a ladle to pick up a mass of undifferentiated parts, or a formless material such as a liquid or sand. Inflating around a part requires inflating a bladder around either the inside or the outside of a part. Magnetic attraction involves using an electromagnet or a permanent magnet to attract magnetic materials such as parts made of iron. If a permanent magnet is used, some type of mechanical device, such as a pin, must be used to eject the part from the gripper once it has been carried to its destination. Vacuum attraction works best on smooth-surfaced materials. It can be used to handle difficult, unwieldy parts such as automobile windshields. Sticking to a part is accomplished by equipping the gripper with a sticky surface. Once again, a mechanical device is required to eject the part from the gripper when it should be released.

Things to consider in deciding what type of gripper a robot should use include: the weight of the part; the gripping distance from the center of gravity of the part; the coefficient of friction between the part and the gripper; the gravitation force (G) involved in moving the part, including acceleration and deceleration; the angle of the gripper's fingers to each other; the necessary safety factor; and the part's characteristics (such as composition, fragility, rigidity, shape, surface, and magnetic qualities).

CALCULATING GRIPPER PAYLOAD AND GRIPPING FORCE

The maximum payload that a manipulator can handle is usually identified by the manufacturer, and sometimes a gripper may be given a payload specification by its manufacturer. When both are given, the lower payload specification of the two is the one that must be followed. While the gripper specification tells exactly how much payload the gripper can hold, the manipulator specification is less straightforward. Since a manipulator may be sold without either a wrist or a gripper, the payload specification may simply represent the amount of weight that can be attached to the end of the manipulator as sold. Thus, if the manipulator can handle 60 pounds, including a 10-pound wrist and a 6-pound gripper, the gripper can handle only 44 pounds ($60 - 10 - 6 = 44$). Of course, the manipulator might include a wrist, in which case the manipulator's payload specification only needs to be adjusted for the weight of the gripper.

Calculating the force that needs to be exerted by a gripper to do a certain job involves considering the following factors: the angle at which the part is grasped and moved; the thickness of the part; the width of the gripper jaws; the distance from the center of gravity of the part; the coefficient of friction between the gripper and the part to the center of force (the center of the contact area of the jaws); the acceleration with which the part is to be moved; and any required safety factor. Not all these factors are relevant to every problem.

The angle of gripping varies from 0° for gripping and moving in a vertical plane to 90° for gripping and moving in a horizontal plane.

The thickness of the part, the width of the jaws, and the distance from the center of gravity of the part are needed to calculate the **torque** that exists when a part is picked up at a place other than its center of gravity. The **center of gravity** of a part is the point where its mass seems to be concentrated—that is, the point where the part is balanced.

The **coefficient of friction** measures how efficiently the gripper holds the part. If the part and the gripper surface are both rough, the coefficient of friction may be greater than 1. If the gripper surrounds the part or if the gripper runs a fork through holes in the part, the coefficient of friction can be treated as equal to 1 and thereafter ignored. For most gripping tasks, the coefficient of friction is less than 1.

The **acceleration** or **deceleration** of a part is the rate of change of velocity of the part. It puts added force on the gripper, trying to pull the part away from the gripper. For convenience, acceleration is measured in Gs. Normal gravitational force is a 1-G acceleration. If the part moves upward, the 1 G of gravity must be added to the acceleration. If the part moves downward, the part's acceleration must be subtracted from the 1 G of gravity. The calculated G force is then multiplied by the part's weight to get the total resisting force of the part.

Finally, the **safety factor** is a fudge factor to counteract unaccountable errors or unforeseen factors. A typical safety factor is 2: After all the forces required for a task have been calculated, the final number is simply doubled.

A few sample problems should help clarify the preceding material.

SAMPLE PROBLEM 1.

Determine how much force the jaws of a gripper must exert to hold a part in a vertical plane under the following conditions:

a. The part weighs 20 pounds and is of a nonuniform shape.
b. The gripper's jaws are parallel to each other and are grasping the part by its vertical sides.
c. The part is grasped 24 inches from its center of gravity.
d. The jaws' gripping surface is 4 inches wide.
e. The part is 2 inches thick at the point where it is being grasped.
f. The part is being lifted with a maximum acceleration of 2.5 Gs, including normal gravitational force.
g. The coefficient of friction between the part and the gripper is 0.85.
h. A safety factor of 2 must be included.

These conditions are represented in Figure 5–11.

Answer

First we must determine what information we need to solve this problem. Since the gripper is both grasping and moving in a vertical plane, we do *not* need to know about conditions (c), (d), and (e). The part exerts a stationary force of 20 pounds. Moving at 2.5 Gs, this force becomes $20 \times 2.5 = 50$ pounds.

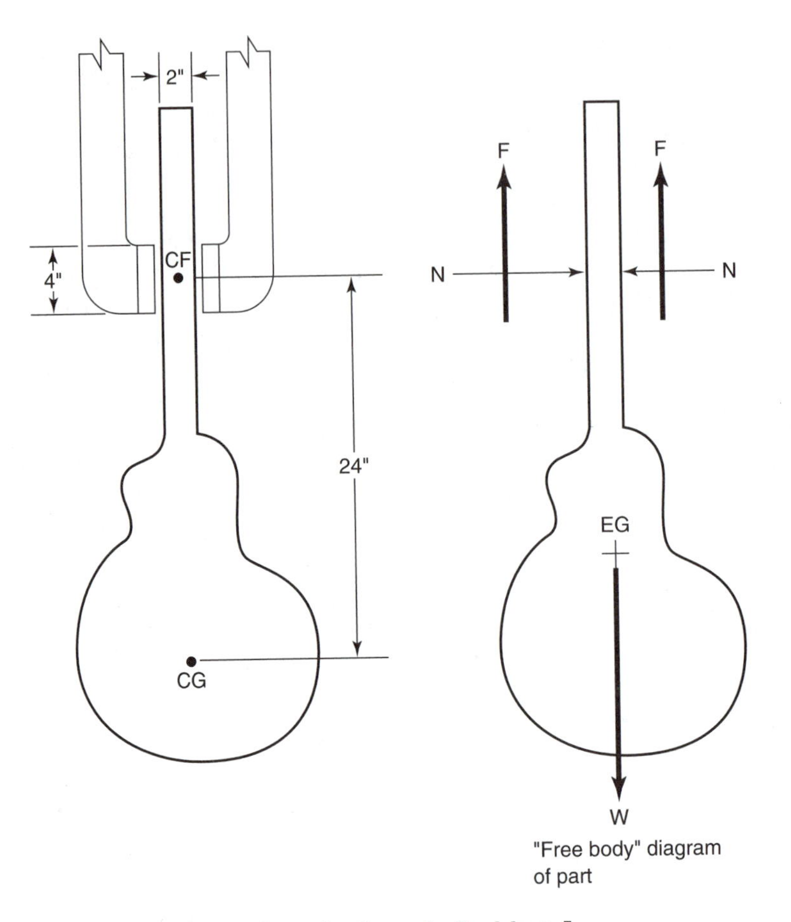

Figure 5–11 Conditions for Sample Problem 1.

In mechanics, the force pressing two surfaces together is called the normal force. The force acting parallel to these surfaces, which tends to prevent any sliding of these surfaces relative to each other, is the friction force. The normal force and the friction force are related through the coefficient of friction as follows:

$$\text{Friction force} = \text{coefficient of friction} \times \text{normal force}$$

$$F = \mu N \qquad \text{or} \qquad N = F/\mu$$

For the problem shown, since the equivalent weight of 50 pounds is opposed by two friction forces (one on each face of the part), each friction force is 25 pounds.

$$N = F/\mu = 25/0.85 = 29.4 \text{ pounds on each face of the part}$$
(note that μ was given as 0.85)

If a factor of safety of 2 is applied, then $N = 2 \times 29.4 = 58.8$ pounds. The force of 58.8 pounds is exerted by the gripper on each of the two sides of the part.

SAMPLE PROBLEM 2.

What force is required if the gripper operates under the same conditions as in Sample Problem 1, but in a horizontal plane? Figure 5–12 depicts these conditions and the stationary forces involved.

Answer

This time we need all the given information. The force exerted by the part while stationary is no longer just its weight, but its weight times the distance between the part's center of gravity (CG) and the center of force (CF). That is, the overall torque (T) on the jaws equals weight (W) times distance (d). The weight equals mass (m) times gravity (G), or 20 pounds. The distance equals 24 inches. The width of the jaws (b) equals 4 inches. The width of the part (p) in the jaws equals 2 inches. The equation for the torque is

$$\begin{aligned} T &= m \times G \times d \\ &= F_2(b/2) + F_1(b/2) \\ &= (b/2)(F_1 + F_2) \end{aligned}$$

Ignoring p,

$$F_1 + F_2 = 2(m \times G \times d)/b$$

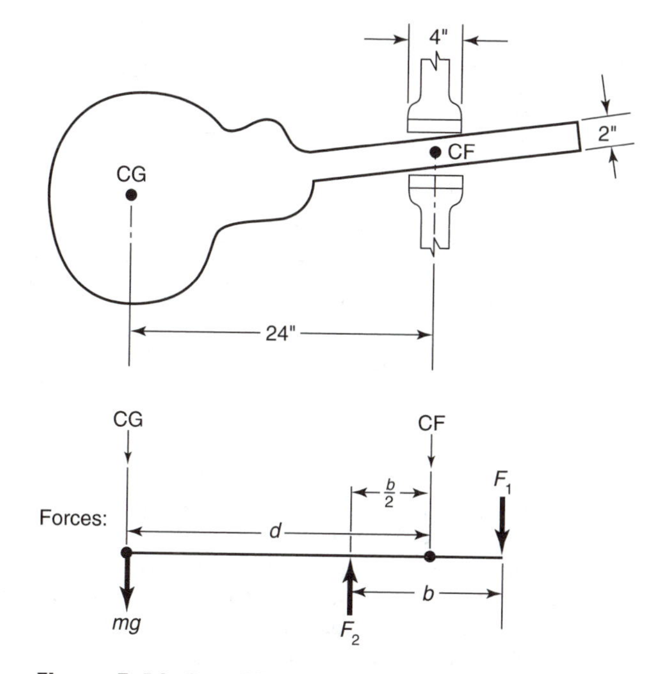

Figure 5-12 Conditions for Sample Problem 2.

Including p,

$$F_1 + F_2 = 2(m \times G \times d)/\sqrt{b^2 + p^2})$$

By Newton's First Law

$$F_1 = F_2$$

In this case, the gripper provides both forces, so

$$F = F_1 + F_2$$

Thus,

$$\begin{aligned} F &= 2(m \times G \times d)/\sqrt{(b^2 + p^2)} \\ &= (2 \times 20 \times 24)\sqrt{(4^2 + 2^2)} \\ &= 214.66252 \text{ pounds} \end{aligned}$$

To factor in the 3 Gs of acceleration, multiply by 3, giving 643.98 pounds.

To factor in the coefficient of friction of 0.85, divide by 0.85, giving 757.62 pounds.

To factor in the safety factor of 2, multiply by 2, giving 1,515.247 pounds.

SAMPLE PROBLEM 3.

What would the required force be if the jaws were 8 inches wide in Sample Problem 2?

Answer

This would change the force equation to

$$\begin{aligned} F &= (2 \times 20 \times 24)/\sqrt{(8^2 + 2^2)} \\ &= 116.4171 \text{ pounds} \end{aligned}$$

This would change the final answer to 821.76 pounds.

The preceding sample problems show the importance of grasping a part as close to its center of gravity as possible. In addition, acceleration and deceleration in the movement of parts should be minimized.

SUMMARY

The robot's end-of-arm tooling may also be known as a gripper, a hand, or an end effector. On an industrial robot such tooling is critical to the robot's effectiveness. End-of-arm tooling can be subdivided into grippers for moving parts and materials and tools for doing work on parts and materials. The manufacture of grippers and end effectors is a highly specialized area of robotics. A built-in remote center compliance (RCC) device can help a gripper overcome slight positional misalignments.

FORMULAS

Required Force for Vertical Grasp

The effective force required to hold a part when grasping it in a vertical direction and moving it in a vertical direction is

$$Fr = W \times G$$

where

Fr = Required force
W = Weight of part
G = Gravitational force

Force with Safety Factor

$$Fs = Fr \times S$$

where

S = Safety factor
Fs = Force with safety factor added in

Torque

$$T = W \times d$$

where

T = Torque
W = Weight
d = Distance at which weight is applied

Weight

$$W = m \times G$$

where

m = Mass of the part

Required Force for Horizontal Grasp

The force (torque) required to hold a part horizontally at other than the part's center is

$$T = (2 \times M \times G \times d)/\sqrt{(b^2 + p^2)}$$

where

b = Width of jaws
d = Distance from CG to CF
p = Thickness of the part

REVIEW QUESTIONS

1. Why is the choice of end-of-arm tooling important to a robotics project?
2. What is an RCC device, and how is it used?
3. What things might a robot hand be able to do that a human hand cannot do?
4. Suggest some new specialized grippers a robot might use that have *not* been mentioned in this chapter. Tell how they could be used.
5. What factors need to be taken into consideration when selecting a gripper for a particular task?
6. A working robot installation is moved from sea level to an altitude of 5,000 feet. What effect will this have on the robot's vacuum grippers' ability to do their task?
7. Suppose that a robot whose manipulator can handle a 50-pound payload is equipped with an 8-pound wrist and a 4-pound gripper. How heavy a part can the robot lift?
8. Calculate how much force the jaws of a gripper must apply to hold a part in a vertical plane under the following conditions:
   a. The part weighs 60 pounds and has a nonuniform shape.
   b. The gripper's jaws are parallel to each other and grasp in a vertical plane.
   c. The part is grasped 20 inches from its center of gravity.
   d. The jaws gripping the surface are 15 inches wide.
   e. The part is 1.75 inches thick at the point where it is being grasped.
   f. The part is being lifted with a maximum acceleration of 4 Gs including normal gravitational force.
   g. The coefficient of friction between the part and the gripper is 0.65.
   h. A safety factor of 2 must be included.

   Figure 5–13 depicts the situation.

Figure 5–13 Conditions for Question 8.

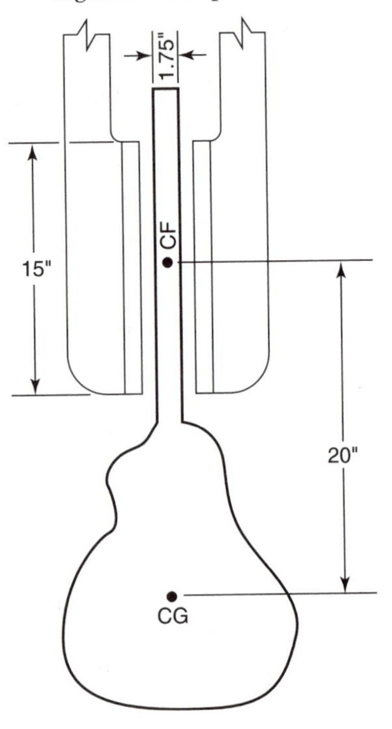

9. What would the required force be if the gripper operates under the same conditions as in Question 8, but in a horizontal plane? Figure 5–14 represents this situation.
10. What would the required force be if the jaws in Question 9 were 8 inches wide?
11. Why might it be necessary to buy a robot more than one gripper when a robot is used to load and unload parts from a machine?
12. Why might a robot need several different grippers during its life?
13. Why is it not practical to buy all the grippers a robot will ever need at the time the robot is purchased?
14. Why use a hydraulic-powered manipulator arm with an electrically powered gripper?
15. Use a computer that is connected to the Internet and try to find the names of several companies that specialize in making end-of-arm tools or grippers for industrial robots.

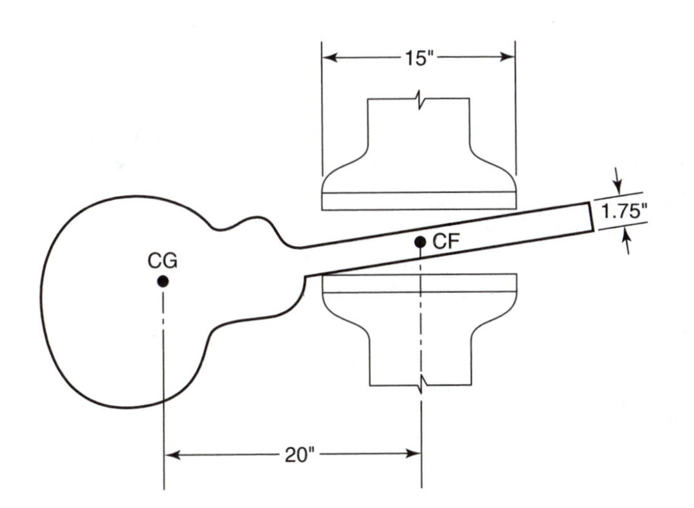

Figure 5–14 Conditions for Question 9.

6

SENSORS

OVERVIEW

The addition of sensors to a machine changes it from a dumb machine into an intelligent machine. This active area of research in robotics shows great promise for improving the industrial robot. In the movies, robots always seem to have better sensors than humans and to be able to do superhuman tasks. Yet, the industrial robot of today has poor sensor capabilities.

But do industrial robots need such sensors? What would they accomplish if they had better sensors? Will robot sensors resemble human sensors or be entirely different? Are robots limited to using the types of sensors humans have? What devices can be used as sensors for robots? These are some of the questions discussed in this chapter.

OBJECTIVES

When you have completed this chapter, you should be familiar with:

- The characteristics of internal and external sensors
- The ways robots use internal and external sensors
- Some devices that can be used as sensors for robots
- How better sensors will expand the range of tasks that robots perform

KEY TERMS

Absolute-readout encoder
Analog signal
Analog-to-digital converter
Artificial skin
Back lighting
Bit
Binary digit
Capacitance detectors
Conductive foam
Conductive paint
Conductive rubber
Differential driver
Digital signal
Diode-matrix camera
Direct-readout encoder
Electrical noise

External sensors
Fiber optics
Governor
Gray code
Gray scales
Hall-effect device
Haptic perception
Incremental encoder
Inductive detector
Interlocks
Internal sensors
Lead screws
Level converters
Limit switches
Line-scan camera
Mechanical stops
Membrane sensors
Modem
Parity check bits
Phoneme
Piezoelectric device
Pixel
Point-to-point wiring
Potentiometer
Proximity detectors
Pulse-reading shaft encoder
Range finding
Resolver
Rigid flat coil
RS-232
Sensor
Shaft encoder
Shielded wires
Signal error detection
Silhouette
Speech recognition
Speech synthesis
Strain gauge
Synchro
Thermistor
Thermocouple
Transducer
Voice synthesis

THE ROLE OF SENSORS

Humans adapt to changes in their environment in part by using **sensors** to detect what is going on in their environment. The ability to adapt to particular surroundings is one definition of intelligence.

Adding sensors to industrial robots could greatly expand their versatility while decreasing the mechanical tolerances they must have. Sensors can furnish a robot with information about its own condition and position, along with information about the outside world; without sensors, a robot requires an almost perfect world in order to do its task. A robot doing small-parts assembly can either use very expensive mechanical parts designed to reach precise locations or use sensor feedback to correct for mispositioning of less expensive parts. Unusual conditions, such as damaged or missing parts, cannot be handled by a robot without sensor information. However, robot sensor devices need not necessarily be located on the industrial robot itself; for some applications, an external device is sufficient.

CLASSES OF SENSORS

Sensors for robots can be divided into three classes: internal sensors, external sensors, and sensors such as interlocks that fall somewhere in between.

Most sensors are some type of **transducer.** Transducers convert one form of energy to another form of energy. The ear, for example, is a transducer that transforms sound energy into electrical signals, and the eye converts light patterns into electrical signals.

Internal Sensors

Early industrial robots did not have internal sensors. Humans programmed them by setting up **mechanical stops** and cams that enabled them to perform a desired task. The robots did not need to know their exact position; they only needed to run a joint up against some mechanical stop that physically halted their motion. These robots could not make any adjustments for misaligned parts or manipulators, and they moved about just as busily with nothing in their grippers as with the proper part. Lacking sensors, they were deaf, dumb, and completely senseless. Such robots worked in an open-loop system of control, without any feedback.

Early electrically powered robots used stepper motors to position their manipulators. By counting the number and phase of the pulses given to the motor since it was last at its home position, the controller could keep track of the position of the manipulator.

The earliest on-board sensors used by robots were **limit switches.** The robot simply extended along an axis until it activated a limit switch that removed power from that part of the circuit and thus halted progress along the axis. The positioning of the limit switches required a mechanical setup.

The first complex sensors used by the industrial robots used feedback information to ascertain the present condition of the gripper and all the changeable axes. This sense is known as **haptic perception,** and it is the robotic equivalent of the human sense of kinesthesia, which allows a person to know the whereabouts of various appendages and extremities without having to locate them visually. The information is fed back from the same muscles and nerves that move the limbs. Therefore, human movement is governed by a closed-loop control system: Each motion command receives information on how well the motion was carried out, enabling the person to make adjustments for slight misalignments in its limbs.

An industrial robot's internal sensors use mechanical, electrical, electronic, and hydraulic devices to obtain feedback information on the position of various movable devices. Many of these devices, called closed-loop servo-controlled systems, produce some type of feedback signal when a device is misplaced. As the device nears its proper location, the error signal diminishes.

One mechanical device that has been used as a control for over a century is the **governor,** which uses adjustable spinning weights to control the rotational speed of devices such as generators and motors. When the rotating device begins to turn too fast, the weights spread out farther from the axle, increasing the drag on the device and thus slowing it down. Then, as the device slows down, the weights drop closer to the axle, decreasing the drag on the device and allowing the device to pick up speed. On a gasoline or diesel engine, the governors monitor the engine's rotational speed. They decrease the fuel to the engine if it is rotating too fast, and they increase the fuel to the engine if it is rotating too slow.

Some hydraulic cylinders are equipped with **lead screws** that can detect the position of the cylinder shaft by counting the revolutions of the screw as the shaft is extended or retracted.

For rotational joints, a **shaft encoder** can be used to detect very fine rotational movement. The expensive **direct-readout encoders** or **absolute-readout encoders** can give the degrees of rotation from 0.000° to 359.999° directly. These readings are expressed in terms of switch openings or closings and thus are already in a digital form that can easily be entered into the robot's electronic memory and interpreted by the robot's controller. Alternatively, the readings can be expressed as lights turning on and off in an optical encoder.

Figure 6–1 shows a sketch of a simplified direct-readout shaft encoder. Notice that each ring of the disk has twice as many divisions on it as does the ring immediately inside it, and that only one division mark change is encountered at a time. This produces an unambiguous count value with an accuracy of plus or minus one

0 0 0 0 0
0 0 0 0 1
0 0 0 1 1
0 0 0 1 0
0 0 1 1 0
0 0 1 1 1
0 0 1 0 1
0 0 1 0 0
0 1 1 0 0
0 1 1 0 1
0 1 1 1 1
0 1 1 1 0
0 1 0 1 0
0 1 0 1 1
0 1 0 0 1
0 1 0 0 0
1 1 0 0 0
1 1 0 0 1
1 1 0 1 1
1 1 0 1 0
1 1 1 1 0
1 1 1 1 1
1 1 1 0 1
1 1 1 0 0
1 0 1 0 0
1 0 1 0 1
1 0 1 1 1
1 0 1 1 0
1 0 0 1 0
1 0 0 1 1
1 0 0 0 1
1 0 0 0 0

(a)

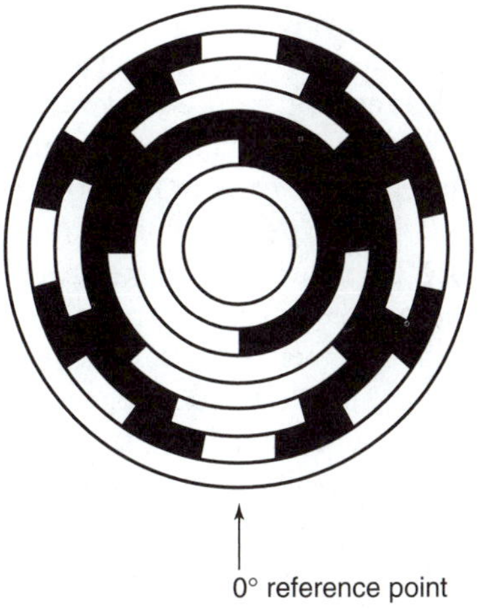

0° reference point

(b)

Figure 6–1 Direct-readout Shaft Encoder.
(a) A direct-readout shaft encoder gives the value of its present orientation. Each revolution is divided into 32 parts. The value is usually expressed as a gray code value; that is, as the shaft rotates, only one bit of the value changes at a time. (b) Gray code counts for the numbers 0 to 31.

division. This unambiguous counting method is known as a **gray code,** as opposed to normal binary codes that switch several positions at a time. The number of circles of division an encoder can use is limited by the detector circuit's ability to detect the on and off markings (which are made of a conducting material). Through the use of fiber optics, it is possible to detect smaller optical marks than can be detected by electrical brushes. As a result, you can always determine where the direct-readout encoder is just by reading its present position value.

The less expensive **pulse-reading shaft encoder** or **incremental encoder** just emits a pulse for each increment of shaft rotation. External counter circuits keep track of the number of pulses and thus the present position of the shaft. This encoder may use one or two circles of windows, with the second set of windows shifted 90° from the first set. By comparing the signal from both sets of windows, you can tell the direction in which the encoder is turning.

An alternative method of determining the direction of the encoder is to use two separate sensors, 90° apart, to detect the single set of windows. Figure 6–2 depicts a pulse-reading encoder with a single set of windows and shows how detectors (LEDs) can be positioned to detect direction from their phase relation. Notice that the pulse-reading encoder has no beginning or ending position; it simply emits pulses as it turns. Of course, if the external counting circuit should happen to miss a count or lose power temporarily, there would be no way to find out what the shaft's position was. You would have to move the joint back to some mechanical stop or home position in order to reestablish a meaningful position to count from.

The price of pulse-reading shaft encoders varies, depending on the number of counts produced per revolution. Encoders that produce up to 600 pulses per revolution may cost as little as $55 each in quantity purchases, whereas encoders that produce up to 50,000 pulses per revolution may cost as much as $500 each. A pulse encoder can also be used to calculate speed and acceleration. Speed is measured by counting the number of pulses per time period, and acceleration is measured by comparing the number of pulses accumulated in two different time periods. Figure 6–3 shows the single-window encoders of the Qume Sprint printer and Hero I robot and the double-window encoder of the Rhino robot. Figure 6–4 shows the low-cost pulse encoder used in the Big Trak gear box.

Another device for measuring small shaft rotations is an electronic **strain gauge,** which in turn may use a piezoelectric device or a Hall-effect device. **Piezoelectric devices** are crystalline materials such as quartz or ceramic that produce electricity when they are distorted. The greater the distortion, the greater the amplitude of the electricity. Modern phonograph record players use piezoelectric devices to convert the mechanical motion of the moving needle into electrical energy. To detect a joint's movement, the piezoelectric device is twisted. The farther the joint moves, the greater the distortion and the larger the piezoelectric voltage. A **Hall-effect device** is a semiconductor device that carries an electrical current within a magnetic field. The current through the device varies as the magnetic field is rotated in relation to the semiconductor's current flow.

Another type of strain gauge is a **rigid flat coil** of wire. As the diameter of the wire changes in response to a twisting distortion, its electrical resistance changes by

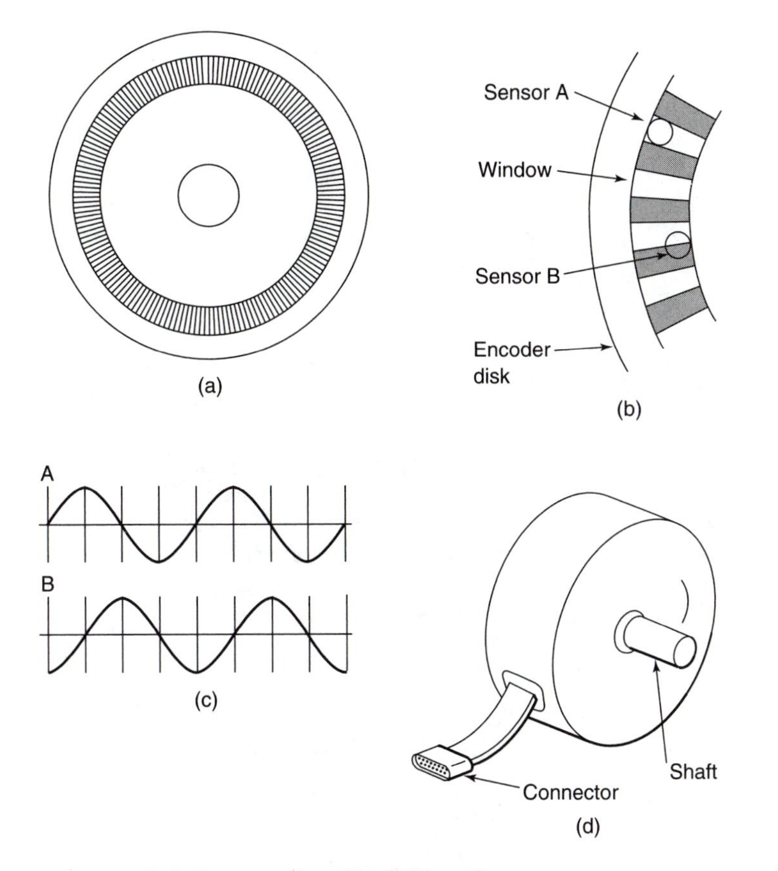

Figure 6–2 Pulse-reading Shaft Encoder.
The pulse-reading shaft encoder outputs one pulse for each increment of rotation. The direction of rotation can be determined by using two sensors that are 90° out of phase with each other, which is half a window apart. (a) The encoder disk. (b) Sensors. (c) Output of sensors A and B for a clockwise rotation. (d) The assembled encoder.

up to 1 percent. If a bridge circuit were used to detect this resistance change, it might register a voltage difference of 0.03 volt. Since these voltages are so low, strain gauge elements must be placed near the bridge and amplifier circuit. Successfully transmitting sensor signals to the robot controller requires careful planning to avoid electrical and electronic noise; otherwise, the sensor information becomes useless.

The rigid flat coil may be wire-bonded to a plastic strip or foil on a printed circuit board. The size of these strain gauges may be from 0.008 to 4 inches in length, with total resistances of 30 to 3,000 ohms. Rigid flat coils can also be used to determine the force exerted by a gripper holding a part. Figure 6–5 shows a single-axis bonded-wire strain gauge and a two-axis printed-circuit strain gauge. Figure 6–6 shows a bridge and an amplifier circuit for the strain gauge.

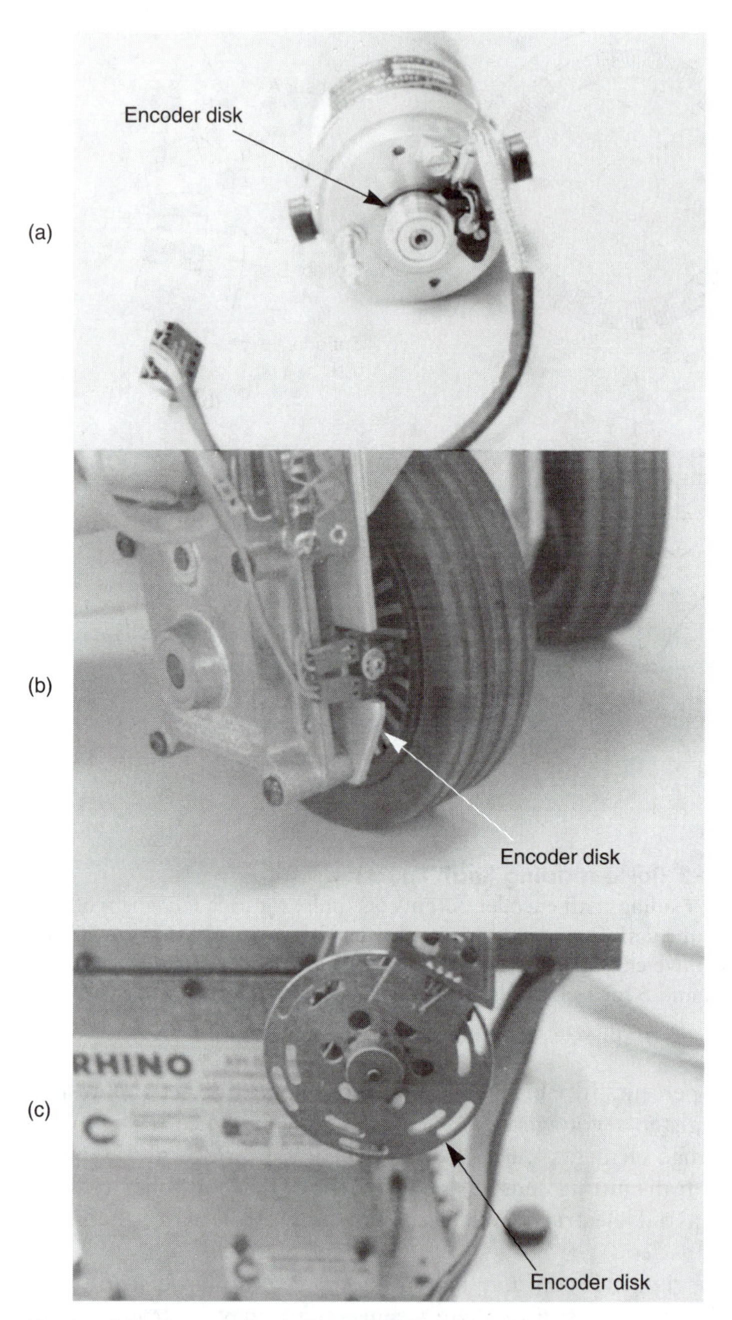

Figure 6-3 Qume, Hero I, and Rhino Encoders.
(a) Single-window encoder disk on the Qume Sprint printer. (b) Single-window encoder disk on the Hero I robot. (c) Double-window encoder disk on the Rhino robot.

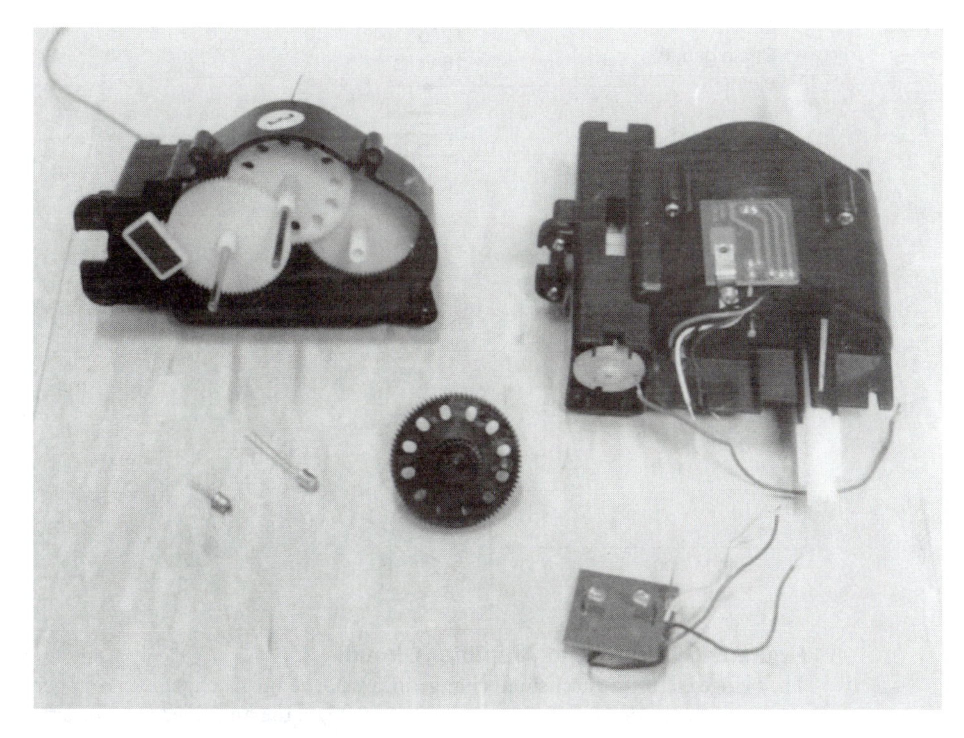

Figure 6-4 Big Trak Encoder.
Upper left: The inside of half of a Big Trak gear box. Upper right: A completely assembled gear box with the encoder board attached. At center: Figure shows the gear that the encoder board mounts against and the LED shines through to create pulses. Lower left: An infrared light-emitting diode (LED) and an infrared phototransistor used in the encoder. Lower right: The encoder board.

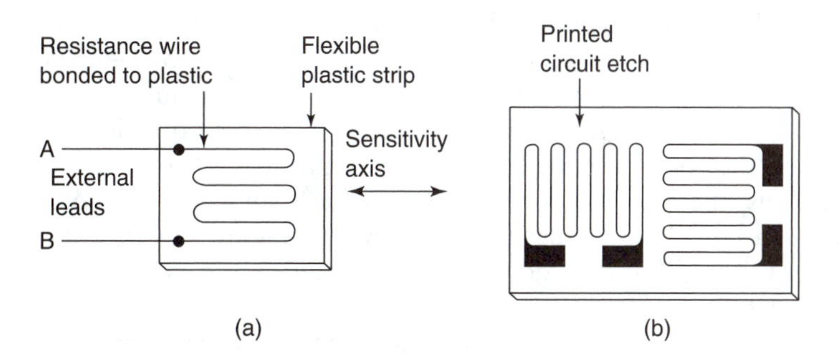

Figure 6-5 Flat Rigid Strain Gauges.
(a) A bonded-wire strain gauge that uses resistance wire bonded to a flexible plastic strip.
(b) A printed-circuit etch or foil strain gauge with two elements. If it is mounted so that the stress is only in one direction, the other element can be used for temperature compensation, replacing the balance adjustment of a bridge circuit.

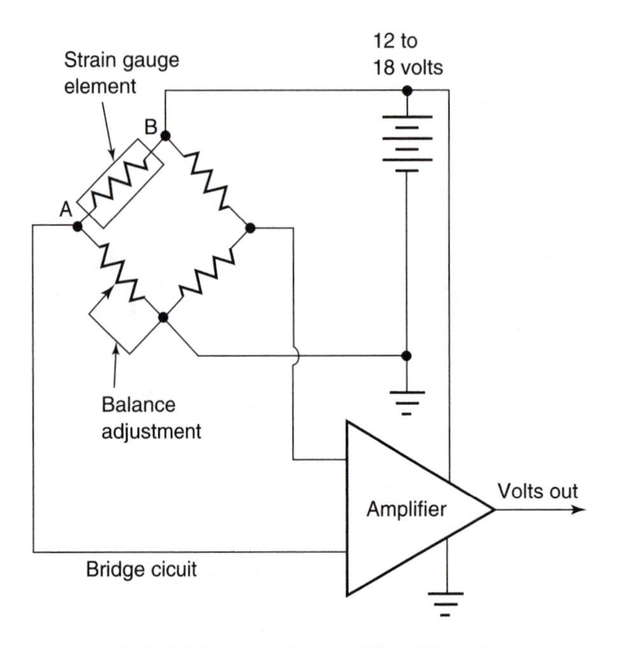

Figure 6–6 Bridge and Amplifier Circuit.
The bridge circuit turns a small change in resistance into a change in voltage. Since the output of the bridge circuit has such a low voltage, the amplifier must be situated near the bridge circuit.

An inexpensive but less accurate movement sensor is the **potentiometer,** a variable resistor. Figure 6–7(a) and (b) shows the physical construction of the potentiometer. It has a circular resistance strip and a movable contact that touches this strip. Most volume controls on radios and televisions are potentiometers. If a voltage is applied across the resistance strip, the voltage at the tap will be proportional to the tap's position along the strip. Figure 6–7(c) and (d) shows the electrical symbol and circuit. For measuring small amounts of angular motion, such as the joints of a robot, the shaft of the potentiometer is connected directly to the rotating part, as shown in Figure 6–7(e). For measuring small linear motions, the shaft is connected to a linkage similar to that of the tuning dial on some radios. This arrangement is shown in Figure 6–7(f). Linear potentiometers are available, but they do not have the motion life expectancy of the circular potentiometers.

Another device for angular positioning is the **synchro** system. Synchros, which have been popular for remote-control positioning of antennas, are motorlike devices with a multiwinding stator and a rotor. They can transform an electrical input into an angular output or an angular position into an electrical output. When two synchros are connected, the rotors both line up at the same angle. One synchro is connected as the master and is adjusted by hand to the desired angle. The other,

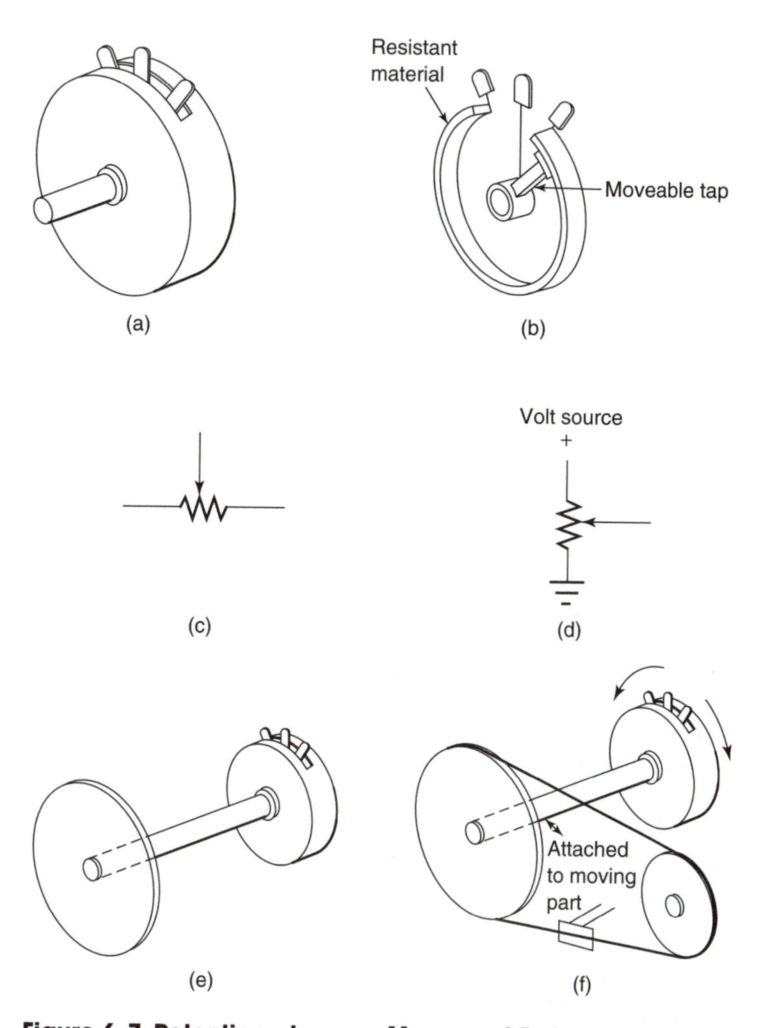

Figure 6–7 Potentiometer as a Movement Detector.
If a movement shifts the setting of a potentiometer, the resulting change in resistance can be used to detect the movement. (a) External physical appearance of a potentiometer. (b) Internal physical appearance of a potentiometer. (c) The electrical symbol for a potentiometer. (d) How to connect the potentiometer so that a resistance change will output a voltage change. (e) Rotational movement detector. (f) How to convert linear movement into rotational movement of the potentiometer.

the slave, then adjusts itself to match the master's angle. Figure 6–8 presents diagrams of a synchro, and Figure 6–9 shows two synchros connected to work as a control system.

A **resolver** produces an output voltage that is proportional to the product of the input voltage and the sine of the rotor angle. Thus, synchros are types of resolvers. Direct-readout encoders may also be used as part of a resolver.

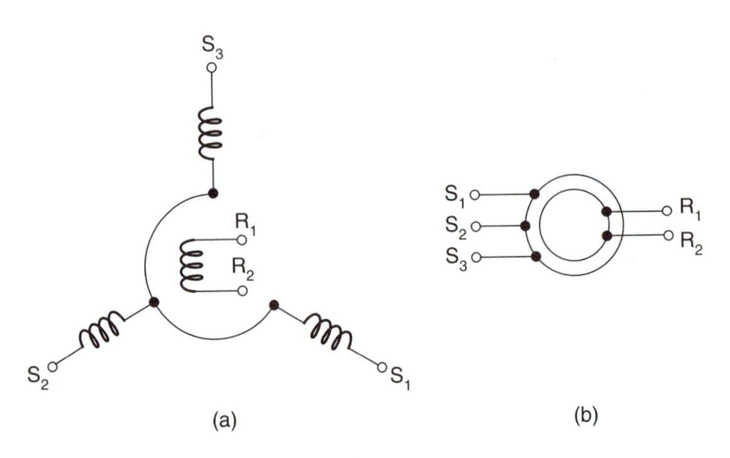

Figure 6-8 Diagram of Synchro.
(a) The detailed symbol for the synchro, with its two circuits. (b) The simplified symbol.

External Sensors

Early industrial robots depended on others for seeing that things were in the proper place at the proper time in their world. If the object the robot was supposed to grasp and move was not where it belonged, the robot would perform the grasping motion anyway and carry a nonexistent part to the prescribed destination.

One of the earliest external sensors for robots was a microswitch or pressure sensor that could detect when the robot had something in its gripper. These were a simple type of touch, or tactile, sensor. A weight or strain gauge attached to the

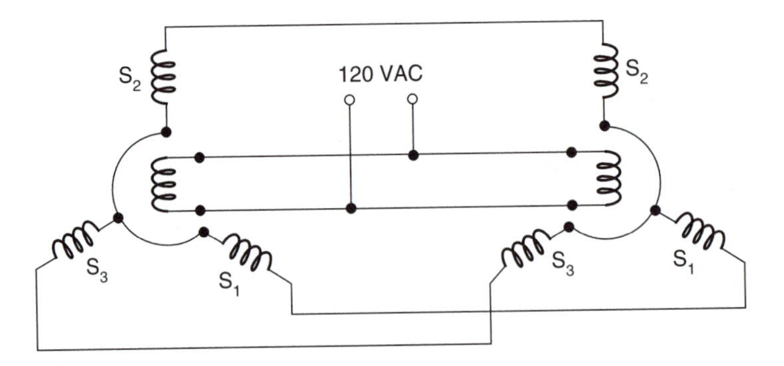

Figure 6-9 Connection of Synchro.
Two synchros are connected in such a way that when the master synchro (on the left) turns its center winding, a phase shift is detected by the slave synchro (on the right), causing it to rotate to the same angle.

gripper enabled the robot to determine whether it had picked up all or only part of a piece. In the latter case, the robot would stop what it was doing and call for human assistance. A third device that could be used to detect whether a part had entered the gripper consisted of a light source on one side of the gripper and a light beam detector on the other side of the gripper. An object entering the gripper would break the light beam, and the robot would become aware of its presence. Figure 6–10 shows two of these simple external sensors.

Another useful sensor for robots working with automated machinery or conveyor belts was an interlock microswitch signal that notified the robot when a machine cycle was finished or a part was in the proper position for the robot to handle. The robot would wait until it received this signal before trying to do its work, thus protecting it, for example, from sticking its gripper in a closing press and damaging its gripper or the press.

Early robots were dangerous to humans who accidentally got in their way, since they had no sensors for detecting human presence. As a result, fences or light beams and detectors had to be installed around the work area to protect humans from robots. The inclusion of ultrasonic and infrared sensors on subsequent robots allowed the robots to sense when something was wrong and stop before hitting an object or person.

The senses of hearing, speech, vision, and touch for robots are still in the experimental stages. Development of the sense of voice recognition (hearing) is mainly intended to make it easier for humans to give commands to the robots when training them to do a specific task. Speech synthesis, or voice, will enable a robot to communicate warnings to humans when something is wrong. Vision will make self-orientation and proper grasping of parts easier for the robot, since it can be used to correct for small errors of positioning or alignment. The sense of touch will help a robot tell when it is gripping a part and what the orientation of the part is.

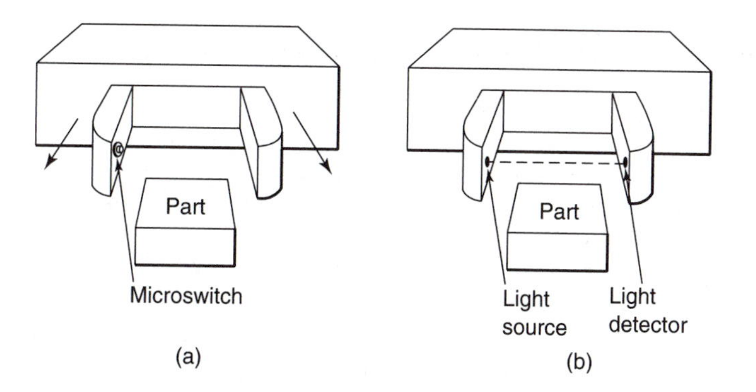

Figure 6–10 Simple External Sensors.
(a) A microswitch being used to detect whether something is in the robot's gripper. (b) A photoelectric device being used to detect whether something is in the robot's gripper.

Interlocks

Interlocks are devices that do not allow an operation to be performed until certain conditions exist. They are very old devices that are used to protect the unauthorized person from harmful conditions. For example, most television sets have an interlock that requires the back to be on the television in order for the set to receive power from the power cord. If the back is removed from the set, the power cord is also disconnected. Many radios have a similar power cord interlock, and most microwave ovens require the door to be shut before the cooking power will turn on. Another example is the deadman switch on trains.

Some interlocks, rather than cutting off power, may sound an alarm or (in the case of a robot) stop the device's motion without switching its power off. Burglar alarms, for example, may be triggered by the moving of a magnet on a window or door.

An interlock may also be used to protect a robot. For example, the airflow in the robot's controller might have to reach a preset level before power can be applied to the controller's electronics. Hydraulic pressure may have to reach some minimum level before hydraulic actuators are allowed to move. The temperature of the hydraulic fluid may have to remain below a certain level if the hydraulic system is to be used. The robot may have to receive an electrical interlock signal indicating that the press has finished its cycle and is open, before the robot can reach in and unload a part from the press.

While many interlocks are electromechanical, completely mechanical interlocks do exist. Some automatic pistols, for example, have mechanical pistol-grip interlocks that prevent the trigger from being pulled unless the grip is squeezed first.

Depending on how they are used, interlocks can be viewed as internal sensors or as external sensors, since they can be described as measuring the state of a device or as warning of an intruder.

INTERFACING NATURAL SIGNALS TO DIGITAL CONTROLLERS

Interfacing sensors to robots involves wiring a stationary device to moving parts. Early robot manufacturers had trouble with sudden emergency stops made by their robots. The cause was eventually traced to wiring that had worn through, causing intermittent breaks in the connections as the robot moved. Sensor manufacturers should profit from these mistakes in designing wiring for sensors on their robots.

All of the electronic strain gauges mentioned thus far produce an **analog signal.** That is, their output voltage may be any of a continuous range of voltages, as opposed to being one of a stepped series of specific voltages as is the case with a **digital signal.** Analog signals cannot be understood by the normal on/off type of digital input circuit. Therefore these signals need to be converted from analog to digital information before being sent to the modern digital controller of a robot. This is done by an **analog-to-digital converter.**

The simplest type of analog-to-digital converter establishes a detection point at some specific level. All values below the critical level are considered to be a digital 0, and all values at or above the point are considered to be a digital 1. By this means all possible analog values have been reduced to a single bit (short for *bi*nary digi*t*). A bit is the smallest piece of memory used in a modern digital controller; it can only hold a value of 0 or 1. In some applications, such as in determining whether the robot's battery level is at an acceptable level, this simple converter may provide a sufficient interpretation of the analog voltage level. The warning or idiot lights on an automobile's dashboard work on this principle.

If the robot can afford to use eight binary bits to hold the digital version of the analog value, an analog-to-digital converter can be used that divides the analog values into 256 separate zones (since $2^8 = 256$).

SENSOR AREAS FOR ROBOTS

Most robot sensors fall into one of several general areas: vision, touch, range and proximity detection, navigation, speech output, speech input, or smell.

Vision

Present-day industrial robots use vision to locate and orient parts. When robots move out of the factory and into the office and home, they are likely to need vision to orient themselves to these surroundings and to handle more complicated tasks.

Robots can be given vision through the use of a television camera. Visual information gathered by the camera can be converted into digital information and stored in the robot's memory. Unfortunately, however, the robot does not know how to interpret most of the visual information it receives. One of the earliest (circa 1962) robots with vision used a single line of a television camera at a specific height above a conveyor belt to detect when a part was in place for pickup.

Most industrial robots with vision use high-contrast television cameras that see things as either completely black or completely white. Shades of gray confuse the robot. The human brain processes its visual information through a smart camera—the eye. The eye preprocesses what it sees (this includes enhancing the edges) before sending it to the brain's vision-processing center. Perhaps what industrial robots need is some way to preprocess visual information before it gets to the vision system's controller.

Establishing vision for robots may seem as simple as adding a television camera to the robot, but it is much more complicated. Human beings use past experiences when trying to interpret a particular sight. Fortunately, industrial robots do not have to understand all the things that humans do when using their vision. In many situations, a robot need only know the outlines of previously recorded pictures of parts, as these appear under controlled, unvarying indoor lighting in a carefully maintained alignment at a predetermined, fixed distance.

All vision systems require at least the following components: an object to be viewed, an illumination source, a vision sensor, an edge detector, a line detector, an object detector, and previously recorded information for interpreting what is being seen.

The Steps of Vision Processing Figure 6–11 identifies the major steps in processing vision. The process starts with an object that is being viewed under some type of illumination. Some of the illumination energy reflects off the object and is received by the vision sensor. The sensor converts the illumination energy into an internal representation that can be processed by the rest of the vision system, and this representation is analyzed for motion and edges. The edges are analyzed for object lines, and the lines are then analyzed for object shapes. Ultimately, the shapes are put together to produce image understanding. Color, texture, and depth may be analyzed to increase understanding of the image. Expectational information may be used to speed up processing of the vision by ignoring alternative interpretations involving situations that are not expected to occur.

Illumination Source The illumination source for natural human vision consists of electromagnetic energy in the visible light range. Radar uses microwave radio waves as an illumination source; sonar uses sound waves. Infrared energy and ultraviolet energy can also be used as illumination sources for machine vision.

High-contrast visual systems usually require placing a high-intensity source of illumination behind the object being viewed. This method, called **back lighting,** only produces a **silhouette** of the object. Back lighting thus creates a visual field of black on white, with nothing between; the subject cannot blend into the background.

Vision systems that have **gray scales**—that is, systems that show shades between black and white—generally require placing the primary source of illumination in front of the object. The light then reflects off the object and back to the vision sensor. This is the type of lighting usually used in taking photographs. With front lighting, the subject may blend into the background, and shadows may hide some edges.

Figure 6–12 shows a ball as it would appear under front illumination and under back illumination.

Vision Sensors Most machine vision systems use some type of television camera as their vision sensor. These may be standard black-and-white or color cameras, and they may use vacuum tubes or solid-state technology. The most common type of vacuum tube for television cameras is the vidicon tube. A black-and-white camera requires a single vidicon. A color camera requires three vidicons—one with a red filter, one with a green filter, and one with a blue filter. The output of each vidicon of a color camera must be adjusted to match the sensitivities of the human eye. That is, when the camera records a "white" scene, red is 30 percent of the signal, green is 59 percent of the signal, and blue is 11 percent of the signal.

The most common types of solid-state television camera use charge-coupled devices (CCD), charge-injection devices (CID), and photo-diode arrays. Solid-state cameras are generally smaller and consume less power than vacuum-tube cameras.

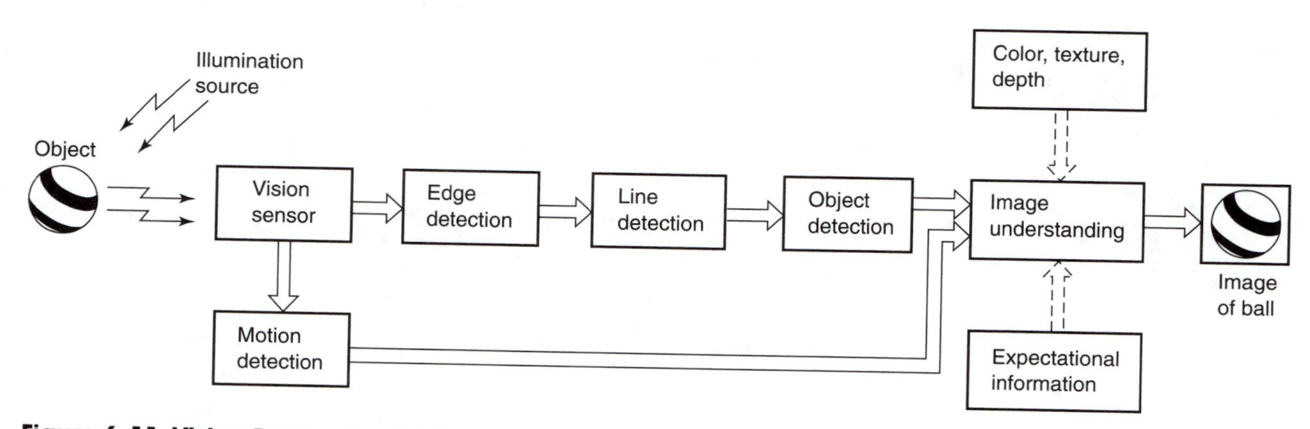

Figure 6–11 Vision Processing System.
A vision system consists of the following parts: an object to be viewed; an illumination source; a vision sensor; motion detection; edge detection; line detection; object detection; image understanding; expectational information; and information on color, texture, and depth.

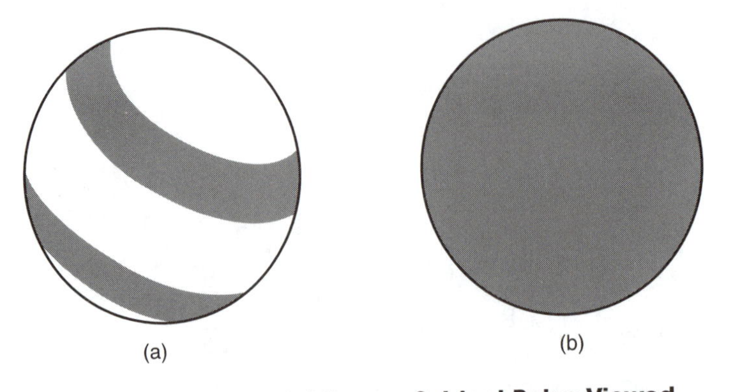

Figure 6–12 Effect of Lighting on Subject Being Viewed.
(a) When a subject is front-lighted, light reflects off the subject and into the vision sensor. This yields the most details about the subject. (b) If the subject is back-lighted, the subject blocks some of the light going to the vision sensor from the illumination source. This produces a high-contrast outline of the subject.

Photo-diode cameras may use linear arrays of photo-diodes (in which case they are called **line-scan cameras**) or two-dimensional area arrays (in which case they are called **diode-matrix cameras**). Photo-diode arrays are made up of individual photo-diodes that change their resistance in proportion to the intensity of the light striking them. They respond faster to light changes than does the vidicon tube.

CCD devices can be bought in linear arrays of 256 to 2,048 cells or vision **pixels** (elements of vision) that see a line of vision at a time. They can also be bought in area arrays of 190 by 244 cells or 380 by 488 cells. CCD cells build up an electrical charge proportional to the light intensity that strikes the cell. This charge is then coupled from cell to cell in the array through application of a clock pulse to the array, which makes each row and/or column of the CCD array act like a shift register. A CCD camera is sometimes called a RAM camera, since the CCD acts like a dynamic RAM memory.

CIDs work similarly to CCDs and can consist of linear or area arrays. The CID has voltage injected into each cell in proportion to the intensity of the light that strikes the cell; but the CID acts like a computer memory and holds the information long enough for it to be read out by a program scanning the CID array. Thus, a CID camera acts like a static RAM memory.

Three-dimensional vision can be achieved by using two cameras for moving and fast-response vision. A single camera can produce the effect of three-dimensional vision if it is shifted between taking pictures. Since the camera must move, however, this is a slower method of achieving three-dimensional vision.

Infrared and ultraviolet cameras work similarly to a normal television camera, except that they are sensitive to different frequencies of light. A single-element infrared camera can be used just for range finding, as is discussed later in this chapter.

Active radar uses its own microwave illumination source—a single-element sensor records responses based on the length of time since the illumination occurred. Inactive radar records the reflected and direct energy that it receives from external sources of microwave illumination.

Active sonar generates its own sound waves and then uses these as an illumination source—a single-element sensor records the responses based on the length of time since the illumination occurred. Inactive sonar relies on existing sound illumination. Like inactive radar, it then records the reflected and direct energy it receives.

No matter what type of vision sensor a machine vision system uses, the output of the sensor must be converted into numerical codes for storage in the system's memory. Figure 6–13 shows how this information might be represented for a back-illuminated object in a high-contrast vision system and for a front-illuminated object in a gray-scale vision system.

Edge Detection Edge detection is the process of finding the boundaries between areas of different intensities of vision energy. An object's edges define its shape.

Most machine vision systems use program processing of visual data to perform edge detection. First the raw visual information is sent to the computer's memory, and then a program processes the data for edges. Since the raw visual information is stored as numbers in memory that are proportional (or inversely proportional) to the light intensity, the actual process of edge detection consists of searching the memory for cells whose values differ from the value of one or more of their neighbor cells. Figure 6–14 shows what the edge-detection data for the ball might look like.

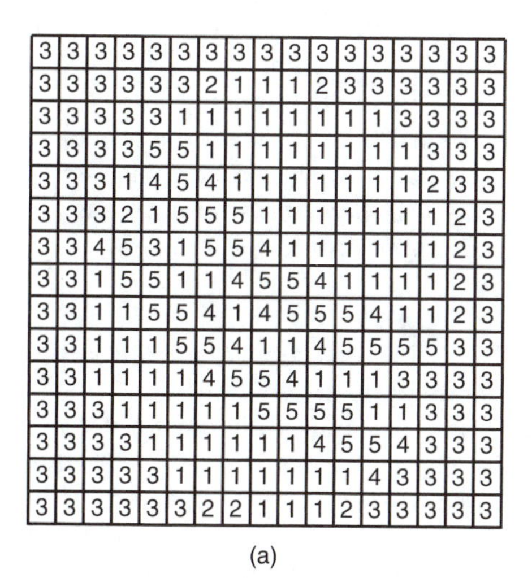

3	3	3	3	3	3	3	3	3	3	3	3	3	3	3	3	3
3	3	3	3	3	3	2	1	1	1	2	3	3	3	3	3	3
3	3	3	3	3	1	1	1	1	1	1	1	1	3	3	3	3
3	3	3	3	5	5	1	1	1	1	1	1	1	1	3	3	3
3	3	3	1	4	5	4	1	1	1	1	1	1	1	2	3	3
3	3	3	2	1	5	5	5	1	1	1	1	1	1	1	2	3
3	3	4	5	3	1	5	5	4	1	1	1	1	1	1	2	3
3	3	1	5	5	1	1	4	5	5	4	1	1	1	1	2	3
3	3	1	1	5	5	4	1	4	5	5	5	4	1	1	2	3
3	3	1	1	1	5	5	4	1	1	4	5	5	5	5	3	3
3	3	1	1	1	1	4	5	5	4	1	1	1	3	3	3	3
3	3	3	1	1	1	1	1	5	5	5	5	1	1	3	3	3
3	3	3	3	1	1	1	1	1	1	4	5	5	4	3	3	3
3	3	3	3	3	1	1	1	1	1	1	1	4	3	3	3	3
3	3	3	3	3	3	2	2	1	1	1	2	3	3	3	3	3

(a)

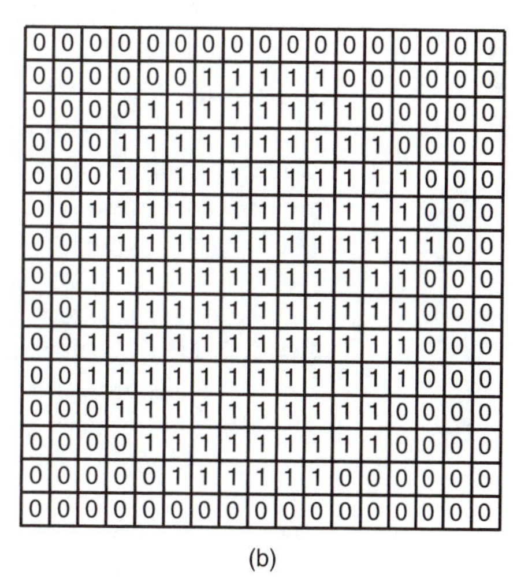

0	0	0	0	0	0	0	0	0	0	0	0	0	0	0	0	0
0	0	0	0	0	0	1	1	1	1	1	0	0	0	0	0	0
0	0	0	0	1	1	1	1	1	1	1	1	0	0	0	0	0
0	0	0	1	1	1	1	1	1	1	1	1	1	0	0	0	0
0	0	0	1	1	1	1	1	1	1	1	1	1	1	0	0	0
0	0	1	1	1	1	1	1	1	1	1	1	1	1	0	0	0
0	0	1	1	1	1	1	1	1	1	1	1	1	1	1	0	0
0	0	1	1	1	1	1	1	1	1	1	1	1	1	0	0	0
0	0	1	1	1	1	1	1	1	1	1	1	1	1	0	0	0
0	0	1	1	1	1	1	1	1	1	1	1	1	1	0	0	0
0	0	1	1	1	1	1	1	1	1	1	1	1	1	0	0	0
0	0	0	1	1	1	1	1	1	1	1	1	1	0	0	0	0
0	0	0	0	1	1	1	1	1	1	1	1	1	0	0	0	0
0	0	0	0	0	1	1	1	1	1	1	0	0	0	0	0	0
0	0	0	0	0	0	0	0	0	0	0	0	0	0	0	0	0

(b)

Figure 6–13 Visual Information in Computer Memory.
(a) The front-lighted ball from Figure 6–12 yields a range of values that are represented in computer memory as numbers from 1 to 5. (b) The high-contrast back-lighted ball from Figure 6–12 can be reduced to values of 1 and 0 in the computer's memory.

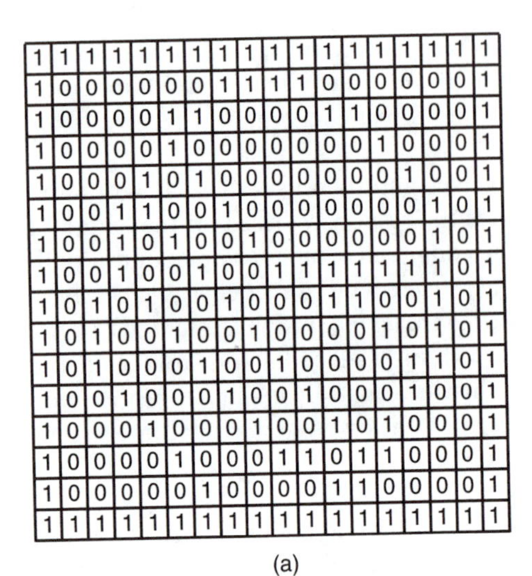

(a)

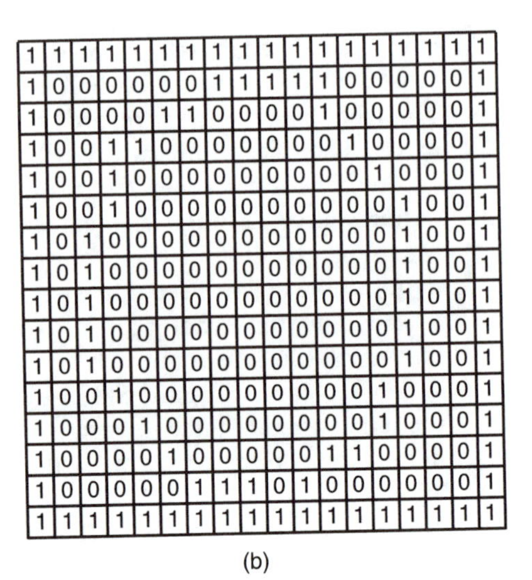

(b)

Figure 6–14 Edge Detection.
Edge detection involves looking for changes in light intensity. (a) This front-lighted figure has visible edges in places other than just around the outside of the sphere. (b) In this high-contrast figure, only the outline of the subject and the edge of the visual field are detectable.

Motion Detection Motion detection is the process of detecting things that are moving in the field of vision. The machine vision system is generally used in a controlled environment where the robot is in no danger, so the system may not be required to perform motion detection. If a machine vision system does have to do motion detection, it generally can be instructed about where to expect the movement. Motion detection is accomplished by comparing the results of two different edge-detection processes. If an edge has moved or a new edge has appeared since the previous edge-detection process was performed, motion has occurred. Figure 6–15 illustrates one method of motion detection.

Object Line Detection Object line detection is the process of connecting detected edges into lines. Because not all points on a line may show up during edge detection, some method must be used to connect disconnected edges into continuous lines. Machine vision systems may use several different procedures on the edge information gathered from edge detection—such as doing vertical lines, then horizontal lines, and finally diagonal lines—in order to complete the task of line detection. If the machine vision system has been instructed about what to expect to see, it can use expectational information to speed up the process. Line detection is an area where much research is going on. Figure 6–16 shows some of the line detection that could be done on the visual information about the ball.

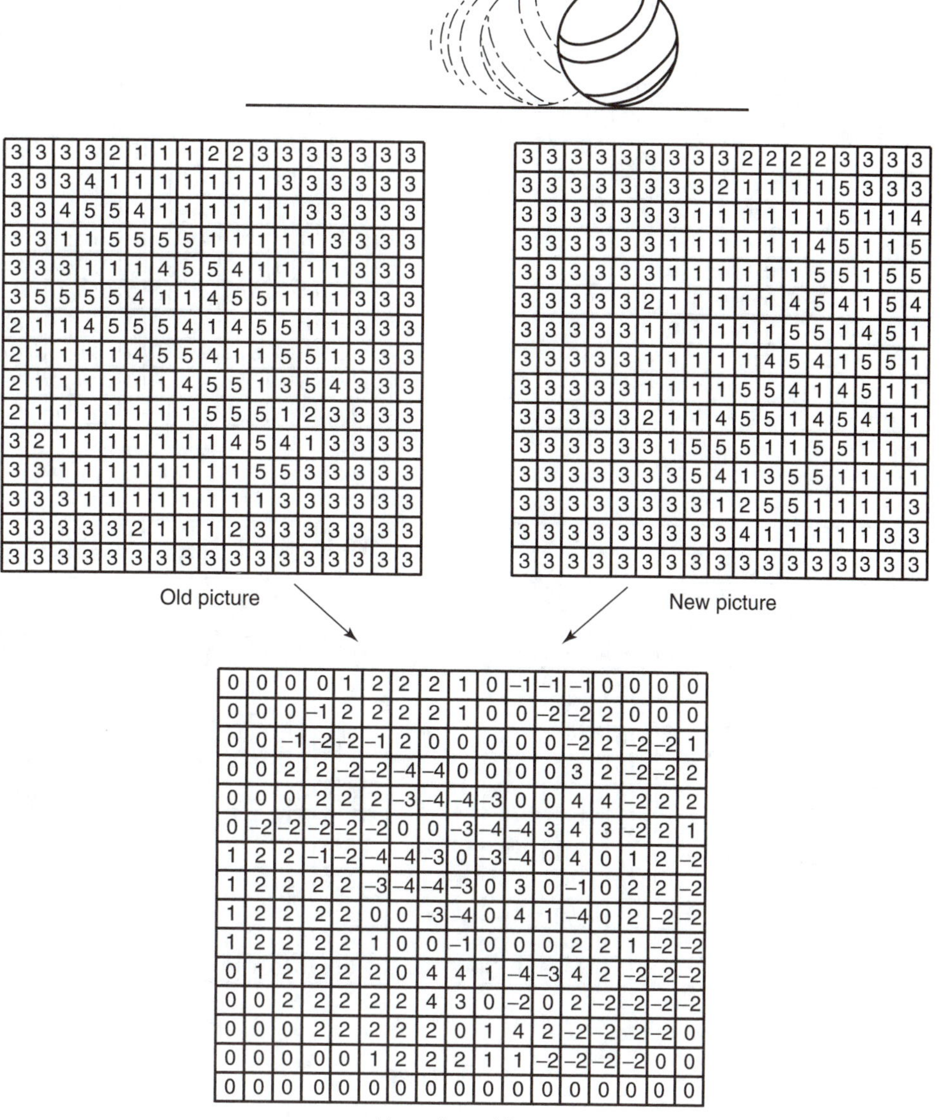

Figure 6-15 Motion Detection.
Motion detection involves comparing an older visual image with a newer one. If the top left number grid represents the older image, and the top right number grid represents the newer image, motion can be detected by subtracting the older image values from the newer image values. The bottom number grid represents the results of this calculation. A minus value represents a cell from which a subject has departed. A plus value represents a cell to which a subject has moved. A zero value represents no change.

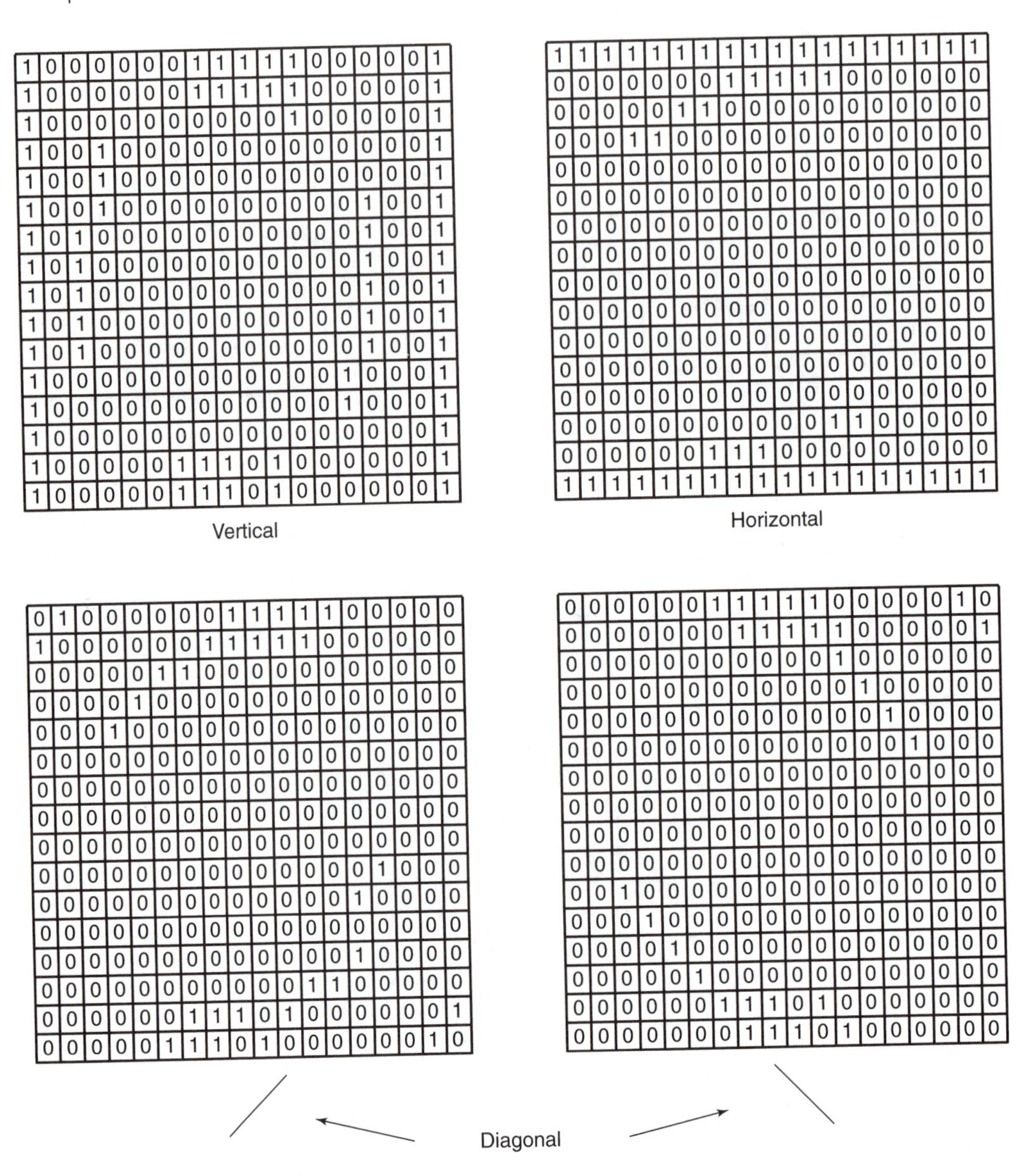

Figure 6–16 Line Detection.
Line detection involves locating line segments in vertical, horizontal, and both diagonal directions.

Object Shape Detection Object shape detection is the process of assembling detected lines to see if they form some known shape. When a single object is involved, the process may be relatively simple—especially if the size and orientation of the expected object are already known. If more than one object is involved, however, they may overlap. This can cause problems in determining which object a common line belongs to. In addition, parts of various lines (except those directly in front) may be missing. Shadows, too, can cause lines to disappear. Machine vision systems need to be taught to recognize objects from only a part of their shape. That is, either the machine vision must be taught to find the closest match or it must be given a reference file of every possible partial shape and size to do its comparing against.

Many features of an object can be used to identify it, including its size, shape, area, perimeter, center, orientation, and Euler number. The Euler number of an object is the difference between the number of objects and the number of holes they have. A uniformly solid object would have a Euler number of 1, while an eye bolt would have a Euler number of 0.

Calculating the perimeter of an object is simply a matter of counting the number of pixels in the lines that make up the shape of the object. This information can be obtained from the shape-detection program.

Calculating the area of an object involves counting the number of pixels on or within the lines that make up the shape of the object.

Machine vision systems often use some type of template or pattern matching to identify different shapes. If the system must be able to identify the shape at different distances, multiple templates are needed. If an object shape is to be identified when only part of the shape is showing, still other templates are needed. Object shape detection is also influenced by rotation of the object and by the angle of view. Figure 6–17 shows one way of doing object shape detection.

Most machine vision systems see the entire object at an unvarying distance and angle, under the same lighting conditions, all the time. If the robot moves around, however, it must learn to identify objects from different angles, under variable lighting conditions, and perhaps from only part of the whole shape. Vision systems will have to be improved considerably to work well under these more difficult conditions.

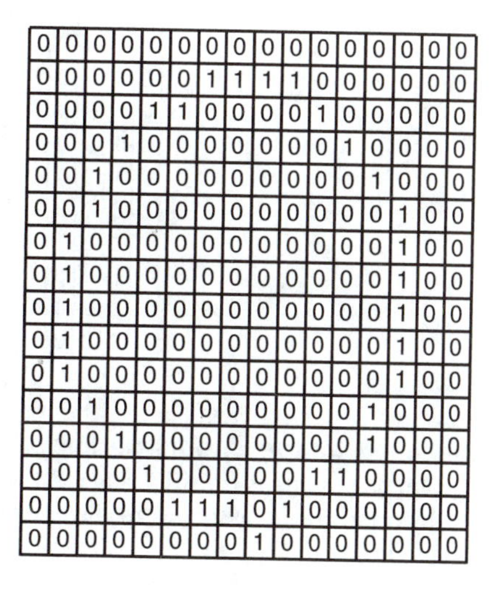

0	0	0	0	0	0	0	0	0	0	0	0	0	0	0	0
0	0	0	0	0	0	1	1	1	1	0	0	0	0	0	0
0	0	0	0	1	1	0	0	0	0	1	0	0	0	0	0
0	0	0	1	0	0	0	0	0	0	0	1	0	0	0	0
0	0	1	0	0	0	0	0	0	0	0	0	1	0	0	0
0	0	1	0	0	0	0	0	0	0	0	0	0	1	0	0
0	1	0	0	0	0	0	0	0	0	0	0	0	1	0	0
0	1	0	0	0	0	0	0	0	0	0	0	0	1	0	0
0	1	0	0	0	0	0	0	0	0	0	0	0	1	0	0
0	1	0	0	0	0	0	0	0	0	0	0	0	1	0	0
0	1	0	0	0	0	0	0	0	0	0	0	0	1	0	0
0	0	1	0	0	0	0	0	0	0	0	0	1	0	0	0
0	0	0	1	0	0	0	0	0	0	0	0	1	0	0	0
0	0	0	0	1	0	0	0	0	0	1	1	0	0	0	0
0	0	0	0	0	1	1	1	0	1	0	0	0	0	0	0
0	0	0	0	0	0	0	0	1	0	0	0	0	0	0	0

Figure 6–17 Object Shape Detection.
Object shape detection involves fitting together all the pieces of lines detected earlier. This occasionally requires filling in missing pixels of a line.

Image Understanding Image understanding, the final step of vision processing, is where the system assesses whether it is seeing the right things. The human vision system does not always have time to complete this step, but in machine vision systems—which work in a very restricted world—this should be done. Figure 6–18 shows image understanding for the ball.

Color, Depth, and Texture Color, depth, and texture provide additional information about a viewed scene.

0	0	0	0	0	0	1	1	1	1	0	0	0	0	0
0	0	0	0	1	1	0	0	0	0	1	0	0	0	0
0	0	0	1	0	0	0	0	0	0	0	1	0	0	0
0	0	1	0	0	0	0	0	0	0	0	0	1	0	0
0	0	1	0	0	0	0	0	0	0	0	0	0	1	0
0	1	0	0	0	0	0	0	0	0	0	0	0	1	0
0	1	0	0	0	0	0	0	0	0	0	0	0	1	0
0	1	0	0	0	0	0	0	0	0	0	0	0	1	0
0	1	0	0	0	0	0	0	0	0	0	0	0	1	0
0	1	0	0	0	0	0	0	0	0	0	0	0	1	0
0	0	1	0	0	0	0	0	0	0	0	0	1	0	0
0	0	0	1	0	0	0	0	0	0	0	0	1	0	0
0	0	0	0	1	0	0	0	0	0	1	1	0	0	0
0	0	0	0	0	1	1	1	0	1	0	0	0	0	0
0	0	0	0	0	0	0	0	1	0	0	0	0	0	0

0	0	0	0	0	0	1	1	1	1	0	0	0	0	0
0	0	0	0	1	1	0	1	0	0	1	0	0	0	0
0	0	0	1	0	0	0	1	0	0	0	1	0	0	0
0	0	1	0	0	0	1	0	0	0	0	1	1	0	0
0	0	1	0	0	0	1	0	0	0	1	0	0	1	0
0	1	0	0	0	1	0	0	0	0	1	0	0	1	0
0	1	0	0	0	1	0	0	0	1	0	0	0	1	0
0	1	0	0	1	0	0	0	0	1	0	0	0	1	0
0	1	0	0	1	0	0	0	1	0	0	0	0	1	0
0	1	0	1	0	0	0	0	1	0	0	0	0	1	0
0	0	1	1	0	0	0	1	0	0	0	0	1	0	0
0	0	0	1	0	0	0	1	0	0	0	0	1	0	0
0	0	0	0	1	0	1	0	0	0	1	1	0	0	0
0	0	0	0	0	1	1	1	0	1	0	0	0	0	0
0	0	0	0	0	0	0	0	1	0	0	0	0	0	0

0	0	0	0	0	0	1	1	1	1	0	0	0	0	0
0	0	0	0	1	1	1	0	1	0	1	0	0	0	0
0	0	0	1	0	1	0	1	0	0	0	1	0	0	0
0	0	1	0	1	0	0	1	0	0	0	1	1	0	0
0	0	1	0	1	0	1	0	0	0	1	0	0	1	0
0	1	0	1	0	0	1	0	0	0	1	0	0	1	0
0	1	0	1	0	1	0	0	0	1	0	0	1	1	0
0	1	1	0	0	1	0	0	0	1	0	0	1	1	0
0	1	1	0	1	0	0	0	1	0	0	1	0	1	0
0	1	1	0	1	0	0	0	1	0	0	1	0	1	0
0	0	1	1	0	0	0	1	0	0	1	0	1	0	0
0	0	0	1	0	0	0	1	0	0	1	0	1	0	0
0	0	0	0	1	0	0	1	0	1	1	1	0	0	0
0	0	0	0	0	1	1	1	0	1	0	0	0	0	0
0	0	0	0	0	0	0	0	1	0	0	0	0	0	0

Previously recorded images or templates

Compare to

0	0	0	0	0	0	0	0	0	0	0	0	0	0	0	0
0	0	0	0	0	0	1	1	1	1	0	0	0	0	0	0
0	0	0	0	1	1	0	0	0	0	1	0	0	0	0	0
0	0	0	1	0	0	0	0	0	0	0	1	0	0	0	0
0	0	1	0	0	0	0	0	0	0	0	0	1	0	0	0
0	0	1	0	0	0	0	0	0	0	0	0	0	1	0	0
0	1	0	0	0	0	0	0	0	0	0	0	0	1	0	0
0	1	0	0	0	0	0	0	0	0	0	0	0	1	0	0
0	1	0	0	0	0	0	0	0	0	0	0	0	1	0	0
0	1	0	0	0	0	0	0	0	0	0	0	0	1	0	0
0	1	0	0	0	0	0	0	0	0	0	0	0	1	0	0
0	0	1	0	0	0	0	0	0	0	0	0	1	0	0	0
0	0	0	1	0	0	0	0	0	0	0	0	1	0	0	0
0	0	0	0	1	0	0	0	0	0	1	1	0	0	0	0
0	0	0	0	0	1	1	1	0	1	0	0	0	0	0	0
0	0	0	0	0	0	0	0	1	0	0	0	0	0	0	0

Present image

Figure 6–18 Image Understanding.
Image understanding is the final stage of the vision process. It requires previous experience with objects identical or similar to the one being observed. In this figure, the present image will be found to match the leftmost template most closely.

Color detection is the process of separating different frequencies of visual energy. Most machine vision systems are not concerned with color. If the part that is the main focus of the vision system differs in color from its surroundings, however, a color filter can be used on the vision sensor to make the part easier to find.

Depth perception, or three-dimensional vision, can be used to combat problems in having a robot pick a part from a bin of overlapping parts. It is also useful for navigating in an unknown area. A moving machine vision system could be taught to use the rate of change in size of a stationary object as a clue to the distance between that object and the vision system. Three-dimensional vision can be supplied by using two sensors or by viewing the object from two different locations with a single sensor. Alternatively, a robot could use sonar or radar as a range or distance finder.

The texture of an object may also be useful in helping to identify it. Is the object solid-colored? Is it flat? Is it wrinkled? Texture can be used to pick patterns out of a visual scene. For example, how can a Ping-Pong ball be distinguished visually from a golf ball? A Ping-Pong ball is smooth and has a single seam that joins the two halves. A golf ball also has a single seam that joins the two halves, but it has dimples all over it.

Expectational Information Expectational information is used to process vision toward an expected goal and to ignore all other possibilities until it becomes obvious that the expected goal cannot be reached. Since machine vision systems often work in a controlled environment, expectational information is extremely useful to them for fast vision processing. On the other hand, a robot that must travel along a corridor and encounter unknown objects in its path will have great difficulty using expectational information successfully. Even though a machine vision system that operates without expectational information is slower, it has the advantage of seeing what is really there!

The expectational goals for a robotic vision system can be entered into the system in one of two ways. The first method is to show the robot the goal by pointing the vision at the goal. The robot then records the image for future reference. This is similar to using voice prints with a speech interpretation system. The other method of entering the goal into a robotic vision system is to describe the picture geometrically. The robot then builds the description into an image.

Limits of Robot Vision Robot vision is not limited to what humans see, since robots can use radar or sonar as an illumination source. Still, some tasks given to robotic vision systems could be handled just as well or better by a robotic touch system. The classic bin-picking problem, where overlapping parts in a bin are to be picked up one at a time and inserted with proper orientation into a machine, might be done in the following steps:

1. The robot grasps a part or parts from the bin.
2. The robot places the part(s) on a table that has an artificial skin surface.
3. The skin locates the orientation of the part(s).
4. The robot is instructed in how to grasp the part with proper orientation for the rest of the task.

Vision input and its attendant problems are covered in more detail in Chapter 17, Vision Systems.

Touch

Tactile sensing, the sense of touch for robots, is needed if a robot is to perform delicate assembly operations. It can be used to recognize parts, determine part positions, determine part orientation, and control force applied to parts. Force needs to be controlled during assembly to prevent cross-threading of screws and bolts.

The sense of touch in the human is greatly underrated. Often, as in screwing together a nut and bolt, touch is more important than vision. When teleoperators were invented to handle dangerous materials, their designers soon found out how important touch feedback is in handling materials successfully. While the early teleoperators could see what they were doing, they had no sense of touch, and this limitation made working with them difficult and exasperating.

The robot's tactile sense is based on the sensing of forces on joints, on the direct contact of external bodies, and on the slippage between the robot and an external body. In humans, the sense of touch in relation to external bodies includes a combination of force and temperature sensing.

When a robot touches something, force is reflected back through each joint. This force on the joints can be measured directly by using strain gauges, or it can be measured indirectly by noting the change caused by the force in the armature current of an electric joint motor, in the air pressure of a pneumatic motor, or in the fluid pressure of a hydraulic motor.

Figure 6–19 shows the "Ding Bot" mechanical bumper sensor, which uses a floating, spring-loaded, pivoting drive wheel. As long as the robot is not meeting

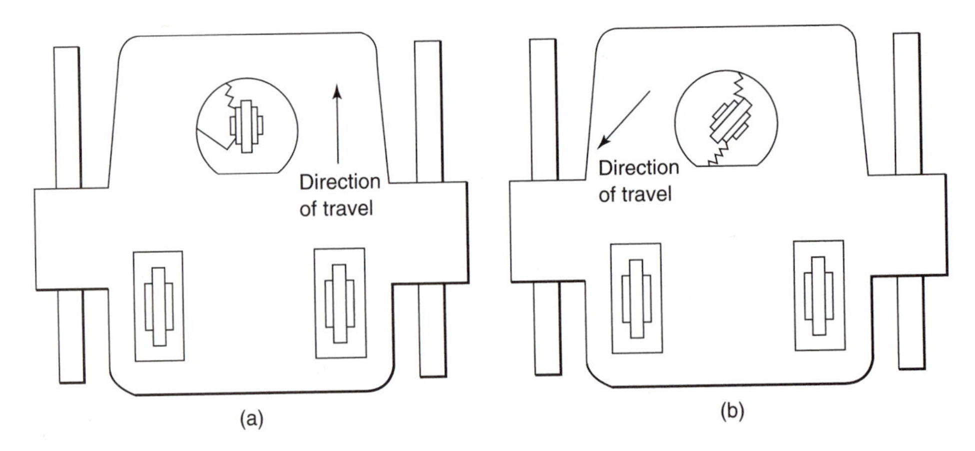

Figure 6–19 "Ding Bot" Touch Sensor.
The "Ding Bot" touch sensor uses a spring to hold the steering straight until an obstacle is encountered. The obstacle resistance overcomes the steering spring resistance and the Ding Bot turns 100° to the left before proceeding. (a) The sensor in normal drive position. (b) The sensor's drive position when an obstacle is encountered.

any resistance, the spring keeps the drive wheel pointed forward. When an object blocks the robot's path of travel, however, the drive wheel rotates 100°, turning the robot away from the object. The spring then returns the robot's drive wheel to a straightforward course.

The force on an external object can be measured directly at the point where the robot contacts the object. If a contact/no contact indication is all that is needed, a microswitch may be used as the sensor. A long-range touch sensor could make use of a whisker attached to a microswitch. If the level of applied force must be ascertained, spring-loaded potentiometers, strain gauges, or pressure-sensitive paint devices can be used. A spring-loaded potentiometer sensor uses a spring to hold the sensor rod fully extended. When the rod touches an object, the force of the touch works against the spring and pushes the rod back a distance proportional to the force. This retraction of the rod changes the setting of the potentiometer and changes its resistance value. A pressure-sensitive paint sensor contains a paint made by mixing piezo-resistant semiconductor powders with an organic material between two metal electrodes. When pressure is applied to the outside surface of one of the electrodes, the pressure is transmitted to the paint layer, which then changes its resistance in proportion to the pressure. Figure 6–20 shows the microswitch sensor, the whisker sensor, the spring-loaded potentiometer, and the pressure-sensitive paint sensor.

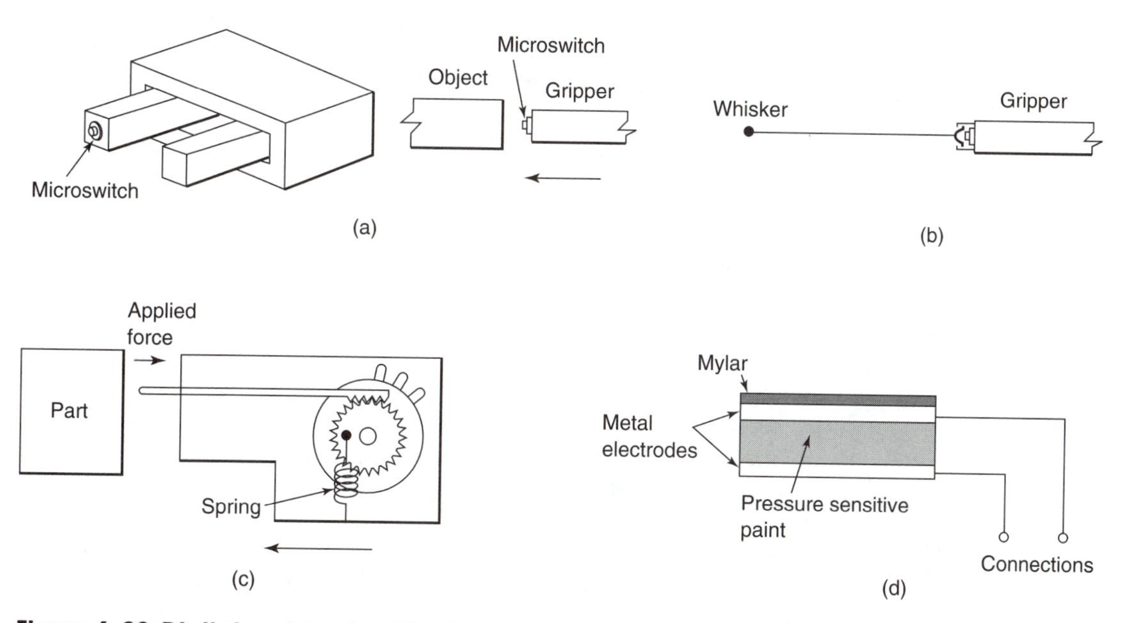

Figure 6–20 Digital and Analog Single-contact Sensors.
(a) A microswitch being used as a short-range digital touch sensor. (b) A whisker and microswitch being used as a longer-range touch sensor. (c) A spring-loaded potentiometer and rod being used as an analog force touch sensor. The potentiometer's setting will vary in proportion to the force of the touch. (d) Pressure-sensitive paint used in an analog touch sensor. The harder the pressure against the paint, the lower the paint's resistance.

Robots can be given touch information through a matrix of strain gauges, a matrix of microswitches, a matrix of crossing bare wires, or a matrix of conductive foam or rubber. The matrix of microswitches is too large to fit on a robot's gripper; however, a robot could use an external device made up of a matrix of microswitches to identify the shape and orientation of a part. The matrix of strain gauges, while smaller than the matrix of microswitches, can measure the forces on various parts of the gripper as it grasps a part. The matrix of conductive foam can take the form of a sheet situated between sets of horizontal and vertical wires; pressing on the foam surface lowers the resistance between the nearest row and column wire intersections. Alternatively, a conductive rubber or foam sheet can be placed on top of a printed circuit board with row and column contacts. Again, pressing the rubber surface lowers the resistance of the row and column contacts in that area. Conductive, D-shaped rubber cords can be overlapped to make a matrix, with the rounded surfaces forming the row and column contacts; pressing near a row and column contact point causes the rounded surfaces to flatten out, increasing the contact area and lowering the resistance of the contact. Figure 6–21 shows examples of conductive foam and rubber matrix sensors. Interpreting pressure data to obtain shapes and part orientations is still highly experimental.

A slip sensor for a robot can be produced by building a roller into the side of the gripper. An object that begins slipping in the robot's grip causes the roller to rotate. The roller may either be part of an encoder built into the gripper or turn a shaft that connects to a sensor on the back of the gripper. A centered, spring-loaded potentiometer or Hall-effect sensor can be used in place of an encoder. Figure 6–22 shows a robot slip sensor.

Touch sensors may contact an object at a single point or at multiple points. The contact may be made directly by the sensor or remotely through some type of linkage.

Single-point Contact A single-point contact device makes contact with an object only once. While this does not reveal much about the size or shape of the object, it does indicate the presence or absence of the object. Such a device can also be used by the robot as an internal sensor to tell when an axis has reached a limit, as in the case of a limit switch or microswitch.

Another device that could be used as a single-point contact sensor consists of two bare wires held apart by a spring; it is sometimes called a **membrane sensor** and is used in membrane keyboards on computers (see Figure 6–23). Its surface is flat, and it is only 1/16 inch thick. There is no reason why such a device could not be built into a gripper face. A strain gauge could also be used as a single-point contact sensor—not only to give a presence or absence signal, but to tell how hard the force on the gripper is. If the robot is trying to pick up metal parts, two bare electrical contacts may be used. When the metal part is contacted, it completes the electrical circuit.

Multiple-point Contact A multiple-point contact device makes (or at least *can* make) contact with a part at more than one point. This enables the device to gather some size and shape information about the part. A multiple-point contact touch

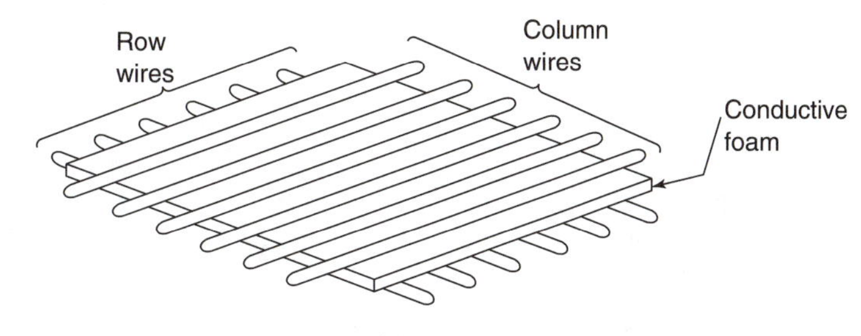

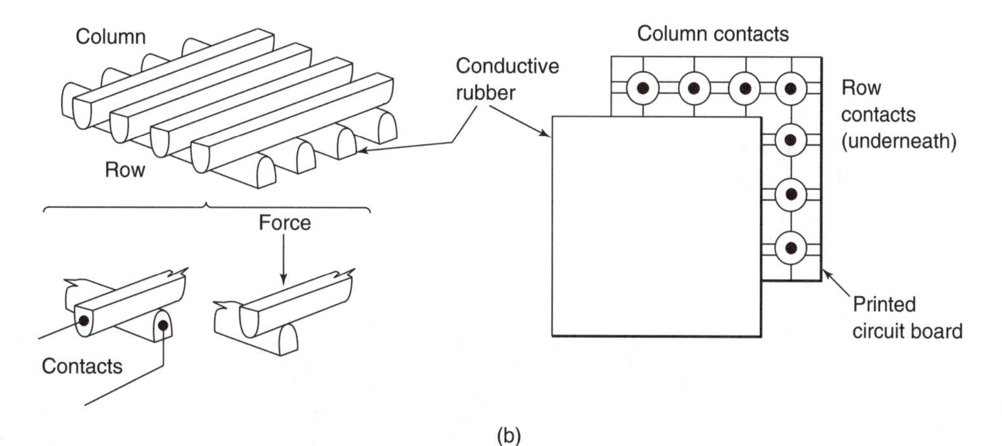

Figure 6-21 Touch Matrix Sensors.
(a) A touch matrix sensor made from conductive foam and laid between row and column wires. It can provide information on the shape of the object touching it, as well as on varying pressure readings at each contact point. The resistance of a row and column junction decreases as the pressure at the point increases. (b) Conductive rubber touch sensor matrices. At left, rubber rods make up one type of touch matrix sensor. The rubber deforms at row-column crossings where pressure is applied. The deforming results in more surface contact and consequently lower resistance readings. At right, a sheet of conductive rubber is placed on top of printed circuit contacts. Pressure on the rubber decreases the resistance between the contacts.

sensor is sometimes composed of a series or matrix of simple single-contact sensors. For example, a matrix of microswitches can be used to get some idea of the shape and heaviness of a part. To make the working surface of the sensor smaller, you could run rods from the switches to the surface of the gripper. As the gripper closed on a part, it would have enough force to activate all contacted switches. A matrix of membrane switches could also be used.

Artificial Skin Artificial skin is a large touch matrix sensor that attempts to imitate human touch. It can sense the shape and various other details about the surface being touched, including the surface's hardness and its texture or smoothness.

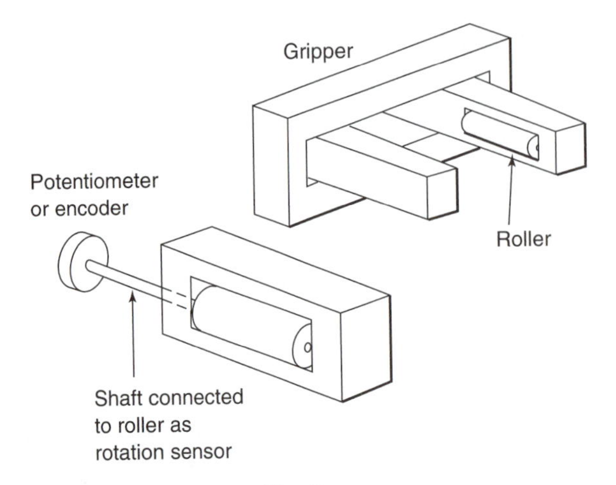

Figure 6-22 Robot Slip Sensor.
A roller built into the side of a gripper can be used to detect part slippage within the gripper. The roller axle is attached to some type of sensor—an encoder, a Hall-effect sensor, or a spring-loaded potentiometer.

Figure 6-23 Membrane Keyboard of the ZX-81 Computer.
The membrane keyboard of the ZX-81 computer is composed of a matrix of bare crossing wires. Spring tension normally holds the contacts open. This is a form of touch matrix sensor.

Artificial skin may also have some pliability, enabling it to develop something of a three-dimensional feeling about a part.

At present, artificial skin is not really available, although small vector and area tactile sensors are available. Some common available sizes of area sensors are 4 by 4, 8 by 8, 8 by 12, 10 by 10, 10 by 16, and 16 by 16 inches. Coarse sensing ability is obtained when sensor elements are spaced 1/10 inch or farther apart. Fine sensing is obtained when sensor elements are spaced less than 1/10 inch apart. The Lord Corporation of Cary, North Carolina, is one supplier of tactile sensors.

Types of Detection There are four types of touch detection: single-point detection, simple edge detection, multiple-point shape detection, and type of surface detection.

Single-point detection is sufficient to tell whether a part is present or not. It can also be used to signal that something or someone is in a robot's path. However, depending on the size and speed of the robot, it may already be too late to avoid a collision.

Simple edge detection can be used to determine how far an object is placed into a gripper or (if the object is small enough) whether the object being grasped is a whole part or only part of the part. Depending on how wide the sensor is, it can tell the width of an object or at least whether it is as wide as the sensor. Used in a vertical line, the sensor can measure the height of an object.

A multiple-point shape detector is a matrix of touch sensors that can determine the shape of one surface of an object. This ability is useful for determining the orientation of a part. If the sensor is fine enough, it can be used not only to inspect the outline of the part but to detect the location and size of holes in the part. A small matrix touch sensor can tell the height and depth of a part located within the grasp of a gripper.

The term *type of surface* refers to such characteristics as hardness and smoothness or roughness. If a surface is rough, is the texture coarse or fine? Is there a pattern to the roughness? Are there holes in the surface? Hardness involves the pliability or give that an object exhibits when pressure is applied to some part of it. Measuring hardness requires the ability of the artificial skin to deform locally and thus apply pressure to an object.

Interpreting touch information is similar to interpreting vision information. Of course, the sense of touch does not require the presence of an illumination source, as vision does. And touch sensor arrays can convey three-dimensional information directly. Detailed coverage of touch sensor interfacing and interpretation is provided in Chapter 18.

Range and Proximity Detectors

A ranging device detects objects situated at some distance from the robot. A proximity detector detects objects in the immediate vicinity of the robot.

Range-finding devices are noncontact devices: they do not have to touch an object to know it is there. Laser, radar, sonar, vision, and infrared devices are used for range finding. Humans use vision, hearing, and smell for range finding.

Proximity detectors can be contact or noncontact devices. Contact devices must actually touch an object to know it is there, while noncontact devices can detect an object without touching it. Humans use vision, hearing, smell, and temperature sensors as noncontact proximity detectors, and touch and temperature sensors as contact proximity detectors.

A stationary industrial robot has little use for ranging devices and might use a proximity sensor only to tell when a part is approaching it on a conveyor belt or when it has a part in its gripper. On the other hand, a mail-delivering cart must be able to detect an unexpected obstacle in its path. Many small hobbyist robots use bumpers that activate microswitches as proximity sensors. This is fine for a small, light, slow-moving robot, but it is unsatisfactory for a heavy or fast-moving robot. In these cases, it is preferable that the robot change course or stop before touching an object.

Proximity sensors have long been in common use. Examples include pressure pads or magnetic detection coils for detecting automobiles at a traffic light, automatic fire sprinklers, the horn on a World War II explosive mine, the bimetallic strip on a thermostat, photo cells for detecting oncoming automobile headlights and dimming your own headlights, and ultrasonic sensors for detecting an approaching customer and opening a store door.

Several types of noncontact proximity sensors are available for robots: magnetic, light, sound (sonar), capacitance, tuned radio circuit, temperature, radiation, and radar. The choice of which type to use should be based on the type of material to be detected and the distance at which the detection is to take place.

Magnetic Detectors A magnetic detector responds to the presence of magnetic materials such as iron, cobalt, some ceramics, and some steels. Some alarm systems use magnetic detectors to signal when a door or window has been opened. The magnet is fastened to the movable surface and a spring-loaded reed relay is fastened to the adjacent fixed surface. When the magnet is near the relay, the relay contacts are pulled closed, completing an electrical circuit. When the magnet is away from the relay, the spring keeps the contacts open and breaks the electrical circuit.

A swinging magnet of the type used in a stud detector is another style of magnetic detector. The electrical circuit in this case can be as simple as a magnet swinging between a phototransistor and its light source. A Hall-effect circuit consisting of a Hall generator and an amplifier can also be used to detect magnetic materials.

Inductive Metal Detectors Tuned radio circuits are one form of inductive metal detector. When a piece of metal enters the inductive field of a tuned radio circuit, it changes its frequency of operation and the current through the circuit. Magnetic materials, such as iron and nickel, increase the inductance by concentrating the field; nonmagnetic materials, such as copper and aluminum, decrease the inductance by scattering the field. The mass of the metal object determines how far the radio circuit is detuned.

The changes in current through the circuit can be used to detect or count metal objects. It is even possible to tell the difference in mass in the objects. A robot could use this detector to locate a metal object on a conveyor belt.

Light Ranging Detectors Several methods of using light as a range or proximity detector are possible. One of the earliest electronic proximity detectors was the photoelectric eye, which consisted of a light-beam source and a beam detector. As long as the detector receives the light beam, nothing is blocking the path between the source and the detector. When something breaks the beam, the detector registers the absence of the beam and the object's presence is detected. A robot gripper with a photoelectric eye can tell when something is in position for its gripper to grasp it. Infrared sources may be used in much the same way as visible light photoelectric systems.

Figure 6–24 shows the use of a photoelectric system for sorting parts. If no parts are present, the far light detector signals that no part is approaching. When a part intercepts this light beam on its way to the part detector, the far detector signals that a part is approaching. If the part is light-colored, light reflects off it to the second detector. This then signals the ram to push the part into the light parts bin. If the part is dark-colored, light is not reflected off it and it is carried to the end of the conveyor belt, where it falls into the dark parts bin.

The light source and photodetector can be mounted on a moving robot in such a way that the light source shines away from the robot and the photodetector detects the reflected light coming back to it. As the robot approaches an object that has a reflecting surface, the level of reflected light reaching the photodetector greatly increases, producing a current change in the device and setting off an alarm.

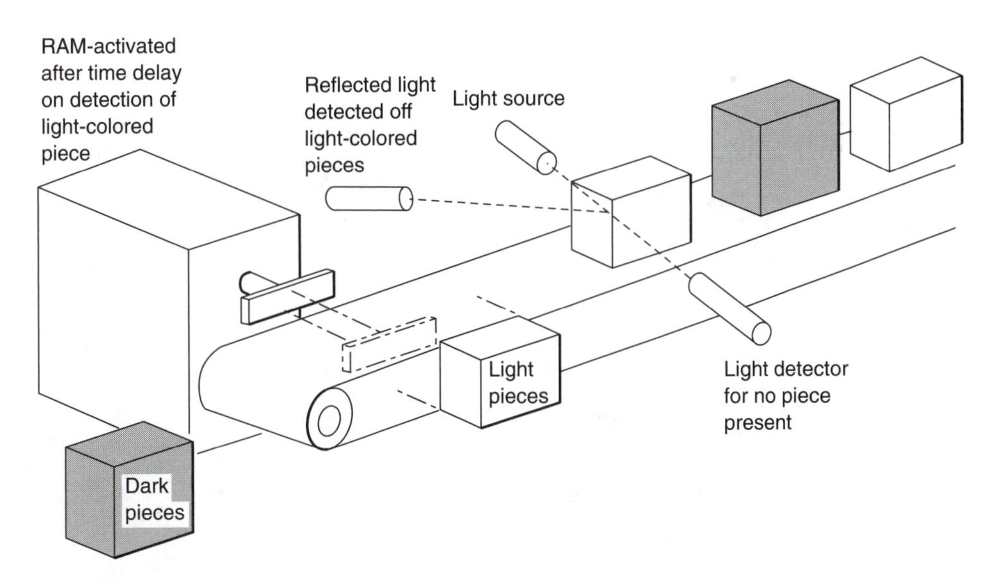

Figure 6–24 Sorting by Means of Proximity Detectors.
Photoelectric proximity detectors can be used to sort certain parts. Here, two types of parts are distinguished by color. Light-colored parts reflect enough light for the second detector to notice and trigger a moving ram. Dark-colored parts do not reflect enough light to trigger the detector.

Now that it knows it is about to run into some other object, the robot may stop or turn to avoid the object. Of course, reflected light works best when there is only one source of light. If two moving robots both use light proximity detectors, one robot's photodetector may pick up light from the light source on the other robot and stop or turn long before the robots are anywhere near each other.

The Avoider robot uses an infrared ranging detector in essentially the same fashion as other robots use photodetectors.

Sound (Sonar) Range Detectors Sonar or sound range finding was developed during World War II. However, it has only recently been adapted for use in consumer products. The fish finder uses underwater sound waves to locate the bottom, fish, and other objects (see Figure 6–25). World War II sea captains would have loved having sonar this precise. Sony has developed an electronic measuring device that uses sound waves in the air to measure distances. The transmitter and receiver in a single unit can be used to measure distances of up to 100 feet. For longer ranges, two units with separate transmitters and receivers are required. An older device of this type is the sonar-operated automatic focus on many cameras.

Robots can be equipped with an ultrasonic range finder system similar to that possessed by bats and by the Polaroid self-focusing camera. This system can tell the robot approximately how far away an object is for distances of 0.9 to 35 feet. The Polaroid ultrasonic range system attempts to overcome the problem of a particular sonar's frequency signal not reflecting from some surface by using four

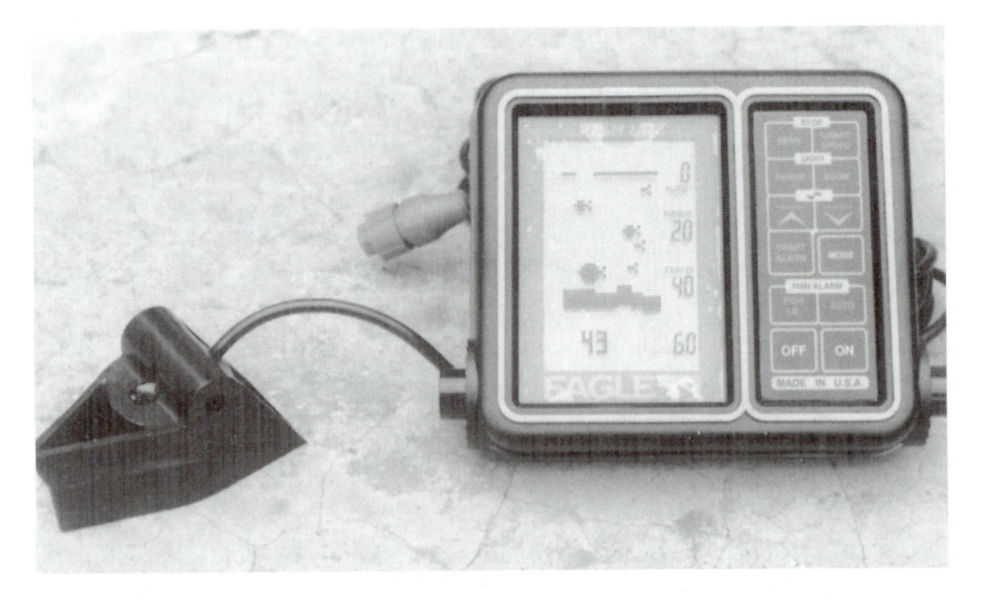

Figure 6–25 Miniature Sonar Fish Finder.
The fish finder uses ultrasonic echoes to detect underwater objects. It emits ultrasonic pulses and measures how long it takes for the echo to return.

different frequency signals—49.7, 53, 57, and 60 kilohertz. A Polaroid ultrasonic range kit can be purchased as an experimental board with a distance readout display from Polaroid Corporation's Commercial Battery Division, 199 Windsor Street, Cambridge, Massachusetts 02139. Figure 6–26 shows the Polaroid kit. The Hero I robot also comes with an ultrasonic range finder (see Figure 6–27). Both systems are relatively short-range, since ultrasonic sound is attenuated very rapidly in air.

An ultrasonic range finder works by measuring the length of time that elapses between emitting a signal and receiving its echo. By dividing the length of time in half and factoring in the speed of sound, you can calculate the distance from the range finder to the object reflecting the sound. Of course, some materials reflect sound better than others, and the angle at which sound strikes an object may prevent the echo from returning to the detector.

Not all sonic proximity sensors use an active sound source. Some devices just quietly listen for sounds. Moving objects give off ultrasonic sound, so nearby

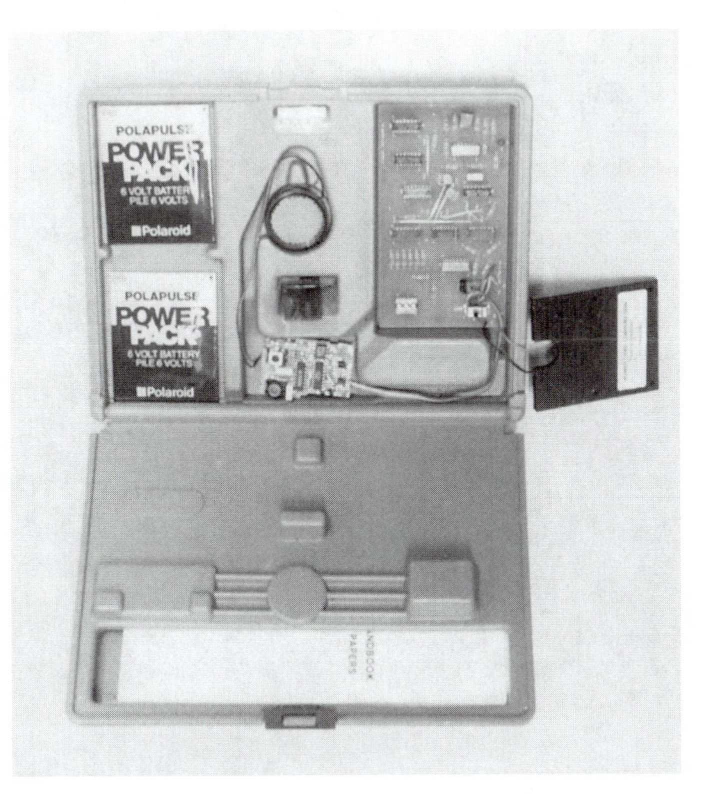

Figure 6–26 Polaroid Ultrasonic Range Finder Kit.
The Polaroid ultrasonic range finder kit emits sound at 49.41 kHz in an effort to detect a wider range of objects. Some materials reflect certain ultrasonic sound frequencies better than others.

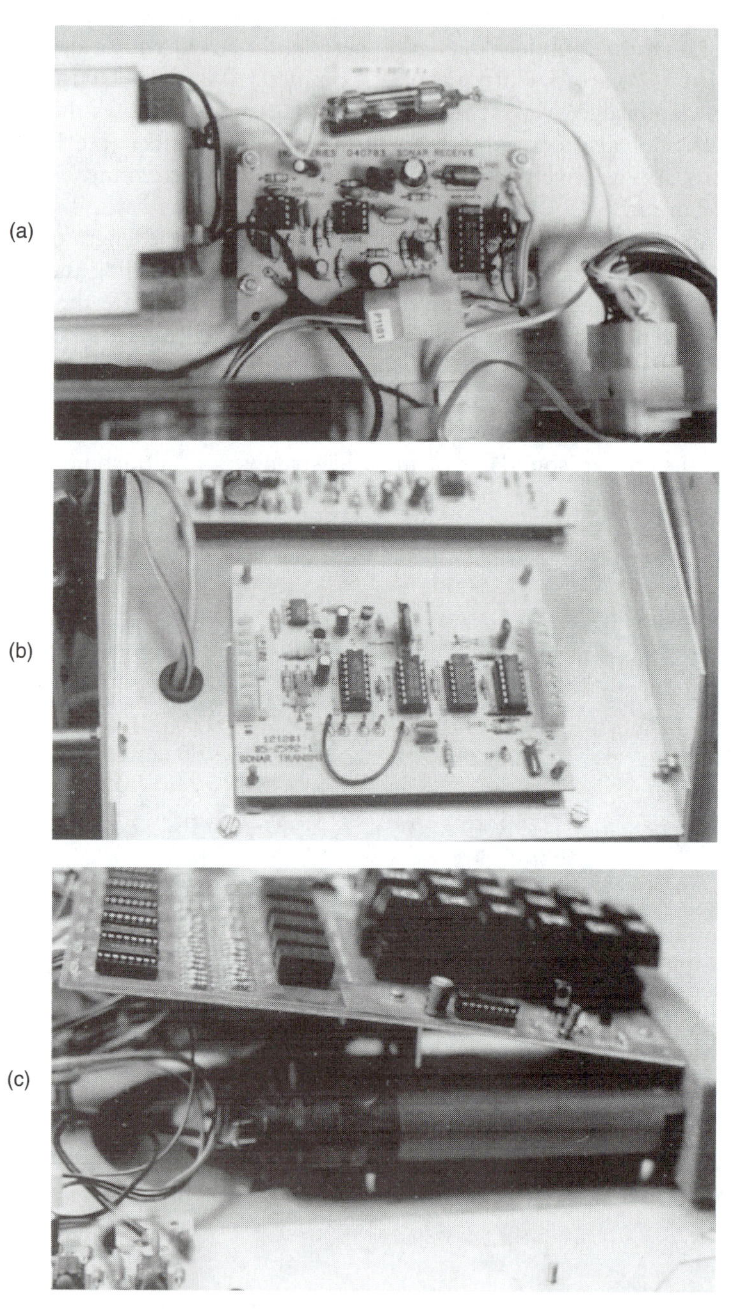

Figure 6-27 Hero I Ultrasonic Range Finder.
The Hero I ultrasonic range finder uses a single ultrasonic frequency. (a) Sonic receiver circuit card. (b) Sonic transmitter circuit card. (c) Sonic transducers.

motion of objects can be detected simply by listening for ultrasonic sounds. Motion detectors and some burglar alarms work on this principle.

Capacitance Detectors The capacitance of an electrical circuit changes as the dielectric constant of the circuit changes. Some electric lamps equipped with a capacitance proximity detector can be turned on and off by having a hand placed next to them. This is because a human hand has a higher dielectric constant than air has. The air near the lamp forms part of the capacitance circuit of the lamp. As a human hand comes near the lamp, it changes the capacitance of the circuit.

Capacitance detectors use changing dielectric constants to detect the presence of a nearby object. Nonmagnetic materials and even nonconducting materials such as plastics can be detected by a capacitance detector. However, capacitance proximity detectors have a very short range of effectiveness—generally, 1 inch or less. Figure 6–28 shows a modern stud finder called a *density meter* that uses a capacitance detector to sense the difference between air and a two-by-four stud behind plaster or wallboard.

Tuned Radio Circuits There are two types of tuned radio circuit proximity detectors: those that use a single radio circuit, and those that use two radio circuits. When operating, a tuned radio circuit is surrounded by an electromagnetic field. If an object enters this field, it detunes the circuit, causing the current through the circuit to change. Such a change can be used to set off an alarm.

A system using two tuned radio circuits is a little different. Both circuits start out tuned to the same frequency, and a difference detector is connected to both circuits. As long as the two circuits remain tuned to the same frequency, no output comes from the detector. One of the circuits is placed at the end of a rod or pole, while the other circuit is placed in a shielded container. When an object interferes with the field of the exposed circuit, it changes the frequency of that circuit. Since the two circuits are no longer at the same frequency, the detector produces a sound equal to the difference between the two frequencies.

Metal detectors can be designed to use either of these two methods. The better metal detectors use two tuned radio circuits.

Since the extent of detuning the radio circuit depends on the type and size of the material, as well as its distance from the circuit, the tuned circuit is not generally able to measure the distance from itself to an object. But on a production line, where the type and size of the material interrupting the field of the tuned radio circuit is known, the distance can be calculated.

A Faraday shield is added to the search coil of a tuned radio circuit metal detector to prevent it from responding to electrostatic fields. Otherwise, the metal detector will respond to your hand's capacitance if the hand is positioned near the coil. Figure 6–29 shows a metal detector.

Radiation Radiation detectors can be used to detect radioactive materials. Such detectors are useful in nuclear power plants and in work with X rays.

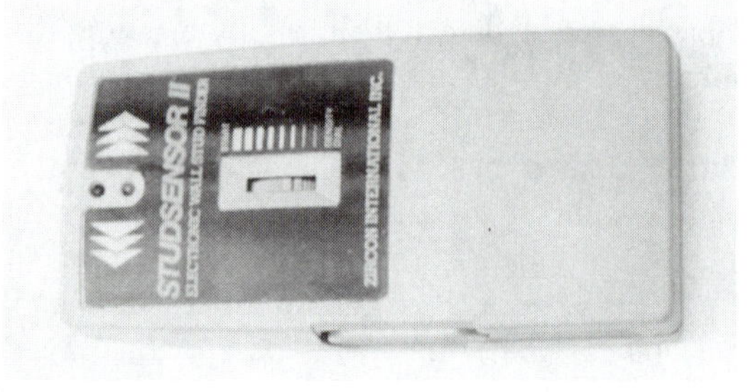

(a)

(b)

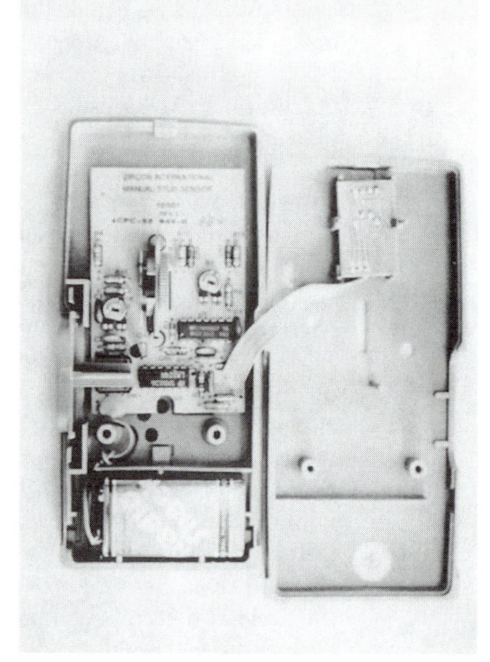

(c)

Figure 6–28 Stud Finder Using a Capacitance Detector.
(a) A modern electronic stud finder that uses a capacitance detector to sense the difference in density in the wall where a stud is located. (b) The foil side of the printed circuit board, showing the capacitance sensor. (c) The component side of the circuit board, inside-mounted in the stud finder.

Radar Radar, short for *ra*dio *d*etection *a*nd *r*anging, was developed during World War II. Since radar signals travel at nearly the speed of light, they can be used to detect objects hundreds of miles away. Although space and missile robots may benefit from using this type of proximity detector, most industrial robots do not need it.

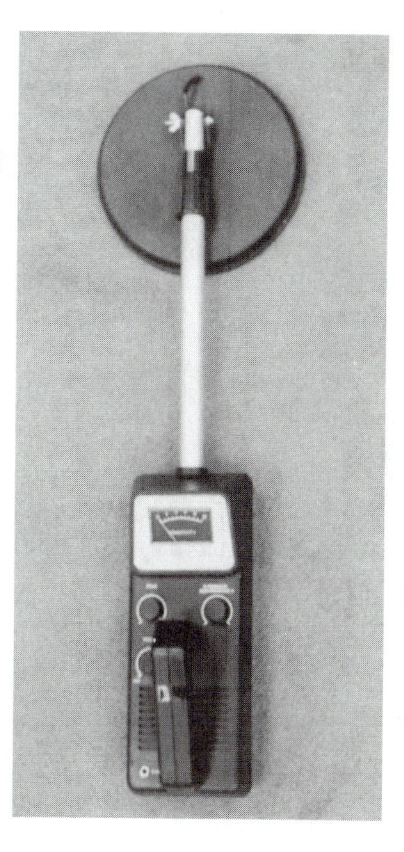

Figure 6–29 Metal Detector.
Radio Shack's Standard Detector uses two tuned radio circuits. The large, flat, circular sensor holds the search coil surrounded by a Faraday shield. The reference tuned circuit and the rest of the detector's tuned circuit are shielded in the case by the handle.

Radar may be active or passive. Active radar works by sending out radio waves and measuring the length of time it takes for the echo to return; the time elapsed is proportional to the distance of the object from the transmitter. Inactive radar just records existing radio waves; it is used for listening to the stars and in the radar-detecting consumer product known as a fuzz buster.

Single-use Sensors Some proximity detectors are intended for a single use. For example, the foil in a glass window must be broken to stop the flow of electrical current and set off the burglar alarm; the Wood's metal in a fire sprinkler head must melt to turn on the water to stop a fire; a sonar buoy is dropped from an airplane or helicopter for a one-time use in detecting a submarine from the air; and electrical and mechanical fuses detect a one-time system overload. The Airco AR-1 robot, for example, uses a mechanical fuse made from fiberboard to protect its welding tool from excessive force. The fuse mounts between the robot's wrist and the welding tool (see Figure 6–30).

Proximity detector interfacing and programming problems are discussed in Chapter 19.

(a)

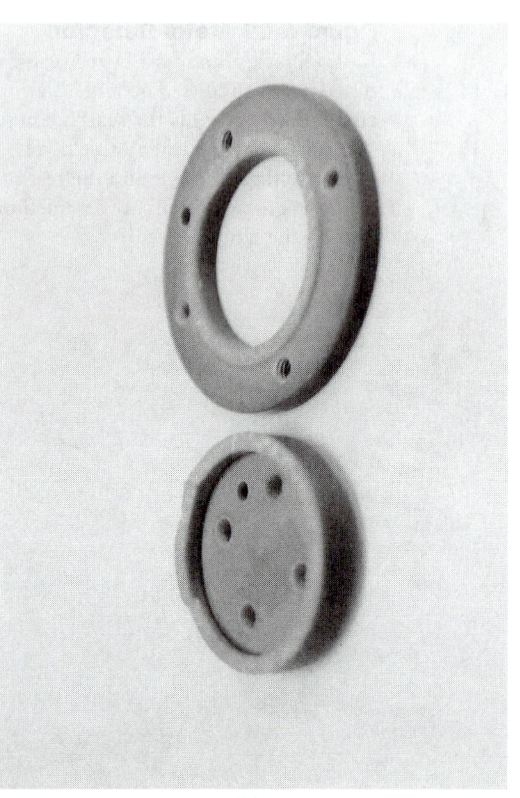

(c)

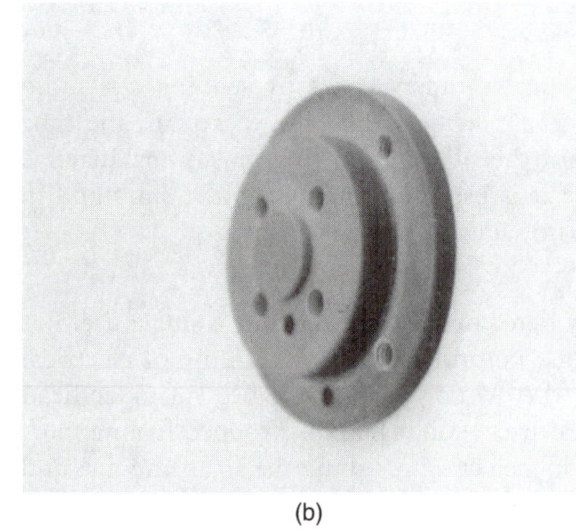

(b)

Figure 6-30 Mechanical Fuse for the Airco AR-1 Robot.
(a) The AR-1 robot's mechanical fuse, attaching the end-of-arm tooling to the robot's wrist. (b) An unblown fuse. (c) A blown fuse. The fuse will break or blow before enough force can be applied to cause damage to the welding head.

Temperature Sensing

Temperature sensors can be used to detect very hot or very cold objects. They are also useful for monitoring air and fluid temperatures inside the robot. Although a robot could use the temperature sensor to monitor the temperature of a furnace, it is much more likely to use the sensor to ensure that it has picked up a properly heated part from the furnace. Under some circumstances, the temperature sensor could be used to detect the presence of humans in the robot's work area.

The two primary temperature sensors are thermocouples and thermistors. In addition, bimetallic strips can be used for coarse temperature sensing, as is done to control air conditioning and heating. Bimetallic strips, which are used in many household thermostats, consist of strips of two different metals fused together. Changes in temperature cause the bimetallic strip to bend mechanically one way or the other in proportion to the temperature. Preset electrical contacts are made if the strip bends far enough (because the temperature has gotten too low or too high).

A **thermocouple** is made up of two dissimilar metals—such as iron and constantan, copper and constantan, or platinum and rhodium—with one end welded together and the rest open. A voltmeter is connected to the open end of the thermocouple. As heat is applied to the welded tip of the thermocouple, a voltage proportional to the temperature of the tip of the thermocouple is read by the voltmeter. Figure 6–31 illustrates a thermocouple.

A **thermistor** is made from a material whose electrical resistance changes at different temperatures. Thermistors are usually made from various metal oxide semiconductor materials. Figure 6–32 shows a thermistor, together with its temperature characteristic chart, and a pencil for size comparison.

An infrared detector can also be used to determine the temperature of a known object that is located at a known distance, under other controlled circumstances.

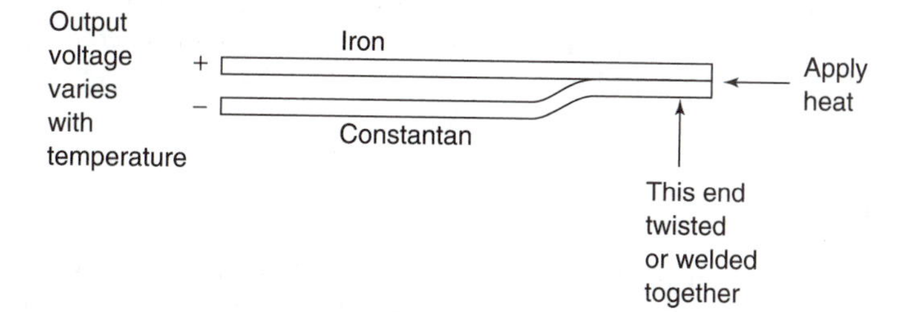

Figure 6–31 Thermocouple.
Thermocouples are composed of strips of two dissimilar metals welded together at one end. Temperature changes at this junction are reflected at the other end of the strips as voltage changes.

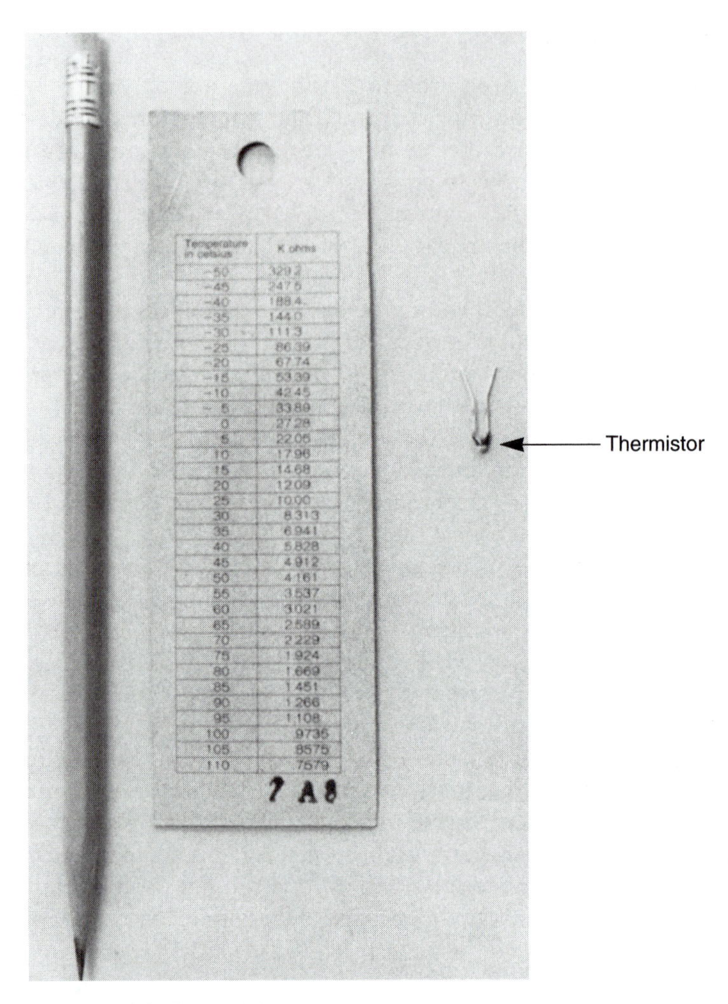

Figure 6–32 Thermistor.
Thermistors are semiconductor materials whose resistance varies with temperature. The chart identifies the resistance offered by this thermistor at various temperatures. The pencil is in the picture for size comparison.

Navigation

The current generation of industrial robot needn't be able to navigate. However, today's automatic guided vehicle (AGV) does. AGVs use either a line follower or a magnetic cable field follower.

A line-following robot uses a simple type of vision to detect the edges of lines (see Figure 6–33). It uses its own light source, such as an infrared LED, to illuminate the line. (High contrast between the line and its background works best.) There are two detectors of the light source, each of which receives only reflected

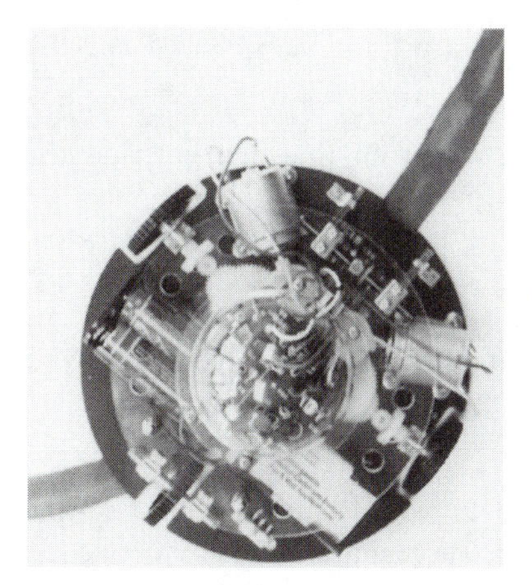

(a)

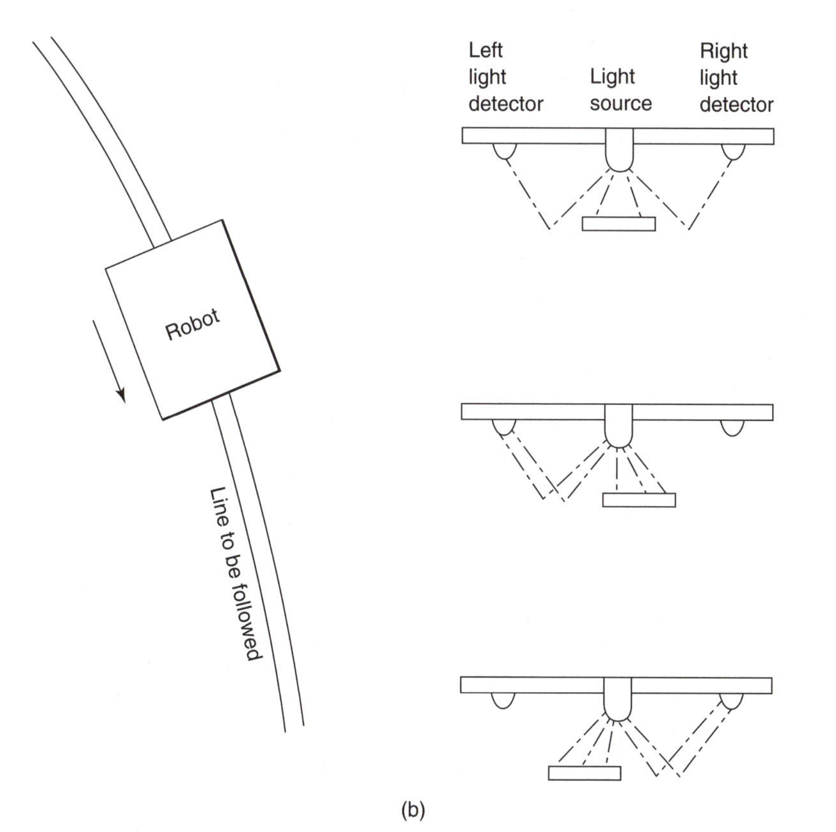

(b)

Figure 6-33 Line-following Robot.
A line-following robot uses a light source (such as an infrared LED) and two light detectors (such as infrared phototransistors) for simple vision.

light. Assuming that the line is dark and the background is light, one of three conditions exists:

1. If the robot is properly centered on the line, both detectors receive the same amount of reflected light.
2. If the robot is off center to the right, the right detector receives more reflected light than the left detector. In response, either the right motor must be speeded up or the left motor must be slowed down.
3. If the robot is off center to the left, the left detector receives more reflected light than the right detector. In response, either the left motor must be speeded up or the right motor must be slowed down.

Magnetic-cable-following robots work similarly to line-following robots, except that a buried cable generates a magnetic field and two detectors on the robot measure the strength of that field. If both detectors measure the same field strength, the robot is properly centered above the cable.

Robots can use ultrasonic ears and an ultrasonic beacon to locate a battery charger or some other home condition. Radio Shack's Robbie Jr. has two ultrasonic ears and an ultrasonic remote control (see Figure 6–34). It can find and come to the control from 10 to 15 feet away if the control is hand-held. If the control is placed on the floor of a school hallway, Robbie Jr. can find it from as far as 30 feet away.

If the course that a robot must follow is known in advance, the robot can be equipped with some type of inertial guidance system. An inertial guidance system uses self-contained instruments to control the path taken by a device.

There is certainly no reason why a robot could not have its own compass. If a robot must navigate, knowing the direction it is facing is a big help. Radio Shack sells a magnetic flux-gate compass.

Experimental computerized navigation systems exist for automobiles. They project a map and show both the car's location and the direction it is facing.

In the summer of 1989, the United States military placed a satellite in orbit that can pinpoint the position of ground and air troops to within 50 feet. In the future, robots could use such a system for industrial and general-purpose applications.

One very simple way to navigate is by touch. Electronic mice (robot mice) contests have been held since the early 1970s. In these contests the mice must navigate an unknown maze. The fastest successful mice have used bumpers to follow a wall until they reach an exit.

Which Way Is Up? If a robot is going to move about, it is helpful for it to know which way is up, so it can avoid falling over. A digital sensor for this purpose could consist of a weighted pendulum arm and two microswitches. Figure 6–35 shows such a sensor. The microswitches are placed so that the pendulum will strike one of them if the robot tilts dangerously to the left or right. A second pair of microswitches and a free-hanging pendulum could be used to warn of excessive forward or backward tilting.

An analog sensor could use a potentiometer with a weighted pendulum arm attached to its shaft. Since the resistance of the potentiometer changes as the

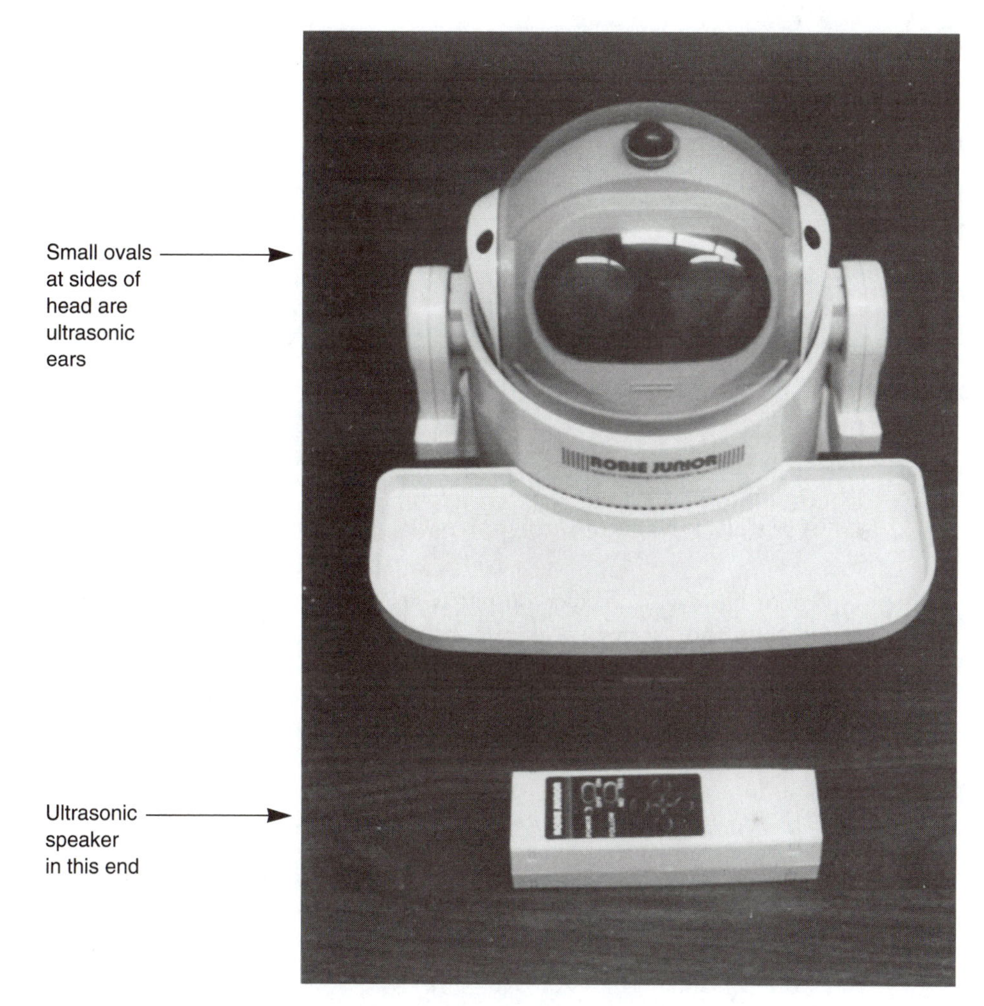

Figure 6–34 Robbie Jr. from Radio Shack.
Robbie Jr. uses two ultrasonic ears to home in on the ultrasonic speaker in its remote control.

pendulum moves, it provides readings that could be calibrated to identify degrees of variation from vertical. Figure 6–36 shows such a device.

Speech Output

Technically, speech output is not handled by a sensor. However, it allows the robot to communicate with us through the additional sensor of our hearing. This can be very handy when trying to get the attention of a busy human. Moreover, voice output is often used in conjunction with voice input to confirm that the robot understands a voice command.

In a noisy factory, a robot using voice output may not even be heard. However, voice output is a valuable safety feature when robots are used in offices or hospitals, as some already are. The use of robots by disabled people is easier if the

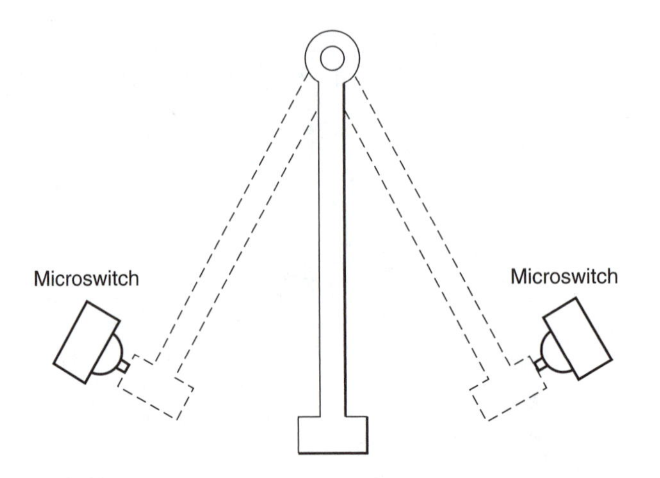

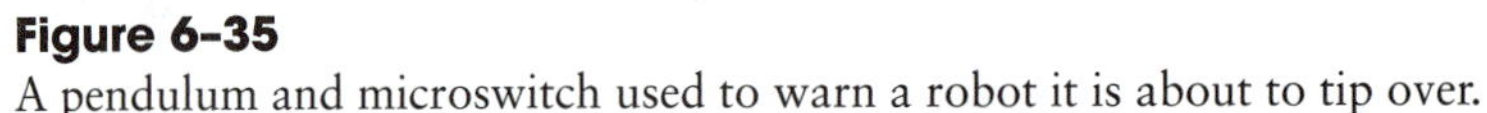

Figure 6–35
A pendulum and microswitch used to warn a robot it is about to tip over.

robot has a voice. Voice output is also useful in medical robotics for the disabled people who cannot speak. Stephen Hawking, a well-known scientist in England, has Lou Gehrig's disease; he is unable to talk and cannot move much, so he uses a computer voice to do his talking. His speech synthesizer, from Speech Plus, has a natural-sounding voice.

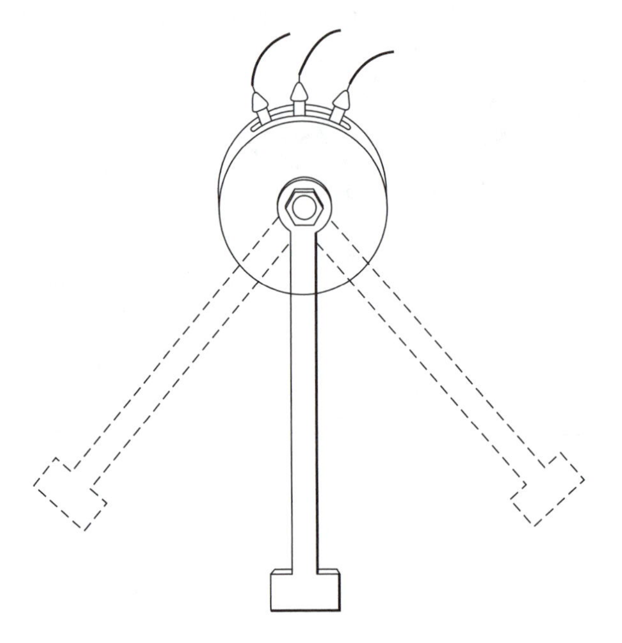

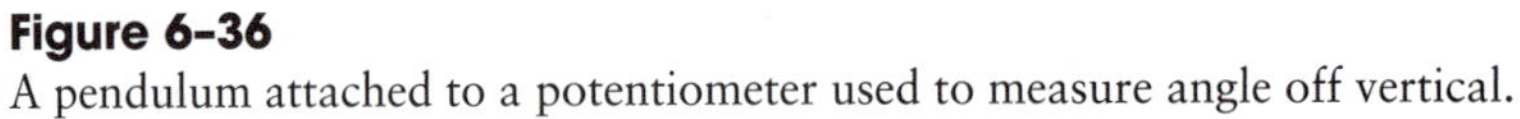

Figure 6–36
A pendulum attached to a potentiometer used to measure angle off vertical.

Voice output, **voice synthesis,** or **speech synthesis** for a robot can be accomplished by three different methods. The oldest method is to have a human record the desired words, phrases, or sentences on individual pieces of magnetic tape. A computer or robot then plays back the appropriate piece of tape to produce the desired message. Because this method plays back real human speech, the individual words sound very natural. The second method of voice output involves having a human speak the words into a microphone. A computer stores each word digitally, and this digital information is then converted back into analog speech one word at a time as the computer or robot speaks. This method tends to sound very much like a human speaker. The third method of voice output constructs speech from stored phonemes for the English language. A **phoneme** is an utterance that cannot be reduced to a smaller unique sound. This method tends to sound very mechanical, because there is no emotion or inflection in the speech. The frequency of the speech can be changed to sound as though it were being said by a man, a woman, or a child, but that is about all the variability possible. Adding an inflection or emotion content circuit to the speech synthesizer would make it much more acceptable to humans.

Speech synthesis through phonemes has been available for robots ever since Texas Instruments Corporation came out with the Speak and Spell educational toy in the late 1970s. The integrated circuit used in the Speak and Spell toy made it possible for most electronic computers and electronic computer-controlled robots to have a primitive voice.

Radio Shack sells a voice synthesizer integrated circuit called an SP0256-AL2 (part number 276–1784) that contains all the logic and memory required to produce every phoneme used in the English language. Additional parts are needed to drive a loudspeaker and to interface the device to a computer or robot. Radio Shack's Robbie Jr. robot, mentioned earlier, has voice output.

Speech synthesis is covered in more detail in Chapter 20.

Speech Input

Using voice input commands is not practical for a robot working in a noisy factory. As robots move into offices, hospitals, and homes, however, voice input will become highly desirable. Physically disabled persons are already making experimental use of voice-commanded robots.

The area of sensory input interpretation is one that will benefit greatly from advances in artificial intelligence. Currently, the interpretation of speech input, vision inputs, and touch information is being intensively studied.

Robots can be given speech input (known as **speech recognition**) through the use of a microphone. This speech information can be converted to digital information and stored in the robot's memory. Unfortunately, the interpretation of speech by the robot is another matter. Most speech recognition systems now in use recognize a very small vocabulary. Ten to twenty words is a large vocabulary for a robot, if it must recognize the words spoken by more than one person. The interpretation of speech by these systems is very slow, and they cannot handle continuous speech.

Speech understanding is covered in more detail in Chapter 20.

Smell

Industrial robots of today do not have a sense of smell, but the more mobile robots of the future may need to have one. Sensors already exist that can sense nitrates and detect explosives. Smoke detectors can react to particles in the air. Exhaust gas analyzers are available for checking car exhaust emissions. The military has sniffers for detecting unauthorized people. And domestic robots could use synthetic noses for detecting a natural gas leak from a water heater or oven, or a freon leak from a refrigerator or air conditioner.

ELECTRICAL NOISE AND ROBOT SENSOR INFORMATION

An industrial robot generally works in an area with a high level of **electrical noise.** This noise is caused by the machinery around the robot and by the robot itself—by the opening and closing of electrical contacts and electronic circuits, and by the starting and stopping of electrical motors. The electrical noise generated takes the form of pulses that last several thousandths of a second and have an amplitude of several thousand volts. The noise can make a light bulb blink, a television picture momentarily change in size, or a radio briefly buzz with static. The light bulb, television, and radio quickly recover from the noise and continue on as if nothing had happened, but the memory and control circuits in a robot's controller respond to signals that are a millionth of a second or less in length. A burst of electrical noise lasts long enough to change the controller's memory and/or mislead it about what it is supposed to be doing. When this happens, the robot will no longer do its task correctly.

An industrial robot almost always has some electrical motors. Many hobbyists have had problems with their robot controller doing unpredictable things every time the manipulator starts or stops a motor. Commercial robot designers have the same problem.

Several things can be done to help eliminate electrical noise within the robot. For one, the power supply for the motors should be separate from the power supply for the controller and the sensors. For another, the wiring for the motors should be kept separate from and routed separately from the wiring for the controller and the sensors. A more expensive and complete solution is to use shielded cables or wiring for all wiring of the controller, the controller signals, and the sensors. The idea is to build an electrical shield around all the sensitive circuits and wiring.

If the controller and manipulator are not contained in the same area, connecting wires must be run between the two devices. The farther apart the two devices are physically, the better antenna the wiring makes for picking up electrical noise from the robot and from surrounding machinery. The noise problem will be greatest if simple **point-to-point wiring** is used. Point-to-point wiring uses individual wires from one point on the circuit to another point on the circuit. The noise is decreased by using twisted pairs of wires—wires run in pairs from one point to another in a circuit, with one of the wires in each pair grounded. The noise is lowest if **shielded wires** are used. Shielded wires require that each wire have a grounded shield surrounding the wire.

Most signals from sensors are transmitted at a low voltage level. The lower the voltage level, the more seriously electrical noise will affect the meaning of the sensor signal. Most computer controller circuits use voltage levels of 0 to 5 volts. The farther a signal travels along a wire, the smaller its voltage level becomes. A 0- or 5-volt signal, after it has traveled along a wire for more than 50 feet, can produce readings of little value. The 0 volts might read anywhere from –2.5 volts to +2.5 volts. For these longer wire lengths, it is more meaningful to convert the 0- and 5-volt signals to larger and easier-to-understand values—such as between ±3 volts and ±25 volts. The transmission end of the signal may be ±25 volts, while the receiver can work down to ±3 volts. At these voltage levels, it is possible to send signals 200 feet or more without garbling. The use of plus-and-minus voltage levels is specified by the **RS-232** signal transmission standard. A **level converter** circuit is required to change the 0- and +5-volt signals to the plus-and-minus voltage signals, and vice versa.

Another method of sending a signal over a distance is to send it as a differential signal. This uses **differential driver** and receiver circuits. The signal is sent over two wires, and the voltage level is not what is sensed on either wire, but the difference in voltage between the two. This type of signal transmission also tends to cancel out electrical noise, since such noise is picked up equally in both wires.

If the robot's controller connects to a far-away computer, it may be appropriate to use a telephone line and a **modem** (short for *mo*dulator/*dem*odulator) to transmit the signals. In its simplest form, the modem converts the controller's 0- and 5-volt signals into two different tones that can be transmitted over a telephone line. The frequency of a tone can still be recognized, even after it has lost much of its voltage amplitude. Tones are less bothered by electrical noise than are voltage levels.

Another form of signal transmission that is unaffected by electrical noise is **fiber optics.** The signals are light beams sent along an optical fiber. The transmitters and receivers for fiber-optics circuits are available in integrated circuits that allow relatively easy use.

When noise in a signal can cause serious safety problems, some type of **signal error detection** should be used. This may mean adding extra information to a signal (in the form of **parity check bits**) to determine whether it has been changed. The more check bits used, the greater the likelihood of catching any errors. An alternative technique for finding errors is to send the information more than once and see whether all copies of the information agree. If they do not agree, the information is retransmitted. A human being reduces the possibility of information errors by sending information over more than one path and then letting the majority consensus settle which version of the information should be accepted.

More details on how to understand sensor information are included in Part Two of this book. In most cases, it is easier to obtain sensor information than to interpret it. Most sensor interpretation is currently done through the software of an electronic computer. In many cases, the robot's controller is not yet powerful enough to do its own interpreting. This is especially true of vision systems.

SUMMARY

Adding sensors to an industrial robot can greatly increase the range of tasks the robot can perform. It also decreases the mechanical tolerances required of both the robot and the robot's environment, since the alignment and placement of parts being fed to the robot become less critical. Sensors allow a robot to adapt to changing conditions so that its world no longer has to be perfectly aligned.

Sensors can be divided into three categories: internal sensors, external sensors, and interlocks. Internal sensors tell a robot the position of its various joints and report other conditions such as fluid pressure and temperature. External sensors tell the robot what is happening outside. Limit switches, potentiometers, piezoelectric devices, temperature detectors, and encoders can be used as internal robotic sensors. Limit switches, photoelectric devices, vision sensors, speech recognition devices, proximity detectors, and range finders can be used as external robotic sensors. Interlocks are used to protect both humans and robots. They may possess features of both internal and external sensors.

If robots are to use sensors in the midst of the electrical noise of a factory, care must be used in transmitting information from them to the robot's controller without interference. Fiber optics or differential drivers may be needed to get the signals to the controller. If the sensor information is being shared with a production computer, the likelihood of noise problems is increased. Many sensor devices give an analog signal, while the controller uses digital signals; consequently, analog-to-digital and digital-to-analog converter circuits may be required.

REVIEW QUESTIONS

1. What are some advantages and disadvantages of a robot's having sensors?
2. What purposes do internal sensors serve in a robot?
3. What purposes do external sensors serve in a robot?
4. What sensors can be used to measure rotational motion?
5. Which is easier—gathering sensor information, or interpreting sensor information?
6. What devices could be used to give a robot a sense of touch?
7. What devices could be used to give a robot a sense of vision?
8. What devices could a robot use to measure temperature?
9. How does electrical noise affect a robot's ability to receive and interpret information from its sensors?
10. Compare the electrical noise levels on direct point-to-point wiring, twisted-pair wiring, and shielded cable.
11. What methods can be used to detect signal error?
12. How can interlocks make a robot's work area safer?
13. Compare absolute and incremental encoders.
14. What are the differences between internal and external sensors? Which type of sensor gives a robot the higher level of intelligence?
15. What sensors can be used for robot navigation, and how does each work?
16. How can photoelectric devices make a robot's work area safer?
17. Suppose that you want to sort blocks of different heights automatically. How could this task be done with whisker

sensors? How could it be done with photoelectric sensors?

18. What device can be used to convert analog signals into digital signals?
19. What is a pixel?
20. Suppose that you would like a robot to be able to tell how far a part is inserted in its gripper. What sensing device might allow it to do so?
21. Why is vision easier to install on a teleoperator (a remote-controlled device) than on an industrial robot?
22. What sensory devices can be used to protect a human who enters a robot's work area?

7

APPLICATIONS FOR INDUSTRIAL, BUSINESS, AND DOMESTIC ROBOTS

OVERVIEW

The present-day applications for robots are much broader than most people realize. By taking time to look at some of these applications, you may be inspired to find additional uses for robots. While most of this chapter deals with industrial robots, other types of robots will be mentioned.

OBJECTIVES

When you have completed this chapter, you should be familiar with:

- What areas are already using robots
- What areas are presently experimenting with robots
- What areas could use robots

KEY TERMS

Arc welding
Die casting
Drone
Electronic assembly
Finishing
Forging
Inspecting
Investment casting
"Lost wax" process
Machine tool loading/unloading
Orthosis
Parts transferring
Plastic molding
Prosthesis
Robovan
Small parts assembly
Spot welding
Spray painting

USING ROBOTS

While the emphasis in robot development is on industrial robots, factories are not the only place where robots are used. In business offices and elsewhere, robots serve as mail delivery carts, promotional or show robots, laboratory assistants, hospital orderlies, and window washers. Domestic robotics is still in a development stage, but it may one day include baby sitters, companions, and security robots. Some nonindustrial applications use industrial robots, while others require specially designed robots.

In general, industrial robots are best used for jobs that are dirty, dull, dangerous, or difficult—the types of jobs that humans do most poorly. In addition, industrial robots can be used for tasks that need to be done repeatedly, but not often enough to justify the use of automated equipment.

INDUSTRIAL APPLICATIONS OF ROBOTICS

The industrial robot has already found many applications in industry: die casting, forging, investment casting and other foundry work, machine tool loading and unloading, parts transferring, spray painting, small parts assembling, finishing, plastic molding, spot welding, arc welding, machining, electronics assembling, and inspecting. No one industrial robot can work well in all these application areas, any more than a single human worker can. Particular robots have special characteristics that make them more efficient for use in their own style of work. Size, strength, speed, cost, and accuracy all influence where the robot may find work. A robot designed to handle 1-ton parts would probably not do well assembling small parts; a small pneumatic-powered pick-and-place robot simply cannot handle 1-ton parts or perform spot welding.

Die Casting

Die casting involves forcing nonferrous metals into dies under high pressure to form parts of a desired shape. Die casting was one of the first areas to benefit from the use of industrial robots; robots have been involved in die casting since 1961. Die-casting machine and molten-metal feed equipment are examples of specialized automated machine tools.

A typical die-casting task involves unloading a part from the die-casting machine, quenching the part, and then disposing of the part on a conveyor belt or into a bin. The robot needs only a single gripper shaped to handle the specific part being produced. The gripper must be able to handle hot parts and not be affected by the quenching fluid. In some die-casting operations, the robot may run the part through a rough trimming machine after unloading it from the die-casting machine.

Other possible robot task cycles in die casting could include any of the following:

1. Alternately unloading two or more die-casting machines
2. Unloading, quenching, trimming, and disposing of the part
3. Unloading the die-casting machine and preparing the die for the next casting cycle (This would require using another gripper or attachment to spray the die.)
4. Loading an insert into the die-casting machine and unloading the finished part

Using robots in die casting has made several positive contributions to the industry. First, since the die-casting plant atmosphere makes the job of working at loading and unloading of the die-casting machines unpleasant and hazardous for human workers, the use of robots has reduced the need for health and safety equipment. Second, the robot gives more consistent timing to the loading and unloading of the die-casting machines, and this can lead to a productivity gain of up to 20 percent over human-attended machines. Third, the more consistent timing improves the quality of the parts, resulting in an increase in usable parts of up to 15 percent. Fourth, given that it is becoming difficult to find persons willing to work under the unpleasant conditions of a die-casting plant, robots are helping to meet the employee shortage.

Assigning robots die-casting tasks has required very little change in the existing equipment when robots are substituted for human workers.

Forging

Forging involves the shaping of metal objects by hammer blows or presses. The old-time blacksmith's job included forging parts such as horseshoes, nails, and tools. Generally the metal is heated to a plastic state before the hammering process begins. The atmosphere in the forging environment is hostile to human workers. Placing parts in furnaces to heat them to temperatures ranging from 1,000° to 1,600°F, then moving the hot part to a 1,500-ton forge press, and finally removing the part from the press is unpleasant and dangerous. The parts themselves may weigh up to 360 pounds each and may require several human workers and a hoist to move a single piece. Human workers need protection from the heat and the dirt, and from getting caught in a closing press. Robots use special high-temperature grippers to handle the hot parts, plus press interlocks and limit switches to prevent them from getting caught in the presses.

The ovens and forging machines in many industrial operations are specialized automated machine tools. This makes it relatively easy to supply signals to the robot telling it when a part is heated to the right temperature. Similar signals can be supplied to indicate when the press is ready for a new part and when the press has completed a cycle. The robot can position parts very accurately in the presses, and the use of robots in forging operations again removes humans from hazardous conditions, as well as giving more uniformity to the forging process.

The robot's tasks in forging vary greatly, as do the parts made. The task may consist of waiting for the furnace to signal that a part is hot and for the press to signal that it is ready for another cycle, then taking the part from the furnace to the press, and then going back and quenching the gripper while waiting for the next cycle of furnace and press signals. Or the robot may be asked to take a length of hot metal out of a furnace, place it in a machine where it is formed into a link of chain, remove the link, hold it while another machine welds the link closed, and finally go back to work on the next link.

Investment Casting

Investment casting uses an expendable pattern from a die to produce a single-use mold. This mold is used to make detailed and often delicate castings. The pattern is often made from wax, in which case the process may be called the **"lost wax" process.** The steps involved in the investment casting process include the following:

1. Make an expendable pattern from a die.
2. Form a mold by coating or "investing" the pattern with a refractory material.
3. Remove the pattern from the mold, generally by melting the wax.
4. Fire the refractory material of the mold.
5. Cast the metal part.
6. Break away the mold.

The industrial robot has been successfully used in the mold-making step of investment casting. The mold is made by repeated immersion of the pattern in a slurry of fine refractory (difficult to fuse) materials. The materials must be evenly distributed and dried between each immersion step. Up to six immersion steps may be required for each mold. The finished mold must be uniform in thickness for best results in the casting phase. Spinning the mold after each immersion helps to remove excess materials and spread the remaining materials out more evenly. Human workers tend to find the mold-making process boring, and their work is inconsistent. The robot uses the same timing and spin force on each immersion of each mold, producing very consistent work and good-quality molds. A special wrist for the gripper allows the robot to spin the pattern in continuous circles; a human worker cannot spin the molds as evenly as the robot can.

Machine Tool Loading and Unloading

Many manufacturing processes use machine tools to work on different parts. These machines require that a raw part be loaded in, the machine cycle started, and the finished part removed. In many cases, the raw part must be positioned very accurately in order for the part to be properly machined. Many such machines are involved in metal-cutting processes that take a long time to complete. Typically, the loading and unloading phases are much shorter than the machining phase. Because a robot can be given a longer reach than a human worker has, a single robot may be able to service two or more metal-cutting machines.

A machine-loading task for a robot might consist of picking up a raw part from a pallet or conveyor belt and placing the part in the machine tool. The gripper for the robot would have to be designed to fit the raw part. Next the robot could then get another raw part and place it in another machine tool. If the robot is also responsible for unloading the finished part from the machine tool, it may have to use a different gripper due to the change in dimensions of the machined part over the raw part. Depending on the length of the machine cycle, the robot may switch to the second gripper between the loading and unloading steps. Not all machine loading can be done by robots as yet, because of the precision placement needed for some parts. As robot sensors

Figure 7–1
This is a FANUC Robotics M-500 robot unloading a sheet metal press. *Note:* To save floor space, the robot is mounted on an overhead beam. (Photo courtesy of FANUC Robotics North America, Inc.)

improve, however, robots will be able to perform more of the machine-loading jobs. Figure 7–1 shows a FANUC Robotics M-500 robot unloading a sheet metal press. Note that, to save floor space, the robot is mounted on an overhead beam.

Parts Transferring

Many manufacturing jobs involve removing parts from pallets and placing them in bins or on conveyor belts—or removing parts from bins and conveyor belts and placing them on pallets. This type of work is known as **parts transferring.** Parts may also be moved from one conveyor belt to another.

Stacking and unstacking parts can be a boring job for human workers, and bored workers tend to make mistakes. Since robots do not have feelings and do not get bored, they can do the same task hour after hour, day after day, in exactly the same way. Once you teach a robot where to locate each part on a pallet, it can load or unload the parts properly and consistently from then on.

Picking up parts from a conveyor belt is more of a problem for the robot. The robot must receive some kind of timing signal from the conveyor system to tell it when a part is ready to be picked up, or it must be equipped with sensors of its own to see or find the part. If the part is metal and the conveyor belt is nonmetal, a metal-detector sensor might be able to locate the part. If the part is not uniformly shaped, either the robot must be given some means of determining the orientation of the part on the conveyor belt or the part must be placed on the conveyor belt in a proper orientation.

Picking up parts from a bin requires sensors that are still in the experimental stage for robots. In particular, vision systems for robots have not solved the problem of looking at overlapping parts. Moving parts from one conveyor belt to another conveyor belt without regard to part orientation is a much simpler task for a robot.

Despite the part orientation problems, robots are already being used to transfer parts successfully. As their sensor capabilities improve, robots will be able to perform more and more part transfer tasks. Figure 7–2 shows a FANUC Robotics LR Mate 100 robot transferring parts between two conveyor belts.

Spray Painting

Spray painting makes use of quick-drying paint, usually applied as a finely divided air and paint mixture. The spray-painting environment constitutes a fire hazard and is also dangerous to human workers' respiration. Because of the fire hazard, hydraulic-powered manipulators are popular for spray painting: their low electrical control voltages pose no fire risk. In addition, the control unit and power supply for the hydraulic manipulator can be housed outside the painting booth.

Figure 7–2
This robot is a FANUC Robotics LR Mate 100 robot using a vacuum gripper to transfer parts between two conveyor belts. (Photo courtesy of FANUC Robotics North America, Inc.)

An early paint-spraying machine was built by Pollard in the 1930s. Today, this machine would be called an industrial robot. It was programmed by leading a jointed arm through a desired path while a magnetic signal was recorded for later playback. The signals were recorded on a grooved disk resembling a phonograph record. These recorded signals were analog or varying-level signals, as opposed to the digital or number signals that are used by modern robots.

Spray painting requires not only that a robot be able to move from point A to point B, but also to follow a predetermined path at a predetermined speed very closely. If the path to be followed is anything other than a straight line, the spray-painting robot must have a continuous-path controller. A human worker spray-painting an object can see where the paint is going and whether any spots have been missed. A robot spray painter does not have vision, so its orientation is crucial. In many cases, the robot actually paints the part as the part moves down the production line.

A human painter does not paint any two objects in exactly the same way, but a robot painter does. When training the robot to paint an object, the human must be very careful to paint all the surfaces and not to paint an area more than is necessary. Human painters cannot be this careful on every object they paint if they are continuously painting things; they get tired and miss spots or paint some areas more than once—missed spots require touch-up painting later, while painting an area more than once wastes paint. Since the robot paints each object with exactly the same motions, the paint job it produces is always of equal quality and no paint is wasted; in many cases, the savings in paint costs achieved by using a robot painter more than repays the price of the robot.

Spray-painting robots are not limited to applying paint. They have also been used successfully to apply the glue or other adhesives for attaching vinyl tops to automobiles. Anything that can be applied as a spray can be applied by a spray-painting robot.

Advantages of using spray-painting robots include the following:

1. Humans are removed from a hostile environment.
2. Less energy is needed for fresh air requirements, and the need for protective clothing is reduced.
3. The quality of the painting is improved, reducing rework and warranty costs.
4. Less paint and other materials are used.
5. Direct labor costs are reduced.

Small Parts Assembling

Assembly of small parts is already being done in many instances by automated machine tools. To be cost effective, however, automated machines need to process a volume of at least 1 million parts per year, over a market life for the product of at least three years. An industrial robot can work with smaller volumes of parts and then be retrained to do some other task when the market life of the product being assembled ends.

Small parts assembly generally involves parts weighing no more than a few pounds. Electrically powered robots can handle these small parts with very accurate positioning of their manipulators. Using remote center compliance (RCC) devices in their grippers helps them overcome any minor part misalignment that may occur.

While small parts assembly by robots is still in its infancy, many large manufacturing companies are extensively researching improvements in the use of robots to assemble small parts. Unimation Corporation makes the Unimate 500 robot for use by General Motors in doing small parts assembly work. Many of the subassemblies used in making an automobile—alternators, carburetors, dashboards, taillights, and transmissions—rely on the assembly of small parts. The Unimate 500 robot occupies approximately the same workspace as a human and can handle parts weighing up to 5 pounds. General Motors has determined that 90 percent of all car parts weigh less than 5 pounds. Unimate robots are used on the General Motors project known as PUMA (for Programmable Universal Machine for Assembly). The robots are equipped with a computer interface that allows them to understand a special high-level computer language called VAL that simplifies their training.

Robots have successfully assembled heater control panels, aerators for faucets, terminal blocks for the telephone company, and IBM PC computers. Specialized robots have worked in class 10 clean rooms doing semiconductor wafer processing and computer hard-disk assembly. Because the robots are cleaner than human workers in the clean room environment, and since any particle contamination on a semiconductor wafer damages part of it, the yield from wafer processing is higher when robots do the work than when humans do it.

Figure 7–3 shows the control unit, teaching pendant, and manipulator for the Unimate PUMA robot. Figure 7–4 shows a closeup of a Cincinnati Milacron robot

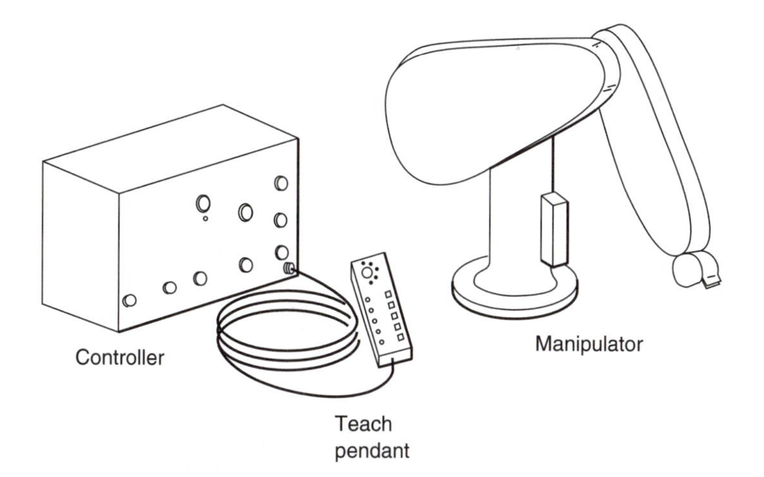

Figure 7–3 Unimate PUMA Robot.
The Unimate PUMA robot series was designed with the help of General Motors. It is a small robot that can handle parts weighing up to 5 pounds. It occupies the same workspace as a human worker and is intended to do small parts assembly for automobiles.

Figure 7–4 Robot Driving Screws.
This robot is equipped with a power screwdriver at its end-of-arm tooling. It is demonstrating driving screws into an assembly. (Photo courtesy of Cincinnati Milacron, Cincinnati, Ohio)

driving screws into a part during an assembly process. Figure 7–5 shows a FANUC Robotics A-520i robot assembling a value body housing.

Electronic Assembling

Large-volume assembly of electronic parts is already being done by automated machine tools. Most of this work consists of automatic insertion of electronic components into printed circuit cards. Robots could be used to assemble smaller-volume parts. Although less efficient than special-purpose automated machines, robots can have changes more easily made to their tasks and as such can support minor design changes. Robots can also handle the installation of finished subassemblies into bigger assemblies.

Figure 7-5
This robot is a FANUC Robotics A-520i robot assembling a value body housing. (Photo courtesy of FANUC Robotics North America, Inc.)

Robots are already used in the following electronic assembly areas:

1. Mounting of components on PC boards
2. Wafer handling
3. Soldering operations
4. Winchester disk drive assembly

Finishing

After the initial machining step in the manufacturing process, many parts require some type of **finishing** step. This step may include trimming, deburring, sanding, heat-treating, and packaging.

The trimming process may be done by placing the part in a trimming machine, or it may require hand-cutting or buffing with the aid of a machine. Robots can feed parts into a trimming machine and can also handle cutting and buffing devices. In some cases, the robot or a human must hold a part against a deburring device such as a grinder, a cutting torch, cutting shears, or a rotating tool.

Rotating and belt deburring and sanding devices can be handled by both humans and robots. Using a robot to sand and deburr requires having the robot closely follow a predefined path and carefully control the amount of pressure it applies to the tool. Robots with continuous-path controllers can follow the path successfully. Strain gauges can be used to control the amount of pressure they apply to the tool.

The length of time that a tool needs to be applied to a surface in order to deburr and smooth it depends on the amount and type of material that needs to be removed from the surface. Since the size and location of burrs on a part are not consistent, the deburring process is not as compatible with robot workers as with human workers. Using automated machines to make the parts and using robots to weld the parts make the location and quantity of burrs or seams more consistent. Robots do not have the tactile sense to recognize fine differences, but they can do much of the work and they do reduce the exposure of humans to flying particles produced by grinding.

Sanding creates another hazardous atmosphere for humans that can be made less dangerous through the use of robots. A continuous-path robot with good internal position sensors and pressure sensors can be used to follow a path, sanding or grinding off any raised surfaces it encounters along the way.

Heat treatment of parts can be handled by robots, too. In this work, robots are used mainly to move parts from conveyor belt to conveyor belt and to load, unload, or quench parts.

Plastic Molding

The **plastic molding** process is similar to die casting, except for the different materials and lower temperatures involved. Plastic molding works with thermoplastics, which can be formed or molded when heated. These plastics are fed into extrusion molding machines, injection molding machines, blow molding machines, or thermoforming machines to form parts of a desired shape; all these machines are specialized automated machine tools. Feeding the raw plastic material into a heat chamber, heating the material, and molding the material are all done by automation. After the molded part has cooled, the part must be removed from the molding machine, have the sprue trimmed off, and be packaged; and if the molding process uses inserts, these must be placed into the machine before each molding cycle. These steps can be handled by human or robot workers.

A typical robot task in plastic molding includes unloading parts from one or more molding machines, moving the parts through a trimming machine, and disposing of the parts on a conveyor belt. Another robot can pick up the parts from the conveyor belt, run the parts through a packaging machine, and deposit the packaged parts on a second conveyor belt. A third robot can pick up the packaged parts and stack them on pallets.

Spot Welding

Welding joins two parts by fusing them together at a spot or along a seam. The heat needed to fuse the parts may be produced by electric arcs, electric currents, or gas torches. Welding creates unpleasant fumes, painfully bright light, and the possibility of being struck by bits of hot metal. Robots used to perform welding reduce the amount of safety equipment needed and diminish the safety hazards to humans.

By 1969, General Motors was using robots to do some of the spot welding on car bodies, a task that requires a robot to distinguish among several different body styles and perform the appropriate welds for each style. Each spot weld requires the robot to perform four steps. It must:

1. Squeeze the welding device on the joint.
2. Weld the joint.
3. Hold the joint while it cools.
4. Release the joint.

The welding tooling must also be able to get to hard-to-reach places at any angle. A typical sequence of steps for a robot doing spot welding is as follows:

1. Determine the style of the car.
2. Move from the home position to the work position for the first group of welds.
3. Perform this group of welds.
4. Move to the work position for the next group of welds, and so on, until all groups have been done.
5. Return to the home position.

The movement from the work position for one group of welds to the work position for the next group can be very complicated.

Arc Welding

Arc welding generates the heat necessary to fuse metals by drawing an electric arc between the metals to be welded and an electrode in the form of a welding wire. Arc welding can be done in a normal atmosphere for ferrous metals but may require an inert gas atmosphere for metals such as aluminum. The use of inert gas in arc welding is known as inert gas welding.

The following steps are involved in inert gas welding:

1. Flood the weld area with inert gas.
2. Feed welding wire to the joint while the electric arc is drawn across the joint.
3. Stop feeding the welding wire.
4. Stop the arc after all the welding wire is melted.
5. Continue flooding the area with inert gas until the joint is cooled.
6. Stop the gas flow.

Not only can robots be used in arc welding, they can make more consistent welds than human workers. Figure 7–6 shows the model $T^3$746 robot from Cincinnati Milacron, set up to do arc welding. Figure 7–7 shows a FANUC Robotics ARC Mate 100i robot doing arc welding.

Flame cutting torches can also be used by robots to cut iron and steel plates. The cutting torch first heats the metal to a cherry red color and then directs a

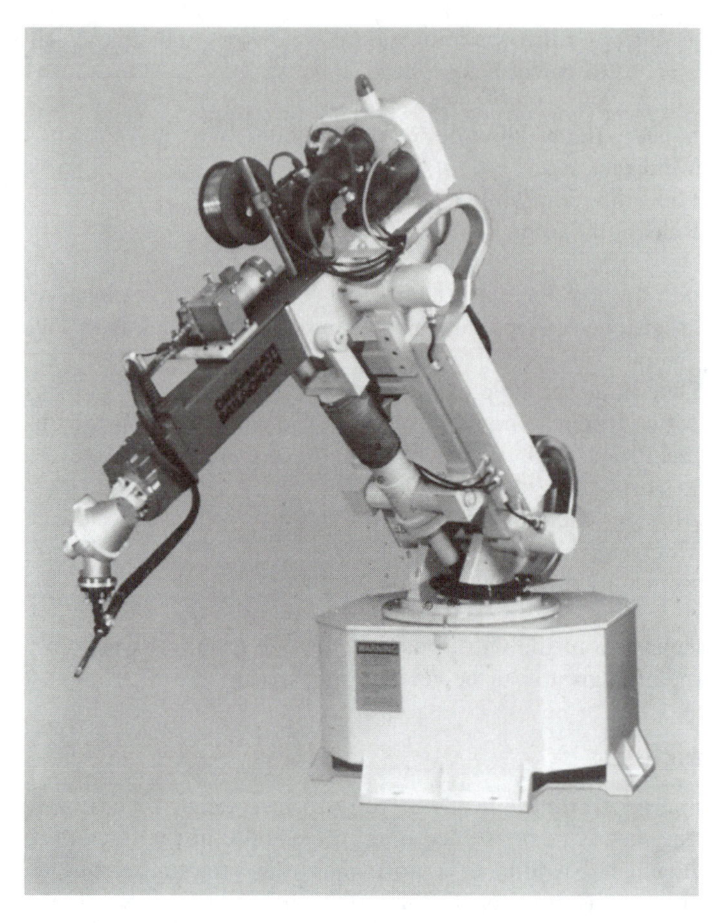

Figure 7-6 Welding Robot from Cincinnati Milacron.
This picture of a general-purpose arc-welding robot shows the reel of welding wire toward the top part of the arm. (Photo courtesy of Cincinnati Milacron, Cincinnati, Ohio)

closely regulated jet of oxygen onto the heated material, causing the metal to burn or oxidize away. Training a robot to do flame cutting does not differ greatly from training it to do arc welding.

Machining

Machining of parts can be done by humans, by automatic machine tools, and by robots. Increased use of robots for machining processes depends on their being able to get better touch and vision sensors. Robots can be used to drill prealigned holes in parts; this requires using drill templates and RCC devices. A robot equipped

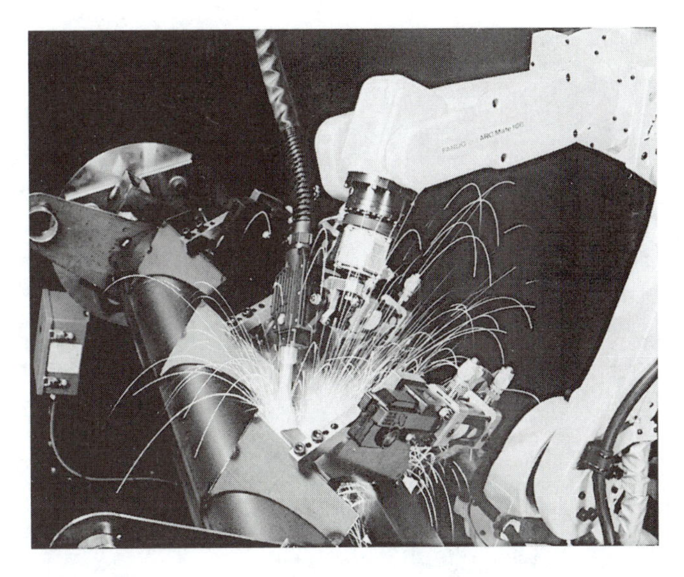

Figure 7–7
This robot is a FANUC Robotics ARC Mate 100i robot doing arc welding. (Photo courtesy of FANUC Robotics North America, Inc.)

with a saw as its end-of-arm tooling could be used to cut stock into predetermined lengths for machining. Figure 7–8 shows the EOA Systems Incorporated EOA CNC AeroDrill end effector. This end effector is used for drilling close tolerance aircraft parts.

Inspecting

While a robot does not have the visual and tactile abilities of a human being, the robot could be used to operate gauges and other measuring devices to inspect parts. Since the robot does not get bored or tired, it is more consistent in its work than a human inspector.

Some television camera systems can be used by robots for inspecting. These cameras, which present digitized vision information to the robot, can either be stationary or mounted on the robot's manipulator. The Eyecom-camera scanner, from Spatial Data Systems, can inspect to tolerances of up to 0.0001 inch when provided with the proper lenses and working conditions (see Figure 7–9).

A robot equipped with some type of weight scale could be used to check the weight of parts or of packaged products. Underweight parts might indicate missing or defective portions. Underweight or overweight packages might indicate a defective packaging machine.

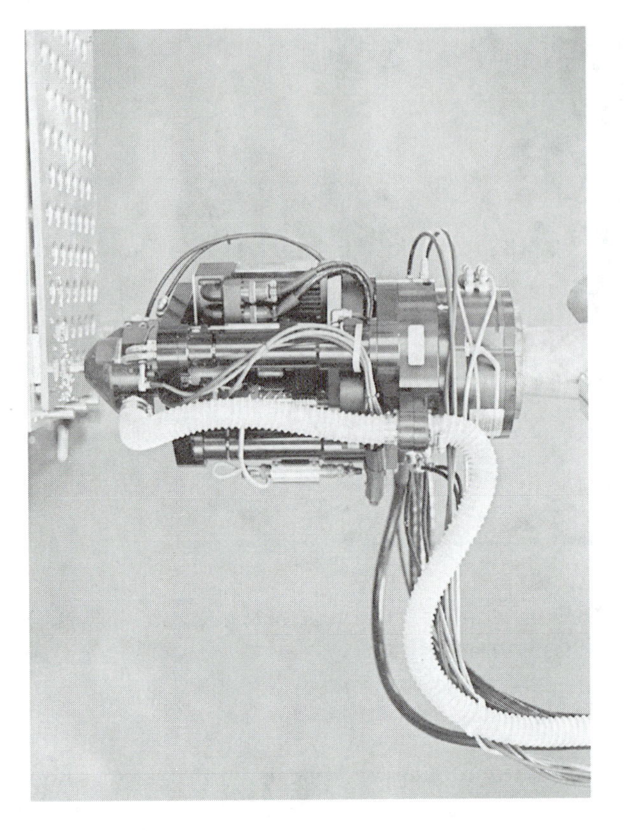

Figure 7-8
The EOA Systems Incorporated EOA CNC AeroDrill. It is used to drill close tolerance aircraft parts. (Photo courtesy of EOA Systems Incorporated.)

Automated Warehouse

Automated warehouses have been in use since the late 1970s. Such warehouses look like unfinished skyscrapers, five or more stories high. They use two types of nonindustrial robots: an AGV or robovan, which can weigh over 1 ton; and a mole, which rides on the AGV and can hop off to load or unload it. The robots operate under computer control, and the warehouse itself does not require human workers.

BUSINESS USES OF ROBOTICS

Businesses currently use robots in a number of areas other than manufacturing. Special fields of application include outer space, under the ocean, medicine, the office, promotion, and the field of art.

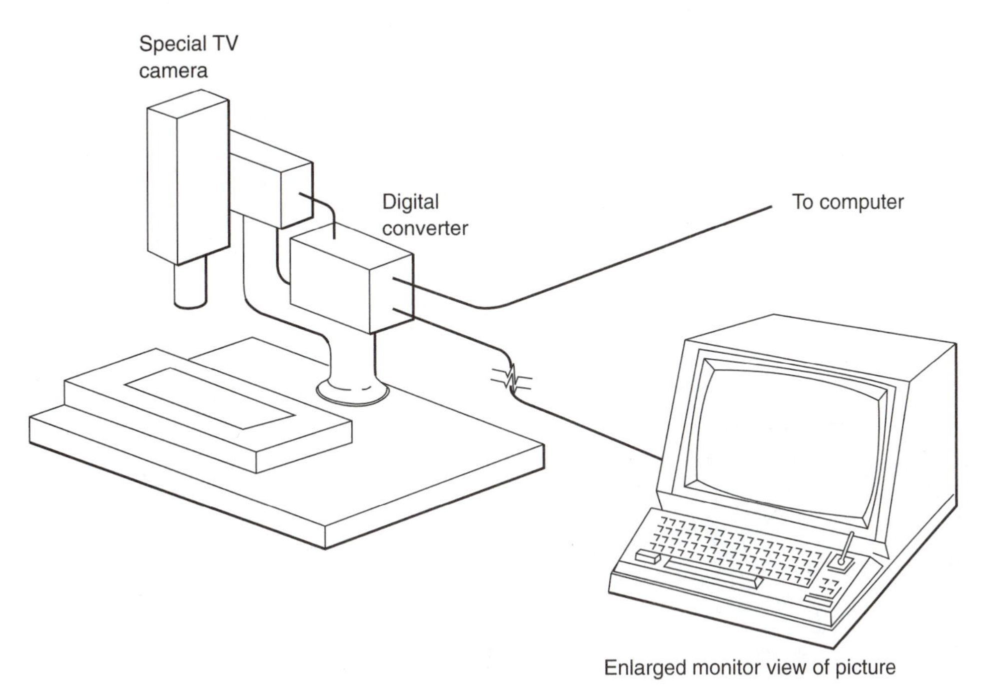

Figure 7-9 Parts Inspection Using a Television System.
With the lighting controlled and the part always situated at the same distance from the camera, this robotic vision system can do very precise inspections.

Outer Space

Outer space is a hostile environment for human beings, so robots are used to do much of the space exploration. *Viking 2* landed on Mars; the *Voyager* probes investigated Saturn and Jupiter, then left our solar system for places unknown; *Galileo* flew by Jupiter; and the Mars *Rover* explored Mars. Unlike the *Viking,* the Mars *Rover* made most of its own decisions, rather than depending on humans back on Earth for commands.

In 1984, the space shuttle proved itself capable of retrieving and repairing satellites. In 1986, Russia placed the MIR Space Station in orbit around Earth. As of this writing, MIR has been in orbit for twelve years. During 1997, the MIR Space Station had many difficulties. It is becoming increasingly important that robots be developed to perform repairs on satellites and spacecraft.

On July 4, 1997, NASA landed the Mars *Pathfinder* Lander and microrover *Sojourner* on Mars. These are outer space robots. Figure 7–10 shows the Mars *Pathfinder* Lander with the microrover still onboard it. Figure 7–11 shows the *Sojourner* rover.

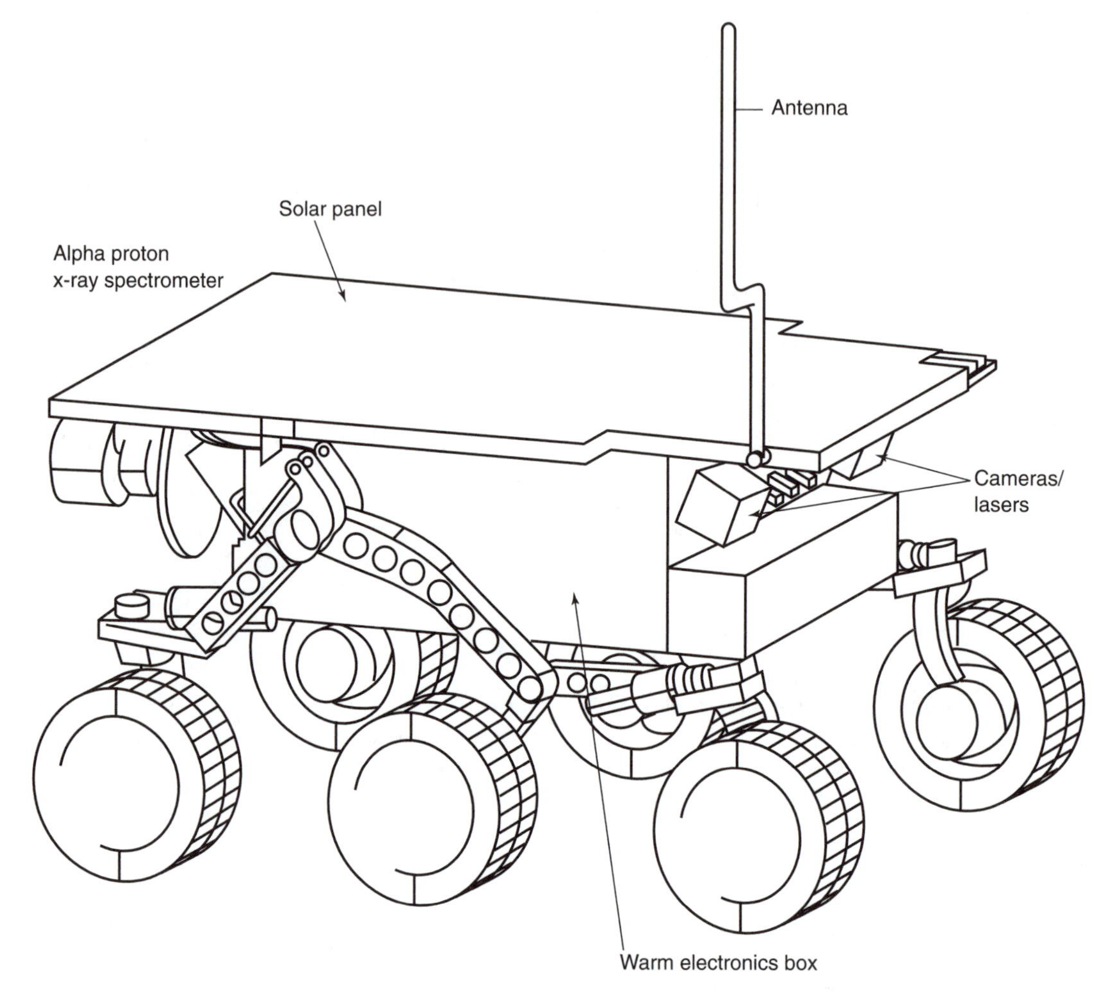

Figure 7-10
Mars *Pathfinder* Lander and microrover. (Sketch courtesy of NASA.)

Under the Ocean

Exploring the ocean depths is very hazardous to humans. People have enclosed themselves in machines and used remote-controlled manipulators to aid in such exploration. Using short-range remote-controlled robots is of limited value in this environment. Researchers are trying to develop free-moving self-controlling robots for use under the ocean in exploration and ocean-floor mining. At present, pipeline companies on land use robots to inspect pipelines from the inside. A modified version of this type of robot could be used to inspect pipelines under the ocean.

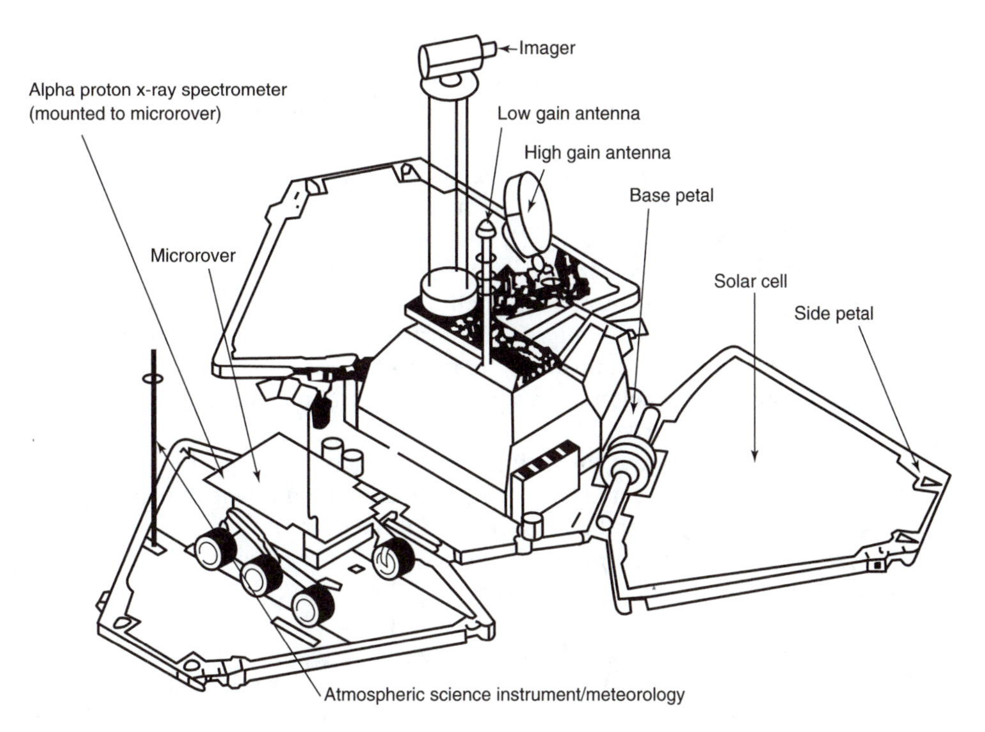

Figure 7-11

Mars *Pathfinder*'s *Sojourner* rover. (Sketch courtesy of NASA.)

Attempts are currently being made to develop artificial gills for robots. These would allow the robot to use a diesel engine for some of its power, thereby extending the robot's working range.

Medicine

Robots are already in use in various areas of medicine. "Resusci™-Anne" was the first of a series of mannequin aids in the cardiopulmonary resuscitation (CPR) training area. Other mannequins in the series include Little Anne™, Baby Anne™, Resusci™ Junior, CPR Cathy I™, CriSis Manikin™, and the Brad™ and Paul Adult Torso Training Manik. The Anne human CPR simulator is credited with training over 150 million persons, worldwide, to do CPR. Through the use of built-in movements and reactions, these mannequins give realism to training that previously could only be talked about. Before resuscitation or CPR is attempted, these simulators have no pulse or heartbeat and are not breathing. Even the eyes' pupil reflex action can be stopped. As resuscitation proceeds, various reactions begin to revive. These reactions increase with proper resuscitation until the mannequin reaches a "state of consciousness." The simulator measures how efficiently the various resuscitation operations have been performed and prints these results out in graph form. The graphs may show, for example, that a trainee pressed too gently or too hard during the heart massage.

A Zymark Corporation robot is now working as a medical laboratory assistant, analyzing the vitamin, mineral, and caloric content of foods. Robots are also starting to be used to analyze and handle dangerous fluids such as blood.

One robot has even been used to perform brain surgery. Dr. Yik San Kwoh, of the Memorial Medical Center in Long Beach, California, has developed a robot for use in stereotactic neurosurgery—a type of surgery that requires precise placement of probes through a small hole in a patient's skull so that information can be gathered by a computer. The alternative procedure is to remove a large portion of the patient's skull through a craniotomy. The robot can position a probe or delicate instrument much more accurately than a human can. The robot was used in actual surgery in April 1985.

Robots are also being used to cut the hole in the bone for an artificial hip joint replacement. They are more accurate than their human counterparts and give increased probability of a successful operation.

A variation on the industrial robovan serves as a hospital orderly, delivering meals and medicine on demand.

Prosthesis The field of prosthesis development is closely related to robotics, if indeed it is not robotics. **Prostheses** are devices that replace a missing or damaged part of a human being. Early prostheses included the peg leg, the hook hand, and the glass eye; more recent prostheses are the artificial heart and artificial limbs that are controlled by the person's brain.

With the use of microprocessors to monitor neurons in the spinal cord, it will be possible to create cybernetic or robotic arms and legs that can be attached to the stub of a person's amputated limb. The artificial limb would then be moved by the same signals that were used to control the original limb. At present, the best cybernetic arms fit over the stump and have sensors that monitor the electrical activity of the muscles in the stump. However, the pressure feedback signals normally received from the fingers as they grasp an object are difficult to simulate.

Some humans born without a hand or arm are now being equipped with artificial (robotic, bionic, cybernetic, or prosthetic) limbs.[1] About 50 children in the United States who were born without a hand have been fitted with bionic hands. The hand and part of the arm fit over the end of the incomplete arm the child was born with. Microprocessors in the bionic hand sense the muscle signals at the end of the arm. Since these children never had a hand on this arm before, some ingenuity is required to teach these children to control the new hand through their own thought waves. To help teach the children how to use the new hand, a remote-controlled toy car is connected in parallel with the bionic hand. The thoughts needed to cause the hand to close also cause the toy car to go forward. The thoughts needed to open the bionic hand also cause the toy car to go backward. Thus, as the child learns how to control the car, he or she also learns how to control the new arm and hand.

[1]"Six-year-old with Bionic Hand," *PM Magazine* (July 6, 1982).

It may soon be possible to have paramedical robots. These robots will assist in giving medical treatment to certain patients. Show or promotional robots are already being used to work with abused children. Future robots will be able to work with patients who have mental problems that require constant monitoring, or lonely persons who need a companion for conversation.

Orthosis Orthoses are devices added to a human to help overcome a nonfunctional or incompletely functional part. Hearing aids and glasses are types of orthoses. Using electronic transducers under the skin for hearing-impaired persons or those who work in high-noise areas is another type, as is the pacemaker for a heart. Thus orthoses can be internally mounted, externally mounted, or both. For example, a mechanical frame may be externally attached to the fingers of a hand to move the hand for a quadriplegic, while another part of it may be implanted and connected to the muscles so that the person's own nerves and muscles provide the power for motions.

In June 1983, Nana Davis, a paraplegic, walked to the podium to receive her B.S. degree in elementary education. This act was accomplished with the aid of a purse-size computer that fed impulses to twenty-four electrodes attached to her legs, and with the aid of two persons at her elbows.

Quadriplegics and even more severely injured persons are finding orthoses useful both for communicating through eye movement and for controlling machines—including robots. The new eye-motion sensors are a godsend for persons who can only move their eyes.

Art

Robots are showing up in the field of art both as artist and as art form. One Japanese company uses a computer vision system and an industrial robot to paint customers' pictures. The vision system takes a television picture of the customers, digitizes it, and transmits instructions to the robot on how to reproduce the information on paper. The robot uses a pen as an end effector to draw the portrait on the paper.

San Francisco has a robotic art show at least once a year. Here the robot is the art. This show is automated and, while it takes months to set up, only lasts a few minutes. If the theme of the work is squares, you might see a robot car moving on square wheels or boxes tumbling into the audience.

The Office

In the office, robots now do windows. At least, one very specialized robot does windows on a glass skyscraper. In addition, robot mail carts deliver mail in some companies. In factories, similar robots known as robovans are used to transport parts. These robots follow lines on the floor or buried cables. Home robots equipped with a vacuum cleaner inside them can vacuum at a party without running into guests or furniture. Can it be long before there are robot vacuum cleaners at the office?

Currently, there are some companies already using robotic vacuum cleaners. Hospitals have started to use Helpmate robots to deliver drugs, food, medical records, X rays, and laboratory reports. Robots from CyberGuard Corporation are working as guards for some companies; these robots can detect fire, monitor air quality, detect intruders, assess situations, respond to alarms, transmit video, and do threat assessment.

Promotion

Promotional robots are already big attractions at electronic and computer shows. Because of their continuing improvement, show robots will remain an attraction. Robots get people's attention and, therefore, have selling potential. At present, robots appear at shows primarily to pique people's interest and to hand out business cards. Perhaps if these robots actually made sales presentations, people would stay attentive during the presentation.

Sheep Shearing

In Australia, experiments are being conducted on using robots to do sheep shearing. Sheep come in different sizes, and they move during shearing. First, a sheep is placed in a stall. Measurements are taken at the forehead, the top of the front legs, and the top of the back legs. Shearing starts at the head and moves back toward the tail. Clippers move within millimeters of the skin, but no nicks or cuts occur. At present, the robot can only shear the left side of the sheep. Robot sheep shearing is definitely the shearing method of the future.

Fast Food

The fast-food industry already uses computers to save time and money. Now it is considering the use of robots and automation. The frequent turnover of staff in fast-food restaurants means that management must keep training new employees. Cooking french fries requires good timing for good results; as soon as the electronic timer on the deep-fat fryer goes off, the fries must be removed from the cooker. The prototype-robo-fries robot developed by Burger King is all electric; an earlier design used pneumatics, but an air supply is not readily available at most franchise outlets.

Sports

The sports world has not escaped automation and robotics. Automated pitching machines throw balls for batting practice in baseball. Other machines are used to practice returning serves in tennis; tennis also has experimental robotic line judges. There are golf shot simulators for the golfer. Germany is using robots as sparring partners for training boxers. Skiing robots in Japan help teach people correct skiing techniques. And finally, Ping-Pong-playing robots have yearly world championship tournaments.

DOMESTIC ROBOTS

The domestic robot is currently at a very limited stage of development. Only a few electronic pet toys and some hobbyist robots such as the Hero I, Hero Jr., and Hero 2000 are now available. Robot lawn mowers and other domestic robots are not yet available. When a personal or domestic robot finally breaks into this untapped market, the market will show a growth rate similar to that of the personal computer market.

MILITARY ROBOTS

The military is pumping large amounts of money into robotics research and is already using robots in several areas. Robot **drones**—remote-controlled airplanes that can do reconnaissance flights of hundreds of miles—are widely used. Small remote-controlled submarines are produced, as well. Drones can also be used as targets in gunnery practice.

Military missiles are special-purpose robots. Military strategists would like to make small remote-controlled tanks. Without the support equipment for the human crew, tanks could be much smaller, making them harder targets to hit.

The army is already using a robot named Manny who can talk, walk, do pushups, and sweat. He even has a temperature of 98.6°F and simulated breathing. Manny's main job is to try out protective clothing. All of Manny's motions are used to put the clothing through the actual motions that a human would use. The clothes are then checked for leaks.

Finally, the military is using computerized and roboticized simulators for both training and research.

OTHER EXPERIMENTAL AREAS OF ROBOTICS

Experimental areas of robot applications include garbage collection, fast-food preparation and delivery, gasoline dispensing, animal husbandry, nuclear maintenance and cleanup, hospital work, prostheses, household service, companionship for the elderly, dishwashing, fruit sorting, candy preparing and inspecting, mail delivery, meat cutting and grading, grapevine pruning, tree harvesting, swimming pool cleaning, gem cutting and polishing, bricklaying, and library work.

AREAS THAT COULD USE ROBOTS

Several areas of industry and government—mining, police, firefighting, and security—could especially benefit from the use of robots.

Mining

Mining activities are unhealthful and dangerous for humans to do. Coal mining is especially known for its dangers, which include explosions, cave-ins, poison gas, and black lung. In 1960, the Joy Pushbutton Miner was designed and built as an unmanned system for drilling into a coal seam at a grade. This remote-controlled device failed to catch on because of its high initial cost, limited application, and maintenance problems. Even so, it shows that coal mining by robot is not impossible. Most underground coal mining involves the following five types of operations:

1. Undercutting of the bottom of a coal vein
2. Drilling holes for setting explosive charges in
3. Blasting
4. Bolting the roof
5. Hauling the coal

Cutting, bolting, and hauling machines are all in use now, but they still require human operators. Automatic rail-hauling machines are under development. The key missing component is intelligent robots to serve as machine operators. As gold, platinum, and diamond mines reach depths of 10,000 feet, they become too hot for human workers. Robot workers are needed here.

Police

Police departments are just now obtaining remote-controlled robots for use in their organizations. Such robots can be used to approach armed suspects who have barricaded themselves into some place, without risking human officers. They are especially useful in handling bombs. Another possible use for robots is to help the police with public relations.

Firefighting

Fighting fires is a hazardous activity for humans. Remote-controlled robots could be used for fighting fires in dangerous positions that are too risky for human firefighters. These robots could survive some conditions that would kill humans, such as those involving extremely heavy smoke and very low levels of oxygen.

Security

Security patrolling by a single human can be hazardous. Remote-controlled robots or even intelligent robots could perform these dangerous patrols. Robots could move faster and remain alert better than humans, and they could be equipped with infrared and ultrasonic detection devices that are not available for human security personnel. The robot could be instructed to inform a human security person if it found something not right. Meanwhile the robot could use armor or its speed to protect itself.

The military has already developed experimental models of robot security guards.

SUMMARY

While the main type of robot today is the industrial robot used in the factory and foundry, other uses for industrial robots exist and other types of robots can be used for many purposes.

Industrial robots currently perform die casting, forging, investment casting, machine tool loading and unloading, parts transfer, spray painting, small parts assembling, finishing, plastic molding, spot welding, arc welding, machining, electronics assembling, and inspecting. Robots working in business are used in outer space, under the ocean, in art, in promotions, for mail delivery, for window washing, and for vacuum cleaning. In medicine, robots are used as lab assistants and brain surgeons. The home is starting to have pet robots, and soon the home robot will be a useful servant.

Many areas of robot use continue to be under investigation. In particular there will soon be use of robots in mining, police, fire, and security.

REVIEW QUESTIONS

1. Why are most industrial robots used in the manufacturing industry rather than as secretaries?
2. What tasks might a robot without sensors be able to handle?
3. Will robots make more jobs for humans or just take jobs away from humans? Why?
4. Aside from the applications mentioned in this chapter, what other jobs might an industrial robot do?
5. What abilities would have to be added to an industrial robot in order for it to work successfully as a receptionist?
6. Why can a robot do a better job of painting parts, on the average, than a human can?
7. Why are remote-controlled robots (teleoperators) used instead of humans for deep ocean exploration?
8. Why do the robots used for space exploration need to make many of their own decisions rather than being just remote controlled?
9. Why is it desirable to remove humans from painting tasks?
10. How would a robot guard be better than a human guard? How might the robot be a worse guard than a human guard?
11. Use the Internet to find companies that make nonindustrial robots such as vacuum cleaning, delivery, and guard robots.
12. Research a pallet loading problem for a factory. What needs to be known before a robot can be given the task of loading boxes from a conveyor belt to a pallet? Once the research is done, select a robot from Appendix D on robot specifications to do the task.

8

ROBOT MAINTENANCE

OVERVIEW

The best equipment money can buy will soon become ineffective and even dangerous if it is not properly maintained. If regular maintenance is not planned, it will not be done properly. Both preventive, or scheduled, maintenance and emergency maintenance require planning beforehand to produce good results.

OBJECTIVES

When you have completed this chapter, you should be familiar with:

- How maintenance can make or break a project
- How preventive maintenance can reduce the need for emergency maintenance
- Problems that may be encountered in working on closed-loop servo-controlled systems

KEY TERMS

Diagnostic programs
Lubrication requirements
Maintenance manuals
Periodic scheduled maintenance
Preventive maintenance
Recalibration
Recommended spare parts list
Substitution of robots
Troubleshooting procedures

Like other complicated machines, robots need to be maintained. A small amount of time spent at regular intervals on **preventive maintenance** could keep a robot from experiencing a major breakdown. While doing preventive maintenance, you may detect small problems before they can cause large problems in the future. Even with preventive maintenance, robots will occasionally break down. They then require **emergency maintenance.**

A company that is planning to have the robot manufacturer provide maintenance on its robots should determine how fast the maintenance service can be expected to respond to problems.

A company that is planning to do its own robot maintenance must figure out who will train its repair personnel. Hydraulic robots require repair personnel who are skilled in hydraulic maintenance as well as in electronic computer maintenance. Hydraulic robots can be damaged beyond repair as a result of using contaminated hydraulic fluid; electronic controllers can be damaged if grounding procedures are not properly followed in work on microcircuits such as those used for memories and microprocessors. The manufacturer must supply a **recommended spare parts list, lubrication requirements** for the manipulator, optimum lubrication schedules, other maintenance schedules, and **troubleshooting procedures.** The company maintenance people should go over all of these items to make sure that they understand and agree with all of the information.

PREVENTIVE MAINTENANCE

Probably the first area of preventive maintenance involves providing appropriate working conditions for the robot. Issues and conditions that need to be taken into consideration when determining where to place the robot include: human and robot safety, heat, shock, vibration, electrical noise, liquid sprays, gases, harmful particles, and risk of fires and explosions. Key elements of preventive maintenance can be accomplished during the project design phase of the robot project, because many potential problems can be designed out of the system, which is why maintenance people should be involved in the design stage.

The safety of human workers controlling the robot or just passing by the robot's area should be a primary consideration. A relatively slow-moving hydraulic robot can strike an undetected object with tons of force—industrial robots have killed people.

The safety of the robot should be the next area of concern. Present-day robots are not very intelligent and need help in protecting themselves.

If the ambient air temperature is above 120°F, an industrial robot needs some source of cooling air. If the robot is equipped with electronic control circuits, it may require cooling at even lower temperatures. Too cool an ambient air temperature can hamper the robot's ability to move accurately. Sudden or wide variations in temperature can adversely affect a robot's mechanical tolerances. Opening an outside door in the winter will throw a precision robot out of tolerance.

Robots are required to meet tight positional requirements. If they are subjected to shocks and vibrations, these will at a minimum alter their positional accuracy. Severe shocks or vibrations may cause joints or slides to stick and may even damage them.

Electrical noise from a robot's motors and switches or from other electrical machines may cause the robot's controllers to misunderstand what is going on. Electrical noise is especially likely to interfere with the robot's communication with other machines or computers. The robot's controller is also easily disturbed by electrical noise, which can cause its memory to become confused or forgetful. Such interference will result in wrong action by the robot, by some other machine, or by the computer, and may lead to damage to a machine or robot. Electrical noise problems can be prevented or compensated for during the robot's installation stage; if the problem is not recognized early on, it becomes much more difficult to solve later.

Liquid sprays can permeate a factory's air. Some of these liquids can pose health hazards to robots as well as to people, since they may interfere with lubrication or even corrode metal parts. The robot may require some type of protective covering against these sprays. Electronic circuits definitely need protection from them to avoid short circuits. Some gases and solid particles are also harmful to robots. One way to reduce the harmful effects of liquid sprays, gases, and particles is to remove all parts of the robot that do not need to be in the work area, such as the power supply and the robot control unit, to another protected area. The manipulator can be given some type of protective covering.

If a robot has to work in dusty conditions, its joints, keyboard, teach pendant, and electronics will need special protection. The keys of the keyboard must be sealed against the dust; it may be necessary to use a membrane-style keyboard. It may be possible to have the robot controller moved to a less dusty area than the manipulator must work in.

The hazards of possible fires or explosions are of considerable importance to both humans and robots. One important safety measure is to remove all nonrequired parts of the robot out of the work area. A second safety measure is to reduce the possibility that the robot will contribute to the hazard by providing an electrical or mechanical spark. Switches on the control devices of a hydraulic manipulator require very low voltages that do not pose a risk of setting off a fire or explosion. Using hydraulic robots is therefore popular in high-risk tasks such as spray painting. Explosion-proof electrical switches can be used on other types of robots. Oil leaks on the robot or on other equipment should be fixed immediately.

Periodic Scheduled Maintenance

A key area of preventive maintenance is **periodic scheduled maintenance,** which covers everything from inspecting the robot before and during each shift of work to performing major overhauls at prescribed intervals.

The inspection of the robot at the beginning or end of a shift should include testing all safety equipment and looking for signs of wear, leaks of any kind, dirt, and proper lubrication. Both lubrication levels and the quality of the lubricant should be checked. Burnt hydraulic oil can seriously damage a hydraulic-powered robot. Like any other tool, the robot needs to be kept clean and in good working condition if it is to perform at its best. Any problems discovered should be reported to maintenance personnel, and corrective action should be taken as soon as is practical. Filters should be checked to ensure that they are not clogged. Air vents should be clear. Without proper ventilation, electrical and electronic parts do not work properly. If the robot uses magnetic tapes or disks, these should be kept clean, dry, and flat; they should be stored neatly in drawers or cabinets and kept away from magnetic fields such as transformers, video displays, speakers, and metal paper clips.

Hydraulic robots should be checked for leaks, which are messy, dangerous, and expensive. It is not uncommon for a hydraulic system to leak four or five times its carrying capacity of oil in a year. The condenser on a pneumatic-powered robot should be checked; if it is full of water, it should be emptied. The oil reservoir should be checked to ensure that it contains enough oil to lubricate the devices that the air supply serves.

During the shift, the robot should be checked for correct performance. Unexpected motions or noises should be investigated, since they could be signs of an impending major problem. The quality of the robot's work should also be checked. An unusual number of damaged or improperly made parts could translate into maintenance problems with the robot. The robot's warning lights and filter plugs should also be monitored.

The manufacturer will generally give a list of suggested routine maintenance steps for upkeep of the robot. These suggestions will include cleaning surfaces, changing filters, changing fluids, and making certain adjustments and measurements. The maintenance schedule may even call for replacing or overhauling some parts at fixed intervals.

Calibration of Robots

Mechanical movements cause physical wear, and electronic power supplies, sensors, and analog-to-digital converters change in value with age. As these inevitable changes occur, the accuracy and repeatability of the robot's movement decreases. Out-of-tolerance power supplies can produce symptoms suggesting that some other part of the robot is faulty.

Preventive maintenance should include checking for possible accuracy and repeatability problems. When these problems arise, they can be corrected through **recalibration** of the robot—the adjustment or replacement of mechanical or electronic parts.

If the robot uses a mechanical setup for its programming, as is the case with some pick-and-place robots, the recalibration may consist of redoing the mechanical setup. If the robot's programming is based on task leadthrough, it may be necessary to redo the programming. Whatever the setup, if the robot's calibration is not checked and corrected periodically, the quality of the work done by the robot will decrease with use.

The following three types of calibration are all useful:

1. Check to see that the robot goes up against its home stops when the robot is given a command to go home.
2. Check to see that the electrical outputs of the power supplies are within their tolerances.
3. Check to see that the pressure leaving the power supply is within the required limits.

EMERGENCY MAINTENANCE

Regardless of whether the breakdown of a robot holds up a production line, the robot does not make any money for its owner while it is waiting to be fixed. Therefore, it is desirable to get the robot working again as soon as possible. Ease of maintenance and repair should be a major feature of a robot's design. One helpful design arrangement is the use of plug-in modules.

Many robot manufacturers supply **diagnostic programs** for use with their robots. Diagnostic programs are used to determine whether the robot is working properly or not. If the robot is not working properly, the diagnostics try to identify what is wrong with it. The robot should also come with **maintenance manuals** that offer additional hints on problems and solutions.

Ford Motor Company is using an artificial intelligence program to help diagnose problems with its robots. This intelligence takes the form of an expert system. When the expert system diagnostic program is run, it asks the operator for specific information about the problem, guides the operator in determining what testing to do, and finally tells the operator what corrective action to take to fix the robot.

Whatever method is used to get information about what is wrong with the robot, the repair person needs to follow some established procedure. Generally, the diagnosis should be conducted as though the robot were three separate systems: a power supply, a controller, and a manipulator. If the controller has a computer interface and includes the signals that are being sent to the manipulator, it may be possible to use an external computer to check out the robot's controller, with the controller disconnected from the rest of the robot. The robot's power supply can cause problems with the manipulator if it is not working right. Before concluding that the manipulator must be bad, you need to check out the power supply. If possible, the power supply should be checked out while separated from the manipulator. If the robot's controller and power supply are working correctly, the problem probably is in the manipulator.

If the manipulator uses an open-loop nonservo-controlled system, the manipulator can be commanded to move, and measurements can be taken to see how well it carried out the command. If the manipulator has sensors, the sensor readings can be compared with the actual position of the manipulator.

If the manipulator uses a closed-loop servo-controlled system, the servo system's attempts to correct the manipulator's position may mask the manipulator's real problems. A problem with the servo system and a problem with the manipulator may partially cancel each other out. In order to simplify the diagnosis, it is desirable to disconnect the servo correction system. The manipulator can now be commanded to move to a position and be checked for adequate response, while the servo correction system can be checked for its ability to generate correction signals.

Another approach to dealing with the breakdown of a robot working on a production line is to substitute another robot for the bad robot. Then the defective robot can be repaired when convenient. If **substitution of robots** is to be used, the feasibility of the plan should be checked out with the robot manufacturer and specified in the purchase contract. Not all robots of the same model necessarily do their motions in exactly the same way. If the task for the robot requires tight tolerances, the robots must be designed with similar enough tolerances to be substituted for each other.

MAINTENANCE AND SAFETY

Good maintenance makes for a safer robot installation; preventive maintenance ensures safety and can be done with little risk to maintenance personnel.

Emergency maintenance must be performed on a robot that is not working correctly. This is the most dangerous time to be around a robot. If the maintenance person must work on the manipulator while the robot is powered up, a second person should be standing by an emergency stop button in case problems arise.

Performing maintenance on any type of robotic power supply—electrical, pneumatic, or hydraulic—is dangerous. Working with a three-phase, 440-VAC power supply is much more dangerous than working with a 5-volt computer logic supply. A 200-pound compressed air supply may not seem dangerous, but it can put out an eye or cause severe cuts if it contains foreign particles. A loose pressure hose can strike with surprising force. A pinhole-size leak on a 2,000-pound hydraulic system can jettison hydraulic fluid right through your hand; if you walk through it, it can cut you in half. A loose hydraulic hose is both quicker and deadlier than a sword!

SUMMARY

Good maintenance does not happen by accident. It must be planned. A complicated device like an industrial robot must be maintained properly if it is to operate correctly. You can either plan preventive maintenance at a small cost or be prepared to pay a large cost for emergency maintenance later on. Maintenance planning should start at the beginning of the robot project: How will the robot problem be diagnosed? What spare parts need to be stocked? What type of maintenance schedule should be set up for the robot? The robot's manufacturer may be able to help formulate sensible responses to some of these questions. Above all, remember that a robot that is not working right is a safety hazard.

REVIEW QUESTIONS

1. Why should preventive maintenance be a consideration in the planning of the robot's installation?
2. How can periodic inspections help prevent major robot breakdowns?
3. Why do closed-loop servo-controlled systems make diagnosis of a robot problem difficult?
4. Why should each part of the robot system be checked out separately?
5. Why is the calibration of a robot important?
6. Suppose that you are planning to place a robot in a dusty area. What can be done to reduce its maintenance problems?

7. Suppose that you are planning to have a robot do spray painting. How do you protect the robot's power supply, controller, and manipulator from the paint?
8. How are preventive maintenance and safety tied together?
9. What can you tell by looking at the hydraulic fluid?
10. Why is emergency maintenance expensive?
11. Why should only trained robotic maintenance people fix a robot, and not the robot operator?
12. Why should the operator inspect the robot at the beginning of every shift?
13. How will the availability of spare parts and spare robots affect robot maintenance?
14. Where can recommended preventive maintenance schedules for a robot be found?

9

ROBOTICS AND SAFETY

OVERVIEW

Safety is everyone's responsibility. Safety precautions make good business sense, and many precautions are required by law. How can we help robots operate safely? How can humans and technology work together safely? How does the law relate to safety? These are some of the questions that will be answered in this chapter.

OBJECTIVES

When you have completed this chapter, you should be familiar with:

- Why robotic safety is needed
- The importance of planning safety
- The need to "think safety"
- What makes up the area of robotics safety
- How robots may make working condition safer for humans
- Legal aspects of robotic safety

KEY TERMS

Beacons
Human safety
Impact injury
Interlocks
Light fences
OSHA
Pinch injury
Pin injury
Pressure pads
Think safety
Three Rs of safety

The present-day industrial robot is a dumb, senseless idiot. Therefore, safety in robotics must be managed by humans. Robot safety is dedicated to enabling the industrial robot to follow the first part of Isaac Asimov's First Law of Robotics: "a robot shall not harm a human being."

WHEN TO CONSIDER ROBOTIC SAFETY

Safety is the first thing you should think about when working with robots, beginning as soon as a robotics project gets underway. Human safety must be built into a robotics system from the outset, even if humans are never expected to venture into the robot's work area. Until robotics systems can repair and maintain themselves without human help, they need some means of detecting the presence of humans. When repairs are needed, it may not be possible to power down a robotics system while humans work on it, and trying to add on safety features later is costly and may not be effective.

Robotic safety continues to be important throughout the life of a robotics project. For humans, safety is a state of mind. Workers who are safety-conscious will experience an optimum level of safety, while those who are not will increase their risk of injury. A very simple formulation of the need for safety awareness is expressed in the **three Rs of robotic safety:**

Robots **R**equire **R**espect

LEVELS OF DANGER

In matters of robot safety, the safety of humans should come first, then the safety of the robot, and finally the safety of other related equipment.

The most dangerous situation in which a human must work with a robot is when repairing it. To repair a robot, the human has to be in the robot's work cell, and since the robot is not working right, it could make an unexpected movement and injure the human.

The next most dangerous situation in which a human must work with a robot is during the robot's training or programming. Once again, the human may need to be in the robot's work cell. Indeed, if the robot is a point-to-point unit, the human may well have her or his nose inches from the gripper to see or measure that the robot is oriented correctly. Programming a robot is different from programming a computer in at least one aspect. A programming error (programming bug) in most computer programs merely produces a wrong answer, whereas a programming bug in a robot's program results in wrong physical movement. This movement can damage the robot, the equipment and parts in the robot's vicinity, or a person near the robot.

The least dangerous situation in which humans must work with a robot is during the robot's normal operation. However, the robot is still dangerous—especially if it depends on the running cycle of other devices, rather than on its own fixed time cycle—and steps should be taken to ensure that humans stay out of the robot's work cell during normal operation. Even if the robot is working correctly, some of the support equipment may not be. For example, a robot that does not have to work on all models of parts coming down a production line will operate only if it detects the presence of a part that requires its services. One person, who had had over ten years of experience working with robots, was killed at a Kawasaki plant when struck from behind by a robot while trying to adjust a malfunctioning process machine.

Both workers and visitors need to be protected from robots. Robotics installations are fairly new phenomena in many factories, and their operation attracts the attention of other personnel in the plant. These unauthorized observers need to be kept at a safe distance from the robot. Light curtains, chains, and fences can be used for this purpose. If a person crosses the barrier, sensors should detect this and shut off power to the robot. At least one death occurred at a Japanese automobile plant when a visiting engineer was killed by a robot he was inspecting.

ROBOT POWER SUPPLIES

Extra caution should be used in dealing with the robot's power supply. While a pneumatic power supply may look benign, compressed air is always dangerous. An air leak may emit contaminants and oil that can do a lot of damage. If a pressure line breaks loose, it can be very powerful and dangerous.

A leak in a high-pressure hydraulic power supply can punch a hole through a person's hand. A broken pressure line is deadly. And of course, the oil spilled on the floor is hazardous.

Some electric robots use 220- and 440-volt AC motors to get a good power-to-weight or power-to-size ratio. These voltages are lethal. The skin resistance of a human being is at best 2,000 ohms. Damp skin has a much lower resistance. A minimum of 30 milliamperes of current, which can be achieved at an electric potential of as little as 60 volts, can kill a human.

HOW ROBOTS INJURE PEOPLE

Robots can injure people in one of three ways. The first way is through bodily impact—since the robot does not know of the human's presence, it can simply run into the human during its normal motion. The second way is by pinching—this can occur if the human gets caught in either the gripper or the joints of the robot's manipulator. The third way is by pinning the human against some structure—this can happen if the human gets caught between a robot's arm and a machine or wall.

People are most likely to be hit by a robot when they mistakenly think that it is not powered up. Using a flashing red beacon on top of the robot to show that it is powered up helps to alert people to the danger of approaching it. Placing sensors around the robot to warn it of approaching humans is an even better idea. Workers must be made aware that even a slow-moving hydraulic robot can hit a person with several tons of force.

A worker can easily get pinned by a robot while trying to work on a malfunctioning robot or on malfunctioning support equipment if the robot is still powered up and trying to do its task. For example, if the unit that is supposed to feed parts to a robot is not working properly, a worker might try to make adjustments to the part feeder while it and the robot are still running. A miscalculation of the timing of the task can leave the worker trapped between the robot and the feeder as the robot tries to get a new part.

SAFETY AIDS

All workers should be educated about the safety issues involved in working with robots or other equipment. Safety issues will persist as long as people and equipment mix.

Always try to match the robot to the task being done. If the task requires that the robot lift only a few ounces, do not use one that can lift a ton; if the task requires moving only a few inches, do not use a robot that has a 30-foot reach.

Several things can be done to aid in robotic safety. Start by fencing off the robot's work cell. The fence can be electronic, such as a light fence (see Figure 9–1). If anything breaks the beams of light, the power to the robot is turned off automatically. As a second line of defense, use pressure-sensitive pads on the floor around the robot (see Figure 9–2). Any pressure on the pads will again shut off power to the robot.

Place a warning beacon on top of a robot so that when the robot has power applied to it the beacon flashes continuously. Because beacons can fail, however, you still need to check that the power to the robot is off before approaching the robot.

Grippers that use spring power should use that power only to open the gripper. This will ensure that when the power goes off the robot cannot accidentally grab a human.

Brakes can be installed on a robot so that it stops and stays where it is when power goes off. Otherwise a robot will not stop instantly, and hydraulic manipulators may droop all the way to the floor as power is removed.

An audible beeper should be used on mobile robots to warn humans of their presence. This will prevent a robot from sneaking up on someone and causing an injury.

When arc-welding robots are used, shields or curtains should be placed around the welding area to protect passersby from the bright light of the arc, which can damage the human eye.

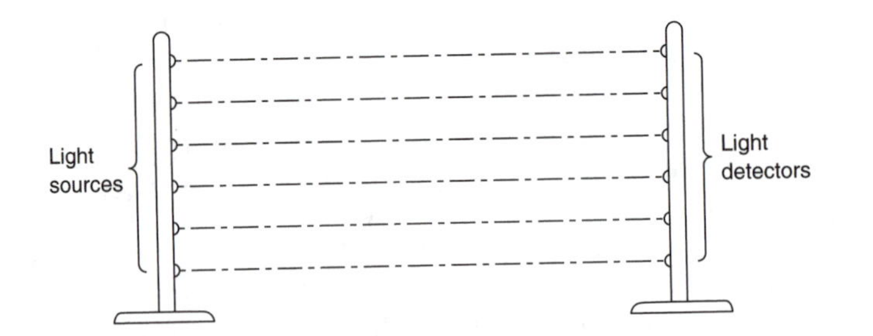

Figure 9–1 Light Fence.
A light fence uses the interruption of one or more of the light beams as a sensor to detect the presence of an intruder. Such an intrusion causes a cutoff of power to the robot.

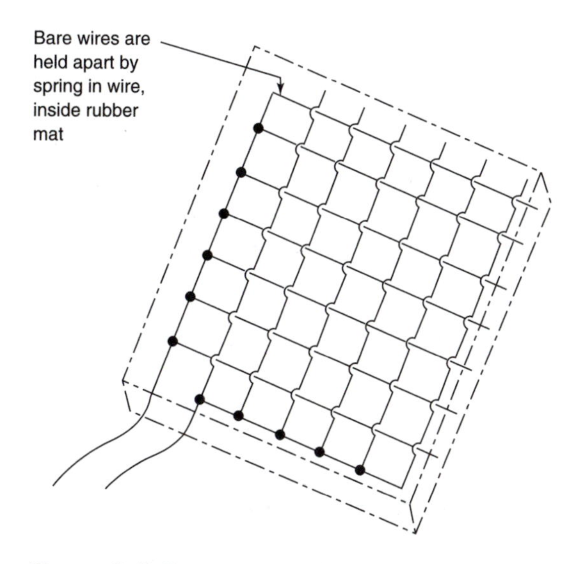

Figure 9-2 Pressure-sensitive Pad.
Stepping on a pressure-sensitive pad closes an electrical contact, which in turn stops the motion of the robot.

To avoid programming bugs that might cause a robot to do damage, programs should be well organized before they are entered into the robot. A robot simulator or CAR system allows you to try out a program on a computer before using it in a robot. Single-stepping a robot through a task program before trying it at full speed may also help you identify bugs before they do physical damage. No matter how the robot's programming is done, great care must be exercised.

Whenever programmers or repair personnel must work with a live robot, they should know where the nearest emergency stop button is. Never let an inexperienced worker go into the work cell, and when possible station a second person outside the work cell next to an emergency stop button while someone is working inside on a live robot.

Deadman's switches might be incorporated into the teach pendant so that the robot stops when pressure is removed from the trigger of the teaching device.

Finally, hard mechanical stops may be used to restrict the work-cell movement of a robot, limiting its chances to make an unexpected movement during a normal cycle. Most industrial robots have provisions for making hard mechanical stops, and most robot tasks do not require the robot to move to all extreme points on every axis.

Robots themselves need protection from the machines they work with. Such protection may consist of interlocks with these machines to prevent a robot from sticking its hand into a closing forge press, or just running into the side of a closed machine.

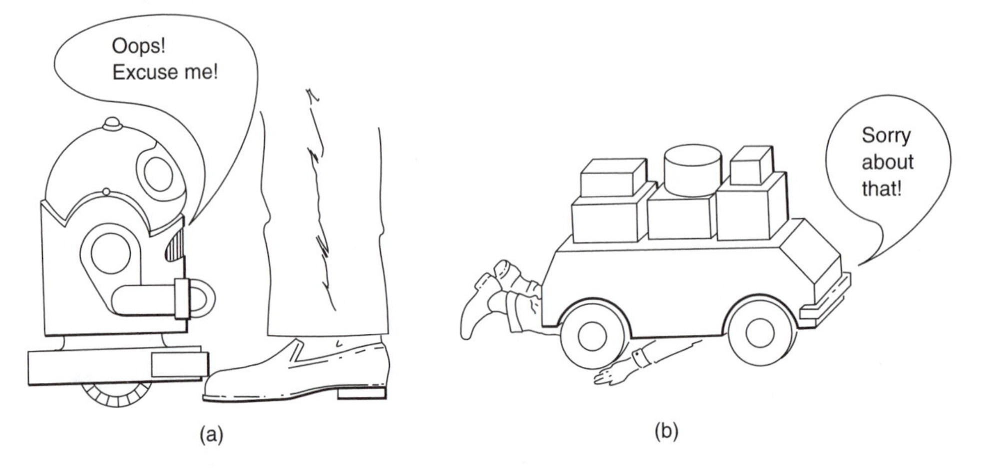

Figure 9-3 Robots Using a Touch Sensor for Safety.
(a) It may be funny when a small hobbyist robot bumps into you. (b) It is not so funny in the factory or office.

Safety devices that work for one type of robot may not be adequate for another type. You need to match the safety device to the robot and its task. As Figure 9–3 indicates, bumpers may be all right as a safety device on a small hobbyist robot, but they will not do much good on a 2-ton robovan. Now that robots are being used to deliver things in hospitals and offices, robots must look out for objects and people in their path! People working in hospitals and offices are not trained robotic specialists. They do not know the dangers presented by the presence of the robots.

One last thing about safety aids: If you are going to have safety aids, make sure they are in good working order and are tested periodically—say, once per shift! Properly maintaining the robotic system is an extremely valuable safety aid.

USING ROBOTICS TO MEET SAFETY REQUIREMENTS

Although much of the subject matter of robotic safety involves protecting humans from robots, robots sometimes may be used to meet human safety requirements by removing humans from unsafe working conditions, such as robots that do spray painting and tend forge presses.

THE LAW AND SAFETY

Taking safety precautions when planning for industrial robots is not only a good idea, it is the law. Both state and federal regulations deal with industrial robotics safety, including the federal Occupational Safety and Health Act of 1970 (OSHA). These laws are aimed at three aspects of the robotic work area: keeping unauthorized persons outside the work cell; protecting workers from fixed machinery; and protecting workers from the robot itself. However, such regulations apply only to the normal operation of the robot; they cannot cover abnormal operations, such as during programming and maintenance of the robot.

Part 1910 of the OSHA standards deals specifically with robotics, including work cell safety and robotic application safety. Nonrobotic fixed machinery is also covered by OSHA regulations; basically, such equipment must carry devices that prevent operators from exposing any part of their body to dangerous parts of the machine. Within a work cell, such protective devices could interfere with the robot's mobility, but the robot does need to be protected from the other machinery.

The latest state and local laws governing robotics installations should be checked at the time the installation is to be done. The latest information on OSHA regulations can be found on the Internet. For example, the U.S. Department of Labor's *Occupational Safety & Health Administration's Technical Manual,* Section III, Chapter 4, titled "Industrial Robots and Robot System Safety," is found at Internet address http://www.osha-slc.gov/TechMan_data/TM41.htm.

A SAFETY CASE STUDY

An automatic warehouse system studied in January 1979 was five floors high and looked like the skeleton of an unfinished skyscraper.[1] Each floor was occupied by two types of robots: robovans, which weighed over 1 ton, ran quietly on rails, and could handle up to 300-kilogram parts; and moles, which rode on the robovans, hopping off to load and unload them. The robovans had whirling red flashing lights and eyes that could read numbers on posts and lanes but could not see humans! The moles had vision that helped them to position loads exactly, but they could not see humans, either. Bumpers would stop a robot on contact, and emergency stop buttons could also be used.

The warehouse area was classified as "unattended" and had signs everywhere that warned "THIS SYSTEM IS COMPUTER CONTROLLED, AND CAN START AT ANY TIME WITHOUT ANY OPERATORS IN SIGHT. ALL UNAUTHORIZED PERSONS KEEP OUT FOR YOUR OWN SAFETY." The warehouse originally did not even have walkways for humans in it.

The day-to-day operation of the warehouse did not prove to be unattended—a fallen part might block an isle; a robot might stop short of its mark; or the computer inventory might not match what was really there. In any of these cases, a human operator would have to climb into the maze and check things out. Walkways were added, but no upgrading was done on the warning devices.

On January 25, 1979, a worker who was checking the inventory was struck in the back of the head by the number reading eye of one of the robovans. The worker died. He had not heard the robot coming, and its bumper—which would have stopped it—never came into contact with him.

If the robot had had an audible beeper, this accident might not have occurred. If the warehouse had had interlocks that shut off the robots in any section of the

[1]John Grant Fuller, "Death by Robot," *Omni* (March 1984) 6:44–46.

warehouse where a human operator was at work, the accident definitely would not have happened.

SAFETY FEATURES OF A ROBOT WORK CELL

Figure 9–4 shows a diagram of a robotic work cell with some safety devices in place. While it does not show every possible safety device, the drawing does give a general idea of how safety devices may be used. Pay special attention to the emergency stops, fences, interlocks, and mechanical stops.

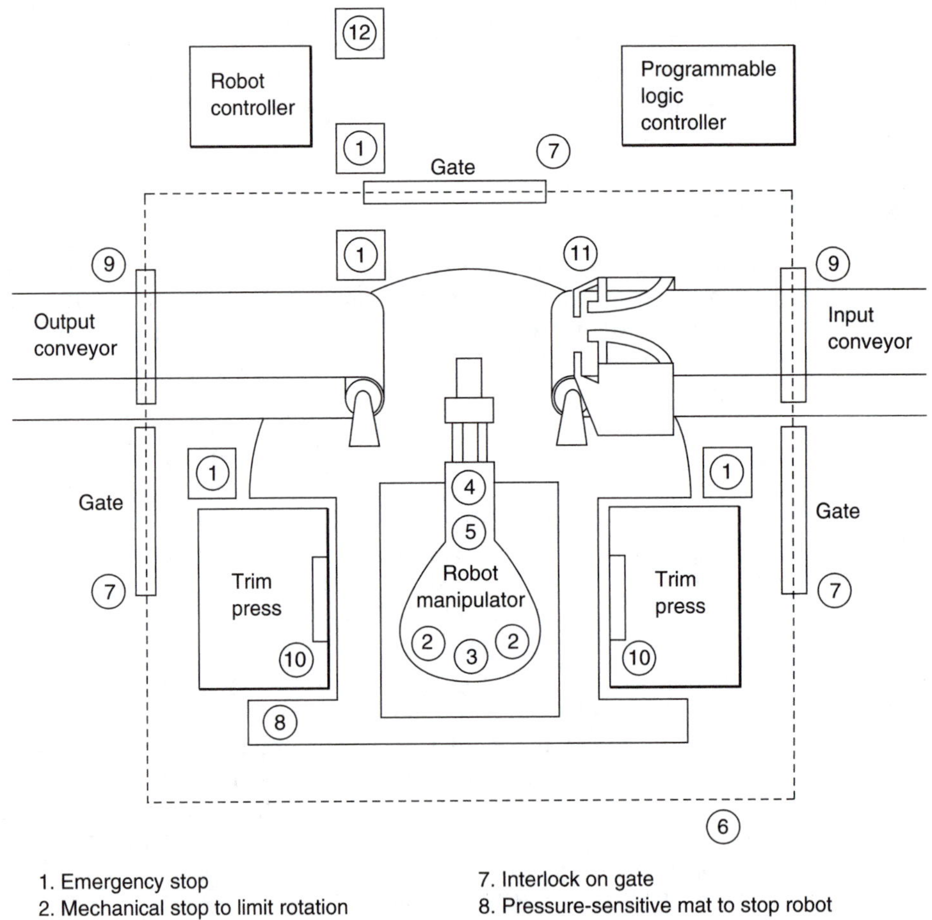

Figure 9–4 Safety Devices in a Robotic Work Cell.
The safety devices in this drawing are identified by number.

SUMMARY

Human safety is a function of safety consciousness. The key to safety around robots is to remember the three Rs of robotic safety: *R*obots *R*equire *R*espect. Even slow-moving robots can be very dangerous. They can hurt humans through impact injuries, pinching injuries, and pinning injuries. The best way to protect humans from robots is to keep the humans away from working robots. Safety devices that can be used include beacons, interlocks, light fences, and pressure-sensitive pads.

The robot is most dangerous when it is being repaired, since it is not working right and is therefore unpredictable. The robot is next most dangerous when it is being programmed or trained. As when during repairs, this condition requires that a human be near the robot. Robots can sometimes be used to make working conditions safer for humans by replacing the humans on some dangerous jobs.

The most important safety rule is to keep untrained personnel away from robots. Appropriate training of persons who will be working with robots is a vital part of any robot safety program.

REVIEW QUESTIONS

1. When should safety issues begin receiving attention in a robotics project?
2. Why is a broken robot more dangerous than a working robot?
3. Why don't present-day robots follow the First Law of Robotics?
4. What are some types of safety equipment that a robot can use?
5. How are robots sometimes used to improve human safety?
6. What precautions should be taken when work must be done on a malfunctioning robot?
7. List the different power sources used for robots and the safety considerations for each source.
8. What is OSHA? Why is it important?
9. Access the Internet, then find and make yourself a copy of OSHA's "Industrial Robots and Robot System Safety."
10. Access the Internet, then look up articles about robotic safety.

10

ARTIFICIAL INTELLIGENCE

OVERVIEW

Artificial intelligence is a branch of computer science that deals with the automation of intelligent behavior. Research in this area is already helping industrial and other types of robots toward becoming "intelligent."

What is artificial intelligence? How will it help progress in robotics? What is an expert system? What is natural language understanding? What is vision interpretation? These are some of the questions answered in this chapter.

OBJECTIVES

When you have completed this chapter, you should be familiar with:

- What artificial intelligence is
- How artificial intelligence may help improve the industrial robot of today
- How to distinguish among the three levels of artificial intelligence
- How artificial intelligence can help a robot become more flexible
- Some current uses of artificial intelligence computer applications
- How a simple expert system works

KEY TERMS

Alpha-level intelligence
Animal
Artificial intelligence
Automated factory
Batch mode
Beta-level intelligence
Built-in intelligence
ELIZA
Expert system
Fuzzy logic
Gamma-level intelligence
Heuristics
Hexapawn Educable Robot
Inherited intelligence
Intelligence
Intelligent tutoring systems
Knowledge diagram
Learning
Natural language understanding
Phonemes

Real time
Sensory input interpretation
Speaker dependent
Speaker independent
Speech interpretation
Speech recognition
Turing's definition
Vision interpretation
Voice prints
Voice recognition

The following definition of the present-day industrial robot describes its intelligence level:

> A robot is a one-armed, blind idiot with limited memory which cannot speak, see, or hear.[1]

The present-day robot cannot make many decisions and receives little or no sensory input. Therefore, the robot has little ability to adjust to changing conditions. Many of today's robots make no attempt at all to adjust. If a part is not where it belongs when the robot tries to pick it up, the robot simply moves empty air. Some of today's robots can call for help if the part is misplaced, but none can search for the missing part. Artificial intelligence may well be the answer to the robot's problem of not being able to adjust to circumstances.

WHAT IS ARTIFICIAL INTELLIGENCE?

When a person or an animal performs some task or act that seems to have required intelligence, we call it an intelligent act. When a machine performs some act or task that seems to have required intelligence, we call it an **artificially intelligent** act. **Intelligence** has been defined as both the ability to learn from experience and the ability to adapt to a surrounding environment. The field of artificial intelligence deals with trying to get machines to duplicate various acts of intelligence that people can perform.

Studying intelligence in humans and animals helps us understand better how to build intelligence into machines. Conversely, studying machines that demonstrate intelligence helps us understand humans and animals better. In other words, "Instead of looking inside to see what crossed the gap when man was created, why not build a machine that behaves like a man and see what must be put into it to make it work?"[2]

Intelligence may be demonstrated in several ways:

1. Learning from experience
2. Adapting to changes in an environment
3. Receiving and processing vision, speech, and other inputs to obtain specific information
4. Making decisions based on the input of information
5. Generating actions and/or memories from the decisions

[1]James Rehg, *Introduction to Robotics: A System Approach* (Upper Saddle River, N.J.: Prentice-Hall, 1985), p. 7.
[2]B. F. Skinner, "The Machine That Is Man," *Psychology Today* (April 1969): pp. 20–25, 60–63.

Learning is one of the most obvious areas of intelligence. A less obvious area involves **built-in,** or **inherited, intelligence.** Most robots of today work exclusively with built-in intelligence, although it may be loaded into their memory at the time they are trained.

The biggest boost robots will get from artificial intelligence is in the area of sensor input and interpretation. Speech recognition, vision, and touch are the sensors most needed by robots at this time.

Artificial intelligence is probably best known for its advances in the fields of expert systems, vision processing, image understanding, speech recognition, speech synthesis, and natural language understanding. Some of these areas are already finding applications in commercial computer products and robotics. Many people think of artificial intelligence as being preoccupied with trying to teach machines to think, but artificial intelligence is also concerned with how human beings think. The use of intelligent computer tutoring systems allows the computer not only to tell a user that an answer he or she has given is wrong, but to pinpoint where the person's thinking process was in error.

Three Levels of Artificial Intelligence

In an effort to establish more precision in the study of intelligence, researchers have defined three different levels of intelligence: the alpha level, the beta level, and the gamma level.

Alpha-level Intelligence The **alpha level** of intelligence is defined as a nonlearning (and probably nonlearned) intelligence. Intelligence at this level does not allow past experiences to enter into its decision making. A demonstration of this level of intelligence can be observed in the present-day hand-held electronic chess-playing game. This game will always repeat its errors, no matter how often it makes them.

All levels of intelligence start from this built-in or nonlearned level. At best, industrial robots of the first and 1.5 generations use alpha-level intelligence to perform their various tasks.

Beta-level Intelligence The **beta level** of intelligence is defined as possessing an elementary learning capability. Because this level of intelligence can learn from its mistakes, past failures affect its decision making. A simplified demonstration of beta-level intelligence can be observed in the **Hexapawn Educable Robot (HER).** HER will not choose the same losing move in the game of hexapawn twice, and thus it could be said to learn from its mistakes or past experiences. Industrial robots of today do not have this ability, nor do they require it. A domestic robot that is to do work around the home, however, may well be required to learn its way around the house and thus show some beta-level intelligence.

Gamma-level Intelligence The **gamma level** of intelligence is defined as having the ability to make generalizations from past experiences. This is also known as **heuristics.** Computer programming attempts to achieve this level of intelligence have not been very successful, but this area of intelligence is as yet not very well defined. A program for gamma-level intelligence would at a minimum have to keep track of averages and other statistics and to draw general conclusions from them. Even for humans, drawing conclusions based on statistics is not an exact science.

Built-in Intelligence

The many chess- and checkers-playing hand-held electronic games use built-in intelligence—intelligence that was placed in their memories during their manufacture. They do not learn, although they may use a random number generator to choose among moves of equal importance, giving the appearance of a variable style of playing the game. These machines use mathematical formulas to calculate the value of each alternative move, and they make use of memorized opening and ending positions. This is also the intelligence method used by present-day industrial robots.

Human beings call their built-in intelligence *instinct*. The human autonomic nervous system uses built-in knowledge to control heartbeat, breathing, temperature, digestion, reflexes, ability to learn, and the movement of information between the brain and the senses and muscles. Humans cannot survive without this built-in intelligence.

Early automated devices, from clocks and windup toys to very complicated electronic numerical control machines, depended on built-in intelligence for their functioning. Some early robotlike machines depended on built-in intelligence for their functioning, too. The mechanical windup clock, which uses weights or springs and gears, has been around for centuries. The intelligence for keeping track of the time was built into the gears. The modern digital clock has its intelligence built into electronic circuitry. A clock can be thought of as a special-purpose robot whose task is to keep the correct time; some electronic clocks can even accept spoken commands asking what time it is, responding through a speech synthesizer with the time.

Alan Turing, a British mathematician, proposed what is now considered a standard test for machine intelligence. **Turing's definition** of artificial intelligence can be summarized as follows:

1. Place a person in a room with two computer terminals. One of the terminals is connected to a terminal in another room, with another human at the keyboard. The other terminal is connected to a computer in another room, with an artificial intelligence program loaded in its memory.
2. Have the person try to carry on two typewritten conversations over the terminals.
3. If at the end of fifteen minutes, the person cannot tell which terminal connects to the human and which terminal connects to the computer, the computer has demonstrated intelligence.

The Doctor Program The program **ELIZA** was written by Joseph Weizenbaun to show how a relatively simple program could carry on a somewhat meaningful conversation with a human. The DOCTOR version of ELIZA pretends to be a

Table 10-1
Interactive Conversation Between ELIZA and a Human Being

WHAT IS YOUR PROBLEM?
i hate computers!
DO COMPUTERS WORRY YOU?
yes!
ARE YOU SURE?
yes!
PLEASE DON'T REPEAT YOURSELF!
why not?
WHY DO YOU ASK?
say, are you a computer?
WHAT MAKES YOU THINK I AM A COMPUTER?
i don't know.
ARE YOU SAYING NO JUST TO BE NEGATIVE?
of course not!
ARE YOU BEING A BIT NEGATIVE?
what makes you think i am being negative?
DID YOU COME TO ME BECAUSE YOU ARE BEING NEGATIVE?
do you always answer a question with a question?
WE WERE DISCUSSING YOU. . . NOT ME!
i want to talk about you!
OH, I . . . !
you are a computer, aren't you?
YOU'RE NOT REALLY TALKING ABOUT ME, ARE YOU?
yes i am!
SAY, DO YOU HAVE ANY PSYCHOLOGICAL PROBLEMS?
no, but you do!
WE WERE DISCUSSING YOU . . . NOT ME!
i give up!
WHAT DOES THAT. . .

psychotherapist. The program was prepared as a demonstration of artificial intelligence on the party-game level, not as a finished work of artificial intelligence. This program can also be run on the personal computer.[3] The conversation in Table 10–1 between ELIZA and a human was carried on with the ELIZA program executing in a Radio Shack TRS-80 Model I computer. The human responses are given in lowercase letters.

[3]John Krutch, *DOCTOR—Experiments in Artificial Intelligence for Small Computers* (Indianapolis: Howard W. Sams, 1981).

The DOCTOR program of ELIZA works by syntactically tearing apart the statements made by the human and rearranging them as a response to the human. ELIZA not only does not learn, it does not understand a word the human says. The ELIZA program might just as well be firmware or built-in intelligence, since it does not change. An improved version of the ELIZA program could be useful as part of the intelligence for a robot bartender, in that it would be able to listen and react to customers telling it their problems. Since some people are helped just by having a chance to tell someone—or in this case, something—their problems, ELIZA the bartender could be a real help to humanity.

Learning

A better-recognized area of intelligence relates to the ability to learn. This ability is generally assessed on the basis of what one has learned. The study of dogs by Ivan P. Pavlov shows **learning** emerging as a result of repeated cause-and-effect events upon the test animals. In the 1960s, Frederick W. Chesson constructed a Robot Rover that exhibited Pavlovian conditioned-learning responses and extinction responses.[4]

Hexapawn Educable Robot (HER) The process of learning from one's mistakes is demonstrated by the Hexapawn Educable Robot (HER) (Gardner 1962). HER is given all the possible moves for the player who moves second in the game of Hexapawn. The rules are implied, in that only the legal moves are given to HER, although HER does not know a good move from a bad move. A human plays one side of the game and goes first. The game of Hexapawn (six pawns) is played on a three-square by three-square board, using six pawns. At the beginning of the game the pawns are lined up on opposite ends of the boards. White moves first. Like a chess pawn, a pawn in this game can move forward one square or take an opponent's pawn on the diagonal. A player wins by accomplishing one of the following three things: getting a pawn to the third row (the opposite end of the board); capturing all of the opposing player's pawns; or blocking the position so that the opponent cannot move. Figure 10–1a shows the starting position for the game of Hexapawn. Figure 10–1b shows the positions after White makes its first move of moving its center pawn forward one square. Figure 10–1c shows the position of the game after Black, or HER, responds by moving the left pawn past White's center pawn. Figure 10–1d shows the position after White takes Black's pawn and White wins the game by reaching the third row! HER now learns from the mistake of its last move, which resulted in a lost game, and removes that move from its list of possible moves.

The HER robot is composed of a series of matchboxes. Attached to the top of each matchbox is a diagram of the Hexapawn game position after an odd-numbered move has been made (first, third, or fifth), with different-colored arrows

[4]Frederick Chesson, "Robot Rover or Build Your Own Pavlovian Pooch," *Creative Computing* (June 1979): pp. 62–64.

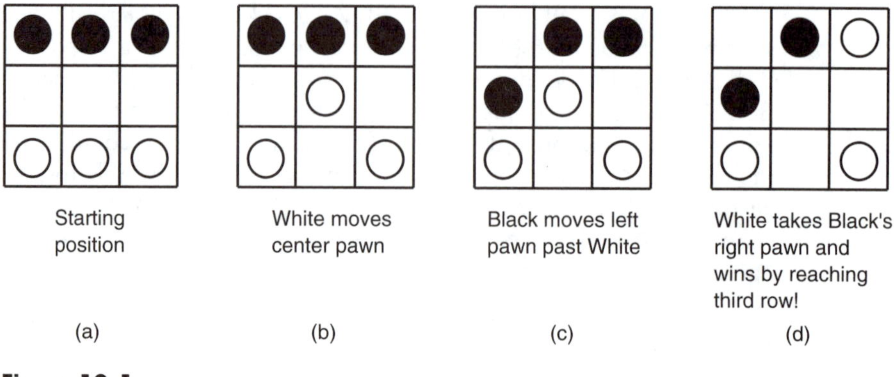

Figure 10-1
A Sample Game of Hexapawn.

pointing out HER's legal moves. You can save about 30 percent of the moves by only taking the right side move of any symmetrical position. This limits the first move to two choices, instead of three. With these limitations, HER can be made from twenty-four matchboxes. Inside each of HER's boxes are different-colored beads—one bead for each arrow on the lid of the box showing a possible move.

When it is HER's turn to move, perform the following steps:

1. Find the box with the present position on it. It helps to number all the second-move boxes with a 2, all the fourth-move boxes with a 4, and so on, and to line up each set of boxes for a particular move in a row.
2. Shake the box and randomly select a bead.
3. Make the move indicated by the arrow that is the same color as the bead.

When a game of Hexapawn is completed, determine who won. If HER lost, remove from its list of possible moves the last move it made (by removing the bead from the box). This is how HER learns from its mistakes—by not being able to make the same bad move twice. HER is unaware that it has won or lost the game; it only knows that it cannot find a legal move with which to continue the game. It takes HER about eleven losses to become a perfect Hexapawn player.

The game of Hexapawn has been programmed to run on the personal computer by several different persons.[5]

If you would like to see two matchbox robots play each other at Hexapawn, try constructing the Hexapawn Intelligent Matchbox (HIM). HIM, which has all the odd-numbered moves of Hexapawn, requires twenty-three matchboxes to construct, using the same guidelines as for HER. HIM requires approximately six losses to become a perfect Hexapawn player of the White pieces; that is, after that many losses, it learns to resign at its first move. (White has a losing game in Hexapawn.)

[5]David H. Ahl, *BASIC Computer Games* (Morristown, NJ: Creative Computing Press, 1979).

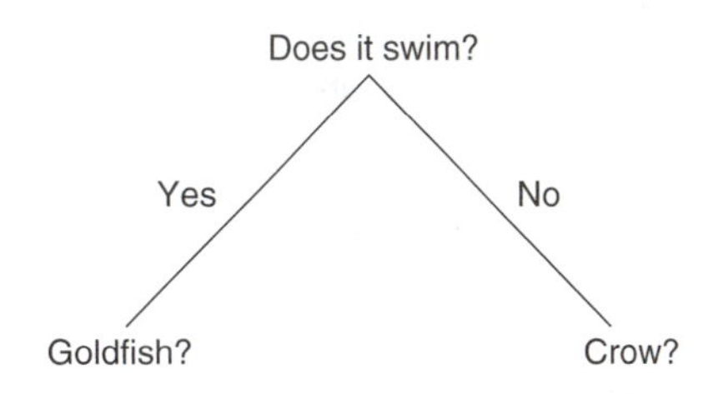

Figure 10–2 The Animal Program's Beginning Knowledge About Animals.
The Animal program starts out knowing only one question, which it uses to distinguish between two different animals.

Unfortunately, HIM usually beats HER only about three times and thus does not complete HER's education.

Donald Michie designed the Matchbox Educable Naughts And Crosses Engine (MENACE) to play tic-tac-toe. MENACE requires 300 matchboxes.[6] The number of boxes could be reduced by recognizing symmetry and by rotating positions.

The Game of Animal Another type of intelligence consists of building one's knowledge on the basis of previous knowledge. The game of **Animal** demonstrates this type of artificial intelligence learning. In our version of Animal, the program's beginning knowledge of animals consists of "Goldfish" and "Crow," along with the question "Does it swim?" The program knows that a goldfish swims and that a crow does not. This knowledge is diagrammed in Figure 10–2.

The computer begins by asking the human player to think of the name of an animal so that the computer can try to guess it. If the computer cannot guess the name (because it isn't in its vocabulary), the computer asks the human player to enter the name of a new animal. The player is then asked (first) to type in a question that would distinguish the new animal from the last animal the computer guessed, and (second) whether the answer to this question is yes or no for the new animal. The computer then saves up all the information given to it and uses the information in future guesses of an animal.

To accomplish this, the computer must integrate the new knowledge into the appropriate open slot in its bank of previous knowledge. If the player enters the command "list" in response to the question "Are you thinking of an animal?" the computer will provide a list of the animals it now knows. This game has been programmed for the personal computer.[7]

Table 10–2 shows the results of a sample session with the program Animal. The program's responses are given in all capitals, while the human player's responses are given in lowercase letters. Figure 10–3 shows what the program's final animal knowledge diagram has become. Caution should be used when trying to educate the naive program Animal. Like a small child, it is trusting and gullible and believes whatever it is told. If care is not taken, its **knowledge diagram** could end up looking like the one in Figure 10–4.

[6]Donald Michie, *Trial and Error* (Penguin Science Survey, Vol. 2, 1961).
[7]David H. Ahl, *BASIC Computer Games* (Morristown, NJ: Creative Computing Press, 1979).

Table 10-2
Sample Run of the Animal Program

```
ARE YOU THINKING OF AN ANIMAL?
  yes
DOES IT SWIM?
  no
IS IT A CROW?
  yes
WHY NOT TRY ANOTHER ANIMAL.
ARE YOU THINKING OF AN ANIMAL?
  yes
DOES IT SWIM?
  no
IS IT A CROW?
  no
THE ANIMAL YOU WERE THINKING OF WAS A?
  horse
PLEASE TYPE IN A QUESTION THAT COULD DISTINGUISH A HORSE
FROM A CROW.
  does it have four legs
FOR A HORSE THE ANSWER WOULD BE?
  yes
ARE YOU THINKING OF AN ANIMAL?
  list
ANIMALS I ALREADY KNOW ARE GOLDFISH   CROW   HORSE
ARE YOU THINKING OF AN ANIMAL?
  yes
DOES IT SWIM?
  no
DOES IT HAVE FOUR LEGS?
  no
IS IT A CROW?
  no
THE ANIMAL YOU WERE THINKING OF WAS A?
  snake
PLEASE TYPE IN A QUESTION THAT WOULD DISTINGUISH A SNAKE
FROM A CROW.
  does it have scales
FOR A SNAKE THE ANSWER WOULD BE?
  yes
ARE YOU THINKING OF AN ANIMAL?
  yes
```

(continued)

Table 10-2 *(continued)*

DOES IT SWIM?
  yes
IS IT A GOLDFISH?
  no
THE ANIMAL YOU WERE THINKING OF WAS A?
  whale
PLEASE TYPE IN A QUESTION THAT WOULD DISTINGUISH A WHALE FROM A GOLDFISH.
  does it breath air
FOR A WHALE THE ANSWER WOULD BE?
  yes
ARE YOU THINKING OF AN ANIMAL?

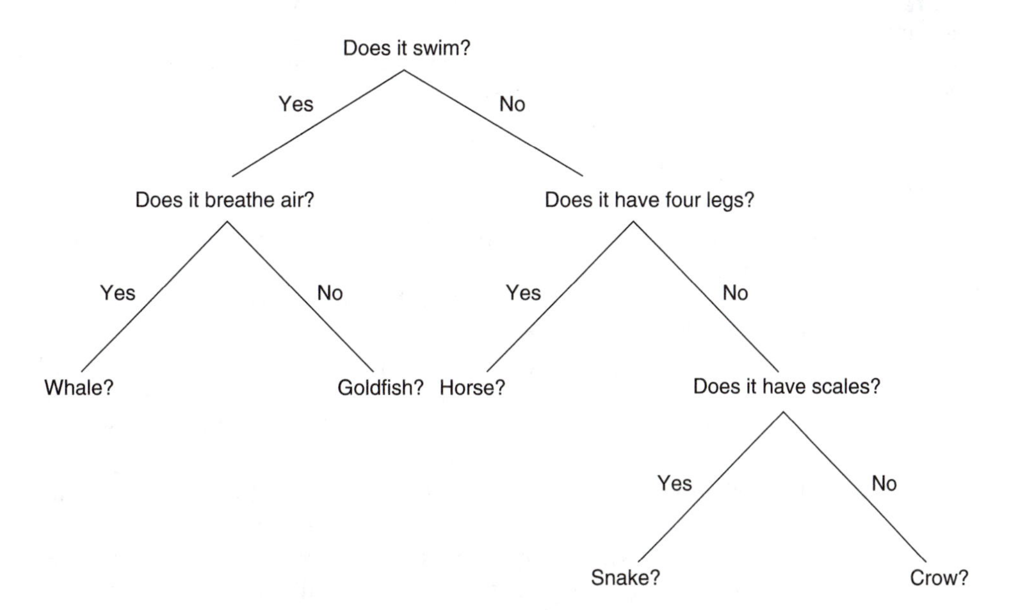

Figure 10-3 The Animal Program's Ending Knowledge About Animals.
At the end of the session described in Table 10–2, Animal knows four questions, which it uses to distinguish among five animals.

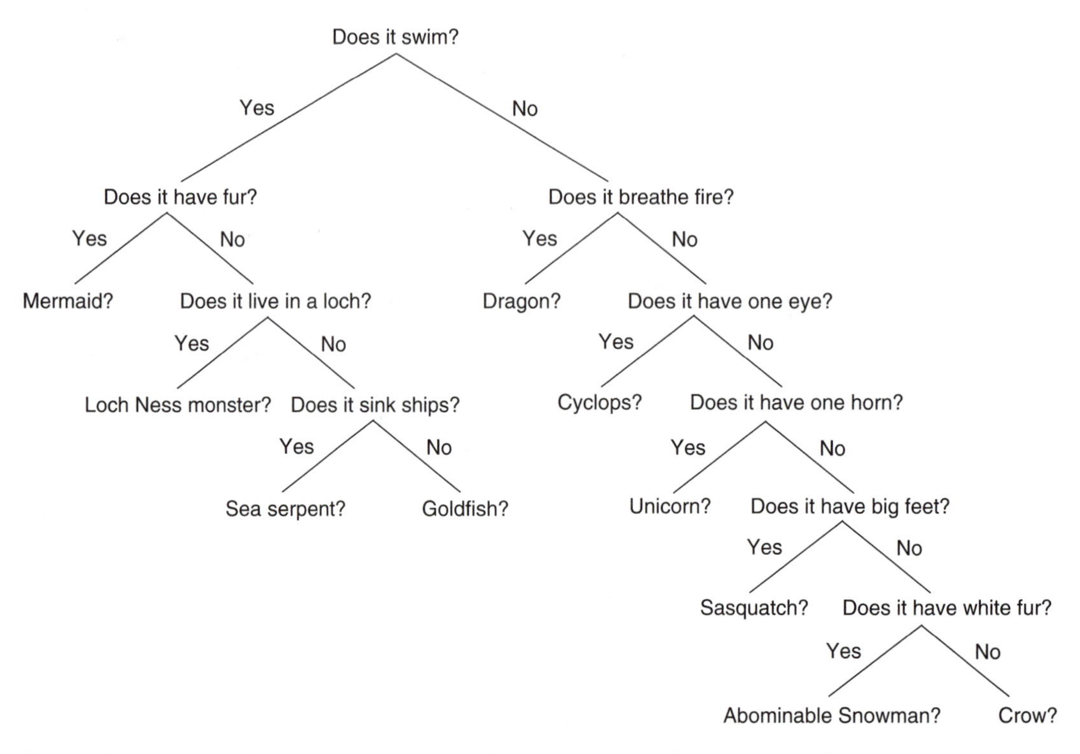

Figure 10–4 The Animal Program's Educational Results from an Unreliable Instructor.
The Animal program's knowledge is no better than the instructor who is teaching it. In addition, Animal does not know when the instructor is just fooling.

EXPERT SYSTEMS

An **expert system** is a computer program containing information that represents a condensation of some small part of human knowledge. An older, manual form of an expert system is the troubleshooting table given in a user's manual. A complete book of tables and procedures for troubleshooting a computer system is also known as a diagnostic decision logic table (DDLT).

A robot could use an automated expert system to distinguish between different parts that it may pick up off an assembly line. Robotic system engineers are already using expert systems to decide which robot is best for performing a particular task and to help troubleshoot broken robots. Expert systems will soon be used to schedule factory production and maintenance around disabled or busy robots.

An expert system consists of four parts: They are an inference engine, a knowledge base, a database, and a natural language interface. The inference engine, or rule interpreter, matches facts and rule conditions; thus, it is the controller of the system. The knowledge base contains the rules. The database collects the facts given to the system by the user. These facts then guide the inference engine in deciding what rules to apply to the problem. The natural language interface allows the user to communicate with the expert system in a natural language such as English, French, or German. The expert system operates in one of two modes: the knowledge acquisition mode, which is used to enter rules into the system, and the consultation mode, which the user refers to in solving a problem.

The logic behind the game of Animal is the same as the logic used in an expert system. When the Animal expert system asks for and receives answers to questions like "Does it swim?" or "Does it breathe air?" it is operating in the consultation or user mode. When the Animal system makes requests like "Please type in a question that would distinguish a snake from a crow," it is operating in the knowledge acquisition mode. Thus, in the game of Animal, the computer is attempting to construct or add to its expert system of animal classification. Figure 10–5 shows a diagram of an expert system for the game of Animal.

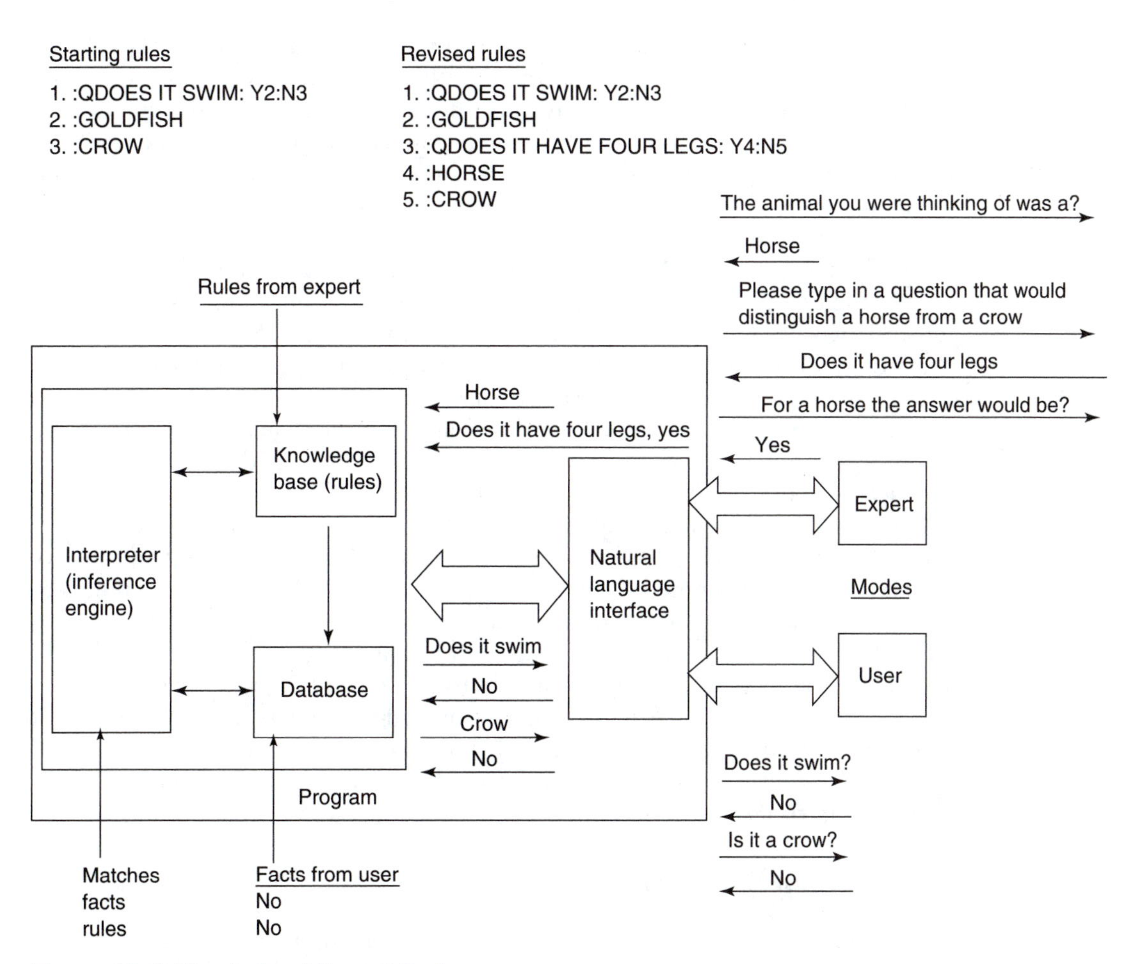

Figure 10–5 The Animal Expert System.
The Animal program is a type of expert system. It operates in the user mode while working its way through the knowledge base. It operates in the expert mode while learning new facts.

Once such a system of information is established, other users could work with it to learn to distinguish between different animals. A similar system could be constructed to help students distinguish between different states of the United States.

Digital Equipment Corporation is using expert systems to help its personnel place computer orders for customers. Expert systems are either in use or under construction in the fields of medicine, publishing, geology, and construction.

The expert system for the game of Animal works with complete and certain information. The MYCIN expert system, constructed in the mid-1970s by Stanford University, was the first program to work with uncertain or incomplete information. MYCIN's purpose was to diagnose and prescribe treatment for spinal meningitis and bacterial infections of the blood.

LEARNING TO UNDERSTAND HUMAN LANGUAGES

Two areas of artificial intelligence deal with understanding human languages: voice, or speech, recognition, and natural language understanding. **Voice, or speech, recognition** deals with converting the spoken language of humans into recognizable patterns within the computer. **Natural language understanding** involves accepting commands in a human language, such as English or Japanese, and translating these commands into the computer's own machine language. A voice recognition system or a keyboard may be used to get the words into the computer; then the natural language understanding program finds the meaning for the words.

Computer machine languages are very precise in meaning, while natural human languages tend to be inexact. The same English sentence conveys different meanings to different persons. In fact, the same person may give different meanings to the same sentence under different circumstances. Part of the reason for the inexactness of natural languages is that the meaning of the sentence is influenced by the context around the sentence. Unfortunately, the computer does not necessarily have access to (or insight into) this context.

One of the immediate goals of natural language understanding is to build a 1,000-word speaker-independent system. This goal will tax the processing capacity of today's biggest computers, which can perform only 200 million operations per second. An ultimate goal for natural language understanding is a 10,000-word vocabulary, with continuous voice input and output. It is estimated that this will require approximately 20 billion instructions per second of computer processing time. Should a person desire to build a limited-vocabulary speech-recognition system for use in a fighter aircraft, with noise levels of up to 115 decibels and acceleration of several Gs, the computer would only need to process approximately 40 million instructions per second.

INTERPRETING SENSORY INPUTS

Sensory input interpretation is an area in which artificial intelligence is deeply involved. Research into how the human robot hears, understands speech, and interprets vision continues to amaze the researchers and to give them a greater appreciation for the human ability to process masses of information every second.

Most humans depend on vision for feedback information on many tasks that do not actually require vision. A person who is learning to drive a car, for example, tends to look down at the shifting of the gears instead of continuing to watch the road. Experienced drivers, on the other hand, can shift gears by feeling, without looking. They do not even need to look at the speedometer to know when to shift gears, since they can tell this by the sound of the engine. Humans also have feedback devices in their arms and hands, in the form of pressure (or feeling) and position (or kinesthesia).

Most of today's industrial robots do not have vision and feeling for feedback, although many do have positional feedback. Primitive vision systems are just moving out of the artificial intelligence laboratory and into widespread use in industrial robots, even though the first industrial robot with a single-line television system was installed back in the early 1960s. Learning to handle even simple vision in **real time**—that is, as the event is actually taking place—requires a large amount of computer processing time and memory. Artificial intelligence research is continuing to develop more efficient ways to do vision processing, at the same time that faster computers are being developed.

Speech Interpretation

The interpretation of speech by human beings may seem rather simple, but it probably uses even more processing power than is required for vision interpretation. Early voice recognition worked on **voice prints**—digital recordings of short pieces of speech. These prints are analyzed for the length of a word and for the frequencies used in the word. Each voice print is as distinctive as a fingerprint, and each must be memorized for every person from which the computer or robot receives voice input. As the number of words in the vocabulary grows and the number of persons using the system increases, the amount of memory required for storing the voice prints expands geometrically. Meanwhile, however, the time needed to search the voice prints for a match goes up exponentially. Thus voice print systems are impractical for handling large vocabularies.

Another style of voice recognition makes use of phonics to recognize words. This requires storing all the known phonemes of speech in a particular language. **Phonemes** are the smallest units of speech that distinguish one utterance from another. Next, each word that is to be included in the vocabulary must be represented in these phonemes. When voice input is received, the input must be broken down into a string of phonemes, with the computer making its best guess at matching the utterances to the phonemes. The computer must also find the breaks between each word.

At this point, the computer takes the string of received phonemes and compares them to known word-phoneme patterns. If all else fails, the computer may have to ask for the meaning of a new word or for repronunciation of the word. If it is a new word, the computer needs to add the phonemes and definition to its dictionary. Again—although the basic sounds for the dictionary are much smaller than for voice prints—as the dictionary grows, the computer's speed of understanding necessarily decreases.

Human beings use phonics in their speech recognition system. They increase the speed of their speech processing by assuming what they expect to hear. This technique occasionally leads to misguesses as to what is coming, and the hearer may have to ask for that part of the speech to be repeated. In fact, all of a person's past experiences are available for reference in speech processing.

Vision Interpretation

Vision interpretation is another area with high processing requirements. Human beings do their vision processing in stages. The human eye does the edge-detection processing and probably screens out everything but the new visual data from being sent on to the brain at all. This process gets fast results by assuming that the eye is seeing what it expects to see. Here again, humans depend on their previous experiences to understand what they are currently seeing. Humans also have the ability to partition their field of vision into several areas, or windows of vision. Generally the central field of vision gets the most attention; it is referred to as the center of one's attention.

The most primitive organ of vision for a living creature uses a single-light detector as an eye. The Hermissenda marine snail has two eyes, each of which consists of a single light detector. It uses its eyes to move toward the light, where the sea is calmer. The only information this eye can convey is about the level of light striking it. If the information from both eyes is combined, it becomes possible to tell (by comparing the light levels received at each eye) which direction to turn toward to find brighter light.

Like a Hermissenda snail, a robot with a single eye could, with the aid of an analog-to-digital converter and a simple search program, turn its eye toward the brightest source of light. It could then track the light source. It is also possible to make a line-following robot that uses two such eyes, spaced apart by just under the width of a line. In this case, the analog-to-digital converter would not be needed. The robot could simply use a comparator integrated circuit.

If an electronic eye could detect only the two states of light and dark, the eye could be used with a light source to warn the robot when it was approaching an object. A pair of these eyes could be used to detect the edge between black and white under high-contrast conditions. They could also be used to help locate an object that a robot was trying to find and pick up. If the robot could interpret only light and dark signals from the eyes, it would be able to detect the four conditions shown in Figure 10–6. The condition of both eyes seeing dark would mean that the robot path was headed entirely off the target. The condition of one eye light and one eye dark would mean that the robot's path was headed toward one or the other edge of the target. And finally, both eyes seeing light would mean that the robot's path was centered on the target.

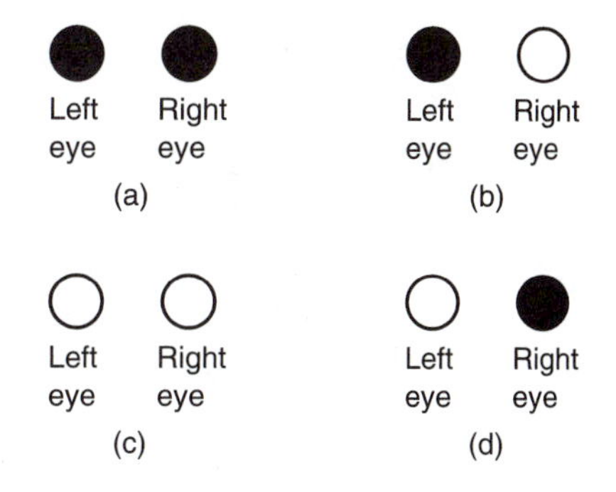

Figure 10–6 Two Two-state Simple Eyes.
Two two-state simple eyes are sufficient to enable a robot to follow a white target on a black background.

Three eyes of the above type could be used on a robot to enable it to follow a white line on a black background. The added eye would give more room for slight steering errors. Figure 10–7 shows the six possible states of these eyes. This vision arrangement would be useful for a robot (such as a mail-delivery robot) that was intended to follow a predefined course.

If the two eyes can detect an additional state of gray light halfway between light and dark, they can sense eight possible conditions (see Figure 10–8).

Most vision input sensors for industrial robots use some type of television camera. The signal from the camera is then digitized and placed in computer memory for processing.

Based on experimentation, it has been estimated that a robot vehicle traveling at a walking pace and trying to move down a hall with unknown obstacles in its path must manage about 1 trillion instructions per second to process its vision. Currently, no robot can do this type of processing. In his 1981 Ph.D. thesis at Stanford University, Hans P. Moravec mentions a visually guided robot rover developed by the Stanford AI Lab. The robot is about the size of a card table and is designed to work its way through a cluttered space. Its eyes are a three-dimensional

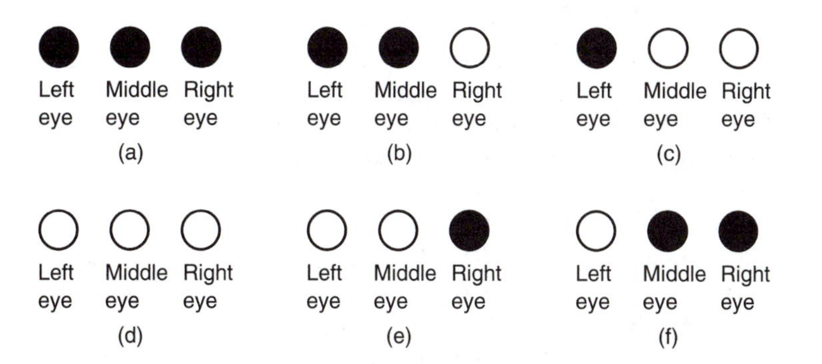

Figure 10–7 Three Two-state Simple Eyes.
Six possible states of three two-state simple eyes, giving a robot even more control in following a white-on-black target.

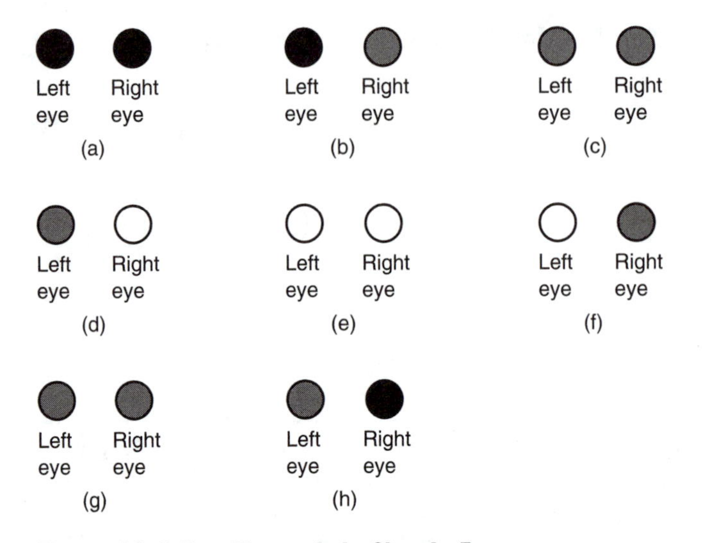

Figure 10-8 Two Three-state Simple Eyes.
Eight possible conditions of two three-state simple eyes, giving the robot still more flexibility in following a white-on-black target.

television system. The cart moves 1 meter at a time and then stops to look around and analyze its surroundings. It requires 10 to 15 minutes to process what it sees before taking the next step. Consequently, the robot takes about 5 hours to travel through a 20-meter course. Indoors, with good lighting, the cart is very successful at avoiding obstacles. Outdoors, where the lighting casts shadows, the cart is less successful and only avoids about two out of every three obstacles.

INTELLIGENT TUTORING SYSTEMS

If a computer can understand how humans think, why can't it understand where human thinking goes wrong? This is the premise for the articles in the book *Intelligent Tutoring Systems.*[8] Some primitive attempts were made at intelligent tutoring in the programmed textbooks of the early 1960s. Based on a person's wrong answer, the book would refer the person to a section of knowledge that had not yet been mastered. But the book did not pretend to know why the person had not mastered the material. Our present-day computer systems are able to store much more information than a book possibly can, and the same computer program and information can be shared by more than one person at a time.

In many cases where faulty logic is responsible for a person's erroneous answer to a problem, it is possible to tell from the person's answer what went wrong

[8]Sleeman and Brown, *Intelligent Tutoring Systems* (New York: Academic Press, 1982).

in the problem solution. Thus, a math instructor can tell a student what step was faulty in an answer if the student gives a step-by-step solution. In high-school algebra, most problem errors come from the misuse of a known rule in some new situation. Often, the same error is made by more than one person, since misusing the same math rule in the same way will produce the same wrong answer. By analyzing the wrong answers, it becomes possible to determine where a rule was misapplied, without having to see all the steps of a person's solution. Programs have been written that can find these misapplied rules and then not only point out what the mistake is but explain how it can be corrected. These programs can be used in any subject that follows a logical application of rules.

WAYS THAT ARTIFICIAL INTELLIGENCE MAY HELP ROBOTS

It is one thing to be able to build a sensor, such as a television camera for vision input, and another thing to be able to interpret the information received from that sensor. Giving the robot improved sensory input and the ability to interpret those inputs will expand the range of tasks that the robot can perform and will simplify the job of training the robot for the tasks. The sensors for vision, tactile sensing, and voice recognition are all receiving intensive study.

At present, robots work in a real-time world where they must interpret sensory input in time to use the information in their present task. Artificial intelligence programs work at a batch-level mode. A **batch mode** of operation means that a device processes the information at its own convenience, which means not right away. Consequently, it cannot process the sensory input in time to use it in the immediate task. By the time a present-day artificial intelligence program could process the vision input to warn a robot that it was about to fall off a cliff, the robot would already have fallen off the cliff and hit the bottom. Artificial intelligence devices must become faster and less expensive if they are to give practical senses to robots.

At the present time, a self-navigating robot requires hours to navigate a cluttered path that a human can navigate in seconds to minutes. It is just that difficult to process and interpret vision system information at this time. The controller of the robot needs to increase greatly its speed of operation and size of its memory. Robots need to build up past experiences to make better use of their intelligence. The problems will decrease with time.

An industrial robot with speech recognition would be easier to train or program; a very limited vocabulary would be sufficient for training, teaching, or programming the robot. For example, a list of key words for a robot to understand might consist of the following: *ahead, all, behind, close, enter, hand, higher, home, left, lower, open, right,* and *run.*[9]

[9]Jorg R. Jemelta, "Designing a Reliable Voice-Input Robot Control Language," *Robotic Age* (February 1984): 30–34.

Flexibility

As research in artificial intelligence provides robots with new and improved sensor capabilities, the robots will become less dependent on their environment's being perfectly controlled. A misplaced or wrongly oriented part could be taken into account or corrected by the robot. The robot would no longer need to know the exact location of things for each phase of a task, but just their general position. The sensors of the robot could be used to adjust the robot's readings of the precise positions, as necessary. This would also allow the robot to do much more complicated tasks in the field of parts assembly. In addition, the robot's training would be simplified, since the mechanics of taking minor corrective actions would not need to be included in the training.

With simplified training techniques and improved sensory skills, the robot would be able to perform easy assignments around the home, such as mowing the lawn and taking out the garbage. House painting and nursemaid monitoring are other possibilities.

All of these new uses for the robot will require improved sensors and increased memory capability. Any advances in these areas will give the robot more flexibility than it currently has.

Artificial intelligence will enable the robot to work with incomplete and uncertain knowledge, help management select the right robot or machine for the job, and allow robots to diagnose unexpected problems and work around them.

Fuzzy Logic

One way artificial intelligence will change the robot's controller is by allowing it to handle problems that do not have exact solutions. These require the controller to work with **fuzzy logic.** Fuzzy logic problems deal with situations that have several gray areas and may have several reasonable solutions. The objective is to find the best of these possible solutions. Many of the problems people deal with require solutions that are only close enough. The use of fuzzy logic will greatly increase robots' flexibility.

Automated Factory

The ideal **automated factory** takes in raw materials at one end and outputs finished goods at the other end, without the aid of human labor. Humans will provide their special talents of reasoning, designing, and decision making, rather than manual labor. They will set up the supply of raw materials, tell the factory what is to be produced, and arrange for shipping the finished goods.

The automated factory will consist of production equipment, robots, and material-handling equipment—all computer controlled. In fact, most of the equipment will have built-in computers. The management within the automated factory will be expert systems. Supervisor computers will schedule the equipment computers, while a central management computer will schedule the supervisor computers. The company's human management in turn will control the central management computer. The expert systems will allow the factory to route production around defective machines, or to route secondary production through less efficient equipment if primary production already occupies the best equipment.

All the computers and equipment used in the automated factory must have a higher level of intelligence than is now available for computers, robots, and equipment.

At least in its early stages, the automated factory's ability to produce any product at will is quite dubious. More likely is a factory that can produce just one product or one closely related family of products.

Currently, between 5 and 15 percent of most manufacturing processes requires the use of humans. This portion of these processes does not lend itself to robotics or automation. However, reevaluation and redesign of various products could lead to the elimination of this nonautomatable work. Such events will not happen, though, as long as it is more economical to use humans than to use machines for some product processes. The bottom line for manufacturing remains maximum possible profit.

SUMMARY

Most industrial robots of today have poor sensors or none at all. Improving robot sensors will greatly increase the usefulness and flexibility of the industrial robot. Artificial intelligence is concerned with making machines that think and sense.

There are three levels of intelligence: alpha level, or nonlearning intelligence; beta level, or simple ability to learn; and gamma level, or ability to make generalizations from past experiences. Alpha-level intelligence is built-in intelligence and can be demonstrated by a hand-held chess-playing game. Beta-level intelligence is the ability to learn from past mistakes or to build on past knowledge. It can be demonstrated by the game of Animal or the Hexapawn Educable Robot (HER). Gamma-level intelligence is the ability to formulate rules of thumb on the basis of prior experience.

Artificial intelligence is dedicated to making computers, and eventually robots, easier to use by teaching them to understand human languages, also known as natural languages. Research in artificial intelligence focuses on interpreting sensory input for computers and robots; it will lead to improved speech and vision interpretation abilities.

In the foreseeable future, artificial intelligence will yield increased flexibility in robot use. The fully automated factory, which works without human labor, is on the horizon. An important element of this improved flexibility will come from the future ability of robots to work with fuzzy logic—addressing problems with incomplete information and giving answers that are close enough.

REVIEW QUESTIONS

1. Will the robot become a new life form? Explain.
2. What is artificial intelligence?
3. Will artificial intelligence ever exceed human intelligence? Explain.
4. What are some educational uses for an expert system program such as Animal and States?
5. If you had the choice between giving a robot speech interpretation abilities or vision interpretation abilities, which would you choose? Why?
6. What applications might primitive vision help a robot accomplish?
7. Starting from the Animal game knowledge of Figure 10–2, construct the knowledge needed to distinguish a dog, a cat, a chicken, and a monkey. Animal is an expert system builder. What rules does the system contain? What facts does the system receive as it works its way through the rules for distinguishing a chicken?
8. Suppose that your car won't start. List the rules that you might need in an expert system in order to find out what is wrong with your car. In other words, what questions do you need to ask, and how do you respond to the answers for each question?
9. List some of the intelligence that would be needed by a robot guard.
10. What intelligence is needed by an office or hospital, mail and/or food delivery robot?
11. Why would a Mars lander robot need the intelligence to make some of its own decisions rather than letting humans make all the decisions for it from Earth?
12. Does a vacuum cleaning robot need the intelligence to remember a room's layout or just the intelligence to avoid things?

11

CLASSIFICATION OF ROBOTS

OVERVIEW

This chapter discusses the following ways of differentiating among robots: by arm configuration, by type of controller, by type of power supply, by task being performed, by design generation, by type of motion (LERT), and by parts not found in most computers.

OBJECTIVES

When you have completed this chapter, you should be familiar with:

- How to classify robots by several different methods
- What developments are likely to appear in future generations of robots

KEY TERMS

Arm configuration
Controller type
First generation of robots
LERT
1.5 generation of robots
Power supply type
Second generation of robots
Technology used
Third generation of robots
2.5 generation of robots

Robots can be classified by several different methods: arm configuration; shape of workspace (cell); operating method; type of controller; type of power; size; type and number of joints; type of technology; task being performed; period or generation of design; and LERT (*l*inear, *e*xtensional, *r*otational, or *t*wisting) motion.

CLASSIFICATION BY ARM CONFIGURATION

There are five generally recognized arm configurations: rectangular coordinates, cylindrical coordinates, SCARA, polar coordinates, and jointed arm. Chapter 3 discusses and illustrates these arm configurations. Rectangular-coordinates robots are also known as Cartesian-coordinates robots. Polar-coordinates robots are also known as spherical-coordinates robots. Jointed-arm robots are also known as revolute-coordinates, articulate, or anthropomorphic robots.

CLASSIFICATION BY CONTROLLER

There are three basic types of robot controllers: limited sequence, point to point, and continuous path. Limited-sequence controllers use a timer to determine when to activate an axis, and they use mechanical stops or microswitches to mark the end of travel for an axis. Point-to-point controllers have memory to hold the coordinates of each axis at various points in a task, and they use sensors to help reach each point. Continuous-path controllers have larger memories than point-to-point controllers and are able to record many point coordinates per second.

CLASSIFICATION BY POWER SUPPLY

There are three basic types of robot power supplies: electrical, pneumatic, and hydraulic. If the arm of a robot is powered by electricity, it is considered to be an electric robot; if the arm is powered by hydraulics, it is considered to be a hydraulic robot. The wrist and gripper, however, may not necessarily use the same type of power as the robot arm.

CLASSIFICATION BY LEVEL OF TECHNOLOGY

Currently, three levels of technology are used for robots: low-level, medium-level, and high-level technology. This is a very fuzzy way to classify robots, and the definitions used may change at any time. Obviously, a high-technology robot of today may well be a medium- or low-technology robot in ten to twenty years.

Low-technology robots are used for material handling, for machine loading and unloading, and for very simple assembly operations. These robots usually have two to four axes of movement and stop only at the extremes of the axis. The wrist may not add any axis of movement. Low-technology robots are nonservo-controlled and usually have a payload capacity of 25 pounds or less. Pneumatic-powered pick-and-place robots fall in the low-technology class.

Medium-technology robots are used for picking and placing and for machine loading and unloading. These robots have four to six axes of movement, some of which come from the wrist. A medium-technology robot can handle payloads of up to 300 pounds.

High-technology robots can be used for material handling, machine loading and unloading, painting, welding, and for many other manufacturing tasks. These robots may have from six to nine axes of movement and can handle payloads of 300 pounds or more. Most continuous-path controller robots are high-technology robots.

CLASSIFICATION BY TASKS DONE

Classifying robots by the task they are designed to do is similar to classifying human workers by their occupations. Some industrial robots work as painters, welders, assemblers, and material transferers. Others work as brain surgeons and artists. A basic problem with this classification system, however, is that robots are designed to be general-purpose machines, and they can change their task simply by changing their end-of-arm tooling and being reprogrammed.

On the other hand, if you buy a wire-feed flat-welding robot, you will have trouble adapting it to do some other task, because of the support equipment (such as the inert-gas bottles and lines, and the wire-feed mechanism) for the welding task—it cannot even be adapted to do spot welding.

Of course, the less flexible or adaptable the robot is, the more likely it is to be considered a form of hard automation rather than an industrial robot. For example, robovans or automatic guided vehicles may well be called automated machine tools.

CLASSIFICATION BY DESIGN

Classifying robots by their technological generation is similar to how electronic computers have been classified. The first generation of electronic computers used vacuum tubes; the second generation used transistors; and so forth.

Because this system of classification is open-ended, it leaves room for future expansion.

First Generation of Industrial Robots

The **first generation of industrial robots** used fixed-sequence programs. They had to be reprogrammed before they could do another job—although this reprogramming might be through retraining, depending on how advanced the individual robot was. These robots were and are used mainly to perform pick-and-place operations. They were basically deaf, dumb, and blind; that is, they had no sensors. They also lacked the ability to make decisions. All control was through an open-loop controller system, where the output had no effect on the inputs to the system. The simplest of these robots actually had their programming sequence set up mechanically on the basis of mechanical stops, limit switches, and positioning pegs on a timing drum.

1.5 Generation of Industrial Robots

The **1.5 generation of industrial robots,** the state-of-the-art robots of today, have some sensor-controlled actions, giving them enhanced capabilities. These robots are able to make minor positional corrections to adapt to varying situations through the use of a closed-loop servo controller that uses the outputs as part of the inputs.

In many cases, the robot is able to tell when something is wrong beyond its ability to correct the situation. The robot then requests human help.

Second Generation of Industrial Robots

The **second generation of industrial robots** will have hand-and-eye coordination control. Some of these capabilities are now in an experimental stage. The capabilities may include mobility, voice recognition commands, vision, touch, multiarm action with hand-to-hand coordination, flexible hands, and microprocessor intelligence. These robots will be able to make their own decisions, correcting minor misalignments of parts or of the manipulator.

2.5 Generation of Industrial Robots

The **2.5 generation of industrial robots** will have perceptual motor function. That is, they will respond to sensory stimuli for control of motion, rather than depending on prerecorded motion. These robots will only require the sequence of steps that are needed to perform a task; they will then determine for themselves what must be done to accomplish the desired steps.

Third Generation of Industrial Robots

The **third generation of industrial robots** will have the intelligence to handle discrete parts assembly, and they will be able to make decisions on how to do their tasks. These robots will require only generalized information about what task needs to be performed.

THE LERT CLASSIFICATION SYSTEM

The LERT classification system uses the type of motion produced by each robot axis as a basis for classifying the robot. The four basic motions are *l*inear, *e*xtensional, *r*otational, and *t*wisting (LERT). Figure 11–1 shows a linear motion of the type that might be seen on a rack and pinion. Figure 11–2 shows an extension motion such as that which occurs when one part of a robot arm slides inside another part of the arm. Figure 11–3 shows a rotational motion such as that found when a part turns at something other than its center; this is something like the arm bending at the elbow. Figure 11–4 shows a twisting motion, which may be seen when a part turns about its center; this is something like the turning of a human neck joint. For a rectangular robot with a wrist that has three rotational axes and with a gripper that has linear motion, the classification would be L^3R^3L, as shown in Figure 11–5. Figure 11–6 shows how a Rhino robot mounted on an *xy*-coordinate table is classified.

The LERT system allows an additional term in the LERT number to identify the type of controller being used by the robot. For instance, you could have a pick-and-place L^3R^4.

Figure 11-1 Linear Motion.
Linear motion is produced by a part moving along the outside of another part, as in a rack-and-pinion system.

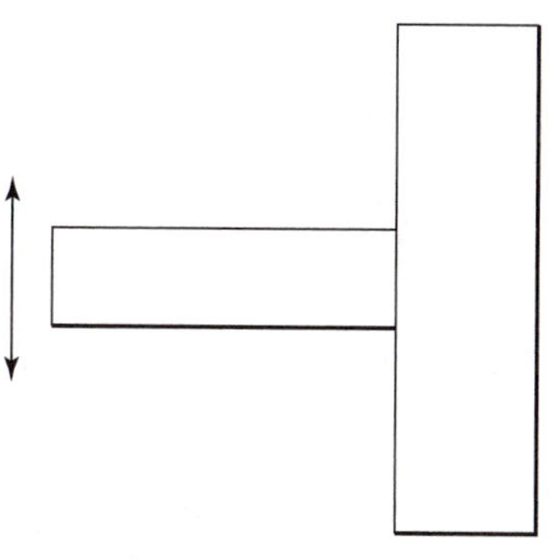

Linear motion

Figure 11-2 Extensional Motion.
Extensional motion is produced by one part moving within another part, with a telescoping motion.

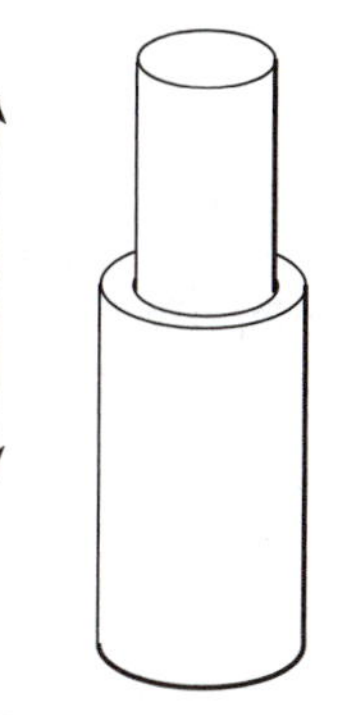

Extensional motion

Figure 11-3 Rotational Motion.
Rotational motion is produced by a part turning about something other than its center.

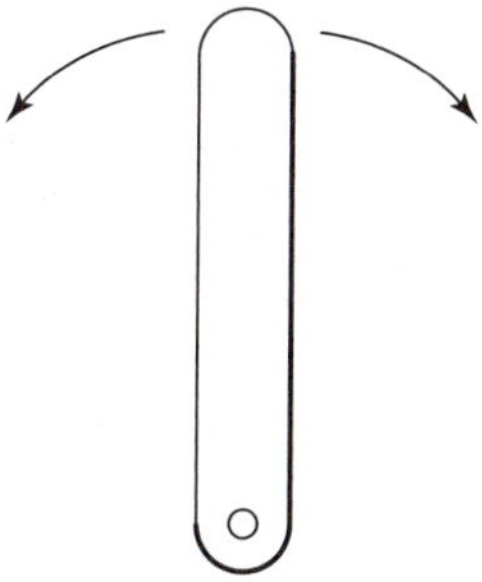

Rotational motion

Figure 11-4 Twisting Motion.
Twisting motion is produced by a part turning about its center.

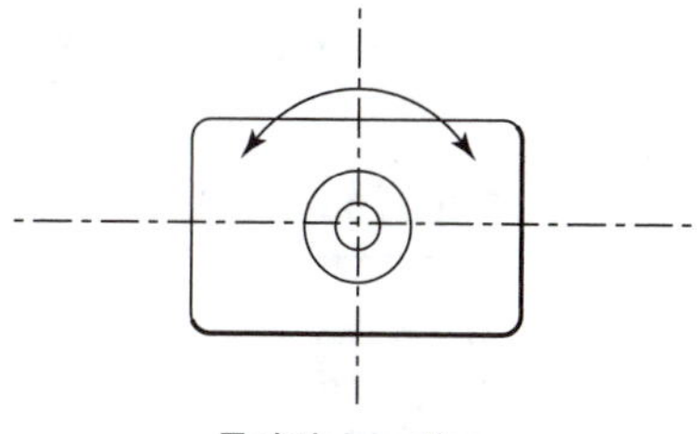

Twisting motion

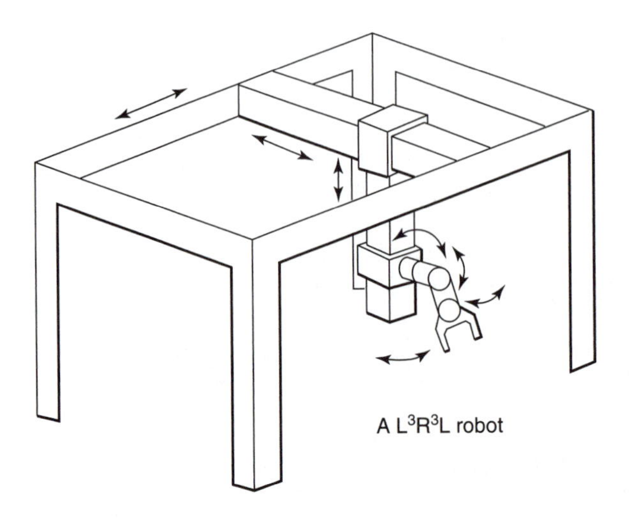

Figure 11-5 An L^3R^3L.
A gantry robot moves with three linear motions, has a wrist with three rotational motions, and has a gripper with one linear motion.

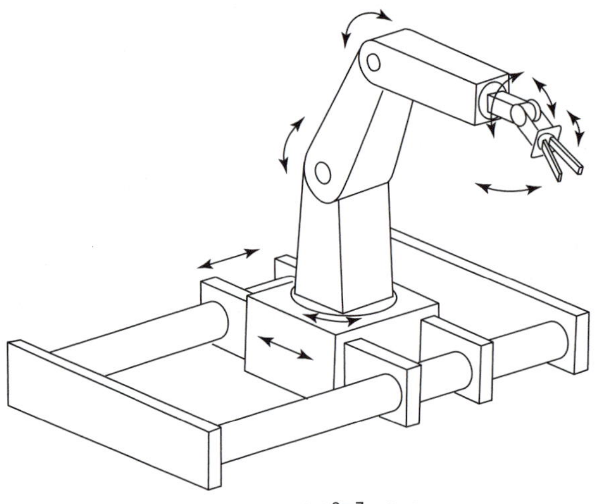

Figure 11-6 An L^2R^7 Robot.
The Rhino robot, mounted on an *xy*-coordinate table, has two linear motions for the base and seven rotational motions for the arm, wrist, and gripper.

SUMMARY

Robots can be classified in several ways: by size, by type of power, by type of controller, by arm configuration, by number and types of joints, by technology used, by task performed, and by period when they were designed. The design period is also known as the generation of the robot.

First-generation robots used fixed-sequence programming. The 1.5 generation of robots (which is where we are now) have some sensor-controlled actions. The second generation of robots will have hand-and-eye coordination. The 2.5 generation of robots will have perceptual motor function. The third generation of robots will have the intelligence to handle discrete parts assembly.

The LERT (*l*inear, *e*xtensional, *r*otational, and *t*wisting) classification system uses the type and number of joints as the basis for classifying the robot.

Table 11–1 shows a comparison of different robotic arms and their reaches, along with their LERT classification. Table 11–2 shows the arm payload versus power supply used for the arm.

Table 11-1
Arm Configuration, Reach, and LERT Classification

Arm Configuration	Reach	LERT Classification
Cartesian coordinates	Only in front of itself	L^3
Cylindrical coordinates	Around itself	L^2R^1
Polar coordinates	Above, below, and around itself	L^1R^2
SCARA (jointed-arm horizontal)	Around itself or around an obstacle	R^3
Revolute (jointed-arm vertical)	Above, below, and around itself, and reach over or under an obstacle	R^3

Table 11-2
Power Source Comparison

Power Source	Payload	Cleanliness
Pneumatic	Light	Semi-clean
Electric	Light to heavy	Cleanest
Hydraulic	Heavy	Dirty

REVIEW QUESTIONS

1. Using the LERT classification system, how would you describe the Hero I robot?
2. How would you classify the Hero I robot by arm configuration? (The Hero I robot is shown in Figure 21–6.)
3. How would you classify the Hero I robot by type of controller? (The Hero I robot is shown in Figure 21–6.)
4. How would you classify the Rhino robot by type of power supply? (The Rhino robot is shown in Figure 21–9.)
5. What is the LERT classification for the FANUC robot in Figure 7–1?
6. What is the LERT classification for the FANUC robot in Figure 7–2?
7. What is the LERT classification for the FANUC robot in Figure 7–5?
8. What is the LERT classification for the Cincinnati Milacron robot in Figure 7–6?
9. What is the LERT classification for the FANUC robot in Figure 7–7?
10. Why would a hydraulic-powered robot be a bad choice for working in a clean room?
11. A robot is to unload a machine in front of itself and deposit the part to a conveyor belt behind itself. What configuration of robot arm could do the task?

12

JUSTIFYING THE USE OF ROBOTS

OVERVIEW

Before any robot gets a chance to prove itself on the production floor, it must prove itself on paper. From an economic standpoint, this involves calculations of the payback period, return on investment, and discounted cash flow.

How is its payback period calculated? How is its return on investment calculated? How is discounted cash flow calculated? How do people react to robots? These are some of the questions that will be discussed in this chapter.

OBJECTIVES

When you have completed this chapter, you should be familiar with:

- Areas that could profit from the use of a robot
- How to justify a robot's cost through payback-period calculations
- How to justify a robot's cost through return-on-investment calculations
- How to justify a robot's cost through calculations of discounted cash flow
- How to justify a robot's use on grounds of improving worker safety
- How people can be expected to react to robots
- Some of the disadvantages of robot labor

KEY TERMS

Depreciation
Discounted cash flow
Future value of money
Internal rate of return
Maintenance costs
Net present worth
Payback period
Return on investment
Savings
Staffing costs
Total cost
Total yearly wage

STEPS IN JUSTIFYING ROBOT USE

The first step in justifying the use of a robot is to decide where and when to put the robot to work. This involves examining the jobs available for possible robot labor. Generally, if a job is appropriate for a robot to do, it should be one that can be done now by unskilled labor. The areas of spray painting, arc welding, and spot welding are exceptions to this generalization.

The next step is to study the relative costs of doing the job by human labor and by robot labor. The final step is to make sure that all other possible alternatives for doing the job are checked out. These include using automation instead of a robot, or making slight changes to the production area, the production equipment, or the parts. Payback period and rate of return on investment are two calculations used to justify the use of a robot from a cost standpoint.

Just because a robot can do a job for less money does not mean that it should do the job. The relations between the robot and human workers, supervisors, and the general public must also be considered.

ASSESSING THE AREA'S NEED FOR A ROBOT

To be appropriate for an industrial robot to do, a job should either be repetitive or be highly hazardous. If a job is dirty, dull, dangerous, or difficult, people will not like doing it and will not do their best at it. If few humans are willing to do the job, it may be a good position for a robot to fill.

Robots do unskilled jobs approximately the same way as a person would, except that the robots may be able to lift more, move faster (on the average), reach a little farther, and handle parts that are too hot or too cold for humans to touch. Robots have been put to work in automotive assembly doing painting, welding, material transfer, and processing. Studies are being conducted into using robots for small parts assembly in the automotive assembly industry, as well. Other areas of successful use include investment casting, assembling, grinding, sanding, inspecting, and machining. Robots are most successful when doing jobs designed with robots in mind. Consequently, just replacing a human with a robot will probably not make the best use of the robot.

Neale W. Clapp of Block Petrella Associates has suggested the following Laws for Industrial Robots:

1. Organizations may not install robots to the economic, social, or physical detriment of workers or management.
2. Organizations may not install robots through devious or closed strategies that reflect distrust or disregard for the work force, for surely they will reap distrust and disregard.
3. Organizations may only install robots on tasks that, while currently performed by humans, require the human to act like a robot—and not vice versa.

JUSTIFYING THE COST OF ROBOTS

There are only three reasons why a company should put in a robot: to increase profits, to do a job that humans are unavailable to do, or to do a job that is too dangerous for a human to do. At the same time, robots must be cost-effective; they must save a company money. W. R. Tanner of Tanner Associates suggests the following rules of thumb regarding robot installation:

1. Avoid extremes of complexity.
2. Make operations orderly and systematic.
3. Remember that robots are generally no faster than people.
4. For short runs, use people; for very long runs, use fixed automation.
5. If it doesn't make dollars, it doesn't make sense.
6. One robot is not better than none.
7. If people don't want it to succeed, they won't let it.

Calculating the Payback Period

To help justify the installation of a robot, it is necessary to calculate the length of time it will take for a robot to pay for itself. This time period is known as the **payback period.** This generally needs to be 2½ years or less; however, the exact requirements for maximum length of payback period vary greatly from company to company. The type of business a company is in also influences the allowable payback period.

The formula for calculating payback is

$$P = \frac{C}{W + I + D - (M + S)}$$

where

P = Number of years for the payback
C = Total cost to build the robot system, including installation
W = Annual salary of workers being replaced by the robot labor, including fringe benefits
I = Savings from robot installation in productivity, quality of work, more economic use of materials, no parking space, no restrooms, less lighting needed, less air conditioning required, and so forth
D = Robot depreciation allowance
M = Robot maintenance costs
S = Staffing costs for the robot, including programmers or trainers, supervisors, operators, and repair persons

The Payback Period (P) The payback period in years (P) equals the total costs of an installed robot (C) divided by the quantity of wages of displaced workers (W), plus savings from the use of the robot (I), plus robot depreciation (D), minus new maintenance costs (M) and new human staffing costs (S).

The Total Cost (*C*) The **total cost** (*C*) to build and install a robot includes the cost of the robot, tool costs, and installation costs. Many manufacturers sell an incomplete, basic robot system for which the purchaser must select certain add-on options (at extra cost) to form a complete robot. All the required options needed by the robot to perform its desired task must be included in *C*. Most types of robot end-of-arm tooling are priced separately, since the robot generally needs a different hand whenever it is assigned a new task. The manipulator, power supply, and control unit are a one-time expense, while the grippers are generally an ongoing expense throughout the life of the robot. Any special modification or other change costs needed for the robot must also be included in *C*. The total cost of the first robot should also include training costs, the cost of the manufacturer's recommended spare parts, and any new test equipment not already used by the company.

Each robot task requires some additional tool costs, including one or more grippers designed for that task, special fixtures for placement of tools and parts, and material-handling equipment. The machines being serviced by the robot may require special changes in order to work with a robot, such as the addition of sensors or switches to synchronize its operations with those of the robot.

The installation process itself involves some costs. These include the cost of moving things about and putting the parts together. Some electrical setup will be required for the electrical power outlets needed by the robot. There may even be additional costs for help from the vendor in the form of advice, physical assistance, or the complete installation. If this is the first robot installation for the company, add 20 to 25 percent to the total for unknown problems.

The Yearly Wage (*W*) The **total yearly wage** or salary (*W*) for the laborers who have been displaced by the robot includes wages and all fringe benefits. Robots do not pay social security tax, get vacations, have retirement plans, or receive health insurance coverage. They may, however, require a service contract. But robots do not get overtime pay nor require the same level of supervision as human workers do.

Savings from Robot Installation (*I*) The installation of a robot can bring **savings** (*I*) through increased productivity and better-quality work. The latter improvement results in fewer rejected (and subsequently reworked) parts and better use of materials. There is no need for parking spaces, rest periods (coffee breaks), pauses for socializing, strikes, or work slowdowns.

Robots increase productivity by not slowing down due to fatigue, by being more consistent in the quality of their work, and by being able to work more than one shift at little added expense. Robots save on materials through more consistent work habits than humans. In the job area of painting and other spray-surfacing applications, for example, robots have been known to pay for themselves within the first year of use just through savings in paint and other spray-on materials.

Other areas of cost savings that can be achieved through the use of robots include investment tax credit, reduced safety equipment costs (such as shoes, glasses, and gloves), and reduced environmental control costs—especially within painting booth areas. If some or all of the robot's costs can be deferred until some time in the future, this saves money, since future money is cheaper than present money.

Depreciation (*D*) Like all other pieces of capital equipment, the robot can be depreciated by one of several depreciation methods. The company accountant can tell you which method is best for your business. Included in the robot's **depreciation** (D) should be the concomitant depreciations for grippers and other special tooling for the robot.

Maintenance Costs (*M*) The **maintenance costs** (M) for a robot include the costs for supplies and for replacement spare parts. If the company has only one robot, keeping enough spare parts on hand to fix the robot is very expensive. The more robots the company can spread the spare parts over, the lower the per-unit cost. If the robots are used in critical positions on an assembly line, it may be desirable to stock a complete robot as a spare to be switched into the production line to keep it running. Maintenance costs are one reason for W. R. Tanner's suggestion, "One [robot] is not better than none."

Robot Staffing Costs (*S*) The robot **staffing costs** (S) are the salaries and fringe benefits of all personnel added to program, run, and repair the robot(s). These include the engineering personnel involved in the design and implementation of the robot, the computer personnel hired to program the robot, and the maintenance personnel hired to repair and keep it up.

The engineers' work includes rearranging the production line to accommodate the robot, modifying or adding safety switches and guard rails, and changing machine tools and even the end product to make it easier for the robot to handle it. Unlike hardwired automation machine tools, a robot is a general-purpose machine tool and requires reprogramming when it switches from one task to another. Maintenance personnel are needed to provide both preventive maintenance and emergency maintenance for the robot. Again, some of these per-unit staffing costs go down as more robots are used, since some of the costs are then divided among more than one robot.

SAMPLE PROBLEM 1.

Assume that the following figures are correct. The total cost of the robot system is \$125,000. The salary and fringe benefits of replaced workers is \$24,000, based on \$12 per hour and 250 workdays per year. The savings from using the robot is \$10,000. The depreciation is \$25,000, based on a 20 percent depreciation. The maintenance costs are \$5,000, and new staffing costs are \$24,000. What is the payback period?

Answer

Plugging the values given into the formula for payback period (P), we get

$$\begin{aligned} P &= \frac{C}{W + I + D - (M + S)} \\ &= \frac{125{,}000}{24{,}000 + 10{,}000 + 25{,}000 - (5{,}000 + 24{,}000)} \\ &= \frac{125{,}000}{30{,}000} \\ &= 4.166 \text{ years} \end{aligned}$$

SAMPLE PROBLEM 2.

Suppose, in addition, that the robot is asked to work two shifts, increasing the maintenance costs (M) of the robot by \$1,000 but doubling the workers' salary savings (W), robot's savings (I), and staffing costs (S). What is the payback period?

Answer

In this case, we have

$$\begin{aligned} P &= \frac{C}{W + I + D - (M + S)} \\ &= \frac{125{,}000}{48{,}000 + 20{,}000 + 25{,}000 - (6{,}000 + 48{,}000)} \\ &= \frac{125{,}000}{39{,}000} \\ &= 3.205 \text{ years} \end{aligned}$$

SAMPLE PROBLEM 3.

Suppose now that the robot's new staffing costs (S) increase by only half for the new shift. What is the payback period?

Answer

Now we have

$$\begin{aligned} P &= \frac{C}{W + I + D - (M + S)} \\ &= \frac{125{,}000}{48{,}000 + 20{,}000 + 25{,}000 - (6{,}000 + 36{,}000)} \\ &= \frac{125{,}000}{51{,}000} \\ &= 2.4509 \text{ years} \end{aligned}$$

SAMPLE PROBLEM 4.

Suppose that three robots each work two shifts. Now some of the spare parts cost (M) and some of the robot staffing costs (S) are shared by the robots. If the per-unit figures for M and S, respectively, are \$4,000 and \$30,000, what is the payback period?

Answer

The equation now reads

$$P = \frac{C}{W + I + D - (M + S)}$$

$$= \frac{125{,}000}{48{,}000 + 20{,}000 + 25{,}000 - (4{,}000 + 30{,}000)}$$

$$= \frac{125{,}000}{59{,}000}$$

$$= 2.1186 \text{ years}$$

Of course, if the price of the robot system (C) decreases, the payback period for the robot will decrease, too. A welding robot that cost \$150,000 five years ago now sells for around \$75,000. The payback period also decreases if the robot becomes more reliable, thus lowering maintenance costs (M). And if the robot requires less human help, the staffing costs will go down, again shortening the payback period.

Calculating Return on Investment

Another mathematical way to justify the installation of a robot is to calculate the **return on investment.** The formula for calculating annual return on investment is

$$ROI = 100 \times \frac{W + I + D - [(C/N) + M + S]}{C}$$

where

ROI = Annual percentage return on investment
N = Useful life of the robot, in years (at least 8 years for a robot)

All other symbols have the same meanings they did in the calculations of payback period.

SAMPLE PROBLEM 5.

Assume that the same conditions identified in Sample Problem 1 are again present, and that the useful life of the robot is eight years. What is the percentage return on investment?

Answer

Plugging the values given into the formula for return on investment (ROI), we get

$$ROI = 100 \times \frac{W + I + D - [(C/N) + M + S]}{C}$$

$$= 100 \times \frac{24{,}000 + 10{,}000 + 25{,}000 - [(125{,}000/8) + 5{,}000 + 24{,}000]}{125{,}000}$$

$$= 100 \times \frac{14{,}375}{125{,}000}$$

$$= 11.5\%$$

SAMPLE PROBLEM 6.

Now take the conditions identified in Sample Problem 2. What is the percentage return on investment for these conditions?

Answer

In this case, we have

$$ROI = 100 \times \frac{W + I + D - [(C/N) + M + S]}{C}$$

$$= 100 \times \frac{48{,}000 + 20{,}000 + 25{,}000 - [(125{,}000/8) + 6{,}000 + 48{,}000]}{125{,}000}$$

$$= 100 \times \frac{23{,}375}{125{,}000}$$

$$= 18.7\%$$

SAMPLE PROBLEM 7.

Now take the conditions identified in Sample Problem 3. What is the percentage return on investment for these conditions?

Answer

Now we have

$$ROI = 100 \times \frac{W + I + D - [(C/N) + M + S]}{C}$$

$$= 100 \times \frac{48{,}000 + 20{,}000 + 25{,}000 - [(125{,}000/8) + 6{,}000 + 36{,}000]}{125{,}000}$$

$$= 100 \times \frac{35{,}376}{125{,}000}$$

$$= 28.3\%$$

SAMPLE PROBLEM 8.

Now take the conditions identified in Sample Problem 4. What is the percentage return on investment for these conditions?

Answer

The equation now reads

$$ROI = 100 \times \frac{W + I + D - [(C/N) + M + S]}{C}$$

$$= 100 \times \frac{48{,}000 + 20{,}000 + 25{,}000 - [(125{,}000/8) + 4{,}000 + 30{,}000]}{125{,}000}$$

$$= \frac{100 \times 43{,}375}{125{,}000}$$

$$= 34.7\%$$

Once again, if the cost of the robot system (C) decreases or the maintenance cost of the robot (M) goes down due to better reliability or the robot requires less human help (S), the return on investment will increase.

Calculating Discounted Cash Flow

Discounted cash flow represents the expected value today of a money return expected some time in the future. This is also known as the **future value of money,** or the **internal rate of return.** Its exact calculation, the effects of other things such as tax laws, and how it all relates to robotics is covered by Von Blois.[1] The effect of discounting cash flow is to lower the rate of return on investment or to lengthen the payback period.

Discounted cash flow is calculated by using interest rates in reverse. A robot with a payback period of 2½ years will probably not be set up and working properly for the first six months to a year. Nonetheless, the real payback period needs to include the entire time between when the money is spent and when it is recovered. The future value (FV) of each current dollar invested in the project is equal to $1 plus the yearly interest rate (r), expressed as a percentage, raised to the power of the number of years (n). That is,

$$FV = ar \times (1 + r)^n$$

Table 12–1 shows the results of using this formula at interest rates of from 10 to 30 percent, and for terms of from zero to five years. What do the figures in Table 12–1 mean? They mean that, if the project costs $100,000 and will take three years to pay itself off, and if the money used to finance it would have been

[1]J. F. Von Blois, "Robotic Justification Considerations." In *Robots 6 Conference Proceedings,* March 2–4, 1982 (Dearborn, Michigan: Robotics International of SME, 1982).

Table 12-1
Value of Money Received in the Future

	Interest Rate				
Year	10%	15%	20%	25%	30%
0	1.000	1.000	1.000	1.000	1.000
1	1.100	1.150	1.200	1.250	1.300
2	1.210	1.320	1.440	1.560	1.689
3	1.332	1.520	1.727	1.953	2.198
4	1.464	1.748	2.075	2.439	2.857
5	1.610	2.012	2.488	3.049	3.717

earning 10 percent per year, the project needs to pay back $133,200 to break even. This result is 33 percent higher than the amount calculated by the formulas for return on investment and for payback period. Money is more expensive than it appears to be on the surface.

Net Present Worth

Net present worth (NPW) is a method of showing how money is worth more now than in the future. If one can purchase something now but pay for it later, it costs less (assuming one doesn't have to pay any interest). If the purchase can be delayed without the price rising, it will cost less. Even if only part of the cost can be delayed, it will cost less. The **NPW multiplier** is equal to the quantity of 1 divided by the quantity of 1 plus the discount rate raised to the power of the years delayed in payment. That is,

$$\text{NPW multiplier} = 1/(1 + \text{discount rate})^{\text{years}}$$

The NPW is equal to the cost times the NPW multiplier. The NPW multiplier for paying for something one year from now at a discount rate of 8% is:

$$\text{NPW multiplier} = 1/(1 + 0.08)^1 = 0.926$$

Suppose one wishes to buy a $40,000 robot and can pay for it now or up to four years from now. Table 12–2 shows the NPW of the robot depending on when the bill is paid.

Table 12-2
Net Present Value

Year	0	1	2	3	4
Cost	$40,000	$40,000	$40,000	$40,000	$40,000
NPW multiplier	1	0.926	0.857	0.794	0.735
NPW value	$40,000	$37,040	$34,280	$31,760	$29,400

Improving Workers' Safety Through the Use of Robots

One very important justification for the use of robots is that it can improve workers' safety. Some tasks performed by human workers require the use of extensive safety equipment that interferes with performing the task and yet does not fully protect the worker. A spray-painting booth operated by human workers requires special ventilation, protective clothing, and face masks; but even so, there is a significant hazard of explosions. For repetitive painting tasks, a robot can do the job, excluding the human from the hazardous condition and saving on protective equipment. Robots can also be used to load and unload presses and to handle hot, cold, or otherwise hazardous materials.

DISADVANTAGES OF ROBOT LABOR

Human workers can be laid off during periods of slow production, in which case their labor costs stop. A robot that isn't being used still accumulates depreciation expenses. Robots require a large capital outlay when they are installed, but their costs must be depreciated over a period of years. Human workers represent a pay-as-you-go labor expense and can be completely deducted as a current business expense.

Just because a robot is capable of doing a particular job does not mean that it should be doing that job. The robot still needs to make economic sense.

If robots are installed under the wrong conditions or under the right conditions with the wrong kind of preparation and publicity, the human workers at the facility may feel threatened and lower their own productivity. The introduction of robots can result in the elimination of some existing jobs, forcing job displacement or even unemployment. Companies installing robots should have an enlightened, humane, and far-sighted policy for dealing with job displacement. A possible policy might be to furnish retraining, relocation assistance, or (in a few cases) early retirement benefits. Whenever possible, displaced workers should be given a choice of new job alternatives.

Even though robots can be reprogrammed to do new tasks, some companies buy a robot for a particular project; then, when the project is over, the robot becomes surplus equipment. This wasteful approach is something like laying off all employees at the end of each project and then hiring a bunch of new workers to take their place on the next project. Companies must plan new tasks for the robot to do when it has finished work on its present project.

HOW PEOPLE REACT TO ROBOTS

From literature and the movies, most people get the idea that robots will either develop into friendly and helpful creatures like R2D2 and C3PO from the *Star Wars* movies or reveal themselves to be evil war machines like the Imperial Walkers. To workers concerned about losing their jobs, the robot is likely to seem heartless and malevolent. To workers who expect the robot to help them in their work, the robot is likely to seem friendly and useful.

Children will react to robots much as they react to the electronic computer: with fascination; adults will react to robots much as they react to computers: with fear, distrust, and suspicion.

Being Replaced by a Robot

In 1993, there were 50,000 industrial robots in the United States, while there were approximately 100 million human workers in the United States. While this might appear to be only 0.05 percent of the jobs, and while the number of new jobs created between 1985 and 1992 increased by more than 0.05 percent, most jobs for robots fall into a very small area of the total job market for humans—the area of unskilled manufacturing applications. The number of these unskilled jobs for humans is already very limited and is decreasing.

People have lost jobs to robots and automation for decades. How many pin setters at bowling alleys lost jobs to automatic pin setters? How many farm hands lost jobs to automated harvesters? How many horses were put out of work by the tractor, the truck, and the automobile?

Production workers often dislike changes and suspect that management underestimates their intelligence and common sense. And when it comes to replacing a human worker with a robot, management does tend to underestimate the intelligence and common sense that are needed to do a production task. If management is thinking of bringing a robot into a production job, human workers like to be consulted and to help in the planning. They are especially helpful in bringing to management's attention the not-so-obvious areas of intelligence used in their job.

The less workers know about the robot joining their production line, the more they will resent it and be afraid of losing their job to a robot. Workers who are not included in the planning for the robot will not believe management promises of no layoffs. If anyone is laid off anywhere in the company within a month or two of the arrival of a robot, the workers will blame it on the robot. It is helpful if management can tell displaced workers what their new jobs (preferably with more responsibility) will be.

On the other hand, workers in some settings have been known to adopt the robot as a fellow worker and even send it get-well cards when it is out of commission. Robots should not be used to replace workers already on the job; rather, they should be used to supply the company's future needs for unskilled labor as current human laborers take more attractive jobs.

Workers are afraid of being bumped—that is, displaced from their jobs—by robots. Job displacement may take the form of job loss, being "demoted" to a less desirable job or shift, or getting a pay cut. In the case of an ex-machine operator who is given the new job of robot operator or supervisor, the new job involves more watching than doing, and the worker (who was accustomed to handling parts) may now feel boredom, stress, and even alienation from the lack of interaction with fellow workers. A person's reaction to job shifting must be considered on an individual basis. Some production workers handle the switch to the new job very well.

Unions will accept the use of automation and robots to increase a company's productivity, provided that union job security is not threatened. The new jobs of persons whose old jobs were taken over by automation or robots should be within the domain of the union. Indeed, the robot or automation device should provide union members with enhanced job security and new and better job opportunities. In some countries, employers are required to pay union dues for their robots. In a few cases, employers are also required to pay into the union retirement fund, based on the work done by the robots.

Another method for overcoming popular fear of robots is to issue community news releases about plans for using robots, with news updates as plans and implementation progress. Before actual production use of the robots begins, have a company employee open house to exhibit the robots and demonstrate what they will do.

Production Supervision's Reaction to Robots

Production supervisors keep their jobs because of their success at meeting production schedules. Supervisors do not want anything new and untried to interfere with their ability to meet schedules, and consequently they are skeptical about any new robot proposals. In addition, losing human production workers to robots makes supervisors nervous that their position is now less important or not needed at all. Again, the way to get supervisors on the side of bringing in robots is to consult them during the planning stages.

Engineering and top-level management often tend to overcomplicate and overengineer the installation of robots. Engineers do not always understand the demands of day-to-day production, and they sometimes forget about costs. Engineers like any excuse for using a mini- or microcomputer—even when a simple mechanical fixture or a piece of baling wire could do the job perfectly well. Supervisors need to be kept informed about what oversight is involved in the installation and use of the robots. They may also be able to enlighten the engineer about problems in their production area. The better informed the engineers are about the production task that the robot is to perform, the better they will adapt the robot to its new task.

All alternative methods for doing a job should be considered before committing to do the job with a robot. Perhaps a slight redesign of the work area, the production equipment, or the part is the optimum solution in a particular case.

Pre-Owned Robots

When trying to justify the cost of a robotics system, one must not forget about the **pre-owned robot,** or used robotic market. They offer factory reconditioned robots for sale for less than it would cost for a new robot. Two such used robotics companies are The Robot Company, which can be found on the World Wide Web at robotco.com, and Robotic Technology Resale Ltd., which can be found at robotictech.com.

SUMMARY

The first step in justifying the use of a robot is to identify the task that the robot is to perform. Tasks that are dirty, dangerous, dull, or difficult for humans to do are the easiest to justify having a robot do. The next step is to prove or justify on paper that using the robot is economically feasible—that is, that it will save money over the present method of doing the task. It should also be proved that the task cannot be done less expensively through automation. If all of these things prove out, the robot may be allowed to justify itself on the factory floor.

While all this is going on, management must not ignore the Fourth Law of Robotics: "A robot may take a person's job, but should not leave a person jobless." Any robot installation requires thinking about people and their reactions to the robot. If a company's workers do not want the robot to succeed, it will not succeed. Supervisors and the local general public may also need to be kept informed about the project, along with any unions in the company.

Payback period, return on investment, and discounted cash flow are three mathematical methods that may be used to justify on paper the use of the robot. In general, a payback period of 2½ years or less is required to justify the use of the robot. The return-on-investment rate required is tied to the general return on investment of other activities of the company.

Robot labor is not always advantageous. During slow periods, you cannot lay off a robot and reduce costs. And just because a robot can do a job does not mean that it should do the job. Poorly conceived robotics projects may end up costing more than doing the job the old way.

Workers' reactions to robots depend on whether they think the robots will help them make more money and do less difficult labor or whether they think the robots are eventually going to take away their jobs. Even supervisors are concerned about losing their jobs to robots—as they lose human workers to supervise, they become very insecure about their own jobs.

FORMULAS

Future Value of Money

$$FV = ar \times (1 + r)^n$$

where

FV = Future value after n years of each dollar currently invested
ar = Amount of money received
r = Interest rate that the money could have been earning
n = Number of years

Payback Period

$$P = \frac{C}{W + I + D - (M + S)}$$

where

P = Number of years for the payback
C = Total cost to build the robot system, including installation
W = Annual salary, including fringe benefits, of workers being replaced by the robot labor
I = Savings from robot installation in productivity, quality of work, better use of materials, no parking space, no restrooms, less lighting needed, less air conditioning required, and so forth
D = Robot depreciation
M = Robot maintenance costs
S = Staffing costs for the robot, such as programmers, trainers, supervisors, operators, and repair personnel

Return on Investment

$$ROI = 100 \times \frac{W + I + D - [(C/N) + M + S]}{C}$$

where

ROI = Annual return on investment (in percent)
N = Useful life of the robot, in years (at least 8 years for a robot)

And W, I, D, C, M, and S are as defined for payback period.

Net Present Worth (NPW)

$$\text{NPW multiplier} = 1/(1 + \text{discount rate})^{\text{years}}$$

$$\text{NPW} = \text{cost} \times \text{NPW multiplier}$$

REVIEW QUESTIONS

1. If the total robot and installation costs (C) for the last example payback period calculations in Sample Problem 4 were $100,000, what would the new payback period be?
2. If the maintenance costs in Sample Problem 4 could be halved, while the robot and installation costs dropped to $100,000, what would be the new payback period?
3. How would the change in Question 1 affect the return on investment in Sample Problem 5?
4. How would the change in Question 2 affect the return on investment in Sample Problem 5?
5. Where might a robot installation be inappropriate?
6. Suppose that your father has worked as an unskilled laborer for 30 years for a manufacturing company, and now that company is expecting to put in some robots. How would you advise your father in an effort to keep him from becoming unemployed?

7. If there will be only 110,000 industrial robots in the United States by 2001, and there are 130 million workers in the United States, why are some people afraid of losing their jobs to robots?
8. How can the workers and supervisors in a production area where a robot is about to be installed help the engineers who are planning the installation?
9. What can one do to help get the unions to support the installation of robots?
10. Suppose that a robotics project costs $125,000, the payback period for the robot is three years, and it will take six months to get the robot working properly. If the money paid for the robot could have been earning 17.5 percent yearly, what will a calculation of discounted cash flow reveal to be the real cost of the robotics project? Make an educated guess about the real payback period for the project.
11. What is the net present worth of a $25,000 robot that is obtained now, but not to be paid for until two years from now, if the discount rate is 11%, or 0.11?
12. Why is it necessary to justify the use of a robot from a cost viewpoint when you already know that your competitors are using robots to do similar tasks?

13

THE FUTURE FOR ROBOTS

OVERVIEW

Why would anyone want to guess at the future? How do we study the future? How will robots change in the short-term and long-term future? What effect will robotics have on the job market? These are some of the questions explored in this chapter.

OBJECTIVES

When you have completed this chapter, you should be familiar with:

- The need for predicting the future
- The tools used for predicting the future
- How artificial intelligence will help robotics in the future
- The probable short-term future for robots
- The probable long-term future for robots

KEY TERMS

Automated factory
Cyberphobia
Expert systems
Life form
Recent history
SCAMP
Science fiction
Self-replicating factory
Self-reproducing machines
Shape-memory alloy
Toys

GUESSING AT THE FUTURE

Why bother trying to study or predict the future? Because within five years of graduation from college, at least 50 percent of what an engineer or technologist has learned in college will be obsolete and looking at the future may show us future consequences of current policies and actions, which, in turn, may help us to make informed decisions now.

The future is controlled by events from the past and present, and by our imaginations. Important products of imagination include art, literature, cinema, and children's toys. Our imaginations should take into account both the good and the bad aspects of possible futures, in order to guide our choices now.

I would like to start this look at the future of robotics by quoting from *The Tomorrow Makers:* "I met the tomorrow makers. . . . Among them are the World's greatest scientists, and they are in a desperate race to be first at downloading the contents of the human mind into a computer housed within a robotic body so that we will never have to die. . . . Others expressed a fear that what was now being created was in fact a new species that might soar beyond its human creators to become our evolutionary successors."[1] Whether the future of robotics is a blessing or a curse to humanity depends on the conduct of the individuals who make and use these robots.

Predicting the future is iffy at best. Most prophets tend to underestimate the time that will pass before their prophecies are fulfilled. Even so, predicting is a useful tool.

Among the best ways to predict the future are to study the past, to study science fiction, and to study the toys made for children.

In 1995, there were approximately 66,000 industrial robots in the United States, not including simple limited-sequence material-transfer devices. Recent history shows that industrial robot sales have grown at a rate of 10 percent per year. Computer memory devices have been quadrupling in capacity, becoming one-third faster, and halving in price about every three years. The personal computers of today have more memory and run faster than did the mainframe computers of five to ten years ago.

Jules Verne predicted many of the mechanical inventions of the twentieth century in the middle of the nineteenth century. Hugo Gernbeck predicted many of the electronic devices—including radar and solar energy devices—of the late twentieth century during the early twentieth century. Today, *Buck Rogers, Star Wars,* and *Star Trek* are suggesting ways robots may develop in the future.

Have you looked at how many educational and noneducational toys of today are related to computers and robots? As recently as five years ago, it was estimated that 50 percent of the adult population of the United States suffered from **cyberphobia**—an unreasonable fear of computers. Children are not afraid of computers, however, because they are being brought up with them. The same will be true of children's attitudes toward robots when these become widely available as toys.

Are members of the general public ready to accept robots into their homes? In the United States and in most other industrial nations, the answer is yes! Consider the array of gadgets and appliances that are already welcome: garage door openers, garbage disposals, dishwashers, washing machines, dryers, vacuum cleaners, and many others.

[1]Grant Fjermedal, *The Tomorrow Makers: A Brave New World of Living-brain Machines* (New York: Macmillan, 1986).

ROBOTICS AT PRESENT

Right now, new types of manipulator joints, actuators, and grippers are under development. Japanese researchers are experimenting with **shape-memory alloy,** a new actuator for humanlike hands. A wire made of shape-memory alloy (composed of nickel and titanium) contracts within about 1 second when it is heated. It takes the wire somewhat longer to cool and stretch back to its normal length. Figure 13–1 shows a Space Wings butterfly marketed by Mondotronics, Inc., that is powered by a shape-memory alloy wire. The wire, located at the bottom of the wings, exerts up to 10 ounces of force to close the wings when it is heated. Figure 13–2 shows the DH-101 Robot, made by Toki and marketed by Mondotronics. It has five degrees of freedom—three axes of the arm, one axis of the wrist, and one axis or motion of the gripper—all powered by shape-memory alloy "motors."

At present, robots have little or no judgment and decision-making capacity, and their sensory capabilities are quite poor. Consequently, they are not yet ready to perform most complicated tasks. The advancement of the robot to the state

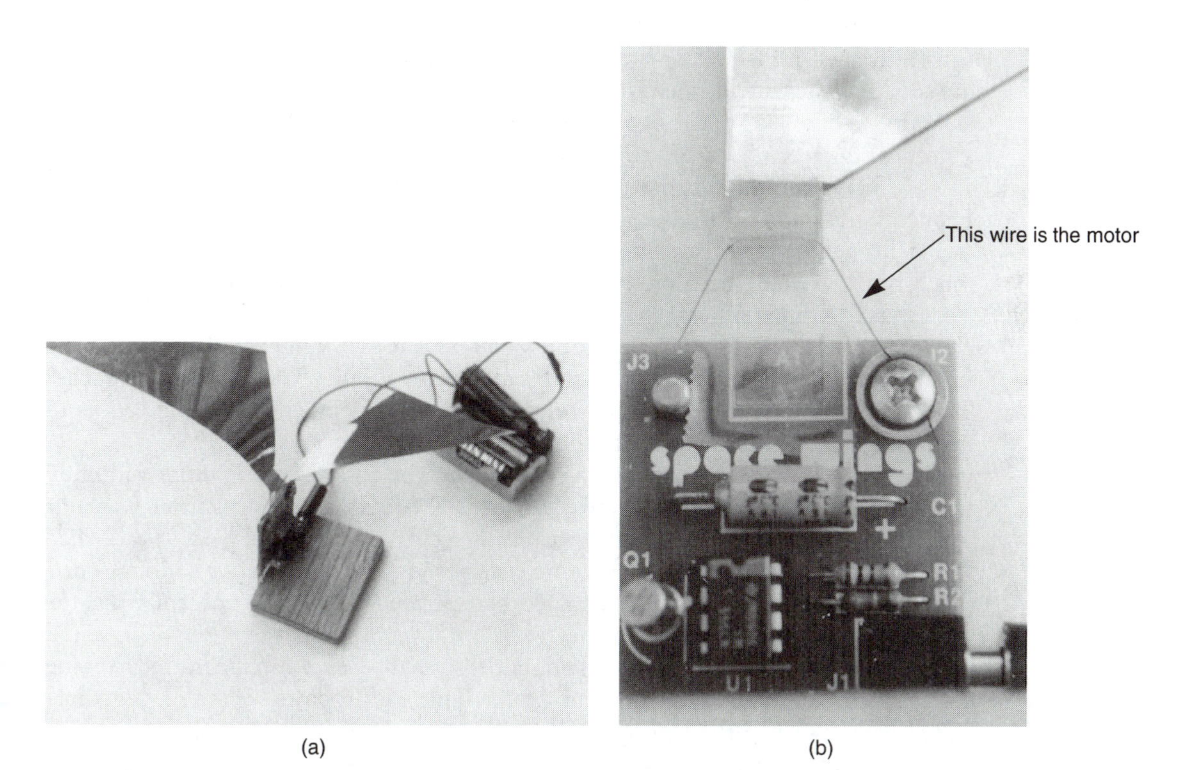

Figure 13–1 Space Wings Powered by Shape-Memory Alloy.
(a) The Space Wings device. (b) Closeup of the power source. The motor for the space wings is the little black wire (0.15 mm in diameter and 6 cm long) at the base. This wire is made of a shape-memory alloy.

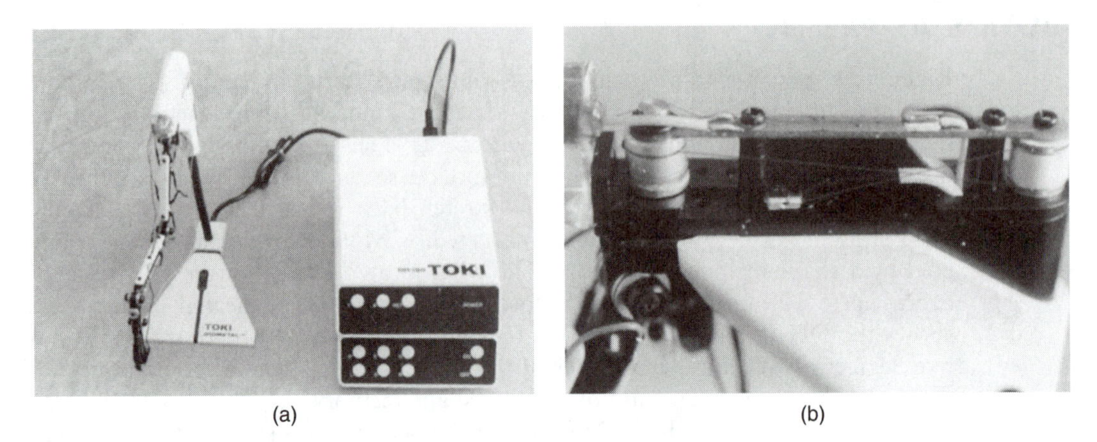

Figure 13-2 DH-101 Robot Powered by Shape-Memory Alloy.
(a) The complete robot. (b) Closeup of wire-controlled joints. All axes of the DH-101 robot are powered by wires made from shape-memory alloy.

shown in such films as *Star Wars* and *Star Trek* must await breakthroughs in the areas of artificial intelligence, voice interfacing, and vision and touch sensors. Considerable research is going on in these areas, and it is only a matter of time before the objects of today's laboratory curiosity become economical enough for use in robots.

Robotics would advance much faster if the future products were needed now. Economics is a major driving force in research and development, as is the government, through its military and space programs. The government was responsible for much of the development of electronics and the electronic computer, for instance. Today, the Atomic Energy Commission is sponsoring research in robotics for advanced nuclear reactors, where robots will have to work in extremes of temperature, humidity, and radiation level and will have to be able to climb over obstacles. The military is hoping to make up for its shortage of personnel by using robots to make the human forces more efficient.

Even in their present state of development, robots are finding more and more jobs. They have already been successfully used as brain surgeons, window washers, and lab technicians. One of their newest uses occurred on February 22, 1988, when ABC began employing robots to run the television cameras for its national news broadcasts. This allows the cameras to be operated by remote control from the control booth by a single person.

The automated robot cook, maid, butler, or gardener is still far from becoming a reality. However, a robot chauffeur can be built with today's technology, if someone is willing to pay the high cost. The navigational computers and sensors necessary for getting around safely in city traffic already exist.

Another desirable robot servant is the robot nursemaid. The story "Strange Playfellow," written by Isaac Asimov in 1940, deals with a robot nursemaid named Robbie. Even with today's limited technology, it would be possible to construct a mobile robot nursemaid that could perform many of the functions of a human nursemaid. It could use a remote-controlled television camera, speaker, and microphone to allow a human to monitor the child while being busy elsewhere. The robot would have its own built-in sensors and be able to move about without running into anything. An infrared object-avoidance sensor is already available for less than $20. A voice recognition and voice synthesis device would enable the robot to communicate with and entertain the child. This would include being able to tell nursery rhymes and stories and even possibly to sing or hum songs. When working with older children, the robot could be equipped with a personal computer for educating and entertaining the child.

As it merges with the electronic computer, the robot is evolving toward the point where it could be viewed as a new **life form.** Such works as the *Theory of Self-Reproducing Automata* by John von Neumann, *The Computerized Society* by James Martin and Adrian R. D. Norman, "Symbiosis and Evolution" by Lynn Marqulis, and *Man and the Computer* by John G. Kemeny give support to this idea. Many space satellites are already able to draw their power directly from sunlight.

Although artificial intelligence is still in its infancy, it has already shown great promise. Work has begun on the development of **self-reproducing machines.** Robert A. Freitas has written an article on the prospects for a **self-replicating factory** to be practical by early in the twenty-first century.[2] Located on the moon and running on solar power, this factory would use the moon's raw materials to reproduce itself and to manufacture additional solar cells, which could then be hauled away for a nearby solar-powered generation station satellite to beam to Earth as electrical power. NASA has proposed a four-part program along these lines, where part one is a robot in a warehouse assembling other robots and part four is a self-reproducing factory the size of a football field.

Robots are currently used in education as tools for teaching various topics. The show robot is very useful for working with abused children[3] whose bad experiences often make it hard for them to trust and talk to adults. Show robots are also useful for working with shy children. Programmable mobile devices such as the Big Trak tank (robot) can be used to teach programming; since they deal with motion rather than with numbers, they are easier for children to relate to. Industrial and educational robots are used to teach applications programming for industrial robots.

Robotics and Artificial Intelligence

Program techniques for artificial intelligence are just now moving out of the laboratory and games area and into practical use. The game of Animal (discussed in Chapter 10) could serve as a prototype for **expert** database **systems.** In the field of medicine, such an area or national database could be used in disease diagnosis to produce a list of diseases (ranked by probability) that match the patient's symptoms.

[2]Robert A. Freitas, Jr., "Building Athens without the Slaves," *Technology Illustrated* (August 1983): 16–20.

[3]Edward L. Safford, Jr., *The Complete Handbook of Robotics* (Blue Ridge Summit, Pennsylvania: TAB Books, 1978).

Digital Equipment Corporation (DEC) has started using two expert systems: XSEL, which is used by salespeople to help them find the right DEC computer system and configuration for a particular customer; and XCON, which is used by salespeople in placing customers' orders to make sure that the correct cables, manuals, nuts, bolts, and so on, are ordered for a particular computer system configuration.[4] SRI International is using an expert system called Prospector to calculate the probability of various mineral ore concentrations at different sites. The new generation of computers currently under development should greatly increase the power of artificial intelligence programs. These programs, when applied to robotics, should greatly increase the new robots' abilities.

The computerized intelligent tutor has been shown to be able not only to tell when a person has made a wrong answer but to explain why the person has made the wrong answer. That is, the tutor can determine from the answer where the person erred logically. When an intelligent tutor is fitted into a home robot, the robot will make an even better nursemaid.

Expert systems are being used to simplify maintenance and diagnostics for robots. In 1986, Ford Motor Company demonstrated a system for troubleshooting and repairing an ASEA S2 robot controller.[5] Reis Machines, Inc., introduced its SERVICE expert system (an IBM PC-based system) for fast repair and adjustment of Reis robots, in April of 1987. Expert systems are also being used to program robotic work cells and to route products dynamically between work cells.

For years, people have predicted that robots are about to experience a market growth rate similar to that of the electronic computer. This has not happened, mainly because robots must compete with humans for their jobs. Through 1980, the size of the human work force in the United States expanded each year and was able to meet most of the nation's labor needs. In 1984, the growth rates of the work force and the population began to slow. Soon the size of the work force will actually decline. When this happens, it will become much more difficult to find workers to fill various, less-attractive jobs—jobs that robots are ready and waiting to fill. It is now estimated that the United States Robotics industry should see a 30 to 35 percent growth rate each year in the next few years.

In the 1960s, it was feared that electronic computers would reduce the number of white-collar jobs, such as for accounting clerks. Instead, the computer has increased the number of these jobs by creating such new occupations as computer operator, computer programmer, and systems analyst. By making more information available, computers have generated an increased desire for information in our society. Now in the 1990s, there is a fear that robots will reduce the number of blue-collar jobs, but it is more likely that robots will cause a total increase in jobs, although some of these may be white-collar jobs. New human jobs such as robot supervisor, robot setup person, robot trainer, and robot repair person are sure to emerge from the widespread use of robots in industry.

[4]Richard Parker, "An Expert for Every Office," *Computer Design* (Fall 1983): 37–46.

[5]Robert N. Stauffer, "Artificial Intelligence Moves into Robotics," *Robotics Today* (October 1987): 11–13.

SHORT-TERM FUTURE OF ROBOTS

In the next five years, the robot can be expected to find a place in the home. It will start out as a novelty but grow into a useful servant. The British have recently invented a petlike robot that appears to have a personality; it is called **SCAMP** (short for *s*elf-*c*ontained *a*utonomous *m*obile with *p*ersonality). SCAMP expresses pleasure, anger, and friendship, and even its programmers do not know how it will react at any given time.

The hobbyist robot field will experience a growth rate in the near future similar to that of the hobbyist computer in the late 1970s. It is already starting this trend with robots that serve as educational toys, and soon it will advance to doing simple tasks about the home. How fast will the home robot market grow? In 1977, Radio Shack introduced the TRS-80 Model I computer. Before this model was discontinued some five years later, 500,000 units had been sold—more than the total number of computers that had been made prior to 1977. The personal robot has at least the same potential for growth as the personal computer.

Voice-activated robots are now being used experimentally to help the physically disabled. They will soon be generally available for helping disabled people in a multitude of ways.

Robots will benefit from more powerful controllers that will allow them to see, learn, and think for themselves. This will happen because the new generation of controllers will have a large enough memory to handle expert systems from the field of artificial intelligence. Artificial intelligence will allow the robot to leave their perfectly controlled world and enter the human world of chance and confusion. Nonetheless, robot intelligence will always be different from human intelligence.

Smart robots will also be able to protect themselves from accidental destruction by an operator's error or inattention to some danger to the robot. This will be necessary for robots working where people cannot go. For example, a radio-controlled robot might be able to rotate its antenna in an attempt to reestablish lost radio contact with its operator.

New types of robot actuators will also appear. These might rely on shape-memory alloy wire or other types of artificial muscles.

The fully **automated factory** has been under development in Japan for some time and should be completed within the next few years. Such factories will use robots in areas that require flexibility.

The Japanese government is also working on the fifth generation of electronic computers in which thousands of microcomputers will work in parallel. It is hoped that this design will greatly increase the intelligence capacity of artificial intelligence programs. Companies in the United States are working on fifth-generation computers, as well. Such computer designs will also lead to more intelligent robots.

Most industrial robots purchased in the next five years will be used as parts of larger manufacturing cells, rather than for stand-alone jobs. The number of robots in United States industry can be expected to triple in the next five years, and by the year 2001 there will be over 117,000 industrial robots in the United States. Sales of military and personal robots will outnumber sales of industrial robots by 2001. The price of low-end robots can be expected to drop. At the present time, most robot technical personnel are obtained through in-house retraining programs. In the near future, fewer people will have to be retrained as more get their training in college. Some twenty institutes in the United States and Canada currently have robotics options, majors, or degrees.

There will be a short-term problem with reassigning persons who are displaced from simple manual-labor jobs by robots. Some will find more interesting and rewarding jobs, other may be given early retirement, and still others will require retraining at the company's or government's expense. But in time the problem will disappear, since no one new will be joining the work force in these areas.

LONG-TERM FUTURE OF ROBOTS

In the long term, the robot will eventually take over many manual jobs now performed by humans. The robot's intelligence will grow and may someday equal or surpass human intelligence. But while robots may help support human society, they will not replace human beings. The robots of *Star Trek, Star Wars,* and *Buck Rogers* are possible in the next fifty years or so. So is the completely automated robot factory.

The day of the *Six Million Dollar Man* or the *Bionic Woman* is not far away, as medical robotics continues to improve. The artificial body parts of today are expensive and not very functional. Those of the future will be more reasonably priced and more functional. Artificial organs—including hearts, kidneys, ears, and pancreas—will be developed. Robotic limbs and other personal robotic devices will become commonplace.

Robots will free humans for tasks that require learning, insight, fantasy, imagination, creativity, and other forms of thinking. This will create a society much like that of the privileged few in ancient Greece. Even when robots are equipped with artificial intelligence expert systems, humans will still be better at some tasks, such as taking insufficient data and working out the probable result.

Since robots will be much more complex and intelligent, their uses will increase, while testing and checking them out will become much more complicated. One possible solution to robot or computer checkout was suggested by David Popoff, who suggested that the newly manufactured robot be given an IQ test to determine where it could be employed.[6] Again, the intelligence level of these robots may well approach the intelligence level of humans, although it will be a different kind of intelligence. Such levels of intelligence may require education instead of the type of canned programs used to give robots knowledge today.

[6]"The Robot Game: What's Your Robot's I.Q.?" *Psychology Today* (April 1969): 33–36B.

SUMMARY

The robot has an exciting future. In the near term, we can expect the robot to move out of the factory and foundry and enter the domestic and business world. The domestic robot will appear in the home as an electronic pet and soon will develop the ability to perform useful tasks there. The sensory ability of all robots will greatly improve. In the long run, the robot will acquire the capabilities they are depicted as having in the movies and science fiction books. Self-reproducing factories may be placed on the moon or on other planets to help meet our growing need for energy and goods. Medical robotics will be able to produce "six-million-dollar men" and "bionic women." Exciting changes are on the way as robots become the helpers that humans have always dreamed of having.

Figure 13–3 shows the number of industrial robots installed in the United States, for the years from 1981 through 1995. It estimates the number of installed robots for the years 1996 through 2001, based on an annual growth rate of 10 percent a year.

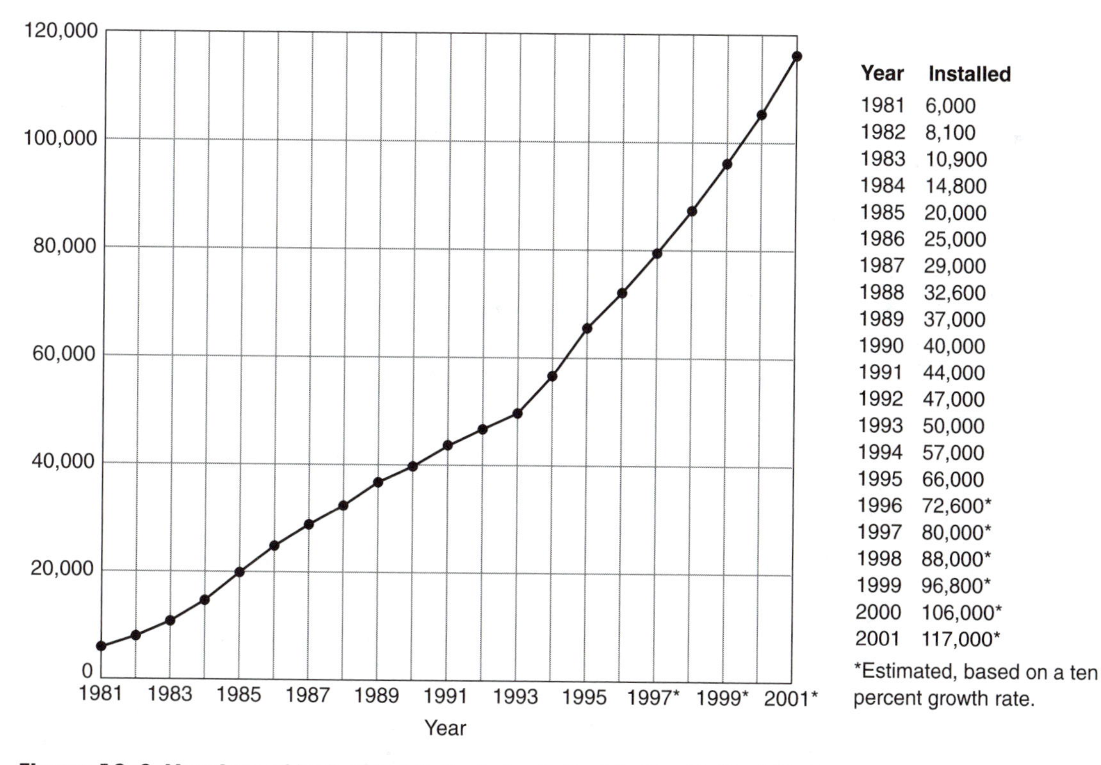

Year	Installed
1981	6,000
1982	8,100
1983	10,900
1984	14,800
1985	20,000
1986	25,000
1987	29,000
1988	32,600
1989	37,000
1990	40,000
1991	44,000
1992	47,000
1993	50,000
1994	57,000
1995	66,000
1996	72,600*
1997	80,000*
1998	88,000*
1999	96,800*
2000	106,000*
2001	117,000*

*Estimated, based on a ten percent growth rate.

Figure 13-3 Number of Industrial Robots Installed in the United States.

REVIEW QUESTIONS

1. Name three tasks around your home that you would like robots to do for you.
2. Name three tasks around the office that you would like robots to do for you.
3. If the preceding tasks were done by robots, how would you use your newly acquired spare time?
4. Is an automated future a blessing or a curse for humankind? Why?
5. How might robots help humans do things that they cannot now accomplish?
6. In the short term, how can we deal with the problem of workers' being displaced by robots?
7. Why should a graduate engineer or technologist be concerned about things that will occur five, ten, or more years in the future?
8. Use Figure 13–3 to estimate how many industrial robots will be installed in the United States in the year 2010 if there is 10 percent growth per year.
9. Starting with the 1995 figure of 66,000 robots installed in the United States, and given that there is an annual growth rate of 15 percent a year, how many robots will be installed by the year 2010?
10. Starting with the 1995 figure of 66,000 robots installed in the United States, and given that there is an annual growth rate of 20 percent per year, how many robots will be installed by the year 2010?

PART 2

ROBOTICS PROGRAMMING PROBLEMS AND TECHNIQUES

This section deals with programming considerations for a robotics system. Chapter 14 considers the operating system requirements for various types of industrial robots and robot controllers. Then Chapter 15 discusses application programming for robots and the programming languages that are used for this purpose. Chapter 16 deals with the steps and tools involved in planning and carrying out a robotics project. Finally, Chapters 17 through 20 deal with present and future programming problems of various sensor devices for robots.

14

OPERATING SYSTEMS FOR ROBOTS

OVERVIEW

Most robot users will never work on the robot's operating system. After all, if the operating system is good, the user can take it for granted. How good the operating system is will determine how easy the robot is to program and use and what limitations the robot has.

What is a robotic operating system? What does the operating system do for the user? What features should be included in an operating system? How will operating systems change in the future? These are some of the questions explored in this chapter.

OBJECTIVES

When you have completed this chapter, you should be familiar with:

- What makes up a robotic operating system
- What a robotic application program is
- The capabilities a robot's operating system must have that most computers do not need
- Differences in operating system requirements for different configurations of manipulators
- Programming languages that can be used for robot operating system programs
- Techniques that can be used to get the application program into the robotic operating system

KEY TERMS

Ada
Application programs
Assembler
Assembly language
BASIC
Binary numbers
Blocks world
C

Continuous-path controller
Control programs
Forth
Home
How
Inputs
Leadthrough
Limited-sequence controller
LISP
Machine language
Mechanical setup
Off-line programming
Operating systems
Outputs
Point-to-point controller
Prolog
ROM
Task commands
Task programming
User's programs
When

TYPES OF ROBOTICS PROGRAMS

The robot needs programs for its controller to execute—just as the electronic computer needs programs to execute—if it is to serve any useful purpose.

Programs for robots can be subdivided into two areas: **operating systems** and **application programs.** Operating systems—which are sometimes called **control programs**—tell a robot *how* to execute some motion such as opening or closing a gripper or moving the manipulator. In other words, operating systems supply the intelligence that enables the robot to receive, understand, and carry out tasks that are given to it. In contrast, application programs tell the robot *when* to do these motions in order to accomplish some specific task. Application programs are also known as **user's programs.** These programs contain positional data, axis velocities, geometric axis moves, gripper instructions, and interface instructions to other devices.

PROGRAMMING NEEDS OF OPERATING SYSTEMS

A robot's operating system consists of programs (generally furnished by the manufacturer) that make the robot easier to use. The more steps and processes of application programming the operating system can perform or support, the fewer mistakes will be made during the application programming.

Operating system programs include the intelligence to move the manipulator, support programs for teaching or training the robot, and a program for communicating with an external computer. Operating systems generally tell the robot how to perform a motion or action, but not necessarily when to do it. Operating systems are usually encoded in some type of read only memory (**ROM**) whose contents are built-in by the manufacturer and cannot be changed. Part of the operating system, however, may be supplied on flexible disks.

In addition to moving the manipulator, operating systems must support other inputs and outputs for the robot. The **outputs** for the robot include the ability to turn end-of-arm tools on and off, to open and close the gripper, to control clamping devices on a machine, and to start and stop a machine. **Inputs** for the robot include the ability to sense various types of information such as a machine completing its cycle, a clamp opening, a clamp closing, a clamp holding a part, or a clamp not holding a part. The robot should also receive inputs about the actual position of the components of its manipulator. In the future, robots will sense more aural, visual, and tactile information than they do now. The operating system must be able to interpret and act on this information.

The complexity and extensiveness of the operating system program's treatment of the robot's motion depend on the type of manipulator and the type of controller used. If the manipulator has a wrist and gripper attached to it, the operating system must be able to command those attachments' positional motions as well as the positional motions of the manipulator. For the sake of simplicity, we will discuss only the positioning of the manipulator at present. With an actual robot, however, the length of the wrist and gripper needs to be taken into account when calculating where the robot is to reach.

The method or methods used to train or transmit applications programs to the robot also influence the programs required in the operating system. Methods of doing application programming include using mechanical setup, using a control panel, using a teaching pendant, using off-line programming, and using task programming.

Home

The points within a robot's work area are defined by their Cartesian coordinates. **Home** for a robot is a known reference position from which all coordinate measurements are gauged. Other things besides the manipulator can also have a home reference, including the wrist, the gripper, the parts and tools being used or moved by the robot, and other machines working with the robot.

Programming Needs That Depend on the Controller

At present, there are three styles of robot controllers: the limited-sequence controller, the point-to-point controller, and the continuous-path controller. Each has unique programming requirements.

Limited-sequence Controller The **limited-sequence controller** is used on the limited-sequence (or bang-bang) robot. In fact, in the United States, these robots are normally called *material transfer devices* rather than robots. The simplest bang-bang robots are pneumatic-powered and have a mechanical controller. Their application programming is done by **mechanical setup;** that is, the endpoints of travel for each axis are established by placing a mechanical stop on each end of the axis, and the activation timing of an axis or of the gripper is fixed by placing pegs as desired in a mechanical timing drum. These mechanical timers are similar to the timers used on dishwashers and washing machines. The main difference in the robot timer is that it is adjustable, while the timers in dishwashers and washing machines are nonadjustable.

The operating system programs for such a mechanical controller are thus built into the mechanical parts. The timing for the robot's cycle is one complete revolution of the mechanical timing drum. The intelligence to move an axis appropriately is built into the mechanism that works the valve controlling pneumatic pressure to the axis. The intelligence to recognize a command to activate an axis consists of a lever that detects a peg in the timing drum and turns on the air pressure. The intelligence capability for receiving the application programs resides in predrilled holes in the surface of the timing drum and places along the axes where a mechanical stop can be fastened.

An electronic controller can be used in place of the mechanical controller on some bang-bang robots. In that case, some of the operating system programs are built into the electronic circuits of the controller.

Figure 14–1 shows a block diagram of a limited-sequence robot operating system.

Point-to-Point Controller The **point-to-point controller** uses some type of electronic memory to record a number of positions for the manipulator. Each position represents a value for each axis and each sensor of the robot. In addition, the operating system must provide ways to move the manipulator to each required position of the task and to determine whether each such position has successfully been reached. This might be accomplished by using stepper motors, in which case the operating system must keep track of the number of pulses that have been emitted and the direction in which the system has commanded the stepper motor to move since it was last set to a home position. If a sensor is used to keep track of the robot's position, the operating system must monitor and interpret the sensor's information.

Figure 14–2 shows a block diagram of a point-to-point robot operating system. This operating system differs from the continuous-path system mainly in the size and speed of each function.

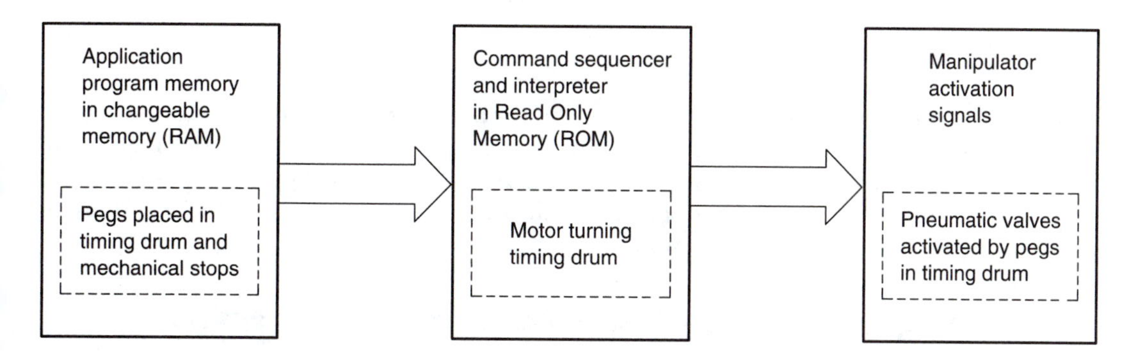

Figure 14-1 Limited-sequence Robot Operating System.
This operating system may be purely mechanical, such as using a motor to turn a timing drum. More modern limited-sequence robot operating systems use electronic circuits and read only memory (ROM). Whether electronic or mechanical, the operating system translates applications program commands into pneumatic value activations.

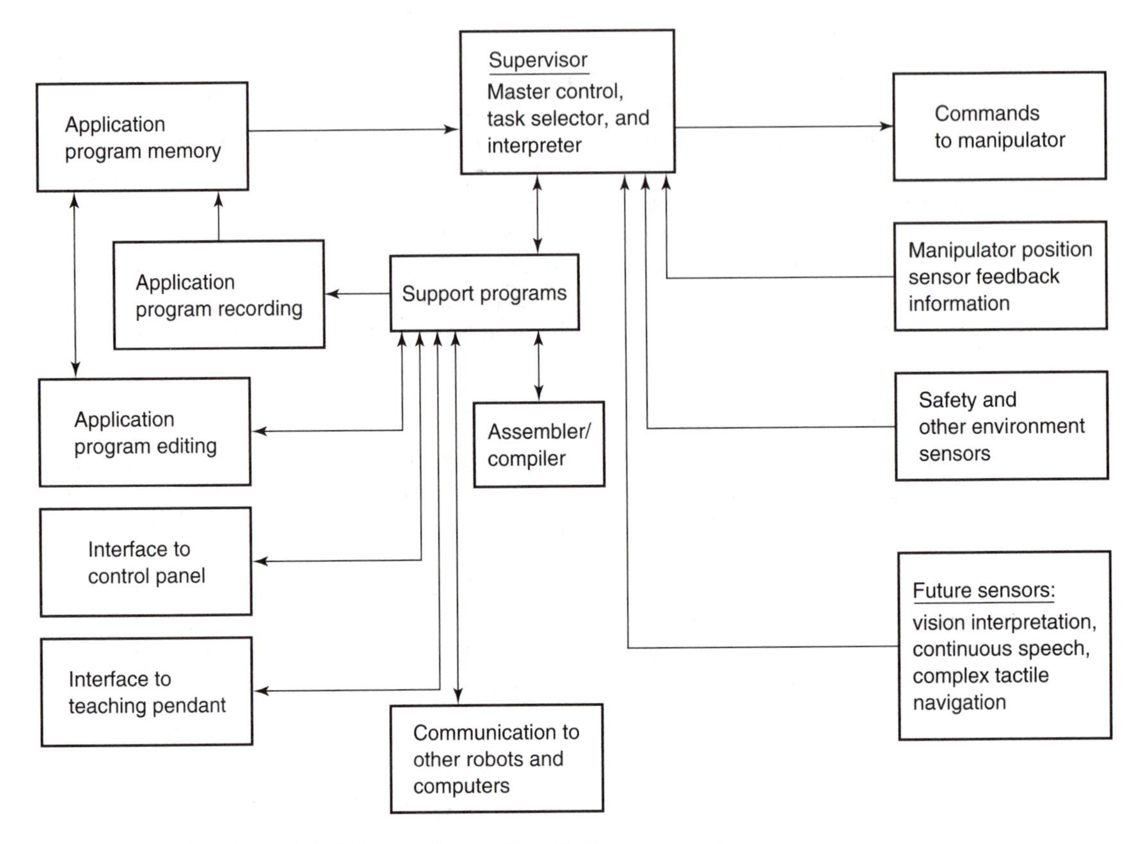

Figure 14–2 Point-to-Point Robot Operating System.
This operating system must support feedback information on manipulator position from a sensor. It must also support control panels and teaching pendants.

Continuous-path Controller The **continuous-path controller** must be able to record and play back the robot's position many times each second. This requires an electronic memory many times larger than that possessed by a point-to-point controller. Most continuous path controllers use sensor information to keep track of position. Figure 14–3 shows the continuous-path robot operating system.

Programming Needs That Depend on Arm Configuration

There are five basic arm configurations for industrial robots: Cartesian coordinates, cylindrical coordinates, SCARA, polar coordinates, and jointed arm. Each configuration has different programming requirements. One could consider the Cartesian-coordinates as people coordinates, or "real-world" coordinates that humans use. These need to be translated into the joint angles for the various configurations of robot arms.

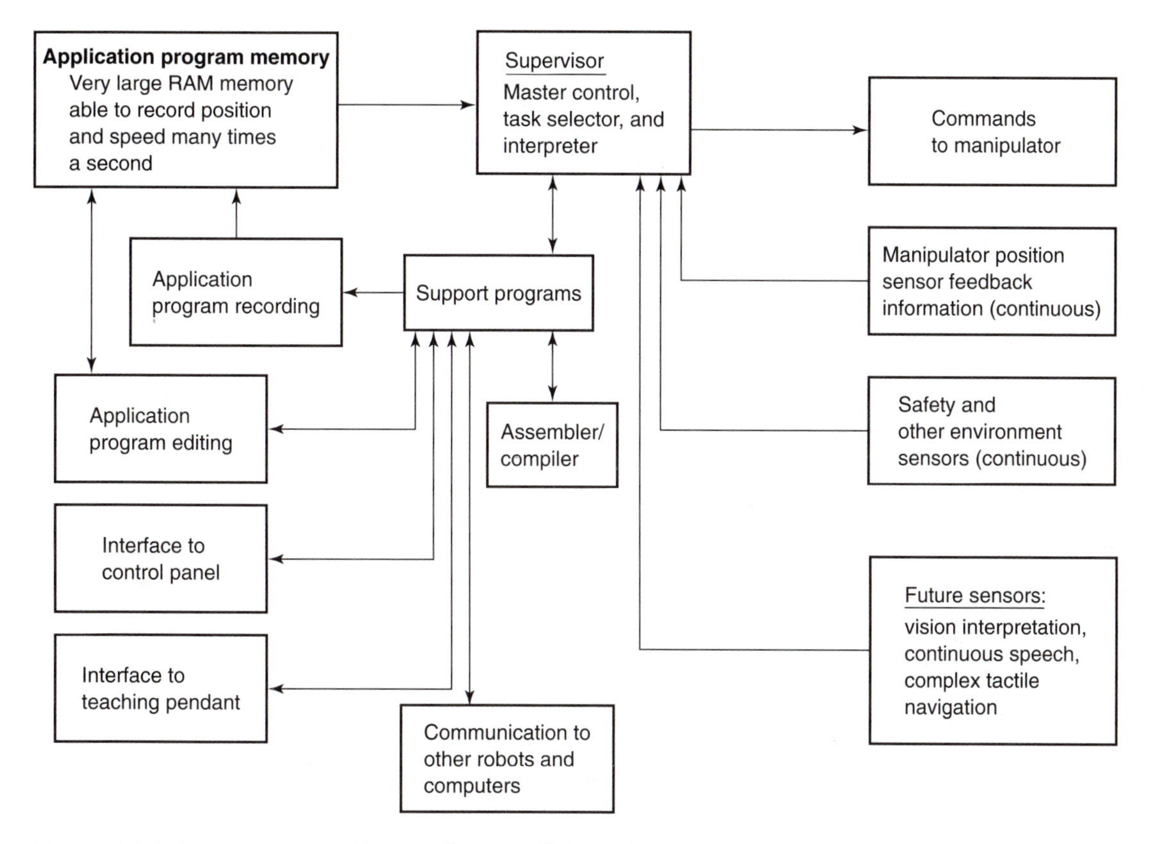

Figure 14–3 Continuous-path Robot Operating System.
This operating system must support continuous feedback information in real time on manipulator position from sensors. It must support playback of a program at the same speed it was recorded. It must also support control panels and teaching pendants and leadthrough programming.

Cartesian-coordinates Robot The operating system for a Cartesian- or rectangular-coordinates robot simply moves the manipulator to specified points on its Cartesian-coordinates axes. The robot can position all three axes at the same time.

The home reference for the Cartesian-coordinates robot could be taken as the lower left front corner of the work area. Assume that the robot to be programmed has a workspace 24 inches square, and suppose that the robot is to reach a point with an x coordinate of 12 inches, a y coordinate of 10 inches, and a z coordinate of 13 inches; such a point could simply be labeled (12, 10, 13). Figure 14–4 shows the workspace for this robot, with the desired point identified inside it. The operating system would command the x-axis of the robot to move out 12 inches from its home position; the y-axis, to move out 10 inches from its home position; and the z-axis, to move out 13 inches from its home position. In this way, the robot would attain its desired position inside the workspace.

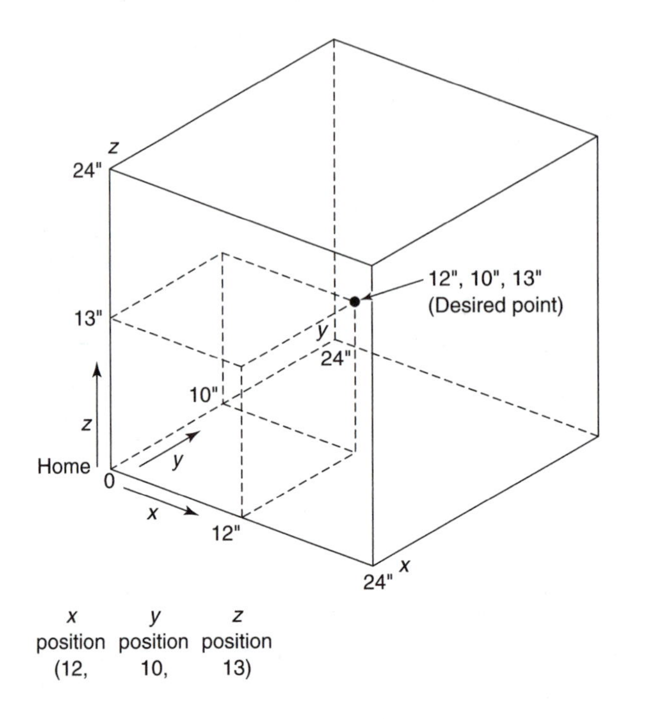

Figure 14–4 A Point for a Cartesian-coordinates Robot.
Points for a Cartesian-coordinates robot are specified as values along three linear axes—*x*, *y*, and *z*. This is easy to visualize and easy to program off-line.

Cylindrical-coordinates Robot The operating system for a cylindrical-coordinates robot must translate the *x* and *y* Cartesian coordinates into corresponding angle and radius position values. The *z* coordinates remain the same. The Cartesian coordinates' home reference point for the cylindrical-coordinates robot could be taken as the bottom center of the work area, with the arm retracted all the way in and aligned with the *x*-axis so that the *z* coordinate would be zero.

Figure 14–5 shows the same Cartesian-coordinates point illustrated in Figure 14–4 (with *x* at 12 inches, *y* at 10 inches, and *z* at 13 inches) within the work area of a cylindrical-coordinates robot. The figure also shows how the *x* and *y* Cartesian coordinates relate to the radius or reach and the angle of rotation for the cylindrical-coordinates robot.

Since the reach value *R* is the length of the hypotenuse of the right triangle whose sides are 12 and 10 inches, *R* is equal to the square root of the sum of 12^2 plus 10^2, or 244, which is 15.62 inches. Now the operating system must use trigonometry to calculate the sine of the angle of base rotation γ. The sine of γ is equal to *y* divided by the radius *R*; and from this value, the angle γ can be found in a sine table. The base angle is thus 39.81°. Finally, the angle is corrected for any minus sign in *x* or *y*.

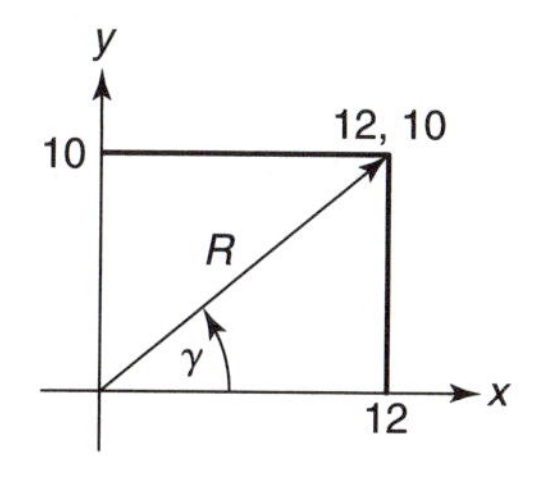

$R = \sqrt{x^2 + y^2} = 15.62"$
$\sin \gamma = y/R = 0.6402$
$\gamma = ? = 39.81°$
If x is negative and y is positive, subtract from 180°.
If x and y are negative, add 180°.
If x is positive and y is negative, subtract from 360°.

Base angle	Reach	z position
(γ,	R,	z)
(39.81°,	15.62",	13")

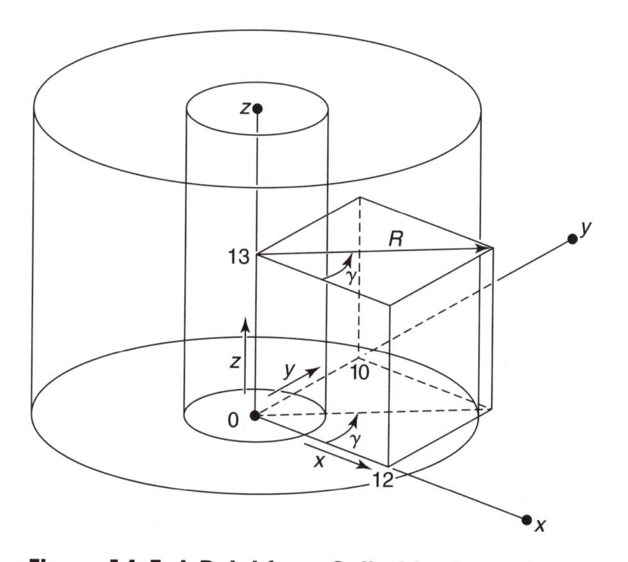

Figure 14-5 A Point for a Cylindrical-coordinates Robot.
Points for a cylindrical-coordinates robot are specified as points defined by two linear axes and one angle. The linear axes are called z and R (for reach). The angle is γ, the angle of rotation. This combination is more difficult to visualize than the one used by the Cartesian-coordinates robot. This makes it harder to program off-line.

As these calculations show, the cylindrical-coordinates operating system must be able to handle trigonometric functions and tables. Since inexpensive pocket calculators can now do these functions, it should not be very expensive to add these capabilities to the robot controller; but if the controller does not have these functions, the robot's programmer will have to do the calculations by hand, which is undesirable because humans tend to make mathematical errors. The final result of changing the x, y, and z coordinates would be to give the following commands to the robot: $\gamma = 39.81°$; $R = 15.62$ inches; and $z = 13$ inches. In point-coordinate form, this would be written as (39.81, 15.62, 13).

SCARA Robot The SCARA robot operating system must translate the x and y Cartesian coordinates into three angles of rotation: the base angle of rotation γ, the preliminary shoulder angle of rotation β, and the elbow angle of rotation α. The SCARA robot's home reference point could be taken as the bottom center of the work area, with the arm folded all the way in and aligned with the x-axis so that the z coordinate would be zero. Figure 14–6 shows these calculations for a left-handed SCARA robot. Since the SCARA robot's elbow can bend either to the left or to the right, these calculations could be worked up into two different solutions.

The angle of base rotation γ and the reach R of the robot are calculated the same way as for the cylindrical-coordinates robot. That is, given our previous Cartesian coordinates, γ remains 39.81° and R remains 15.62 inches. However, the reach information is now used to calculate the shoulder and elbow angles. This, of course, requires that we know the lengths of the upper arm and of the lower arm. Suppose that A, the upper arm length, is 10 inches and that B, the lower arm length, is 10 inches. At this point, we need to check to make sure that the upper arm length plus the lower arm length is equal to or greater than the required reach. If it is not, the robot will not be able to do the task. Here, since 10 inches plus 10 inches is 20 inches, the arm has more than enough length to manage the 15.62-inch reach specified.

Since we now know the length of all three sides of the triangle, we can use the Law of Cosines to calculate the elbow and shoulder angles. The Law of Cosines states that the cosine of an angle is equal to the sum of the squares of the two adjacent sides of a right triangle minus the square of the opposite side, divided by the sum of twice each of the adjacent sides. If an angle has a negative cosine value, the angle is greater than 90°, and we must subtract the cosine table value from 180° to get the real value. If you can use a scientific calculator, the above step is taken out automatically. The shoulder angle β calculates to 38.64°. But since there is no base axis, the complete shoulder angle is equal to 38.64° (β) plus 39.81° (γ), or 78.45°. The elbow angle α calculates to 102.71°. The final command given to the robot would be for it to move to the complete shoulder angle $\gamma + \beta$ of 78.45°, the elbow angle α of 102.71°, and the height z of 13 inches. In point-coordinate form, this would be written as (78.45, 102.71, 13).

Polar-coordinates Robot The polar-coordinates operating system must translate all the Cartesian-coordinates system's axes to its own set of coordinates: a base angle γ, an elevation angle β, and a reach distance R. Figure 14–7 shows the Cartesian-coordinates point of 12, 10, and 13 inches for x, y, and z, within the work area of a polar-coordinates robot.

The Cartesian coordinates' home reference point for the polar-coordinates robot could be taken as the center of its horizontal working plane, with the arm retracted all the way in and aligned with the x-axis so that the z-axis reference would be zero. The figure also shows how the x and y coordinates relate to the radius R and the base angle γ. This yields the same values as were calculated for the cylindrical robot: R = 15.62 inches and γ = 39.81°. Next, R is used with the z

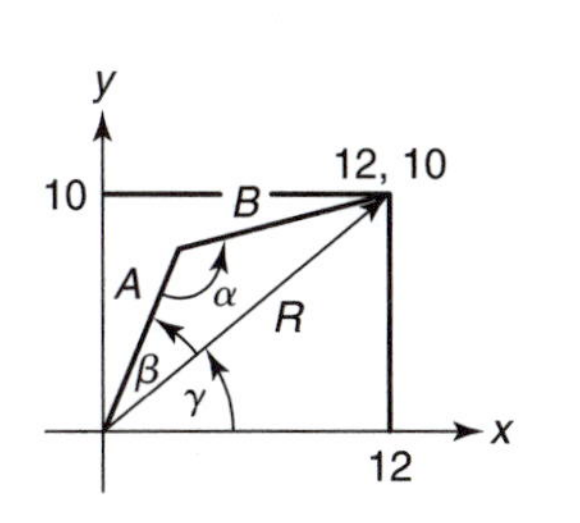

Let $A = 10''$ and $B = 10''$

$R = \sqrt{x^2 + y^2}$

$\sin \gamma = y/R$

$\gamma = ? = 39.81°$

If x is negative and y is positive, subtract from 180°.

If x and y are negative, add 180°.

If x is positive and y is negative, subtract from 360°.

$$\cos \beta = \frac{A^2 + R^2 - B^2}{2(A)(R)}$$

$\beta = ? = 38.64°$

If β is negative, subtract it from 180°

$$\cos \alpha = \frac{B^2 + A^2 - R^2}{2(B)(A)} = -0.23$$

$\alpha = ? = 77.29 + 180 = 102.71°$

If α is negative, subtract it from 180°

Base + shoulder angle	Elbow angle	z
$(\gamma + \beta,$	$\alpha,$	$z)$
(78.45°,	102.71°,	13")

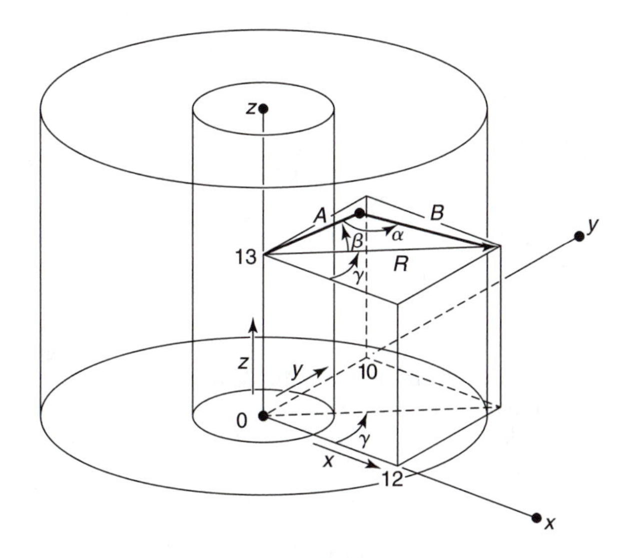

Figure 14-6 A Point for a SCARA Robot.
Points for a SCARA robot are specified as one point along a linear axis and two angles. The linear axis is height, or z. The angles are the base and shoulder angle $\gamma + \beta$ and the elbow angle α. This combination is difficult to visualize and to program off-line. Either a software support program or a scale model of the arm is needed in order to do much off-line programming.

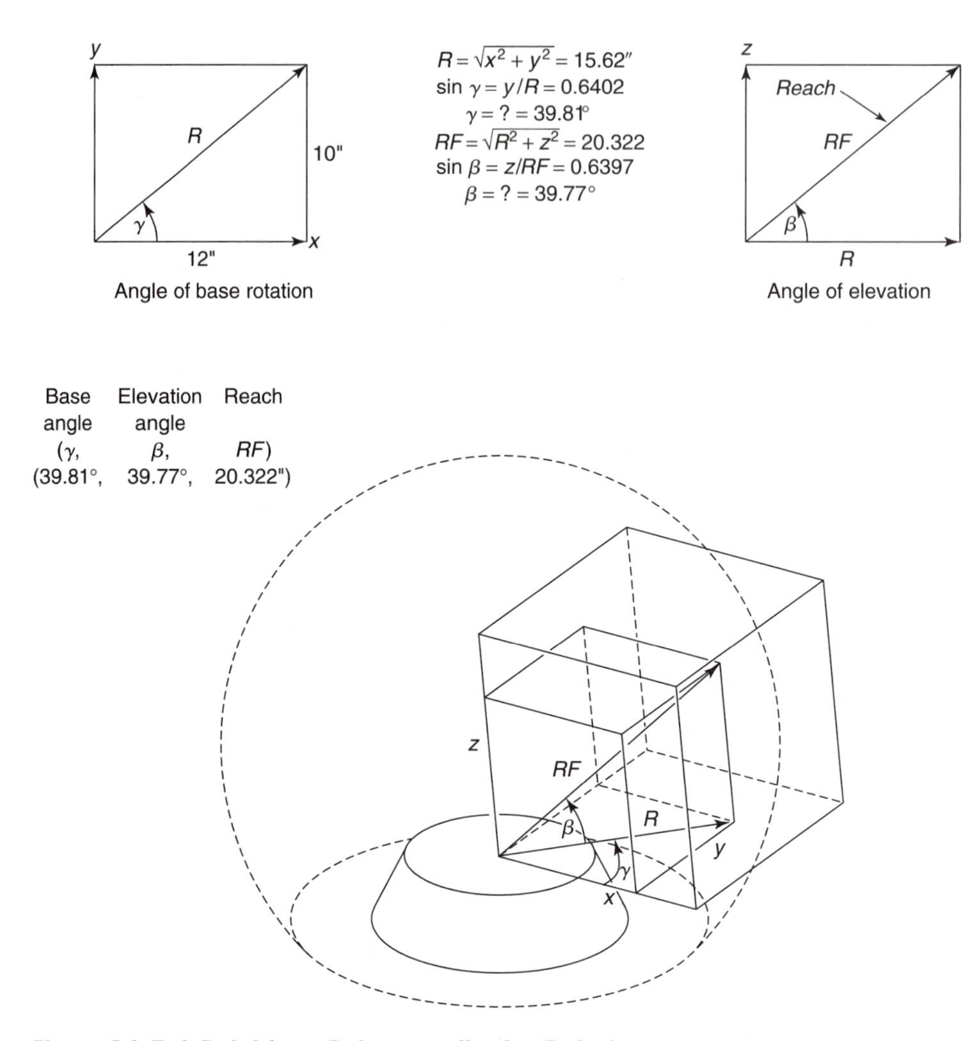

Figure 14–7 A Point for a Polar-coordinates Robot.
Points for a polar-coordinates robot are specified as a base angle γ, an elevation angle β, and a linear reach *R*. This is difficult to visualize. Once again, either software support or a scale-model arm is needed to do extensive off-line programming.

coordinate to relate to a second radius R_F for the robot, along with its additional angle β. These work out to 20.322 inches and 39.77°, respectively. This calculation yields the following final commands to the robot: γ = 39.81°, β = 39.77°, and R_F = 20.322 inches. In point-coordinate form, this would be (39.81, 39.77, 20.322).

Jointed-arm Robot The jointed-arm robot operating system again must translate all of the Cartesian-coordinates system's axes to its own system of coordinates: A base angle γ, a complete shoulder angle (elevation + shoulder angle) β +

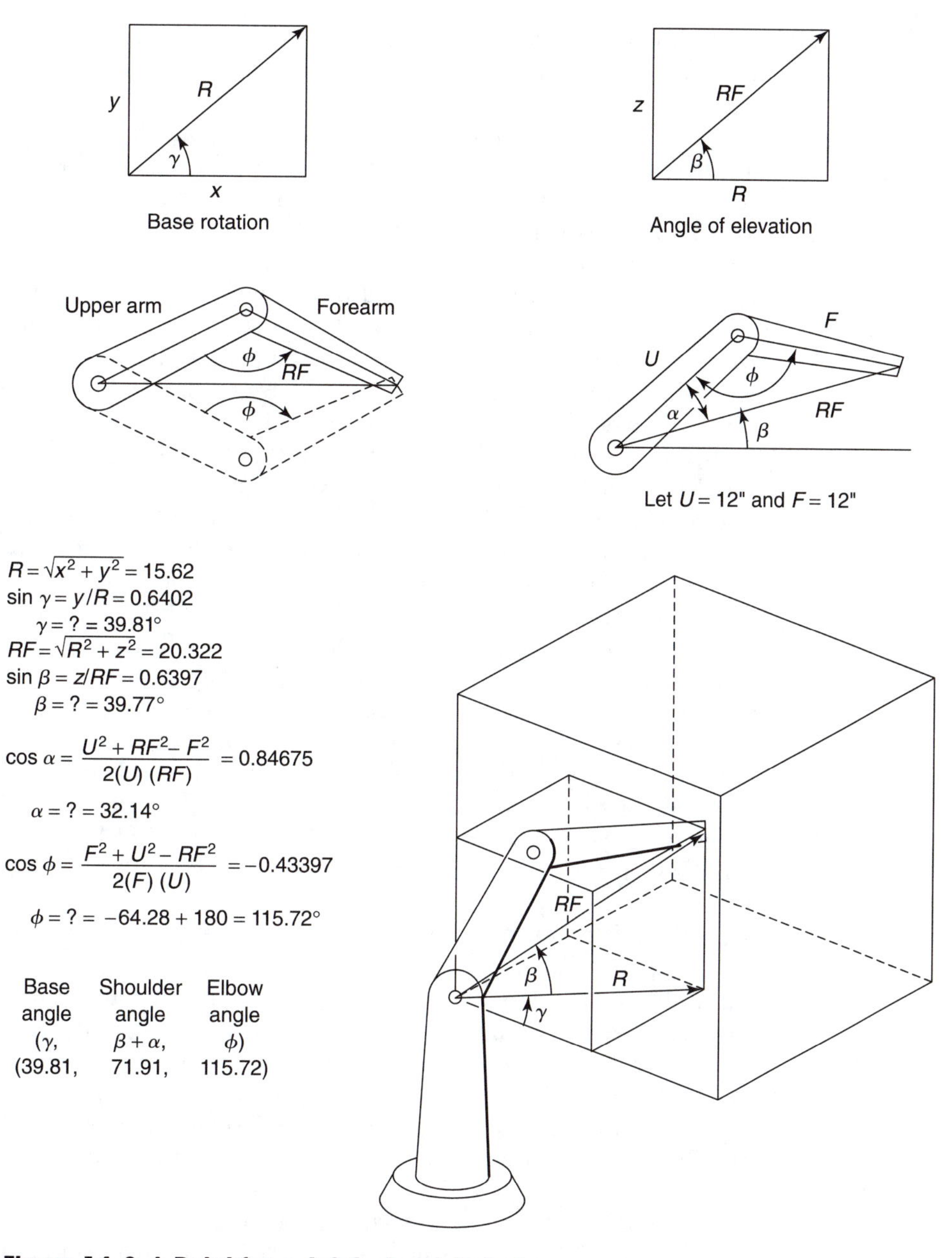

Figure 14-8 A Point for a Jointed-arm Robot.
Points for the jointed-arm robot are specified as three angles: the base angle γ, the complete shoulder angle β + α, and the elbow angle ϕ. It is difficult to visualize and requires software support or a scale-model arm to do extensive off-line programming.

α, and an elbow angle ϕ. Figure 14–8 shows the Cartesian-coordinates point of 12, 10, and 13 inches for x, y, and z, within the work area of a jointed-arm robot.

The Cartesian coordinates' home reference point for the jointed-arm robot could be taken as the center of its horizontal working plane, with the arm retracted all the way in and aligned with the x-axis so that the z-axis reference would be zero. All of the calculations that were done on the polar-coordinates robot—namely, a base angle γ of 39.81°, an effective elevation angle β of 39.77°, and a reach R_2 of 20.322 inches—are used in the jointed-arm system. In addition, it is necessary to calculate the angle between the forearm and upper arm of the manipulator that will allow the arm to reach the proper distance or radius. To add to the complications, the jointed-arm robot may have to choose from among up to four alternative ways of reaching the same point.

Before you can calculate the angles for the shoulder and elbow, you need to know the lengths of the upper arm and forearm. In this case, we will use 12 inches for both the upper arm length and the lower arm length. At this point, we need to check to make sure that the upper arm length plus the lower arm length is equal to or greater than the required reach. If it is not, the robot will not be able to do the task. Here, since 12 inches plus 12 inches is 24 inches, the arm has more than enough length to manage the 20.327-inch reach specified.

We can now use the Law of Cosines to solve for the rest of the upper arm elevation angle α and the elbow angle ϕ. This upper arm angle calculates to 32.14°, which when added to the elevation angle already calculated, gives a complete shoulder angle $\beta + \alpha$ of 71.91°. The elbow angle ϕ calculates to $-$ 64.28°, so we add 90° and get the final elbow angle of 154.28°. Thus, the final values that need to be given to the jointed-arm robot are as follows: $\gamma = 39.81°$, $\beta + \alpha = 71.91°$, and $\phi = 154.28°$. In point-coordinate form, this would be (39.81, 71.91, 154.28).

Other Considerations Up until now, we have been concerned only with defining a point that the manipulator needs to reach. No consideration has been given to the acceleration and deceleration of the manipulator or even to the speed needed to move the manipulator through more than one point. Some applications require the robot to follow a specific path as it goes from point to point in its task. A very sophisticated operating system is needed to move a jointed-arm coordinates robot in a straight line, since this requires coordination of all three axes. The axes of movement in the robot's wrist and gripper must also be controlled.

The development of sensors will greatly increase the demands on the operating system. As vision, speech understanding, touch, navigation, and other senses are added to the robot operating system, the robot will need a very powerful computer and all the help it can get from artificial intelligence programs to deal with the data its sensors gather.

Currently, industrial robots with vision do not process visual information in their own controllers. They just receive the results from a vision processing system, which is a separate computer.

Programming Needs That Depend on Robot Training Techniques

The method used to train the robot (give it application programs) affects the operating system requirements. The limited-sequence controller can receive application programming training only through mechanical setup. The point-to-point and continuous-path controllers may use control panels, teaching pendants, leadthrough programming (lead by nose), off-line programming, and transfer from some other robot or computer to do its training.

Mechanical Setup *Mechanical setup* refers to an arrangement where the placement of mechanical stops determines the movement limits of the robot's axes. The operating system does not keep track of the placement of these stops, nor does it even know that they exist. It responds only to the mechanical signal telling it when to activate the axis. If limit switches are used to control the robot, the operating system must respond immediately to a change of status in one of these "sensor" switches, which determines when to stop the motion along an axis. Mechanical stops may be used on any robot to limit axis movement for safety reasons. But since these stops are not used for guidance in the normal operation of the robot, the operating system is not concerned with them.

Control Panels All robots require some type of control panel, even if it is only used to turn the robot on and off. Most control panels have sensors for over-voltage, under-voltage, low pressure, high pressure, high temperature, and so forth. Most operating systems should monitor and react to these sensors. Control panels for point-to-point and continuous-path controllers may also be used to train and operate the robot. If so, the operating system must monitor all of these controls, be able to accept training commands, and be able to store them for future use. To accomplish this, it must have some method of identifying the location of each axis, and it must be able to execute commands when told to. A control panel that can be used to train the robot can also be used to edit or change the recorded training commands. Chapter 15 presents a demonstration program in which a control panel is used to give the directions to an AR-1 welding robot.

Teaching Pendants The most popular method of programming point-to-point and continuous-path robots is to use a teaching pendant to lead the robot through the task. The operating system requirements for the teaching pendant are about the same as for a control panel used to train a robot. Figure 14–9 shows some teaching pendants. Chapter 15 includes a short Rhino robot teaching pendant program.

Leadthrough Programming Leadthrough programming is the process of moving the robot manipulator through the steps of the task and having the controller record the process. In the case of a spray-painting or welding robot's training, a skilled painter or welder is required to do the leadthrough programming.

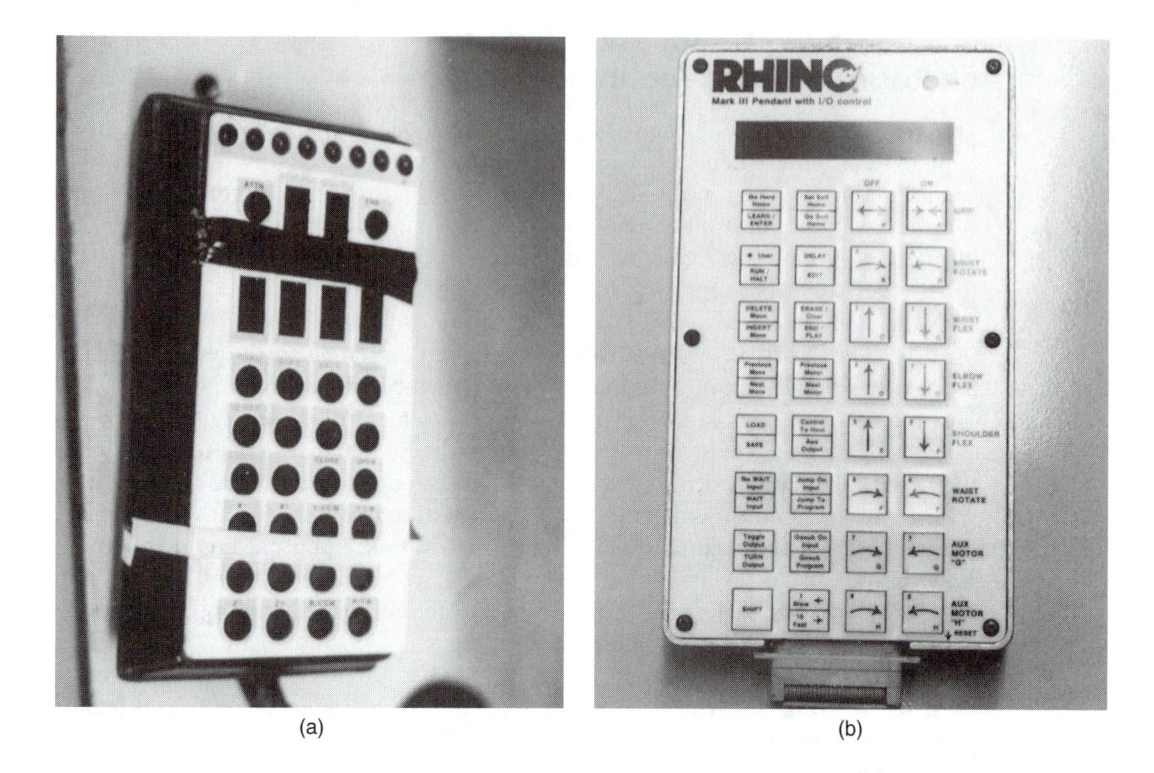

(a)

(b)

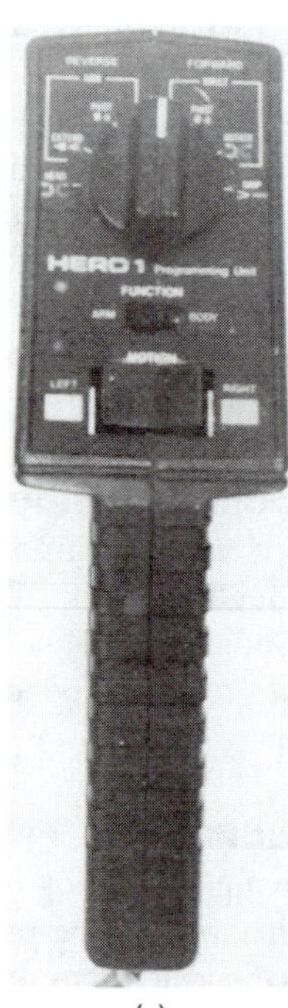

(c)

Figure 14-9 Teaching Pendants.
(a) An IBM robot teaching pendant. (b) The Rhino robot teaching pendant. (c) The Hero I teaching pendant.

Off-line Programming Off-line **programming** is the process of entering robot commands straight into the controller's memory. For a point-to-point robot, which assumes at most only a few hundred positions in the course of its task, the program can be planned on paper and loaded directly into memory. For a continuous-path robot, which passes through thousands of positions in the course of a task, off-line programming is not very practical. The operating system requirements for doing off-line programming are similar to those for editing robot training commands. The capabilities can be expanded to constitute a complete CAR system.

Transfer from Other Robots or Computers If you want each of ten robots to do the same independent task, it is most efficient to train one robot and then transfer the task program to the memories of the other nine robots. To handle this style of training, the operating system must be able to communicate with other robots and computers.

Task Programming In **task programming,** the operating system must plan every detail of how to accomplish a task. This is something like when your boss tells you to clean up the storeroom and you must figure out all the specifics. Task programming for industrial robots does not yet exist, but as the industrial robot becomes more intelligent, it will come nearer to being able to handle task programming.

TYPES OF MEMORY USED IN OPERATING SYSTEMS

Most operating system programs are placed in some type of nonchangeable memory—read only memory (ROM)—by the manufacturer. This allows the operating system to be up and running the moment the robot's power is turned on. Unfortunately, it also makes changing the operating system difficult.

In some instances, only enough of the operating system is in ROM to allow the loading of the rest of the operating system from a flexible diskette. This has the advantage of allowing you to change the operating system simply by running a new or revised program on the flexible diskette.

In hobbyist robots, the operating system's memory may take the form of gears and cams, like those of a windup watch, or wiring and electric circuits. In the industrial pick-and-place robot, memory generally takes the form of wires, limit switches, mechanical stops, and circuits activated by the pegs on a mechanical rotating drum. In the industrial point-to-point and continuous-path robots, memory may take the form of wiring and electronic circuits or the silicon die or chip of a ROM integrated circuit.

LANGUAGES USED IN OPERATING SYSTEMS

The language used to write operating system programs for robots depends on who is writing the programs and what computers are available to run them. People with an electronic engineering background tend to use microcomputers and the program language BASIC. Those who have a solid programming background and use microcomputers may use Forth, BASIC, or possibly even C. Those who have a programming background and use a large-scale computer tend to use LISP if they work in the United States and Prolog if they work in Japan or Europe.

Machine Languages

The **machine language** of a robot is the actual native language the robot's controller understands. In a mechanical pick-and-place robot controller, the machine language consists of the action of limit switches, mechanical stops, and pegs in a drum that make or break mechanical contacts. The electronic computer controller on more expensive robots uses as its machine language a code consisting of a series of ones and zeros known as **binary numbers.** A certain pattern of ones and zeros might signal the controller to add two numbers, while a different pattern might signal it to close the gripper.

Engineers and programmers who write programs in the robot's machine language must be very familiar with every detail of the robot and its controller. Such work is very time-consuming and costly. Consequently, machine language is used to program robots only when they are so new that no other programming language is available or when the robot must respond to commands at maximum speed in order to do a job successfully.

Assembly Languages

An **assembly language** represents a middle ground between machine language and human language. It uses symbolic or convenient abbreviations in place of groups of zeros and ones. Thus, the operation of addition may be written as ADD rather than as, say, 01101011. When working with an assembly language, the computer does part of the housekeeping for the programmer, such as keeping track of the physical addresses of information in memory. Assembly language programs must be run through another computer program, called an **assembler,** in order to get translated into a machine language that the computer can execute.

A robot's assembly language capability is usually determined very early in its development. The robot's controller may well have been designed around an instruction set that already existed as a known assembly language, making early programming of the robot faster and easier. Since assembly language is so closely tied to the robot's machine language, assembly language programs are also used to give the robot maximum response time to execute a command.

High-Level Programming Languages

High-level programming languages use meanings for words that approach their human language meanings. Human languages themselves are ambiguous and remain beyond the level of computer understanding.

BASIC The Beginners' All-purpose Symbolic Instruction(al) Code, **BASIC**, is the most accessible computer language for home hobbyists and others who work on personal computers. It is even more attractive for home robot programming when special robot motion-related commands are added. Forms of BASIC are available for the Hero I, Hero 2000, and Fischertechnik educational robots. Intelledex, an industrial robot manufacturer, has been using a form of BASIC to program its line of robots.

Forth The **Forth** computer language is widely used in controlling the motion of observatory telescopes. It was developed by Charles Moore at the Kitt Peak Observatory in Arizona. The mechanism of an observatory telescope is similar to that of a polar-coordinates robot's manipulator. Forth was designed to run in minicomputers and microcomputers. Its popularity as a control language is due to its adaptability. If Forth does not have an instruction for doing the action you require, you can define a new instruction to accomplish the desired operation. Forth has a drawback of not being very user-friendly. Because it is difficult to understand, it should not be used by novice programmers.

Some early Fischertechnik educational robots use Forth as their operating system.

LISP LISP (short for List Processing) is a computer language developed in the late 1950s at MIT. Most conventional programming languages such as COBOL, FORTRAN, Pascal, and BASIC are designed to manipulate numbers, but LISP is designed to manipulate lists of symbols and is well suited for working with words and phrases. One of the first word-processing programs was written in LISP, and LISP is also used in artificial intelligence work.

There are many versions or dialects of LISP. One, called Interlisp-D, from Xerox's Palo Alto Research Center, has an automatic spelling corrector called DWIM (for Do What I Mean). Since programming began, programmers have been plagued by spelling errors that result in a computer's doing what it is told to do—and not what the programmer means for it to do. In an effort to make all LISP computer programs uniform and usable on any computer that understands LISP, a standard version of LISP—called Common LISP—has been created.

LISP allows a program to assign memory as the program is running, rather than requiring a programmer to assign all memory before a program runs. LISP was designed to run on large computer systems, so it can handle large information bases. Like Forth, LISP is quite adaptable and thus allows the user to create new computer instructions. It has an English language interface for ease of use. A few computers now being designed with LISP as their native language are very efficient for writing artificial intelligence programs.

Artificial intelligence researchers have found that intelligence processes involve many lists, groups, or strings of items. Human languages such as English are composed of lists. Lists of letters make up words; lists of words make up sentences; lists of sentences make up paragraphs; lists of paragraphs make up chapters; lists of chapters make up books; lists of books make up libraries.

In the area of vision processing, humans again work with lists of items. Lists of lights make up patterns; lists of patterns make up shapes; lists of shapes make up objects. In many cases, humans can recognize an object from seeing only some of the shapes that compose it. Similarly, speech processing, smell processing, and most other areas of human intelligence use lists as the basis for interpreting sensory information.

As robots receive more complicated sensor input information, artificial intelligence programs will be needed to help interpret the information. As robots become more intelligent, their operating system will have to support new types of commands, known as **task commands,** rather than the present-day step-by-step directions. With a task command, the robot is told what to do rather than how to do it. The operating system must then figure out the steps needed to accomplish the task.

A LISP program has been constructed to demonstrate task commands in a blocks world.[1] A **blocks world** consists of a table with a series of blocks placed on it. Most industrial robots work in a controlled environment that resembles a blocks world in possessing a very restricted level of complexity. Each block in the blocks world would require a property list of all of its attributes—its shape, its color, the object or objects supporting it, the object or objects it supports, and its location within the blocks world. It is already possible to write LISP procedures to handle a request such as, "Put block A on top of block B." Handling this task would require support procedures that might include the following: PUT-ON, PUT-AT, GET-SPACE, MAKE-SPACE, GET-RID-OF, GRASP, MOVE-OBJECT, UNGRASP, FIND-SPACE, CLEAR-TOP, MOVE-HAND, REMOVE-SUPPORT, and ADD-SUPPORT.

The highest-level procedure for handling the "Put block A on top of block B" request would probably be the PUT-ON procedure, since this procedure would require the least information on the details of the properties of block A and block B. It would need to pass only the names of block A and block B onto the next-level procedures. The PUT-ON procedure could call the PUT-AT and/or the GET-SPACE procedures, and these procedures in turn could call other procedures. For instance, the PUT-AT procedure could call the GRASP, MOVE-OBJECT, and UNGRASP procedures. The lowest-level procedures would need to know all of the properties of the block in order to do the actual work. The procedure PUT-AT could be written in LISP as the following function:

```
(DEFUN PUT-ON (ITEM SUPPORT)
(PUT-AT ITEM (GET-SPACE ITEM SUPPORT)))
```

To use the PUT-ON procedure, you would need to issue a statement like the following:

```
(PUT-ON ⟨item name⟩ ⟨support name⟩)
```

Thus, in the specific case of "Put block A on top of block B," the proper statement would be:

```
(PUT-ON A B)
```

[1]Patrick H. Winston and Berthold K. P. Horn, *LISP* (Menlo Park, Calif.: Addison-Wesley, 1984).

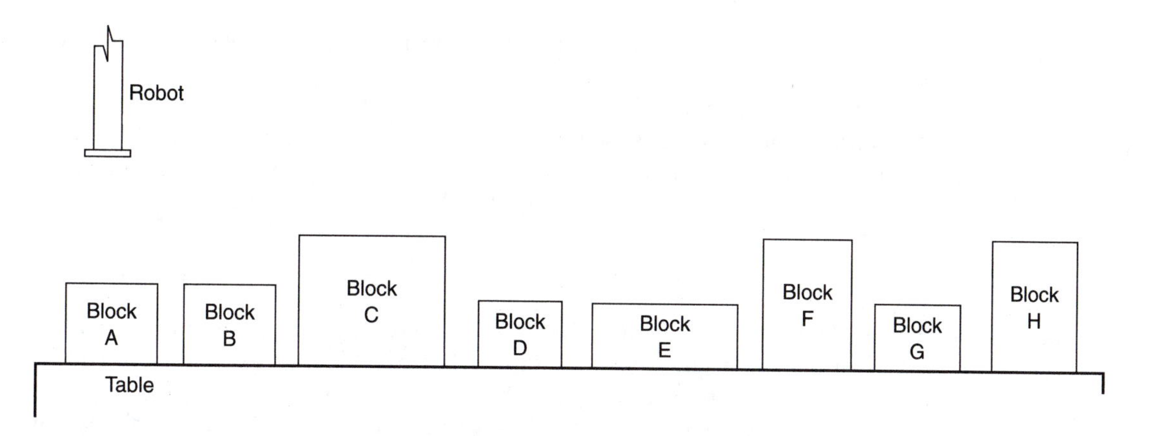

Figure 14–10 Blocks World Before the PUT-ON A B Command.
The blocks in the blocks world start out at preassigned locations on the table.

The PUT-ON command is a task command, because it does not provide any information on how to do the task. The user of the PUT-ON command does not need to know where blocks A and B are or what they are supporting. Figure 14–10 shows a possible starting position for a two-dimensional blocks world. Figure 14–11 shows the changed blocks world after the robot has carried out the PUT-ON A B command.

When PUT-ON makes use of the procedure PUT-AT, PUT-ON transmits to PUT-AT the item and its associated support information—which in this case is A and B. As the robot gets deeper into the job of moving block A, it will need to know some of block A's properties, such as where block A is and whether anything is being supported by the block. But since the PUT-ON procedure did not need to

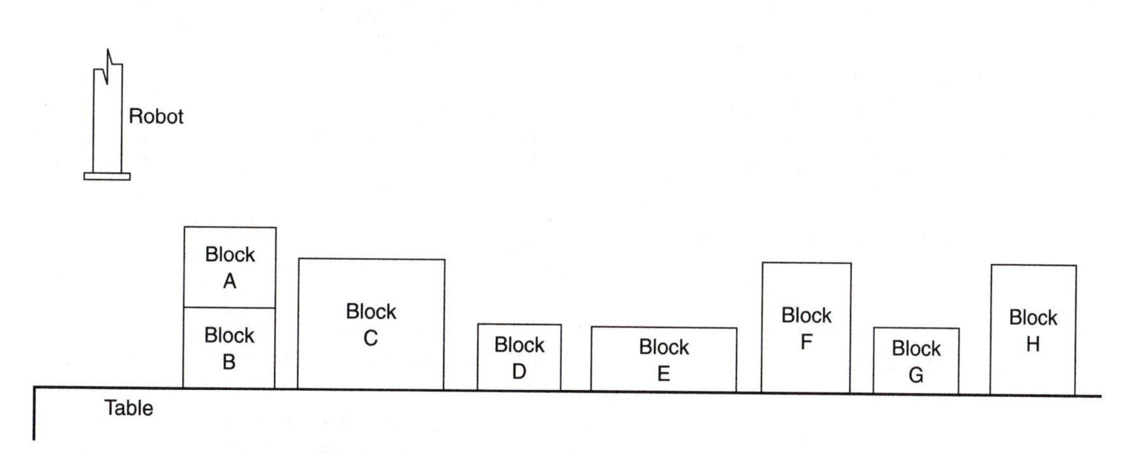

Figure 14–11 Blocks World After the PUT-ON A B Command.
After the command PUT-ON A B has been executed, block A is on top of block B.

possess these details, it is therefore protected from being encumbered by the information.

A person who wishes to give step-by-step PUT-AT commands to the robot needs to know more of the present properties of the blocks. For example, the programmer must tell the robot precisely where to place the block. Getting the robot to place block A on block F, using the PUT-AT command, would require a statement like the following:

```
(PUT-AT' (BF-HOME 1 BF))
```

The BF-HOME term shows that block BF was here when the program was loaded. The 1 identifies the height above the table where the block is to be placed. The BF term represents the supporting object. Clearly, task programming requires a lot less knowledge. The PUT-AT command allows the user to place blocks in places not allowed by the PUT-ON command. This means that the PUT-ON command may not work properly after the PUT-AT command has been used—for example, leading to attempts to crush some of the blocks.

The Oregon Institute of Technology uses a modified X-LISP programming language to program the Fischertechnik educational robot. This makes it easier to tie artificial intelligence programs into the job of robot control.

Prolog The **Prolog** computer language is the main language used by the Japanese in their fifth-generation electronic computer. Prolog was invented in Europe and is a procedure-oriented computer language. It is especially popular for doing symbolic computations such as for relational databases, mathematical logic, abstract problem solving, understanding natural language, architectural design, symbolic equation analysis, biochemical structure analysis, and various areas of artificial intelligence. Prolog can be used to help a robot understand its sensor input information.

A version of Prolog, called Prolog II, has been designed to run on IBM PC computers that have at least 128K bytes of memory. This version should increase Prolog's popularity in the United States.

Writing a computer program in Prolog consists of three stages: declaring some facts about objects and their relationships to each other; defining rules about the objects and their relationships; and asking questions about the objects and their relationships. Prolog is a conversational language. The statement "John likes Jane" would be written in Prolog as follows:

```
likes (john, jane).
```

The relationship is the word *likes*. The objects of the relationship are *john* and *jane*, and their relations are those of actor and receiver, respectively. The period (.) signifies the end of the fact.

A rule in Prolog is used to show that a fact depends on a group of other facts. For instance, if John likes many different persons, this relationship could be summarized by the following rule:

```
likes (john, x).
```

The *x* in the rule can then stand for any or all of the persons that John is shown to like in other fact statements.

Given a database full of facts and rules about John, we are ready to ask the program questions about him, such as "Who or what does John like?" This question could be asked in Prolog as follows:

```
?- likes (john, x).
```

In response to this message, the Prolog program starts at the beginning of the database on John and looks for something or someone that John likes. The first value found as a thing or person John likes is returned as the answer, and the program then waits for the user to tell it what to do next. A sample answer might be as follows:

```
x = jane
```

If we now type in a semicolon (;), the program will look for the next fact that answers the question. This answer, depending on what facts are in the database, might be the following:

```
x = computers
```

The above process can be repeated until the program can find no more facts to use in answering the question. In that case, it returns the following answer:

```
no
```

Like LISP, Prolog is currently used in educational robotics applications, not in industrial applications. But this state of affairs could change soon.

Ada If the Department of Defense has its way, all future programs, including embedded software, will be written in the **Ada** programming language. Embedded software consists of programs built into a computer. Any company that wishes to sell robots to the Department of Defense must write the robot's operating system programs in Ada. Ada is a Pascal-like programming language that in fact includes Pascal as a subset.

C C is a high-level programming language that has features normally found only in machine and assembly languages. This makes C a good programming language for writing operating systems for robots, and this language is gaining in popularity. Some robotics companies (such as Intelledex) that now use BASIC as their programming language for operating systems are switching to C. Certain versions of LISP and PROLOG are written in C—for example, X-LISP. Thus, C may well emerge as both an industrial robotics language and an educational robotics language.

SUMMARY

The programs for a robot fall into two categories: the operating system, which tells the robot how to do something, and the application programs, which tell the robot when to do something. The intelligence of a robot comes from the operating system. Robot operating systems vary from straight mechanical to very sophisticated electronic computers. The capabilities of a robot operating system depend on the type of controller, the arm configuration, and the method used to train the robot.

Many special languages can be used to write robot operating systems: mechanical language, machine language, assembly language, and high-level programming languages such as BASIC, Forth, LISP, Prolog, Ada, and C. Programming through mechanical setup requires the least work by the robot's controller; task programming requires the most. Receiving and interpreting information from sensors increases the system's intelligence needs further.

REVIEW QUESTIONS

1. What functions might a robot's operating system programs have to perform?
2. What factors affect the functions an operating system must handle?
3. Why is it desirable to write programs in a high-level language rather than in the robot's machine language?
4. Compare the Forth and LISP programming languages.
5. How do operating system programs and application programs differ?
6. Given that a robot is to move to Cartesian-coordinates point (11, 12, 13):
   a. Calculate the base angle γ and reach R needed for a cylindrical-coordinates robot to reach this point.
   b. Calculate the shoulder $(\gamma + \beta)$ and elbow (α) angles for a SCARA robot, if the upper arm is 12 inches long and the forearm is 11 inches long.
   c. Calculate the elevation angle β and reach R_2 needed for a polar-coordinates robot.
   d. Calculate the shoulder $(\beta + \alpha)$ and elbow (ϕ) angles for a jointed-arm robot, if the upper arm is 14 inches long and the forearm is 13 inches long.
7. How do the programming requirements for limited-sequence, point-to-point, and continuous-path controllers differ?
8. How does the robot training technique being used affect the required operating system's capabilities?
9. What additional capabilities must be added to a present-day robot controller in order for the controller to support task programming?
10. Given the people coordinates of a point (10, 12, 14):
    a. Calculate the base angle and reach needed for a cylindrical-coordinates robot to reach this point.
    b. Calculate the shoulder and elbow angles for a SCARA robot, if the upper arm is 11 inches long and the forearm is 10 inches long.
    c. Calculate the elevation angle and reach needed for a polar-coordinates robot.
    d. Calculate the shoulder and elbow angles for a jointed-arm robot, if the upper arm is 12 inches long and the forearm is 11 inches long.

15

ROBOT APPLICATION PROGRAMMING

OVERVIEW

When most people refer to teaching a robot, they mean programming the robot to do some task. This is more properly termed *application programming* of the robot. What is a robot application program? What methods are used to teach a robot how to do an application? What languages are used for robot applications programs? These are some of the questions discussed in this chapter.

OBJECTIVES

When you have completed this chapter, you should be familiar with:

- What robot application programming is
- The methods that may be used to give a robot application programs
- Some of the programming languages that can be used in robot application programming
- The steps involved in designing and programming, or training, a robot

KEY TERMS

ACL
AL
AML
Androtext
Application programs
APT
ARMBASIC
BASIC
CAP 1
Compiler
Control panel
Forth
HELP
Inputs
Karel
Ladder logic
Learning
LISP
MCL
Mechanical setup
Operating systems
Operations (op) sheet
Point-to-point recording
PRL
Programmable logic controller (PLC)
PROLOG
RAIL
RAM
ROM

RPL
Scorbase
Task leadthrough
Task setup
Teaching pendant
VAL
Wave

WHAT IS A ROBOT APPLICATION PROGRAM?

Application programs are sequences of instructions to be carried out by the robot to produce some desired objective. An application program might, for example, instruct a robot to remove a part from a machine and place that part on a conveyor belt. Most application programs for robots are written by the user. Getting an application program into a robot's memory is sometimes referred to as training or teaching the robot a task.

Application programs are generally stored in rewritable random access memory (**RAM**) in the robot. When the robot needs to learn another task, its memory is rewritten with the new task instructions.

LANGUAGES USED

The earliest methods for training a robot—mechanical setup, point-to-point path recording, and task leadthrough—did not use word-based languages at all. Nonetheless, each of these methods does represent a programming language.

Many robot manufacturers are trying to come up with high-level computer languages to make training a robot to do a task easier. A high-level computer language resembles human languages, such as English. Examples include Wave, AL, ACL, AML, HELP, Karel, MCL, RAIL, RPL, Scorbase, CAP 1, ARMBASIC, Androtext, VAL, APT, and Ladder Logic.

After a program has been written in a high-level language, it must be translated into machine language by another program, called a **compiler.** Robots are usually unable to run a compiler program, so an external computer must be connected to the system to serve this purpose.

Mechanical Setup

Mechanical setup was the language used to program the earliest robots. A mechanical setup uses adjusted mechanical stops, set limit switches, and the placement of pegs in revolving drums as its application program language. To institute such a program, the programmer must have some mechanical ability.

Early pick-and-place robots relied on mechanical setup for their programming, and some inexpensive robots still use this method.

Point-to-Point Recording

Point-to-point recording was the language used to program the first point-to-point robots. It requires positioning the manipulator at various stages of the task and recording the relevant coordinates at each stage. This is a faster and less mechanical method of giving a robot application training than setting up mechanical stop adjustments.

Task Leadthrough

Task leadthrough, the language used to program the continuous-path robot for an application, can best be described as on-the-job training. It requires that the robot be taken through all the steps of the task at normal task speed. In the case of a spray-painting robot's training, a skilled spray painter would be required to do the initial leadthrough.

High-level Languages

A number of high-level languages are used in application programming. Examples of such languages are discussed individually in the subsections that follow.

ACL The Advanced Command Language (**ACL**) is the robot language that employs a user-friendly conversational command environment. It is a high-level programming language similar to PASCAL. ACL is used by some Yaskaua robots and Eshed Robotec robots such as the SCORA, SCORBOT, and SCORBOT-ER V plus.

APT The Automatically Programmed Tools (**APT**) language is probably the oldest computer language dealing with motion. It was created to make the programming of numerical-controlled machine tools easier. Its basic structure was developed by the Electronic Systems Laboratory of MIT in 1956 and 1957, and the language became practical in 1961 with the release of APT III.

APT refers to both the computer language and the computer system that processes language statements written in APT. The APT computer system takes APT statements as input and produces a punch tape that controls a numerical-controlled machine tool automatically. Figure 15–1 shows the steps of the APT system from the action of the human programmer through the output of tape for the machine. Numerical-controlled machine tools use a Cartesian-coordinates system and require positioning of materials and tools, along with control of tool speed, coolant, and tool changes.

An APT program starts with a PARTNO statement, which identifies a program sequence for making some part. The end of the program is marked by a FINI statement. After the PARTNO statement, the program usually presents a list of geometric definitions, followed by an origin setup that uses a SETPT statement such as:

```
SETPT + POINT/0, 0, 0.
```

This is followed by a list of machine motions that reference the geometry definitions. Finally the machine is stopped by a STOP statement, and the sequence is ended by an END statement.

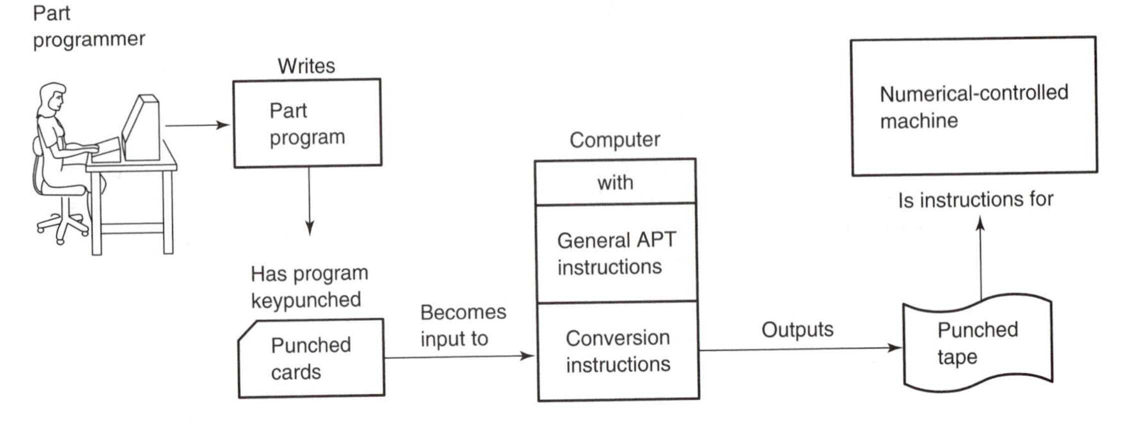

Figure 15–1 Steps in the APT System.
The APT system starts with a human programmer and ends with a punch tape of instructions for performing a task on a numerical-controlled machine.

Among the many APT statement words for describing geometric shapes and actions are POINT, CIRCLE, CONE, CYLNDR, HYPERB, LINE, PLANE, QADRIC, RADIUS, VECTOR, and SPHERE. If an operation can be expressed in geometric terms, it can be written in APT. APT action statements include PICKUP, UNLOAD, LEFT, RIGHT, GOTO, IN, OUT, ON, and OFF. The spindle speed and direction can be controlled with the statement

```
SPINDL/ n, CLW [OR CCLW]
```

where n is the spindle speed in rpm and CLW/CCLW stands for clockwise or counterclockwise, respectively. The feed rate of a tool is controlled with the statement

```
FEDRAT/n
```

where n is the rate in inches per minute. Cooling fluid can be controlled with the statement

```
COOLNT/ cc
```

where cc expresses a release level—flood, mist, off, or on.

Figure 15–2 shows the layout of a possible task for a robot to do using the APT program language. The problem is to have the robot deburr a slot that has been cut into the side of a cylinder. The following APT program listing directs the robot through this task:

```
PARTNO   DEBURR SLOT
         MACHINE/UNIV.1
         OUTTOL/0.0005
SETPT   = POINT 10, 18, 24
C1      = CIRCLE/72, 36, 24, 36
P1      = POINT/41.8231, 18, 24
```

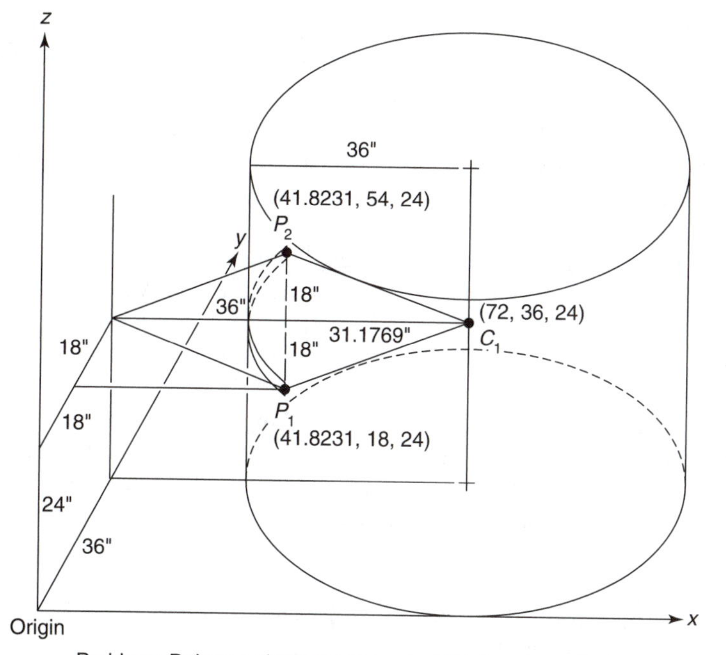

Figure 15-2 Robot Deburring Task Defined for APT.
The geometric layout for the APT task described in the text shows the dimensions of the slot to be deburred, plus the distances to be negotiated by the manipulator arm.

```
P2      = POINT/41.8231, 54, 24
REMARK    DEBURR SLOT
          CUTTER/4
          FROM/SETPT
          SPINDL/600, CCLW
          FEDRAT/RAPID
          GOTO, P1
          FEDRAT/10
          GOLFT/C1, PAST, P2
          FEDRAT/RAPID
          GOTO/SETPT
          STOP
          END
          FINI
```

To formulate an APT program like this one, the programmer must know something about machines and machine-processing techniques. Here, the APT programming requires knowledge of tool sizes, spindle rates, and cutting rates. In every concrete instance, the programmer must understand all of the techniques used in the task to be programmed.

Wave Wave was the first high-level language created for programming a robot. It was developed for research tasks in 1973 at the Stanford Artificial Laboratory.

AL The AL (for Arm Language) high-level programming language was developed at the robotics research center of Stanford University. It is useful for robotics research because it can provide coordination between two arms of a robot.

AML The AML high-level programming language was developed for use with the IBM RS/1 assembly robot. Depending on what source you consult, the acronym AML is short for either A Manufacturing Language or A Manipulator Language. The language is highly structured and has features from ALGOL, APL, and LISP.

HELP HELP is a high-level programming language developed for use with General Electric's Allegro assembly robots. This language can be thought of as a loose extension of Pascal. HELP is useful for controlling more than one manipulator arm from a single controller. It operates with a jointed-arm coordinate system and can support the use of intelligent tactile sensors.

Karel Karel, Karel 2, and Karel 3 are robot control languages used by some FANUC robot controllers.

CAP 1 The Conversational Auto Programming 1 (CAP 1) robot language is used by the FANUC 32-18-T Robot Controller. It uses a user-friendly, step-by-step, interactive approach.

MCL MCL, short for Manufacturing Control Language, is a high-level programming language developed by McDonnell Douglas for the U.S. Air Force's ICAM project. It was used to program a Cincinnati Milacron T3 robot. Basically, MCL is an extension of the APT language. The MCL instruction set includes commands for controlling other equipment in the work cell and for communicating with vision and touch sensors.

RAIL RAIL is a high-level programming language developed by Automatix for use with robots and vision systems. It is an extension of the well-structured Pascal programming language. RAIL can be used in either an off-line programming mode or a walkthrough mode. The following pick-and-place program is written in RAIL:

```
BEGIN
```

;Get the part.

```
OPEN
APPROACH 3.0 FROM LINE2
MOVE LINE2 WITH SPEEDSCHED(1)
CLOSE
DEPART 3.0
```

;Put down part.

```
APPROACH 4.0 FROM PRESS
MOVE BELT WITH SPEEDSCHED(1)
OPEN
DEPART 4.0
END
```

RPL RPL, a high-level programming language developed by SRI, has features of FORTRAN, LISP, and Pascal. It is used in the configuration of automated manufacturing systems and can communicate with intelligence vision systems.

Scorbase The robot programming language used with the SCORBOT robots is called **Scorbase.** It is a PC-based, menu-driven, Windows-based programming language that can be used on Eshed SCORBOT-ER V plus and SCORTEC-ER I robots.

ARMBASIC ARMBASIC is an extension of the hobbyist computer language, BASIC—Beginners' All-purpose Symbolic Instruction(al) Code. ARMBASIC was developed to enable the TRS-80 model I level II microcomputer to control a robot arm. ARMBASIC adds the additional commands of @STEP, @CLOSE, @SET, @RESET, and @READ to BASIC. These commands are used to control the arm of the Microbot Mini-Mover 5. The @STEP command positions all six stepper motors on the arm at once. The @CLOSE command causes the hand or gripper to close until the grip switch is activated. The @SET command sets the robot action delay values and the speed of the arm. The @RESET command zeros out the position registers and stops all movement, thereby establishing a new home position for the arm. The @READ command reads the present values of the position registers, thus identifying how far the robot is from its home position.

Androtext **Androtext** is a high-level computer language developed by Robotronic Corporation to make commanding a personal robot easier. A compiler that translates Androtext into machine language has been written for the Heath Company's Hero I robot. Androtext can be translated by many home computers, including some Commodore computers, the Radio Shack TRS-80 Color computer, and the Atari 800XL computer.

The Androtext language is intended to be understandable to people who have little technical knowledge. It has the usual arithmetic, decision, and control instructions, along with two special classes of instructions just for robots: a set of instructions for setting and reading of sensors, and a set of instructions for operating effectors (also known as the muscles of the robot). The sensor instructions are divided into passive sensors (for feedback information on position of the parts of the robot) and active sensors (for information on the outside world). Similarly, the effector instructions are divided into passive effectors (for control of lights and voice) and active effectors (for controlling moving parts such as the arm).

VAL VAL is a high-level programming language developed for Unimation's Unimate and PUMA lines of robots. It is a loose extension of BASIC. The VAL operating system supports both off-line programming and on-line walkthrough programming of the robot, and it can communicate with tactile and visual sensors. VAL is being used to program the Unimate 500 robot on the General Motor's PUMA project. The VAL language includes the following features:

1. It operates in real time.
2. It can handle programming on the basis of joints in the real world or of tool coordinates.
3. It includes editing abilities such as add, delete, or replace.
4. It has built-in diagnostics for finding out what is wrong with a robot.
5. It has speed scaling so that the operator can run the robot in various gradations of slow, regular, or fast motion.
6. The robot's programming is interactive, so changes can be made while the robot is running.
7. All inputs and outputs for the robot are interfaced.
8. It has automatic approach to and departure from a point.
9. It can compensate for orientation of the workplace.
10. It allows for off-line flexible disk storage of programs.
11. It supports continuous-path operating, with the tool speed held constant.
12. It allows line tracking.
13. It supports sensory perception interfacing and allows override of closed-loop control circuits and loop counters.
14. It supports an RS232C computer interface.

The following is a VAL program for a pick-and-place operator:

```
PROGRAM PICKPLACE
      1. MOVE P1
      2. MOVE P2
      3. SPEED 25
      4. MOVE P3
      5. CLOSEI 0.00
      6. MOVE P4
      7. MOVE P5
      8. MOVE P6
      9. SPEED 20
     10. OPENI 0.00
.END
```

Ladder Logic **Ladder logic** is a programming language designed to be used by electricians. It closely resembles the relay logic that appears on the inside lids or doors of dishwashers and washing machines.

The only robots that use ladder logic programming are those that come without a controller or with a programmable logic controller. But ladder logic is still important to persons working in robotics, because the **programmable logic controllers (PLCs)** that control production lines often are used to interface the robot with the rest of the production line. PLCs are specialized computers that handle electrical inputs and outputs in electrically noisy factory environments. PLCs support connections to AC and DC equipment and sensors, and they work with digital or analog signals.

If a robot without a controller needs to be able to raise, lower, rotate left, rotate right, extend, retract, grasp, and ungrasp its gripper, its ladder logic program will have a rung or circuit for each desired motion. However, if each axis of the robot uses spring return power, the circuits can be reduced to the ones needed for the motions raise, rotate left, extend, and grip. Figure 15–3 shows a sample ladder

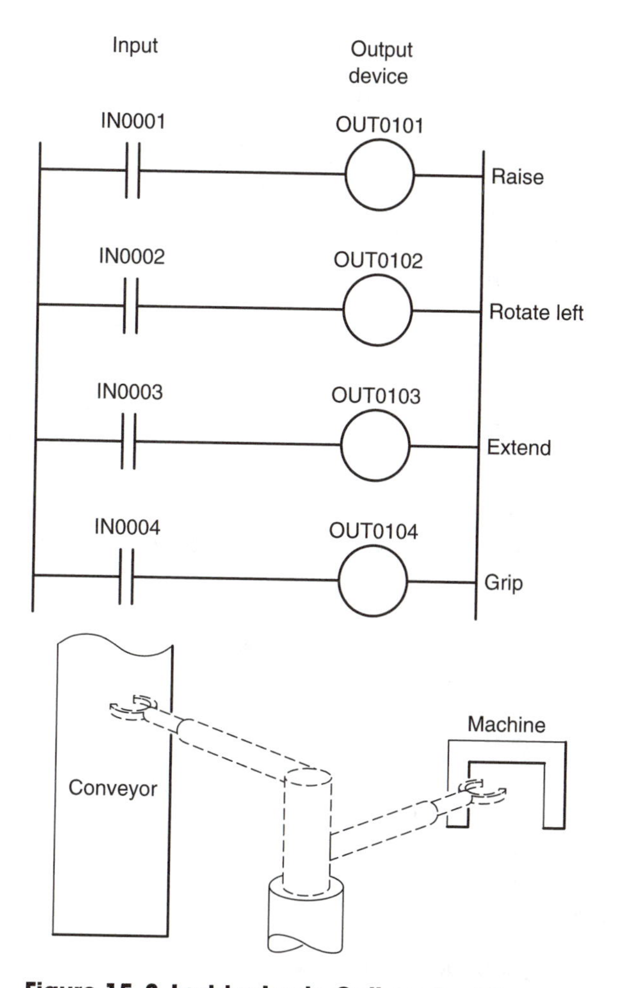

Figure 15–3 Ladder Logic Coil-contact Program.
A ladder logic coil-contact program is similar to a relay logic diagram. A rung or circuit is included for each desired motion.

logic diagram identifying inputs and devices for such a robot. Since a PLC may be equipped with a sequencer control system, it is possible to program the sequence of motions involved in performing the robot's task into the PLC. Figure 15–4 depicts the robot ladder logic diagram and register commands for a robot performing the following sequence of motions: extend the arm; close the gripper; retract the arm; raise the arm; rotate to the left; extend the arm; open the gripper; retract the arm; lower the arm; and rotate to the right.

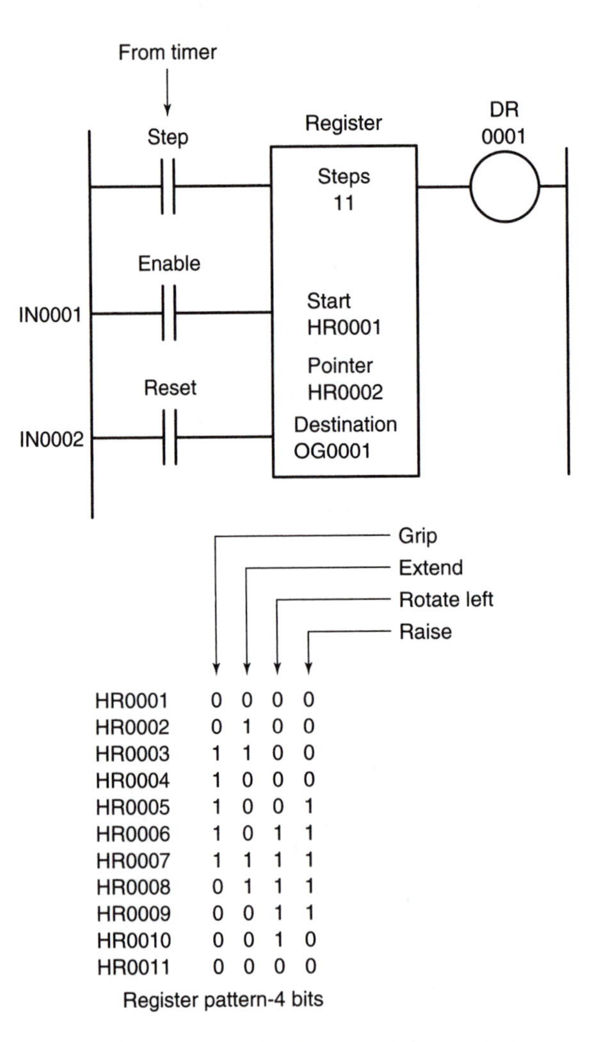

Figure 15–4 Ladder Logic Robot Sequencer.
The sequencing order shown here is for the ladder logic coil-contact program shown in Figure 15–3.

Thorough coverage of ladder logic and programmable logic controllers can be found in John Webb's book *Programmable Controllers: Principles and Applications.*[1]

TEACHING OR LEARNING TECHNIQUES

Teaching, learning, and training—collectively known as robot **task setup**—involve getting the desired sequence of motions into a robot's memory so that it can perform a given job or task. Various techniques can be used to accomplish this, including manual operator control, transfer from another robot, and off-line programming.

Manual Operator Control

Most robots require the use of some type of manual operator control in programming the robot. In the case of simple pick-and-place robots, this control may consist of the manual setting of limit switches and mechanical stops. In the case of point-to-point robots, it may take the form of a teaching pendant for moving the robot to and recording each desired point. In the case of a continuous-path robot, the robot must be led through a task by manual motion.

Mechanical Stops Many pneumatic robots rely on the placement of mechanical stops and electrical limit switches for the programming of their application or task. During the task operation, the robot simply moves a limb until it hits the mechanical stop.

The teaching or training of a simple, open-loop servo-controlled pick-and-place robot consists of two steps:

1. Set the actuator stops at the desired positions. These are usually mechanical stops held in place with bolts. A screwdriver and a wrench are probably the only tools required for this step.
2. Set the indicators for the control system to the sequence in which the actuators are to be driven against the mechanical stops.

The control unit for a pneumatic-powered pick-and-place robot typically consists of a rotating drum with places for inserting pegs to indicate when a particular actuator is to be moved. The pegs not only control the position actuators, but also the opening and closing of the gripper and any required waiting times during the task.

Point-to-Point Control The point-to-point style of controlling a robot involves positioning the robot to each point in its required cycle, in the proper order, and then pressing an enter button. Figure 15–5 shows a possible flowchart of the point-to-point robot programming process. This is just an outline showing the order of things to do. It doesn't list all the details.

[1]Columbus, Ohio: Merrill, 1988.

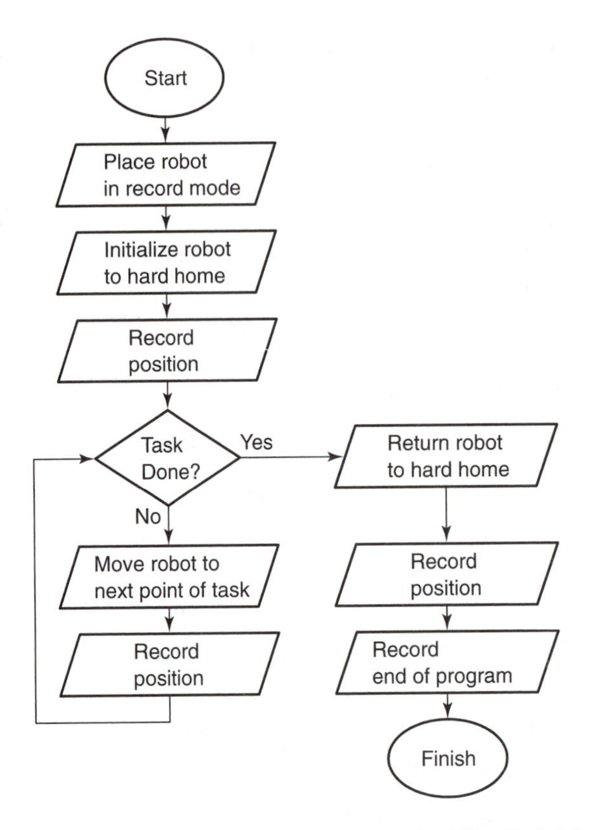

Figure 15–5 A Possible Flowchart of the Point-to-Point Robot Programming Process.

Point-to-point closed-loop servo-controlled robots usually have a **control panel** or hand-held **teaching pendant** for training the robot. The panel or pendant allows the operator to move the robot through its task sequence at reduced speed. Whenever a desired point in the task is reached, the operator (programmer) enters the point into the robot's memory by pressing an enter button (also called a record or program button) on the control panel or teaching pendant. These points then become the robot's program and are stored in an electronic memory inside the robot's control unit. The memory for a point-to-point robot needs to be much larger than the memory for a pick-and-place robot. The path the robot takes between any two points is not programmable, but if only one linear axis moves, the movement occurs in approximately a straight line. If two or more axes are moving, however, the movement may be along a path of any shape.

The panel or pendant also has controls for entering an instruction telling when to open or close a gripper and when to wait for some external signal before resuming movement. Some control panels allow the programmer to enter point information (in the form of Cartesian coordinates) directly into the robot's memory, without having to position the robot manipulator to that point. If a circular path must be traversed by a point-to-point robot, the programmer will have to record many points close together in order to approximate the actual path. The total number of points that can be used in a task is limited by the size of the robot controller's memory.

Figure 15–6 shows a teaching pendant for an industrial robot and another for the hobbyist Movit robot. The Movit teaching pendant has only four function keys plus an enter key: the arrow to the left key controls the robot's left motor; the arrow to the right key controls the robot's right motor; the key with a circle above it controls the robot's light; the key with the solid circle below it controls the robot's horn; and the key with the triangle above it is the enter key. Nothing happens until the enter key is pressed, at which point the state of the other keys is recorded as a robot command. To make the robot move forward for one move time, the programmer depresses both the right and left motor keys and then presses the enter key. The robot will then make the move and record the move in its memory for future use. The

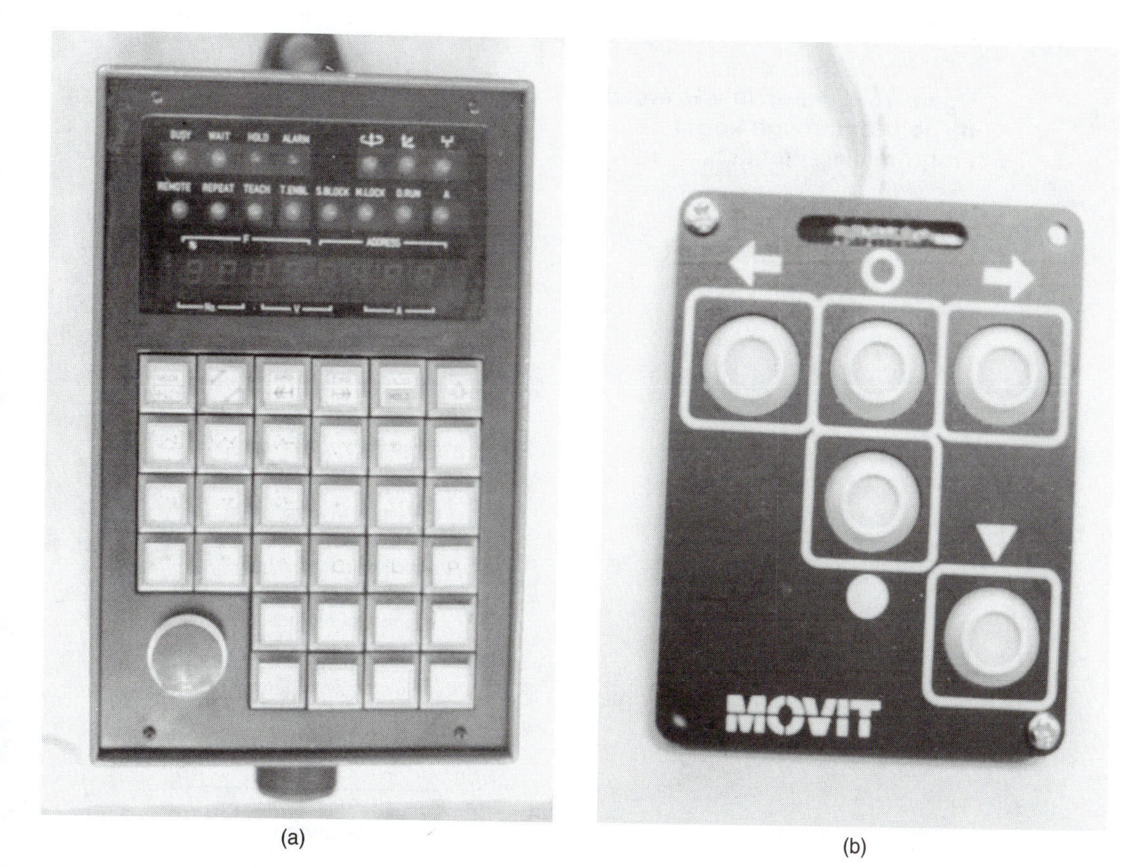

(a) (b)

Figure 15-6 Teaching Pendants.
(a) The controls on an industrial robot's teaching pendant. (b) The simple teaching pendant for the Movit hobbyist robot.

Movit robot's memory can hold 127 commands. The Movit comes with an optional computer interface that allows training commands to be sent to its memory from a computer. Since the interfacing is via the printer port of the computer, any computer language that can normally use a printer can be used to transmit training commands.

Continuous-path Leadthrough Task Most continuous-path robots require having the manipulator led through the desired motion cycle at normal speed while a program button is held down. The control unit records many points per second in its electronic memory as the robot is led through its path. To make it easier to move the robot manipulator at normal speed, most continuous-path robots have hydraulic power systems that are relatively easy to bypass during programming. Any task requiring the activation or deactivation of machines, tools, or sprayguns must have these activations and deactivations performed during the leadthrough programming. Figure 15–7 shows the outline of the leadthrough task programming of a robot.

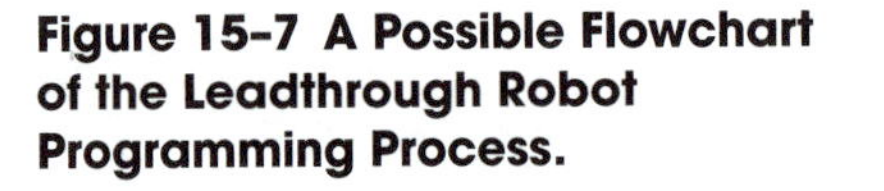

Figure 15–7 A Possible Flowchart of the Leadthrough Robot Programming Process.

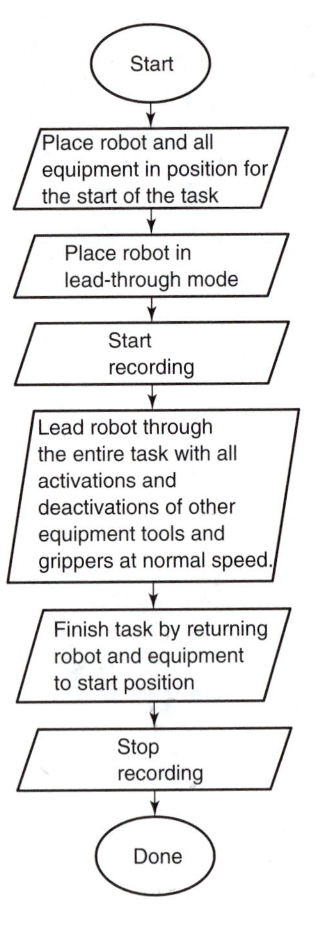

Transfer from Another Robot

Most industrial robots other than pick-and-place robots have a computer interface that allows them to save programs of tasks to the computer's memory and to load a task program from its memory. This allows a robot to build up a library of task programs that is larger than its own memory.

Moreover, if you have a group of identical model robots, one robot can be trained in a task and then can transfer the task information to the computer's memory, from which it can be downloaded to other robots. Due to the mechanical tolerances of the positioning mechanisms of different robots, however, the computer may need to establish correction tables for each robot and translate the task position information for each robot through these. Transferring programs from one robot to another is a very efficient method of programming robots.

Off-line Programming

Off-line programming of anything but a Cartesian-coordinates robot is difficult at best. Usually, a scale-model manipulator or a simulation software program is needed if you are to do much off-line programming for a robot task. The main benefit of off-line programming is that it does not tie up the robot during programming.

Typically, a robot is first programmed for one task by means of task leadthrough. Then the program can be modified off-line to do related tasks. Such tasks might include making adjustments to handle parts of different sizes. Off-line editing of mistakes in programs is also possible.

STEPS INVOLVED IN TRAINING A ROBOT

No matter what programming languages or methods are needed to train a specific robot to perform a specific task, organization is the key to success. The following generalized steps for creating a computer program or solving any other type of problem apply with equal force to training or teaching a robot a new task:

1. Define the problem.
2. Outline the steps of the solution.
3. Express the steps in the language of the computer or robot.
4. Enter the steps as instructions into the computer or robot.
5. Check out the resulting solution.

Listing the Facts of the Problem

In defining a problem that the robot is to solve, you must list the pieces of the problem very specifically. When you understand the details of the problem, your chances of finding a correct solution are greatly improved.

Outlining the Problem Solution

Once the problem has been defined, you must outline the steps involved in solving it. If the problem is solvable with an industrial robot, the solution can usually be acted out by an unskilled or semiskilled human worker. Observing how a human worker performs the task that the robot is to learn will provide a good outline of the steps of the solution. Watch for places where the human makes some type of decision about how to do the task. Notice any unusual steps that the human may occasionally need to take. These decisions and alternative steps may pose a problem for having a robot do the task.

Once you see how a human worker solves the problem, you need to check for possible differences between the robot's techniques and the human's techniques. A robot may be able to perform the steps of the solution differently from (and possibly better than) how the human did them. Any steps that require synchronization with other humans, machines, or robots are especially important. Using a pictorial outline technique such as programming flowcharts may help you visualize the complete processing of the task.

Expressing and Entering the Steps in the Robot's Language

Once you have worked out the outline of the solution, you must translate the steps into the robot's language. This may consist of setting up a set of gears, cams, stops, and limit switches settings, or it may consist of establishing a series of positionings and recordings, or it may consist of walking the robot through all the steps of the solution. Some robots can receive their list of instructions through voice commands, and a few robots allow commands to be entered on a computer in a high-level programming language.

Checking Out the Resulting Solution

After the commands or instructions have been entered into the robot's memory, the program is ready for the checkout step, in which the robot is directed to execute the commands one at a time to verify that each works correctly. When each command by itself checks out, try all of the commands at full speed to see whether all of the commands needed to solve the problem have been included in the program. Human workers can supply missing steps in a prescribed solution as they work. Robots can only perform the commands they are given. If any problems arise, go back to the step of the outline where the error first occurs, and redo all the steps from that point forward.

SAMPLE PROGRAMMING METHODS

Programming the Airco AR-1 Robot, Using Task Leadthrough

The Airco AR-1 robot is a point-to-point flat welding robot (see Figure 15–8). It performs arc welding, either as a carbon dioxide pulse or as an argon/carbon dioxide welder using mild steel or stainless steel wire of 0.045- to 0.0625-inch diameter.

Figure 15–8 Airco AR-1 Welding Robot.
On the near left side of the photo is the robot's manipulator. In the center is the robot's circular worktable. Right of center is the robot controller, with the control panel on top of it. Farther right is the welding power supply. On the far right are argon and carbon dioxide gas tanks.

The controller's memory can hold up to eight programs. A program may include up to 250 points, and a total of 500 points are available for all eight programs. A flat-welding robot is used to join pieces together in a horizontal plane. Because welding is highly labor intensive, welding jobs in batches of three to five pieces can be done cost-effectively with this robot. Unlike most manufacturing robots, which work with other machines in an assembly-line process, the AR-1 robot usually works by itself. Since the AR-1 does not have a teaching pendant, its programming must be performed from the control panel.

Before you can program the robot, you have to decide what task you want the robot to perform. I chose to have the robot weld my initials on a 3-inch by 5-inch by $\frac{5}{8}$-inch mild steel plate.

Once the task has been chosen, the next step is to make a sketch of it. Such a sketch is called an **operations** (or **op**) **sheet.** Figure 15–9(a) shows a full-size sketch of my initials, J L F.

Next the initials need to be divided into roughly linear welding strokes. Figure 15–9(b) shows the breakdown of the initials into strokes. Notice that the horizontal strokes in the middle and last initial stop short of the vertical strokes. This will make the vertical strokes smoother; and since each stroke is about $\frac{1}{8}$ inch wide, the vertical and horizontal strokes will connect without any trouble.

Now we need to pick the order of the strokes. We do not want the welding strokes to splatter on each other, so we will work in an orderly fashion from right

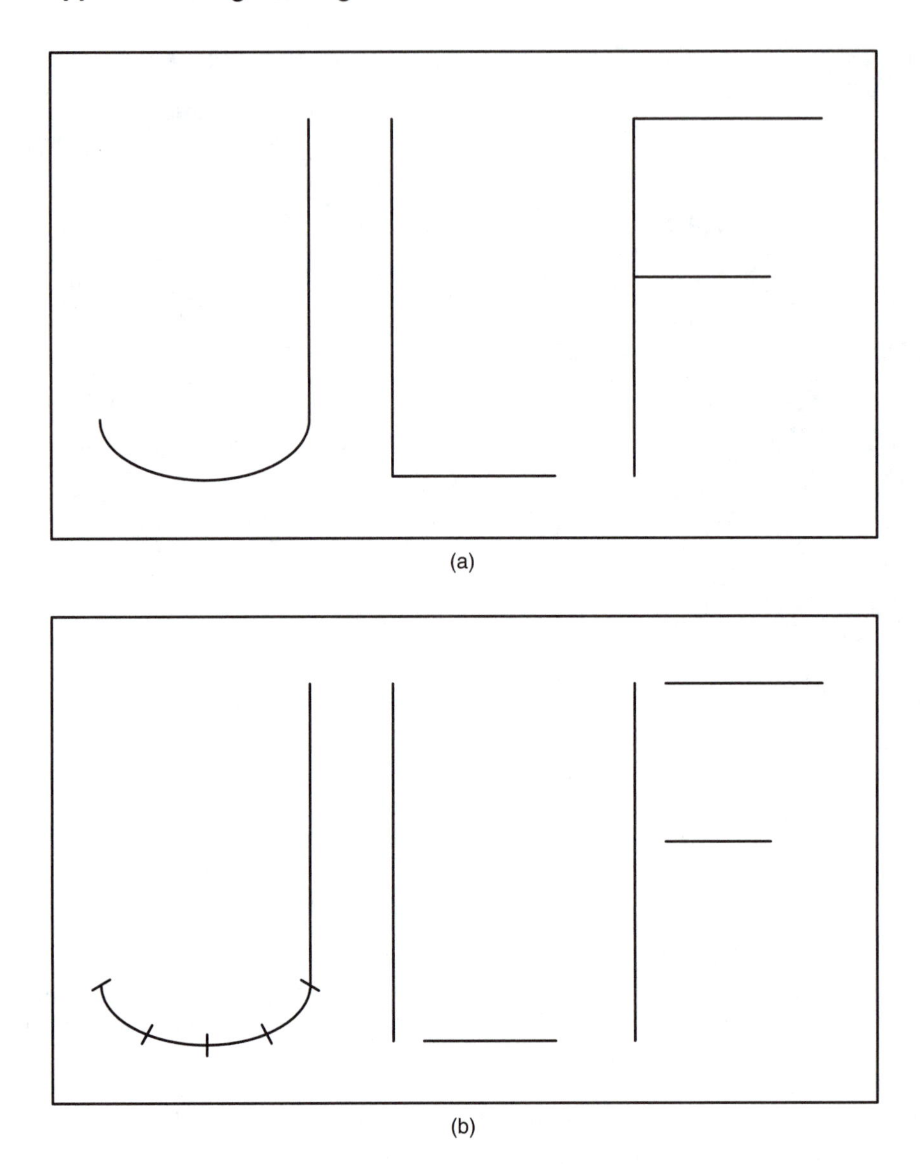

Figure 15-9 Laying Out the Welding Project.
(a) The pattern to be welded by the robot. (b) The same pattern broken into welding strokes.

to left and from top to bottom. We will start by doing the top bar of the F, then the middle bar of the F, and then the vertical bar of the F. Next we will do the bottom bar of the L, followed by the vertical bar of the L. Finally, we will do the J as a continuous series of strokes, starting at the top of the vertical line. As the curved sections of the J are encountered, the welding head will have to be rotated. Figure 15–10(a) depicts the welding strokes with arrowheads showing the direction of the strokes. The numbers identify the order in which the strokes will be made.

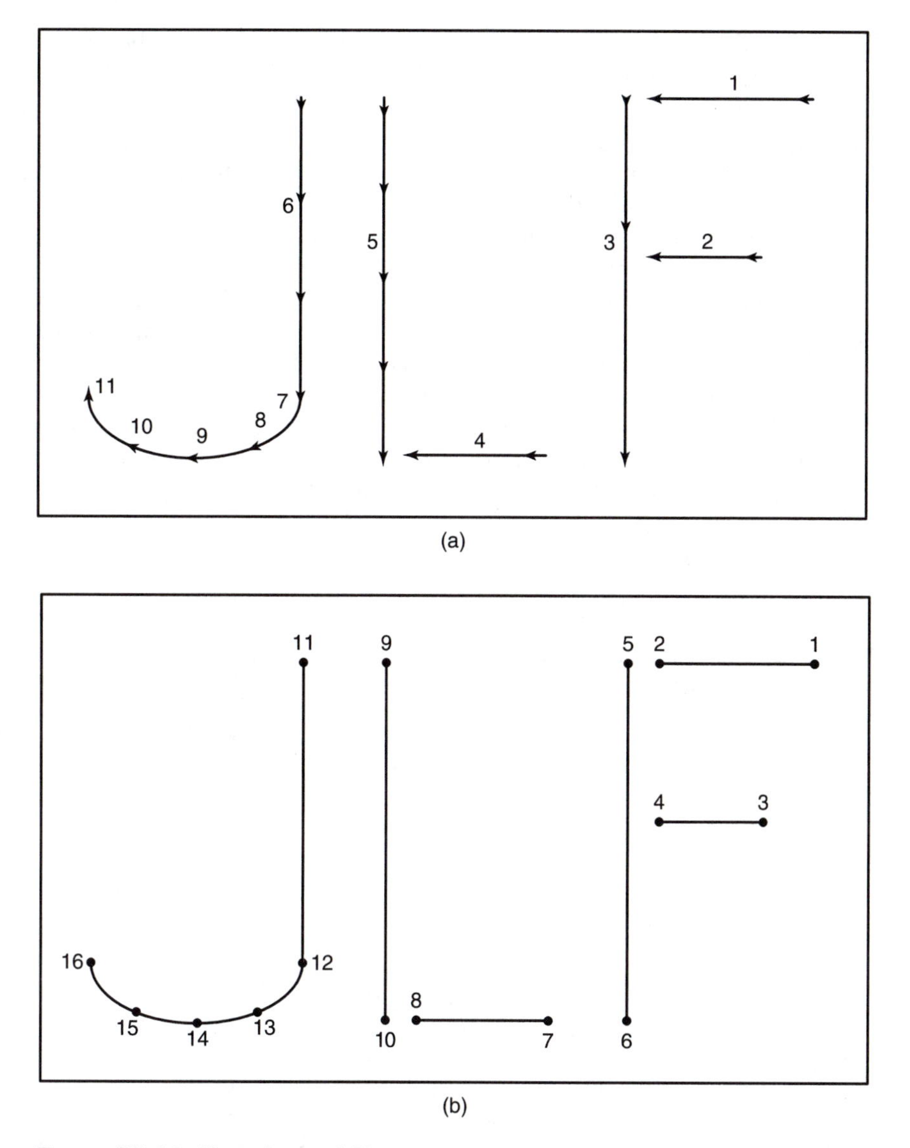

Figure 15-10 More Layout Steps.
(a) The direction of the welding strokes, and the order in which the welds will be done.
(b) Point numbers to identify the start and end of each stroke.

We can now draw a diagram showing the points between which the robot is to weld. Figure 15–10(b) shows these points. The robot will start welding at point 1 and stop at point 2. Then it will weld from point 3 to point 4, from point 5 to point 6, from point 7 to point 8, and from point 9 to point 10. Finally, it will weld from point 11 to point 16, with welding head turns at points 12, 13, 14, and 15.

We now know what we want the robot to do. But before we program the robot, we still need to understand how the AR-1 robot works. A point-to-point robot can move along more than one axis at a time. While it moves in an approximately straight line between two points, it does not have to. Therefore, we must take some precautionary steps to keep the robot from running into the workpiece. First, we will have the robot stop about 1 inch above our workpiece (point 1). Then we will have it move to the beginning of the first welding point (point 2) about 1⁄32 inch above the workpiece. We want the robot to hold at this point, purge the gas lines, and start the arc. The robot should weld to point 3, and then stop the arc. Next, we want to lift the welding head clear of the workpiece to a position about 1 inch above the workpiece (point 4). The robot should then move to welding point 5, where the arc should be restarted. After welding to point 6, the arc should again be stopped. Once more, the welding head is lifted to a position about 1 inch above the workpiece (point 7). The same procedure is repeated until all the welds are made. Figure 15–11 shows all of these points, including the positions above the workpiece.

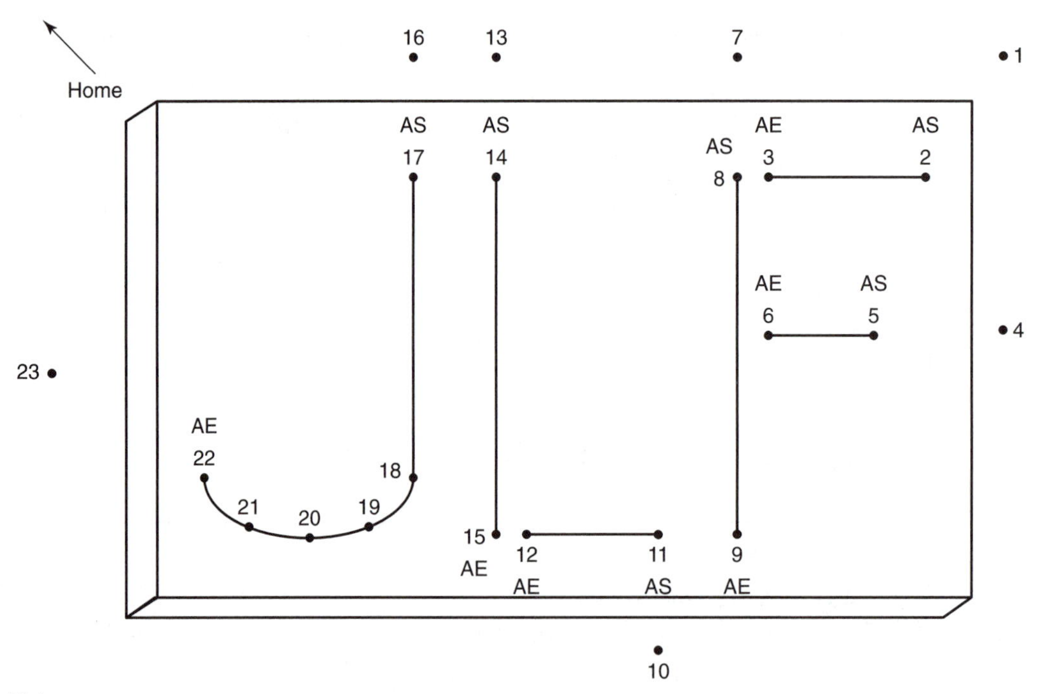

Figure 15–11 Final Programming Points.
This layout includes home, points situated above the work to get the robot clear of the workpiece, and commands for starting and stopping the arc.

We now need to mount the workpiece on the robot's worktable. The robot can only reach about half of the table, but the table has a precision 180° rotation that allows half of the table to be worked on while the other half is being set up. Then the table can be rotated to enable the robot to work on a second part while the first part is removed and replaced with a new part. The workpiece should first be cleaned in a pickling solution and then firmly fastened to the worktable. Remember, you need to have a good electrical connection in order to do arc welding. Finally, we want to place the op sheet of what is to be welded on top of the workpiece.

You are now ready to enter the program into the robot's memory. Programming the robot does not require that the arc controller be turned on. For safety reasons, make sure that the arc controller's power is turned off during the programming phase of this project. These are the steps for entering the program:

Turn off the WELD ON/OFF switch on the interface box. Turn on the power at the line-disconnect switches. Turn on the welding power source. Turn on the circuit breaker inside the power control unit. Turn the POWER key switch on the CRT input terminal unit to ON. Wait for the screen to warm up. On panel B, press the SERVO ON key; then press the STANDBY key; then press the START key. Wait while the robot finds its origin. The z-axis will have to be homed manually by pressing the UP/DOWN switch on the manipulator arm.

On panel A, press F2 for teach mode; then press F1 for program; then press F1 for create; then press F1 for direct mode. On panel B, press 1 for program 1 selected. On panel A, press STORE. This stores the origin position in memory, frees the arm for manual movement, and switches on a green light on the arm. Do not try to move the arm manually unless the green light is on. Use the default values for speed of movement of the arm.

Manually move the arm to point 1 of the layout sketch in Figure 15–11, which is 1 inch above the workpiece. The z-axis movement is obtained by pressing the UP/DOWN switch on the manipulator arm. Then press STORE either on panel A or on the manipulator arm, and move the arm to point 2. Rotate the gripper so that the end of the welding wire points toward point 3. To get the robot to push the welding pool ahead of the wire feed, press ARC START; then press STORE. The HOLD and PURGE commands will be edited in later. Move the arm to point 3; then press ARC STOP; then press STORE. Move the arm above the workpiece to point 4, and press STORE. You have now completed the programming of the first welding stroke.

Move the arm to the beginning of the second welding stroke (point 5). Rotate the gripper so that the end of the welding wire points toward point 6. Press ARC START; then press STORE. Move the arm to point 6; then press ARC STOP; then press STORE. Move the arm to point 7, and press STORE. You have now completed programming of the second weld stroke.

The vertical bar of the F and the two weld strokes of the L are done similarly, and the vertical bar of the J is started similarly. When the arm gets to point 18, however, rotate the gripper to point toward point 19, and press STORE. The arc remains on. Then move the arm to point 19, rotate the gripper to point toward point 20, and press STORE. Move the arm to point 20, rotate the gripper to point toward point 21, and press STORE. Move the arm to point 21, point the gripper toward point 22, and press STORE. Move the arm to point 22, and press ARC STOP; then press STORE. Move the arm above the workpiece to point 23, and press STORE; then press END. This ends the programming of the points of your program.

You can perform a check of the program by pressing F1, then STANDBY, then SERVO ON, and then START for each point. Each press of START will move the robot arm forward one point in the program. Verify that each point is correct, and then press END.

At this stage, you can put in sequencing information such as HOLD on ARC START and PURGE. Proper use of these commands is necessary for good welds. Alternatively, you can press F2 for no sequencing information. Then press END.

Now press F1 for play, press 1 for program 1, press STANDBY, and press START. The robot will now move through the whole program without actually welding anything. Make sure that the robot goes to each point properly. Then remove the paper pattern from the workpiece.

You are now ready for the AR-1 robot to do the actual welding. Turn on the WELD ON/OFF switch on the interface box. Turn on the gas—75 percent argon and 25 percent carbon dioxide. Set the arc current for 200 amps and a voltage of 21 to 25 volts. Obtain and have ready a welding mask; do not attempt to watch the actual welding without wearing it. Press F1 for play; then press 1 for program 1; then press STANDBY. If everything is ready, press START and put the welding mask on. When the program finishes, do not touch the workpiece. It has had 4,000 to 5,000 watts of power applied to it, and it will be hot for some time to come. With gloves and pliers, carefully remove the workpiece and place it in a cooling tank of water. After it is cold, wash it off and clean it up with a grinder.

Figure 15–12 shows the results of two runs of this welding task by a robot. The beginning of the top bar of the F in 15–12(a) is bubbly because I did not purge the gas lines at the start of the welding; the top of the J is short because of bad electrical contacts. The L and the J in 15–12(b) were splattered with welding fragments because the robot was accidentally started in the middle of the cycle, so it welded the J first. This shows the importance of doing the welding strokes in the right order. The width of the welding strokes was probably set too wide to begin with.

It just goes to show you that a nonwelder can program a welding robot's motion but cannot tell the robot how to weld properly. Notice, in any case, how closely the corresponding characters' size, shape, and placement match each other on the two workpieces.

(a)

(b)

Figure 15–12 Two Workpieces from the Welding Program.
(a) The result of forgetting to purge the gas lines before starting the welding is visible in the F.
(b) The result of starting the program in the middle of the program—so that some of the welds splattered onto previously completed welding strokes—is visible in the J and L.

Programming the Rhino Robot

The Rhino educational robot is a point-to-point robot that can be programmed from a teaching pendant (see Figure 15–13). The following sample program shows some of the techniques used in such programming. The task shown is a simple one. The program homes the robot, moves the robot to a part (in this case, a small block), has the robot pick up the part, moves the robot to approximately home, moves the robot back to where the part was first located, has the robot put the part back, and moves the robot to home once more. If properly done, this program could cycle repeatedly as a simple demonstration program.

Before you begin, place the part within reach of the robot, and mark the place so that you can see where the part belongs. After power to the robot and controller has been turned on, the programming proceeds as follows:

1. Use the teaching pendant to set all axes on their limit switches (home).
2. Press the reset button on the end of the teaching pendant. Wait for the robot to finish homing its position.

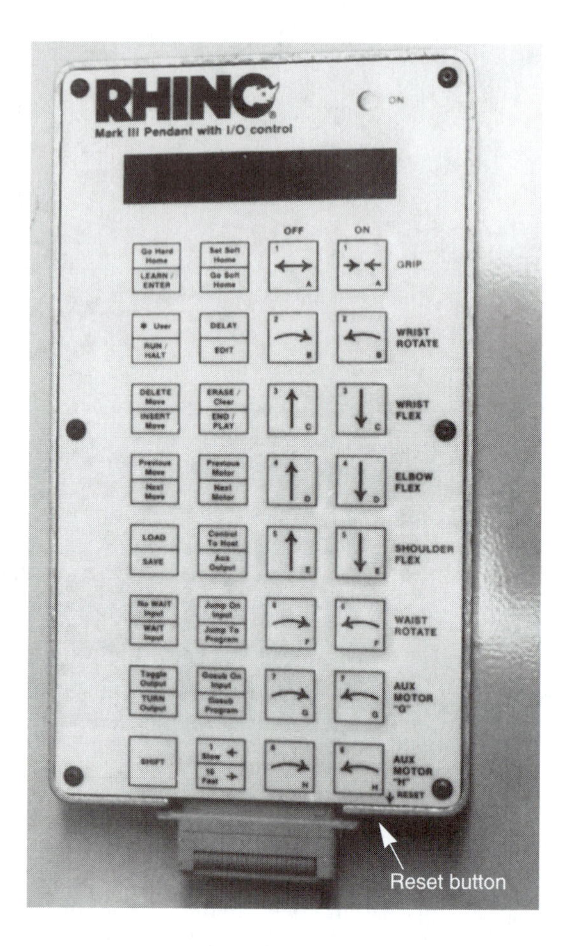

Figure 15-13 The Rhino Teaching Pendant. Refer to this figure while reading the step-by-step programming directions for the Rhino robot.

3. Press SHIFT and GO HARD HOME at the same time. Wait for the robot motion to stop. The robot is now ready to program.
4. Press LEARN/ENTER on the teaching pendant. In response, the display will change from P to L on the pendant.
5. Use the arrow keys to position the robot's hand just above the block that is to be moved. This will require changing more than one axis, and you will have to look at the position very carefully. Press the LEARN/ENTER key.
6. Use the arrow keys to move the hand so that it surrounds the block. Press the LEARN/ENTER key.
7. Press the GRIPPER ON button. Press the LEARN/ENTER key.
8. Lift the block a short distance off the table. Press the LEARN/ENTER key.
9. Reposition the robot at approximately its home position. Press the LEARN/ENTER key. Do not use the GO HARD HOME or GO SOFT HOME keys, since they would open the gripper.
10. Carefully position the robot so that the block is just above its former position on the table. Press the LEARN/ENTER key.

11. Lower the block to the table. Press the LEARN/ENTER key.
12. Press the GRIPPER OFF button. Press the LEARN/ENTER key.
13. Move the robot so that it is just above the block. Press the LEARN/ENTER key.
14. Reposition the robot at approximately home. Press the LEARN/ENTER key.
15. End the program by pressing the GO SOFT HOME key. Press the LEARN/ENTER key.
16. Press END PLAY.

To run the program, press RUN/HALT. The program will then repeat over and over. To stop the program at the end of a cycle, press END/PLAY.

USING THE TEACHING PENDANT TO EDIT RHINO ROBOT PROGRAMS

Sometimes you may find that a step of a long program is missing or is in error. In such a case, it is easier to edit a step or two of the program than to reteach the whole program to the robot. Figure 15–14 shows a possible flowchart for editing a robot program.

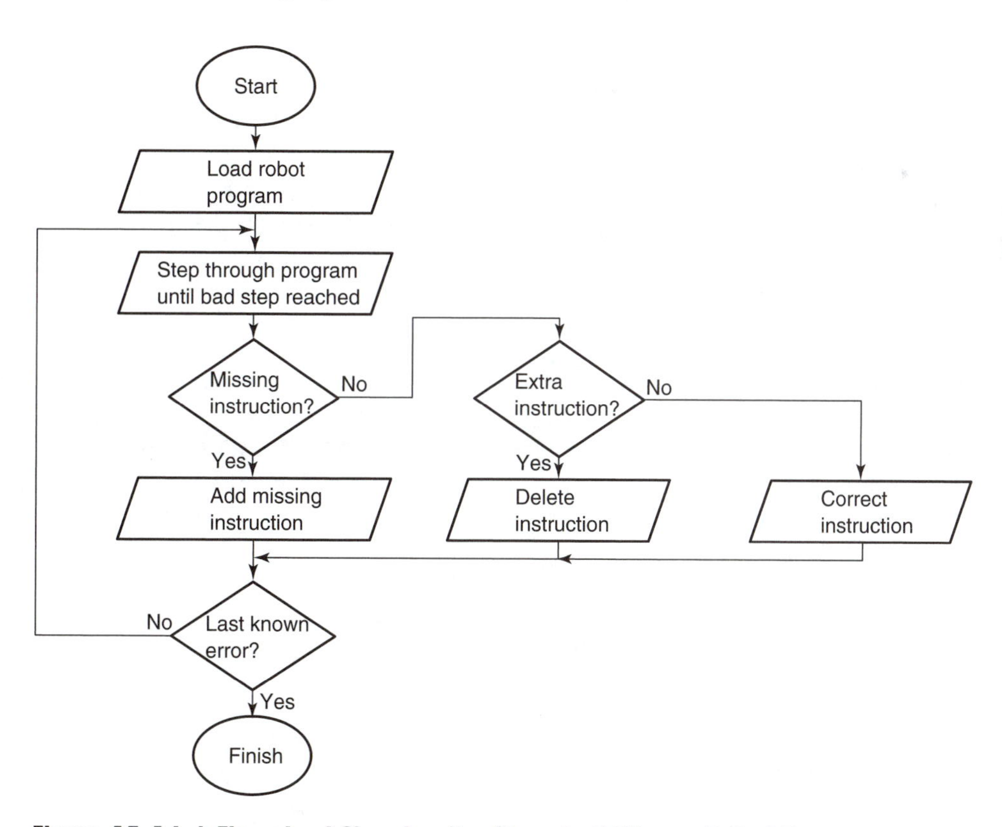

Figure 15–14 A Flowchart Showing the Steps in Editing a Robot Program.

The edit mode of the Rhino teaching pendant can be entered only from the play mode, with a program in the controller's memory. A program can be placed in memory in two different ways: up-loaded from a computer, or created with the teaching pendant in the learn mode.

The edit mode is entered by pressing the EDIT key on the teaching pendant. The teaching pendant's display will then show

```
EStP 1
```

This means that it is in the edit step mode and has just executed the first step.

In the edit mode, you can see a display for every motor or I/O command in any move. To view any motor or I/O segment of a move, perform the following two steps:

1. Press the PREVIOUS MOVE/NEXT MOVE key to get to the desired move.
2. Press the PREVIOUS MOTOR/NEXT MOTOR key to get to the desired motor of the chosen move.

As you step through the program, a specific motor will show a zero in the step section of the display if that motor was not moved. The only exception to this is the A motor (the motor that controls the gripper), which displays either "OPEN" or "CLOSE."

Changing Motor Positions

To edit a specific motor motion, first step through the program to the specific step you want to change. Second, use the PREVIOUS MOTOR/NEXT MOTOR key to find the motor in question. Third, use the specific motor key on the teaching pendant to position that motor as desired. You have now changed the value.

Editing one location does not affect the other locations. The system displays (and remembers) each move as an absolute location from the hard home position.

In the beginning, restrict yourself to changing only motors that have moved—that is, motors showing nonzero values. The trajectory changes in a jointed-arm robot can move in unexpected ways when a zero-value move is altered.

The edited move may be checked by backing the program up several moves with the PREVIOUS MOVE key and then using the NEXT MOVE key to move forward through the edited move to see if the program is doing what it is supposed to. If it moves satisfactorily, you are done. If not, repeat the preceding steps on other moves and motors.

The edit mode may be exited at any time by pressing the END key.

Using the INSERT MOVE/DELETE MOVE Keys

Using the INSERT MOVE and DELETE MOVE keys enables the user to insert a new move between two existing moves and to delete an existing move.

To insert a move, step through the program until you reach the step prior to the point at which you want to insert the move. Then press the INSERT

MOVE key. The display will show EIN, with the next higher move number. Once in the insert mode, enter the move as if you were teaching the robot a point in the learn mode. All keys are active and will be recorded when you press the ENTER key. Multiple moves must be inserted one at a time. After inserting and recording the first move, simply press the INSERT MOVE key again and key in another location.

To delete a move, step through the program until you reach the step you want to delete. Pressing the DELETE MOVE key will delete the move record and move the robot to the next move in the sequence. The next step number will be displayed.

You must be careful when inserting and deleting moves—especially when your application uses I/O. Inserting or deleting a move that changes the state of the I/O may have unexpected results later in the program.

SUMMARY

Application programs deal with getting the robot to do a required task—machine loading and unloading, spray painting, arc welding, or something else. In general, the application programs are kept in the temporary memory of the robot, which changes for each task the robot is to do.

Many programming languages can be used to do robot application programs. These include mechanical setup language, task leadthrough, machine language, and many high-level programming languages. The methods used to program the robot include mechanical setup, point-to-point and continuous-path leadthrough, off-line programming in a high-level language, and transfer of programs from another robot or computer.

Good application programs require careful planning and must include every detail that the robot is expected to perform. A programmer needs to gather all the facts about the task, and then outline the steps of the task. The outline can be expressed in English or in pictorial form, such as with a flowchart or Nassi-Schneiderman chart. The better organized the outline, the easier it will be to convert into a finished application program.

REVIEW QUESTIONS

1. What do a robot's application programs do?
2. Why is it desirable to write programs in a high-level language rather than in the robot's machine language?
3. How does programming a pick-and-place robot differ from programming a continuous-path robot?
4. Compare the APT and VAL languages for programming robots.
5. Why is it necessary to include every detail of a task in a robot's program?
6. How would you program an educational robot to trace your initials with a pencil?
7. a. Starting from a Hard Home command, enter the following Rhino program using the learn mode on the teaching pendant:

```
MOVE 1: F-300 waist
              clockwise
MOVE 2: E 250 bicep out
        D 250 elbow up
```

```
MOVE 3: F 590 waist counter-
              clockwise
MOVE 4: D-300 elbow down
        C 250 wrist in
        B 500 fingers
              clockwise
MOVE 5: F-600 waist
              clockwise
MOVE 6: A CLOSE
        B 250
        C 300
        F 400
MOVE 7: GO SOFT HOME
END
```

b. Save this program under the title A:LAB 2.
c. Press the RESET button on the teaching pendant and load the program you created in part (a).
d. Edit (A:LAB 2) as follows:

```
MOVE 1: change F motor to
        reading of -500
MOVE 2: change E motor to
        reading of 300
MOVE 3: change F motor to
        reading of 400
MOVE 4: no change
MOVE 5: delete (remember
        that at this point
        all step numbers
        will move up by 1)
MOVE 5: no change
MOVE 6: insert F motor move
        to reading of 200
```

e. Save this edited program on disk.

8. Why use the edit mode on a robot program rather than just re-record the entire program?
9. If your school has a robotic work cell set up, go observe it and then draw a sketch of it. List each machine in the work cell and give the functions that it performs. Make a top level flowchart for the overall function of the work cell.

16

CARRYING OUT A ROBOTICS PROJECT

OVERVIEW

Computer software or programming projects are among the most complicated projects that humans can work on. For this reason, it has become necessary to formulate new planning and organization techniques and tools—one of which is the software development life cycle. Comparable techniques are required in robotics projects.

What are some of the details that need to be covered in planning a robotics project? How can the tools for a computer programming project help with a robotics project? Why are robotics projects so complex? These are some of the questions discussed in this chapter.

OBJECTIVES

When you have completed this chapter, you should be familiar with:

- Why robotics projects are so complex
- The steps involved in the robotics development life cycle
- The techniques used to organize a robotics project

KEY TERMS

Automatic parts
Checkout
Customer acceptance
Defining the problem
Design analysis
Design phase
Design requirements
Documentation
Documentation and procedure
Electronic requirements
Fabrication
Implementation phase
Manual parts
Mechanical requirements
Operation/maintenance phase
Origination phase
Outline of project
Procedure/project development
Procedure requirements

Programming requirements
Project development
Pseudocode
Robotics project development life cycle
Robotics system analyst
Software development life cycle
Stepwise refinement
Structured programming
System analyst
System integration
Testing phase
Testing requirements
Top-down structured approach
Training requirements

OVERVIEW OF THE ROBOTICS PROJECT DEVELOPMENT LIFE CYCLE

The steps involved in developing a robotics project are known as the **robotics project development life cycle,** which is patterned in part on the **software development life cycle** for the electronic computer. Each of these life cycles has three main phases: origination, procedure and program development, and implementation. Figure 16–1 shows the robotics project development life cycle, and Figure 16–2 shows the software development life cycle.

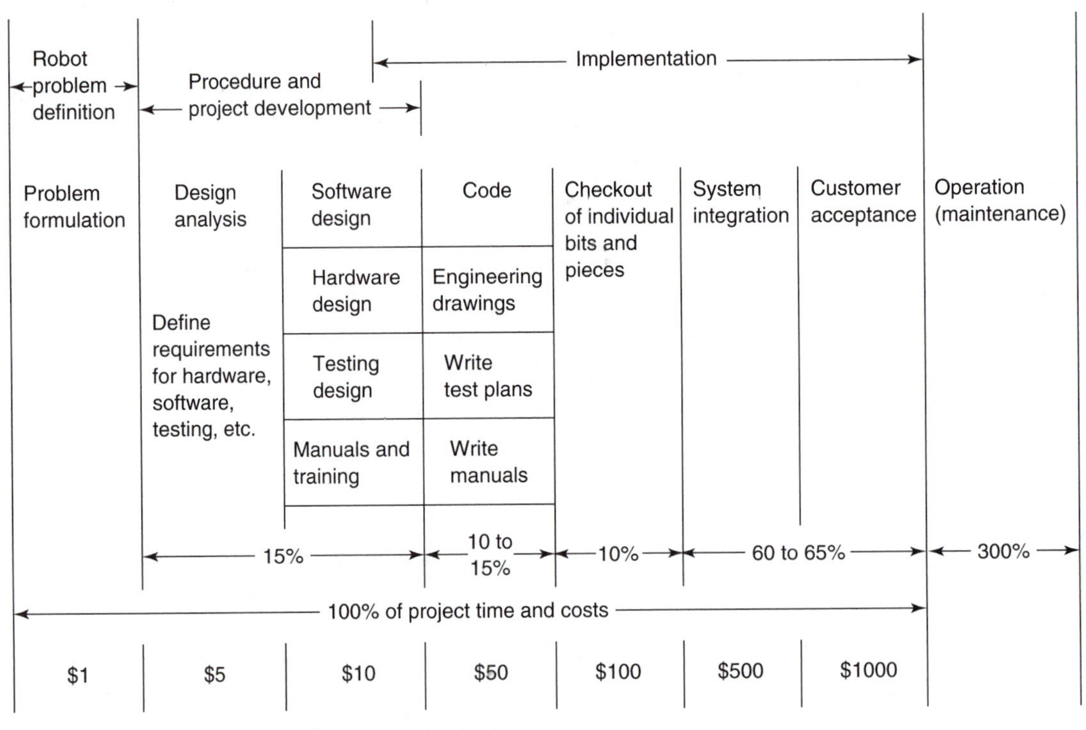

Figure 16–1 Robotics Project Development Life Cycle.
This life cycle is based on the software development life cycle, but it also includes hardware, mechanical, testing, manuals, and training development life cycles.

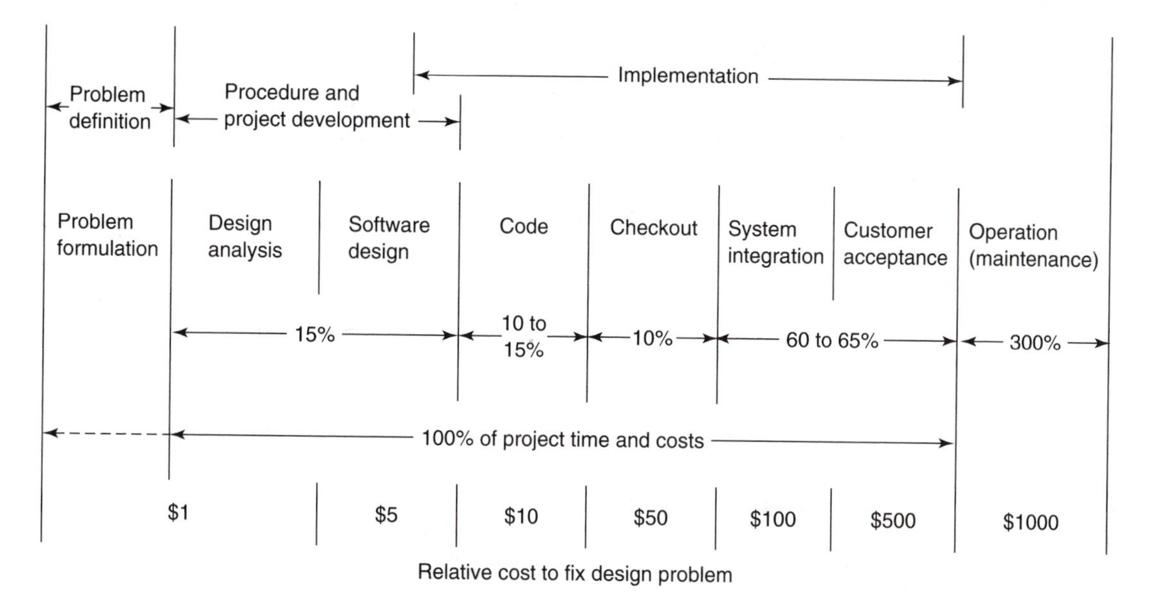

Figure 16-2 Software Development Life Cycle.
This life cycle was designed to help in the organization and planning of software projects, which are very complicated.

ORIGINATION PHASE

The **origination phase** of the robotics project deals with defining the problem or task to be solved by the robot. During this phase, a person called a **system analyst** or **robotics system analyst** develops a functional description of the problem or task and its solutions, after interviewing the customer to gather information. For the software portion of the project, the origination phase consists of defining the problem to be solved by the electronic computer acting as the robot's controller. The results of this phase of the project will constitute a clear and detailed statement of the requirements of the project, but they will not include how to do the project.

Defining the Problem

Defining the problem is probably the most important part of the robotics project. If the problem is misdefined, the solution to the problem will be incorrect. Taking a little extra time to study and define the problem thoroughly can save a lot of time later on in the project. Many tools can help in the problem definition, including interviewing workers, supervisors, maintenance personnel, and engineering personnel.

Robotics System Analysis

To perform a robotics system analysis, a person must be knowledgeable in manufacturing processes, robot mechanics, and computers, and skilled at communicating with other persons. The purpose of the robotics analysis is to determine whether the task *can* be done by a robot and whether the task *should* be done by a robot.

The system analyst's job includes finding out how the job is currently being done—current input information and materials organization, current output information and materials organization, and the methods now being used to get the output or end product from the input or raw materials. The analyst must then obtain a wish list from the customer indicating what the customer wants the robotics system to accomplish.

In order to minimize potential misunderstandings about what is discussed at meetings with a customer, it is useful to tape-record the meeting. The tapes can then be typed up, dated, and signed by all parties present. The system analyst then prepares a preliminary design, in which the task is divided into mechanical, electronic, programming, testing, and procedure requirements, with requirement specifications for all these areas. The system analyst may also prepare summaries of costs versus benefits for several variations of the robotics system.

Outlining Programs

Among the first steps a person performs when writing an essay is to prepare an outline, in order to help organize the material that is to be presented. Doing an **outline of a project** is a creative process. The outline for a robotics project deals with large steps and the ordering of these steps; it is not concerned with small details. When the project has been put into outline form, it is fairly easy to identify missing or misordered steps in the project. Without an outline, it is very easy to overlook some steps and to misorder others. Locating and fixing such problems once the project has been filled with details is difficult.

A good starting point for creating an outline of the robotics project would be to sketch the layout of how the task is currently performed. This could be followed by a sketch of how the robotics system is to perform the job. Finally, the major steps involved in the task could be added to each sketch. Figure 16–3 shows a sketch of how a human worker might service a die-casting machine. Although the steps do not list this, the human is continuously checking the quality of the part. Figure 16–4 shows a sketch of how a robot might service two die-casting machines, because of the robot's ability to work at a peak pace continuously.

Another useful organizing technique consists of making a pictorial outline of the solution. Types of pictorial outlines include flowcharts, structure charts, state diagrams, Nassi-Schneiderman charts, and data flowcharts. Figure 16–5 shows symbols that can be used in pictorial outlines during the origination phase. Most of these symbols represent specific input and output devices for the robot or computer system. Later phases of the project are not concerned with distinguishing among different input and output types.

The system analysis phase of a project is concerned with ascertaining requirements for the project and with dividing the project into various discrete subunits—mechanical parts, electronic parts, programs, and information and material transfer components. Consequently, the system flowchart will try to show all materials and information coming into the system, along with the devices needed to receive these inputs. It will also note what processing is needed on the materials

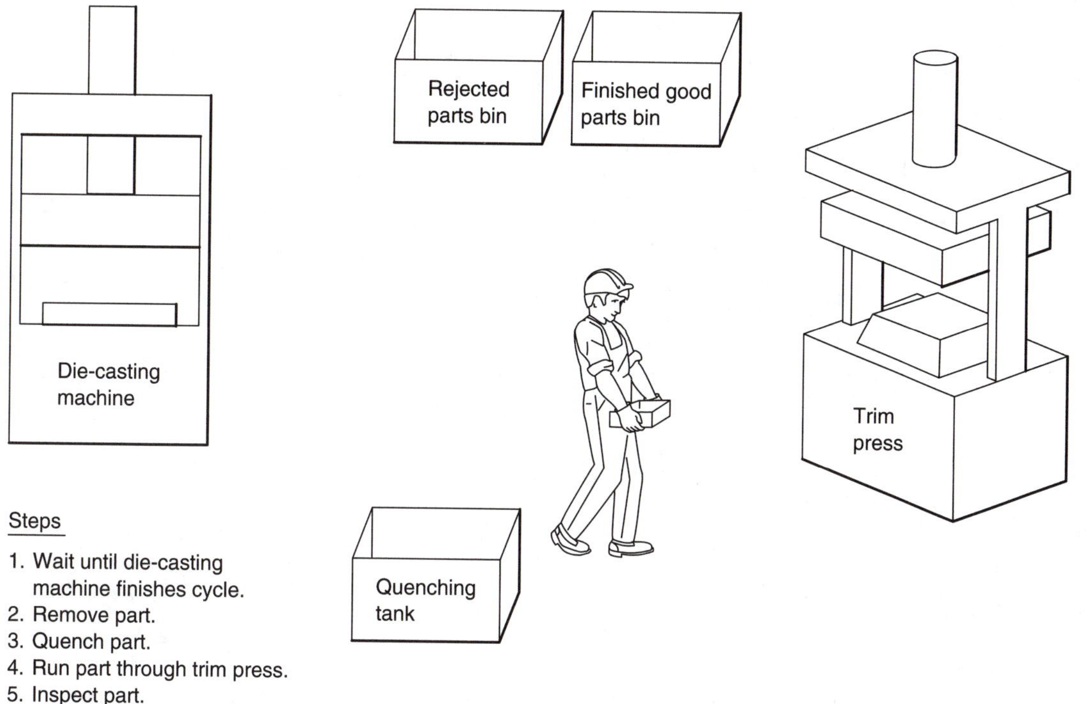

Figure 16–3 Layout for Human Servicing Die-casting Machine.
This layout of equipment would be appropriate for a human performing the task of servicing a die-casting machine.

and information, and it will show the system's outputs and output devices. After the outline of the system has been drawn, the detailed requirements of the hardware and software parts can be specified. These specifications tell what needs to be done—not how to do it. It is extremely important to identify requirements correctly, since a problem cannot be solved unless it is properly understood.

Design Techniques Used

The robotics system designer can benefit from the experiences of designers of other complicated systems such as electronic computers. Two techniques that have emerged from computer work are the top-down structured approach and the stepwise refinement.

Top-Down Structured Approach The **top-down structured approach** starts with a very general goal and slowly works its way into the full detailed design. The technique is similar to top-down programming, which describes a high-level problem

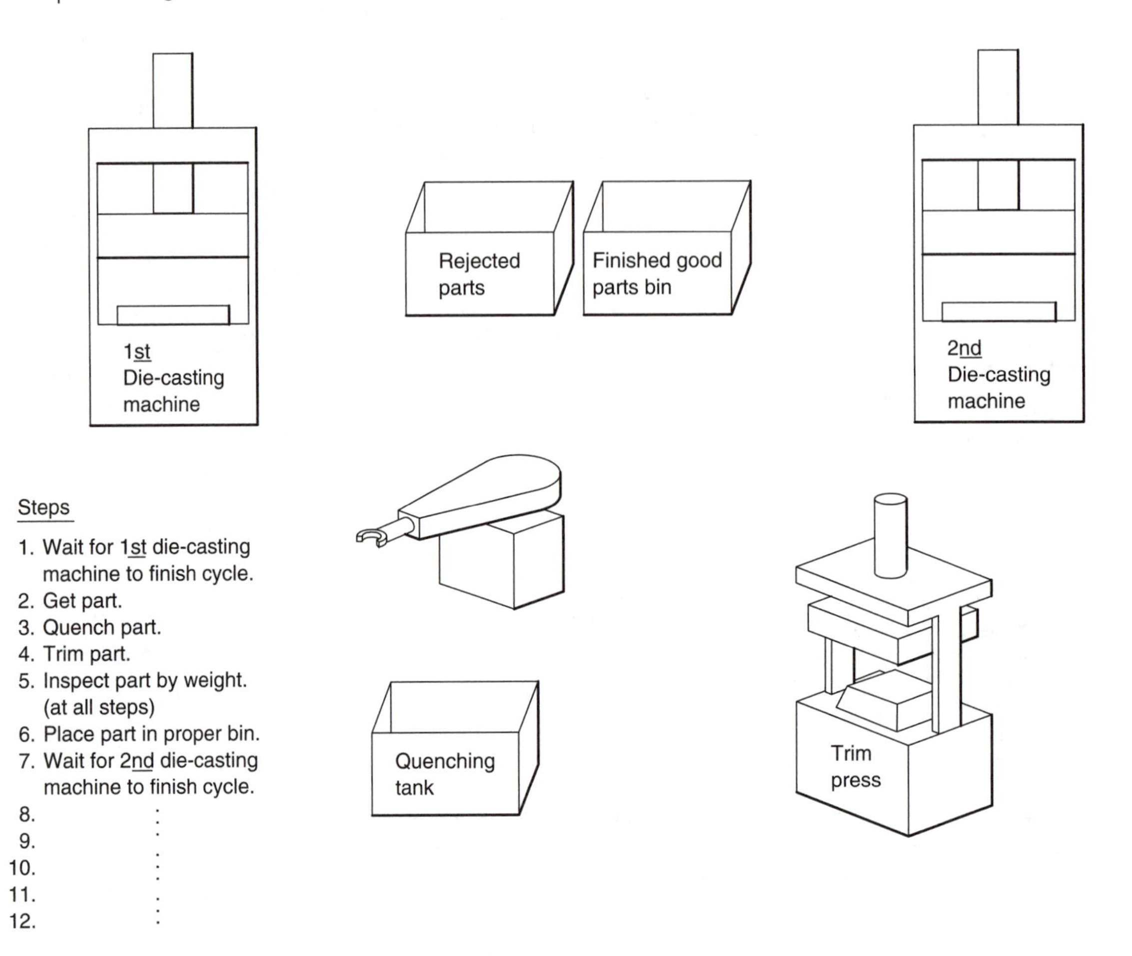

Figure 16-4 Layout for Robot Servicing Die-casting Machine.
This layout of equipment might be appropriate for a robot performing the task of servicing a die-casting machine. Since the robot does not get tired, could it service two machines?

in terms of more elementary subtasks. It uses an enhanced documentation technique over the hierarchical input–process–output (HIPO) technique (in which each block on a hierarchical organization chart lists the inputs to that block, the processing done on the inputs, and the resulting outputs).

A structured design passes through three basic levels of detail. The most general level lists the inputs to the system and the outputs from the system, and it provides a general statement of the action the input undergoes to produce the output. The middle level of detail breaks the action down into subparts consisting of more detailed steps. At this level, the designer determines what steps need to be done and when to do them. The most specific level of detail involves considering how to do each step. Each block at this level represents a single function.

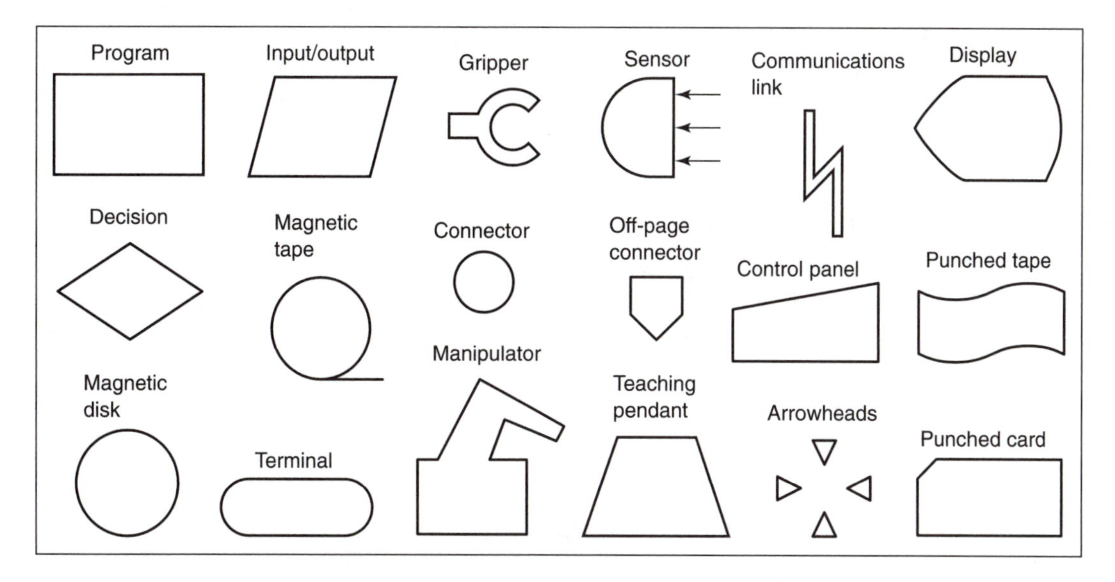

Figure 16-5 Origination Phase Flowcharting Symbols.
A flowcharting template at the origination phase needs symbols for grippers, sensors, manipulators, control panels, and teaching pendants.

The finished structured design can best be shown on a structure chart, which looks very much like a corporate organization chart: people at the top decide what to do; people in the middle decide when each step needs to be done; and people at the bottom decide how to carry out each step when asked to do so. A structure chart also carries labels showing the flow of control and information between different parts of the design.

One visual aid that can be used in constructing a structure chart is a data flow diagram or bubble chart. It consists of a group of lines showing the unprocessed information entering a bubble. The bubble or transaction center identifies the transformations to be performed on this information. Lines leaving the bubble show the new information produced at the transactions center.

Top-down designing allows the control structure of the whole program to be built and tested before the detailed design is completed. This is accomplished by supplying dummy modules during early testing that just print out what their future functions will be. Since most of the problems in a project show up as control problems during system integration, top-down design can isolate these control problems much earlier in the project. The sooner a problem is found, the less it costs to fix.

Stepwise Refinement **Stepwise refinement** refers to the technique of starting from generalities and working toward specifics. Thus, it is the process of defining and expanding each subtask until that part of the problem is solved. This idea works very well in combination with the top-down structured approach.

The basic idea in stepwise refinement is to take a general idea and break it down into steps or parts. If these steps do not give enough detail to complete the design, each of them can be broken down into more detailed steps. The process is repeated until the steps identified are of sufficient detail to allow implementation of the design. This point is reached when all requirements for the design have been defined.

PROCEDURE AND PROJECT DEVELOPMENT PHASE

The **procedure and project development** phase consists of manual parts and automatic parts. The **manual parts** deal with generating the robot's training, operating the robot, maintaining the robot, and checking out the robotics system, including its manuals and documentation. In a large company, the manual parts of development are usually handled by a separate procedure and documentation group. The **automatic parts** of development deal with the mechanical, electronic, and programming aspects of the robotics system.

Documentation and Procedures

The **documentation and procedure** writing should start as soon as the project becomes a working idea, and it should continue to develop as the design of the robotics system develops. Documentation should never be left until the end.

The user's manual for the robotics system is an extremely important component of the system. Without a good user's manual, even a very good robotics system may be useless to the purchaser. The manual should be written with the ultimate user in mind. This person is unlikely to be one of the designers of the system—or indeed a designer, engineer, or programmer at all. Therefore, the text should be clear and thorough, and should not presume that the reader possesses a special level of expertise in dealing with robotics systems.

The documentation for a user's manual should include the following information:

1. A description of the application areas for the robotics system, and a list of its capabilities
2. A description of the input and/or sensory information needed by the robot
3. A description of the output information available from the robot
4. A description of the commands needed to set up and start the robot doing a task
5. A description of the kinds of interaction possible with the robotics system
6. An explanation of all indicators, alarms, and messages produced by the robotics system
7. A discussion, in nontechnical terms, of the performance capabilities and limitations of the robotics system
8. A section on training the robot (which may become a second manual or even a set of manuals as more applications are devised for the robotics system)
9. A good technical documentation of the robotics system, including mechanical, electronic, and programming sections

All procedures and manuals for using the new robotics system should include written narratives with examples of how to do things and with samples of inputs to the systems and the resulting outputs from the system.

Project Development—Analysis and Requirements

Project development—determining the physical and intelligent elements of the automatic part of the robotics system—deals with specifications. This is the process of taking the problem definition and working out the various physical and intelligent requirements needed to solve the problem. The designer must decide on specifications for all of the hardware in the robotics system, including mechanical parts that will give the robot its motions and electronic parts that will make up its intelligence.

Design analysis is the process of looking at the project requirements and then deciding how to meet them. While design analysis deals with the overall requirements of the system, it must also break the design down into various areas such as the mechanical design, the electronic design, the programming design, the testing and maintenance design, and training and manuals.

Then, detailed requirements for each area must be worked out. Once again, these requirements should only tell what specifications or capabilities need to be met. The determination of how to meet these requirements should occur during the implementation phase. The detailed requirements for each area of the project should be related to one or more of the overall system requirements.

Mechanical Requirements **Mechanical requirements** deal with the overall mechanical ability of the robotics system. The manipulator is the part that performs most of the robot's motions. But other mechanical requirements are associated with each physical piece of equipment, including cabinets, control panels, structures, and power supplies.

Electronic Requirements **Electronic requirements** deal with the overall electronic ability of the robotics system. Most of the electronics are located in the robot's controller, but electronic sensors and control circuits are needed throughout the robotics system. Some type of teaching console or pendant is also essential.

Programming Requirements **Programming requirements** deal with the robotics system's overall programming needs. Defining program requirements is the second phase in the software development life cycle discussed earlier in this chapter and shown in Figure 16–2. Also known as the software design analysis phase, this phase is concerned with both the built-in abilities and the learning or task-training abilities of the robot. The built-in abilities programs give the robot the knowledge it requires to move all its joints and control its gripper. If the robot has sensors, program requirements for these must be included. The built-in abilities must also include some provision for enabling the robot to learn a task. Some of the learned task programs are thought out at this stage, too, if for no other reason than to test the robotics system. Figure 16–6 shows symbols that can be used in pictorial outlines during the program development phase of the project.

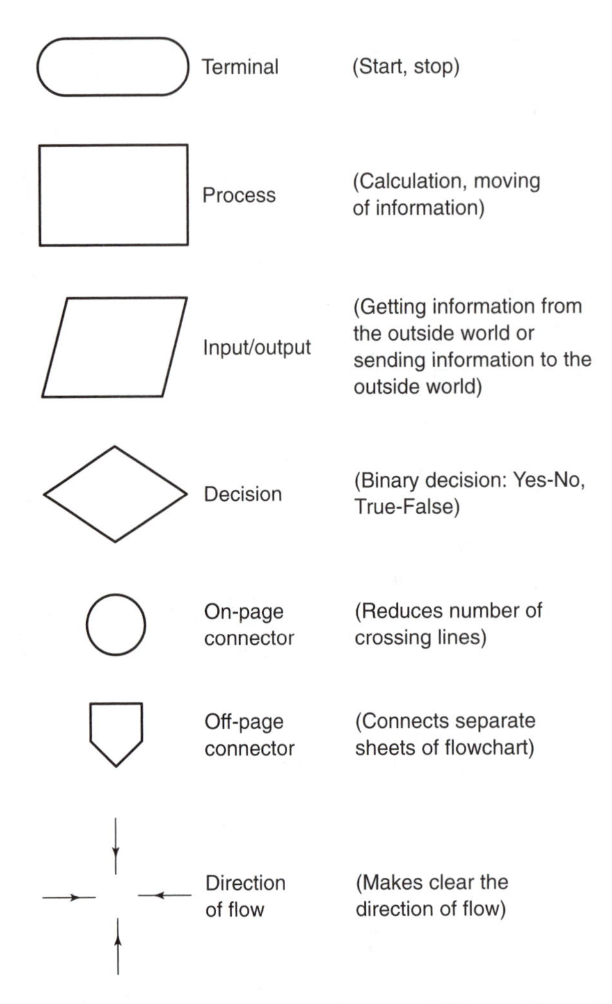

Figure 16-6 Project Development Phase Flowcharting Symbols.
A programming flowchart uses fewer symbols than does the system flowchart.

Testing Requirements Testing requirements deal with all procedures that are needed to verify that the robotics system meets its design requirements. These tests are performed on the design at various stages, but they are concentrated during system integration and customer acceptance. All tests must be referenced to specific design requirements.

Procedure Requirements Procedure requirements might better be called *documentation requirements*. The object here is to define all the elements of documentation and training needed in the system. These include the user's manual, the maintenance manual, any customer training courses, and all the other normal design documentation.

IMPLEMENTATION PHASE

The **implementation phase** sees the transformation of the robot requirements into a working robotics system. The first step is to take each requirement and design the corresponding part of the system in such a way as to satisfy that requirement. This is the actual design step. Next, the paper design must be turned into physical parts, circuits, and programs. This is the fabrication step. After that, the fabricated parts of the system must be checked out—first individually and then as a whole system. This is known as system integration. The final step is to demonstrate the system to the customer for confirmation that the system meets all its requirements. This is the customer acceptance step.

In many projects, the implementation phase accounts for 80 to 90 percent of the time involved in the project and 90 to 99 percent of the budget. Much of this time is wasted as a result of poor problem definition and design requirements at earlier phases of the project.

Design

The design step is different for each area of robotics. It involves designing circuits, parts, or programs to meet **design requirements.** Although most people think of this work as the engineer's main task, in reality the design step generally occupies only 10 to 20 percent of the engineer's time. The design step accounts for 10 to 15 percent of the elapsed time in a robotics project.

Mechanical Design Mechanical design consists of producing mechanical drawings that describe how to fabricate each mechanical part and how these parts fit together.

Electronic Design Electronic design is concerned with making electronic circuit diagrams, logic diagrams, and cabling diagrams for the various control and computing circuits in the robotic system.

Program Design Program design involves writing computer instructions to carry out each program design requirement. This step of writing instructions for a computer is known as coding or writing a computer program. Again, while many people think of coding programs as being central to a computer programmer's job, this is really only a small part of it.

The program design phase of the project shows what actual processing or transformations need to be performed on the input information by the program in order to produce the desired output information. This may also be outlined with a flowchart showing the major steps that must be performed by the program. The outline should be based on program requirements specified during the system analysis phase of the project. The outline should show the inputs provided to the program, the processing steps to be performed on the inputs, and the outputs produced by the program. The actual processing consists of the steps required to get from the given program inputs to the desired program outputs.

Suppose someone asked you to write down the steps involved in taking a test at school. You could represent this information in outline form as follows:

```
Start
Clear off desk
Receive test questions from instructor
Get your name from your memory
Print name on test paper
Have you finished test?
  If yes
    Turn in test paper
    Leave classroom
    Done, Finished, Fini, The End!
  If no
    Read next test question
    Convert visual input into thoughts
    Do you understand questions?
      If yes
        Search memory for answer
        Mark answer on test paper
        Go back to "Have you finished test?"
      If no
        Reread question
        Go back to "Do you understand question?"
```

This style of outlining steps for a solution to a problem, when applied to steps in a robot or computer program, is known as **pseudocode.** It is used by many programmers.

The program flowchart shows the step-by-step time-related organization of a proposed program. The flowchart should be completed before you try to write out the instructions for the computer (the coding phase). Like an outline, it loses its value if it goes into too much detail. Moreover, it should not be related to any specific computer language, computer, or robot; it should simply show the logical steps involved in solving a problem.

The first draft of a flowchart should be limited to showing the major steps or divisions of the problem. This is something like listing only the chapter titles for a book or the section titles for a report in an initial outline. If you are dealing with a large program, this may be sufficient detail for the top-level flowchart.

You might begin by writing out the steps you would have to go through if you were doing the operation manually. Next, take each major step or division of the program and list all of the primary subordinate steps involved in doing it. Each major step was a single block on the top-level flowchart; now each of these steps becomes a page or so of flowcharts at the second level of flowcharting. The process of taking a block at one level and developing a page or so of flowcharts from it at the next level should be repeated until you reach a level at which all required steps are listed and described in sufficient detail to identify every action that must occur in the course of the program. You should not, however, include minute details of how the computer or robot is to do the step. Figure 16–7 shows a flowchart for the outline on taking a test that was presented earlier.

Flowcharts, like any other type of outline, should be well organized and easy to follow. A program made from an outline cannot be any better organized than the outline it is based on. To help produce well-structured outlines and programs, a system known as **structured programming** has been developed. Basically, structured programming allows the designer to use only well-structured program segments or constructs—constructs that have only one entry point and one exit point. Figure 16–8 shows four well-structured constructs: sequence; if/then/else; and two types of loop or iteration.

A sequence outline is simply a step-by-step set of instructions, with no decisions or loops involved. The if/then/else outline allows a decision to be made, based on a question. The then side of the outline is taken if the answer to the question is yes or true, and the else side of the outline is taken if the answer to the question is no or false. The two sides rejoin at the end of this outline construct. The loop or iteration has a sequence of instructions and a test to determine whether that sequence of instructions should continue to be repeated. Figure 16–9 shows two different structures that can be produced from well-structured constructs; therefore, these outlines are themselves well structured.

A second form of pictorial charting is called a Nassi-Schneiderman chart, after its inventors. It provides much the same information as a flowchart but allows the use of only well-structured constructs. It possesses charting techniques for representing only three constructs: the sequence, the if/then/else, and the loop. Figure 16–10 shows the forms for these structures. Figure 16–11 shows a Nassi-Schneiderman chart for taking a test. Figure 16–12 shows a Nassi-Schneiderman chart for making a piece of toast.

Another approach to doing top-down designs is to use structure charts, which resemble organizational charts. Structure charts show the control of a program (what is to be done) at the top levels, and they show the details of the program (how to do it) at the bottom level.

One big problem in program development involves achieving good quality control. To avoid bad feelings on the part of computer programmers whose work is being corrected, quality control must be handled carefully. The goal is to achieve "egoless programming" in which program quality review takes the form of a group project dedicated to finding and solving problems before the system checkout phase begins. The purpose is not oneupmanship or programming elegance. Instead, the focus should be on finding problems that actually interfere with the system's objectives. Do not try to place blame, and do not fight over programming styles.

Testing Design Testing design is concerned with developing techniques and procedures that will be used to verify that a design meets all the relevant design requirements. These tests must be tied to the requirements specification, and they should be designed by a group that is independent of those doing the design of various other parts of the system.

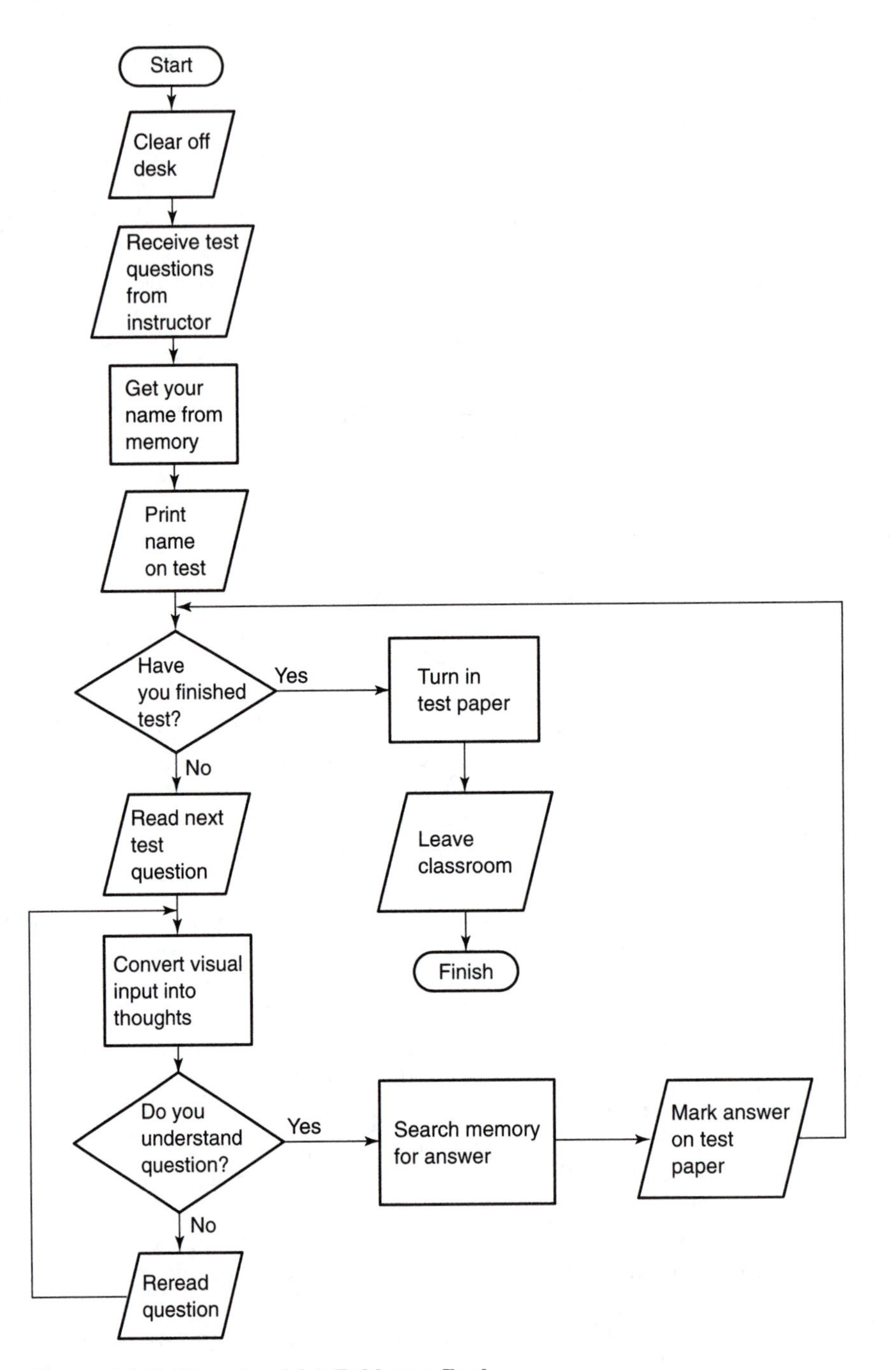

Figure 16-7 Flowchart for Taking a Test.
This flowchart shows the major steps and decisions involved in taking a test. It also shows a possible trap if you don't understand a question.

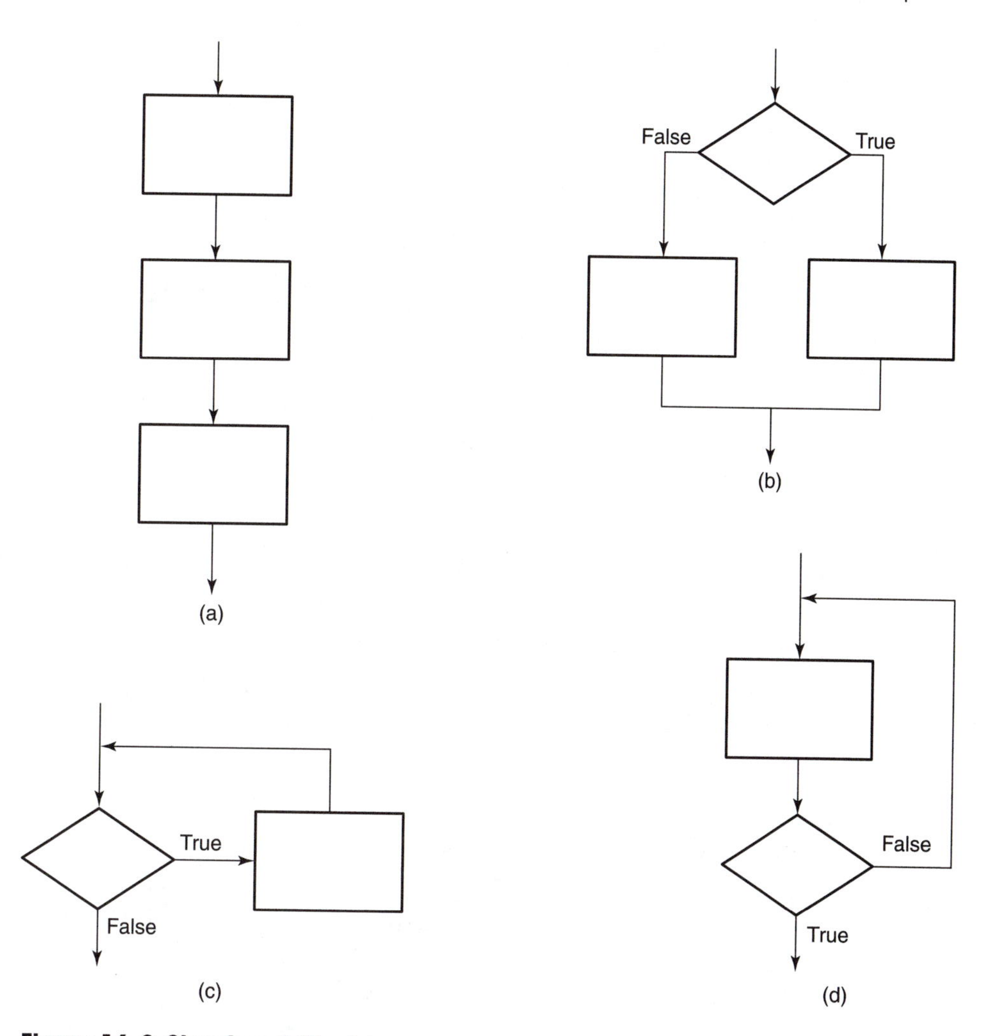

Figure 16-8 Structured Blocks.
Structured blocks are used to break the flowchart into small pieces, each with a single entry and exit, making the flowchart easier to understand. (a) A sequence block. (b) An if/then/else block. (c) A do while block. (d) A repeat until block. Blocks (c) and (d) are types of loop or iteration blocks.

Documentation Design Documentation design is concerned with providing the training, user's manuals, and other documentation for the system. The people in this design group will need to check with the other groups at each stage of the project to verify that all procedures and documentation are being followed and kept up to date. All of the manuals and procedures needed for documentation continue to be refined as the system development progresses.

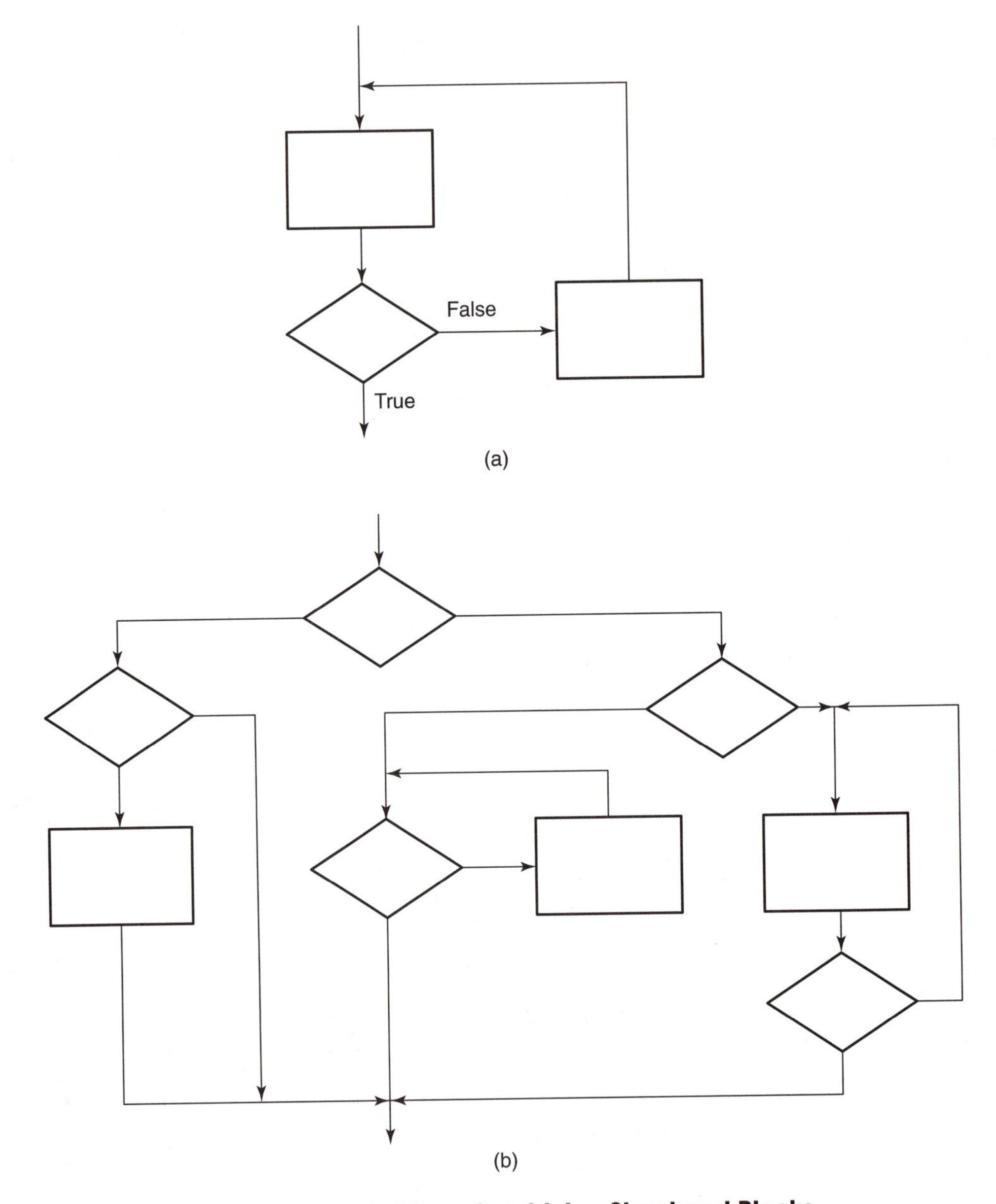

Figure 16-9 Structures Formed from Combining Structured Blocks
(a) A loop exit if end block, produced by a combination of do while and repeat until blocks. (b) An unnamed structure composed of simple well-structured blocks. The picture here is getting confusing; and if a flowchart made up of well-structured blocks is confusing, imagine how confusing things can get when unstructured blocks are used.

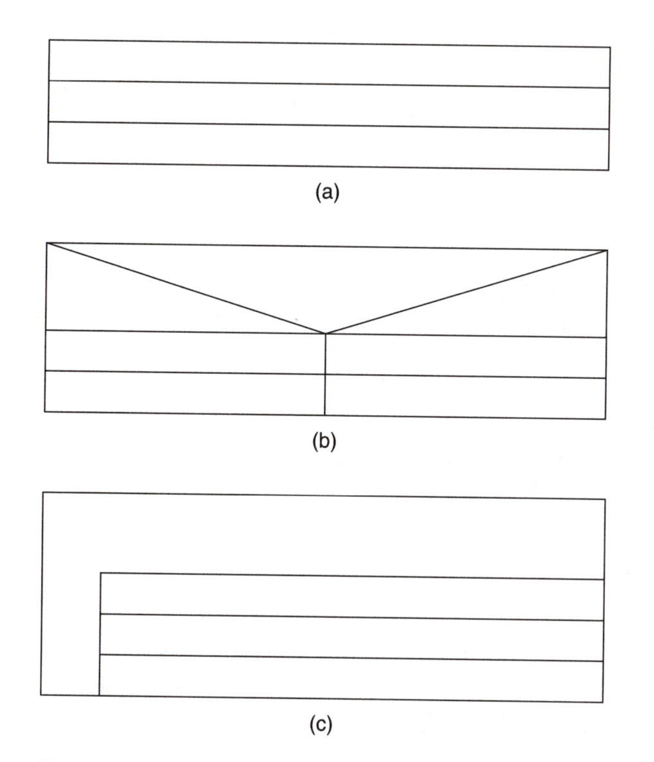

Figure 16-10 Nassi-Schneiderman Chart Symbols.
The Nassi-Schneiderman chart uses only three symbols. (a) The symbol for a SEQUENCE construct block. (b) The symbol for an IF/THEN/ELSE construct block. (c) The symbol for a LOOP construct block. A loop or iteration includes both the DO WHILE and the REPEAT UNTIL constructs.

Fabrication

The **fabrication** step is where the paper designs are changed into physical parts, manuals, and computer programs. This is the point at which you begin to see the physical robotics system come together. Many persons find it a very exciting stage of the project.

Mechanical Fabrication Mechanical fabrication involves transforming the physical drawings of parts into the physical parts themselves, either through purchase or through manufacture. These parts are brought together and assembled into subassemblies. The engineer then must verify that each subassembly does what is required.

Electronic Fabrication Electronic fabrication consists of transforming the circuit and logic drawings into working electronic circuits. The circuits are built or purchased and then interconnected or wired into subassemblies. These subassemblies

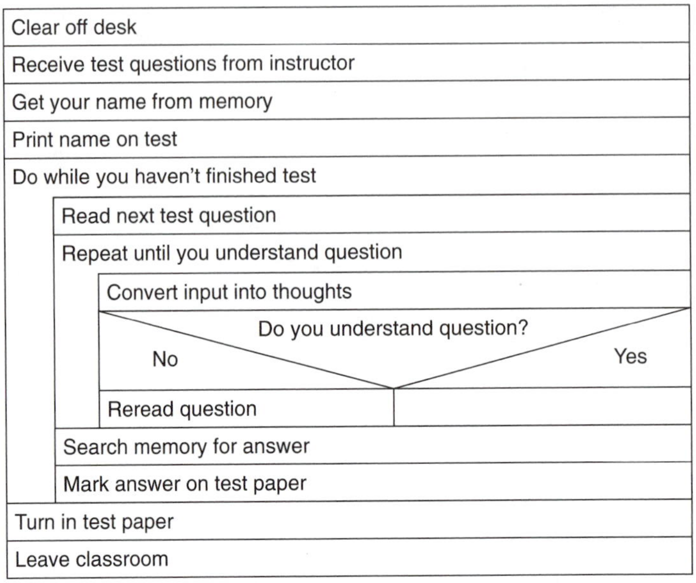

Clear off desk
Receive test questions from instructor
Get your name from memory
Print name on test
Do while you haven't finished test
Read next test question
Repeat until you understand question
Convert input into thoughts
Do you understand question?
No
Yes
Reread question
Search memory for answer
Mark answer on test paper
Turn in test paper
Leave classroom

Figure 16-11 Nassi-Schneiderman Chart for Taking a Test.
This chart presents the same instructions for taking a test as the flowchart in Figure 16–7. It also shows how to nest the three Nassi-Schneiderman symbols.

will be integral parts of the controller, sensors, power supplies, control panels, teaching pendant, and so forth.

Program Fabrication Program fabrication deals with converting the program requirements into instructions for the computer, entering these instructions into the computer, and getting the computer to confirm that it understands all of them. This process is also known as coding the program.

The coding part of the implementation phase consists of converting the outlines of the program into the program language being used to program the robot. This translation step, the simplest part of the implementation phase, may account for 5 to 15 percent of the project's total expenditures of time and money. If the definition and program development phases have been done properly, this step should go relatively quickly.

The coding part of a program is not done until the code has been entered into a computer and the computer has signified that it understands the instructions. This means that no compilation errors have been detected by the computer. Murphy's Law states, "If anything can go wrong, it will." A pertinent corollary to Murphy's Law is offered in Henry Ledgard's book, *Programming Proverbs:* "The sooner one starts to code a program, the longer it will take to complete the program."

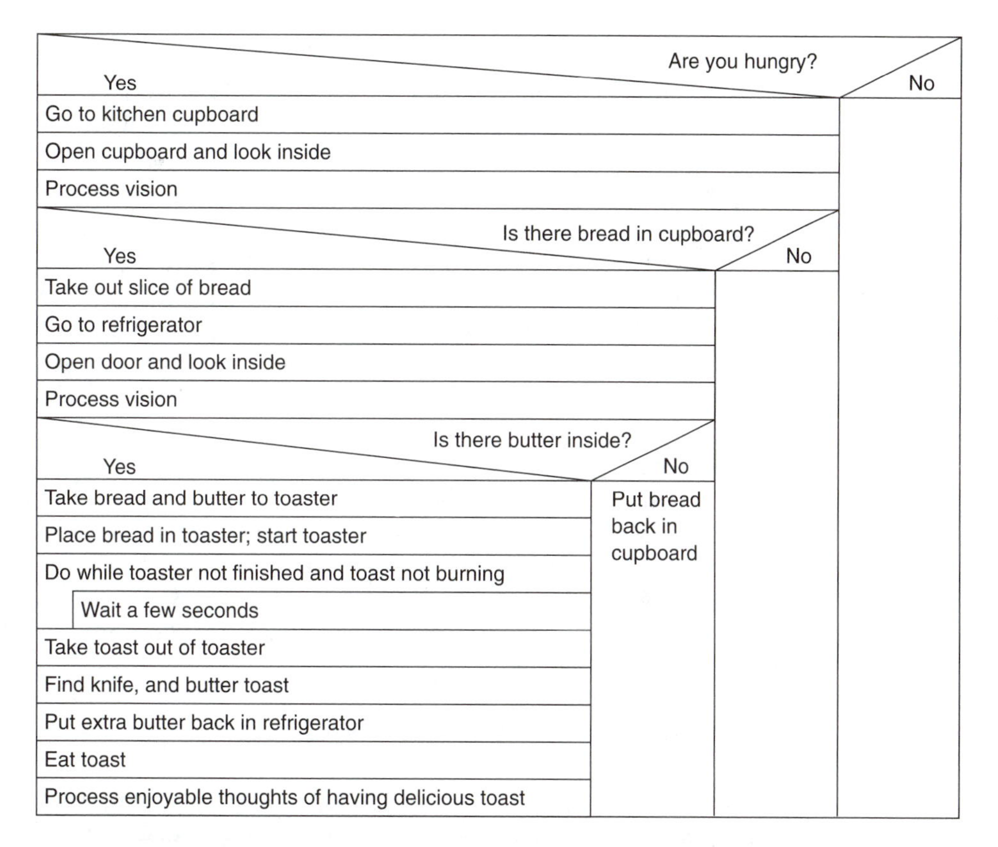

Figure 16-12 Nassi-Schneiderman Chart for Making a Piece of Toast. This chart shows the major steps and decisions involved in making a piece of toast. It also shows that decision constructs do not have to be divided exactly in the middle. If the boxes get too small to hold all the desired information, you can write a page number in them and go to a new page.

Testing Fabrication Testing fabrication is the process of writing tests to verify that each system requirement has been met by the robotics system. It is important that every requirement be tested. To ensure that all requirements are covered, each test should be related to a specific system requirement statement. The designers of these tests must work with the designers of the various other parts of the robotics system.

Procedure Fabrication Procedure fabrication involves writing all the procedures that need to be followed on the robotics project and writing all required manuals. The procedures will state the methods that must be used to document the robotics system. They will also explain how to make changes or updates in the documentation. Manuals that were begun earlier will take on a finished look at this point. However, they will still require some changes as the robotics system goes through checkout, system integration, and customer acceptance.

Checkout

The step of **checkout** is concerned with the inspection of the individual parts of the robotics system by their designers. Each designer is responsible for verifying that his or her portion of the system works as designed.

Mechanical Checkout Mechanical checkout deals with inspection of each of the mechanical subassemblies. Do the parts of all enclosures and cabinets fit and work properly? Does the hydraulic power supply work correctly? Do the parts of the gripper fit correctly, and does the entire gripper move properly? Does the manipulator assemble and move correctly? Every mechanical assembly must be checked out to verify its proper functioning.

Electronic Checkout Electronic checkout involves inspecting each individual circuit to see that it functions correctly. This is the first time these circuits have had power. Checkout consists of applying signals to the circuits and evaluating the resulting outputs from the circuits. It is important to verify the signal levels at interface points between this circuit and other circuits, since these interface points are where most misunderstandings occur. The designers on both sides of the interface need to agree about the interface signals.

Part of the electronics checkout is an electrical checkout. This is concerned with 110-, 220-, or even 440-VAC circuits.

Program Checkout The checkout of the program involves determining whether the program works correctly and meets all design requirements. This checkout is done by executing the program with known input information and investigating whether the results from the program are correct. The information used as input to the program should exercise all paths of the program, not just the normal operation paths. Most program problems occur when processing input values at or near the limit—that is, very large or very small values. The interface points of this program with other programs are potential problem areas, too, since the programmers on either side of the interface sometimes interpret the form of the information differently.

Testing Checkout Testing checkout is a matter of trying out the tests on the robotics system. It must be verified that each test works successfully, meaning that it runs without any problems and that it measures what it was designed to measure. Since the robotics system is not yet complete, most of the checking of tests will consist of having the designers read over the proposed tests. Some of the tests will not pass inspection, and then it must be determined whether the test was designed wrong or whether the robotics system did not meet its requirements. It may take all the system designers working together to determine what is wrong and then to fix the program.

Procedure Checkout Procedure checkout is devoted to verifying that all procedures and manuals are correct. Most of the procedures are already in use and thus have already been checked out. The main focus at this point will be the checkout of the user's manual. The operating instructions for the system must be verified, and so must the training or programming instructions for the robotics system. The method used here is to have someone who is unfamiliar with robotics systems read and explain what the instructions mean.

System Integration

System integration is the process of bringing the various parts of the robotics system together and making them perform as a complete robotics system. The early effort will be spent on subsystem checkout, in which various individual pieces from checkout are grouped together into subsystems and then each subsystem is inspected to see whether it meets its functional requirements. This checkout should reference the original system requirements specifications. Discrete subsystems in a robotics project might include power supplies, manipulators, grippers, controllers, teaching pendants, programs, and vehicles.

The next phase of system integration is to put all of the working subsystems together. This may be done by checking out one subsystem, then adding another subsystem, then another subsystem, and so on, until the complete system is together and working. The unified system is then checked out to verify that it works according to the system requirements specifications. During checkout, care must be exercised to ensure that the robot does not damage itself, especially once the components in place give the robot mobility. Safeguards should also be established to protect the personnel doing the robotics system checkout from unforeseen motions by the robot. After all, there is no guarantee that the robot will work as expected.

The use of testing plans made up during the procedure and project development phase greatly speeds up the task of system integration. It also helps ensure that the testing of the system against the system requirements specifications will be complete.

Customer Acceptance

Customer acceptance of a robotics system consists of successful demonstration of the system to the customer and final acceptance of the system by the customer. This is often considered the final phase of the robotics project, since the developer's time and cost figures for the project are then terminated. By accepting the system, the customer affirms that all the requirements of the system have been met.

Generally, customer acceptance of a system is based on some demonstration that the system works as required. For a robotics system, this might require that the robotics system be installed in the customer's plant and begin doing the job it was purchased to perform. The demonstration might also include demonstrating how easy it is to train and perform maintenance on the robot.

OPERATION (MAINTENANCE) PHASE

The **operation** or **maintenance phase** of the robotics project deals with the useful life of the robotics system. While the official costs of the robotics system may have ended with the customer's acceptance of the system, the robotics system will continue to incur costs throughout its useful life. These costs, which may run to two or three times the cost of the original system, include maintenance costs, personnel costs, retraining costs, additional hardware costs (such as for new grippers), and additional programming costs for improving the robot's capabilities.

Maintenance costs are generally thought of as the costs necessary to keep a robotics system operating correctly, but hidden under maintenance costs are the costs to correct any errors or bugs in the robot's programming or mechanical ability that were not found during system checkout and integration. Some such errors may be traced all the way back to definition errors not noticed during the system design.

Additional programming costs may be incurred if the customer wishes to add additional capabilities or hardware to the system. These additions are sometimes due to a customer's not fully understanding what the robotics system should have been able to do in the first place. In some instances, a more thorough study of the robotics project's present and future uses would have greatly reduced the need for additional programming after the project was completed.

Whatever the reasons for changes, the maintenance costs of the mechanical, electronic, and program parts of the robotics system can be very high. In part this is because program maintenance is almost always done by persons who have no connection with the designing of the program. Consequently, they must figure out how the program currently works before they can change it. This task is difficult enough if good program documentation is available for the programs. But unfortunately, documentation generally is not kept up to date. This is one of the costliest mistakes made in most programming projects, because old documentation is worse than no documentation.

SUMMARY

Robotics projects are very complicated and require special development techniques. One useful tool that can be borrowed from computer projects is the software development life cycle. The robotic programming requirements can use this tool without change, but with a few changes it can be transformed into a robotics project development life cycle that can be used for the whole project.

The robotics project development life cycle consists of three phases: the origination phase, the procedure and project development phase, and the implementation phase. The origination phase consists of defining the program and outlining the problem solution. The implementation phase includes design, checkout, system integration, and customer acceptance. The operation phase covers operation and maintenance of the robotics system.

REVIEW QUESTIONS

1. Why is it important to finish the design of a robotics system before trying to build the system?
2. Why is creating and keeping documentation up to date important in a robotics project?
3. Prepare an outline (flowchart or whatever method you like) explaining how to get up and off to school or work in the morning.
4. Prepare an outline explaining how to make a cup of coffee.
5. Refer to Figure 16–13, the Amazing Maze Problem. Prepare a flowchart that will allow Tovin the Robot to enter, work his way through, and exit the maze safely. Tovin is to enter the maze at the entrance and leave by the exit. The solid lines represent walls. The dashed lines are just a convenient way of breaking the maze into equal-size pieces for easy measurement of progress. Write down any assumptions you need to make.
6. Refer to Figure 16–14, the Another Maze Problem. Prepare a flowchart that will get Tovin the Robot through this second maze.
7. Refer to both Figures 16–13 and 16–14. Prepare a single flowchart that will get Tovin the Robot through either maze.
8. How well will your flowchart from question 7 work if the mazes are inverted?
9. Prepare a general maze flowchart that would work on any maze.
10. Give some advantages and disadvantages of using the organizational techniques:
    a. English outline
    b. Flowchart
    c. Nassi-Schneiderman chart
11. What organizational techniques, such as English outlines, flowcharts, and Nassi-Schneiderman charts, do you like best? Why?

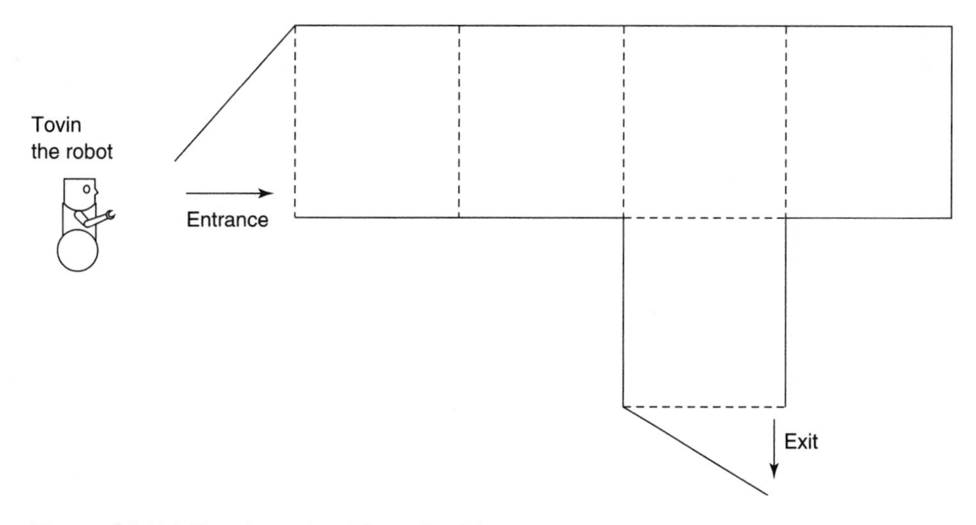

Figure 16-13 The Amazing Maze Problem.
This is a simple maze. (Hint: If Tovin were a person, how would you tell him to get through the maze?)

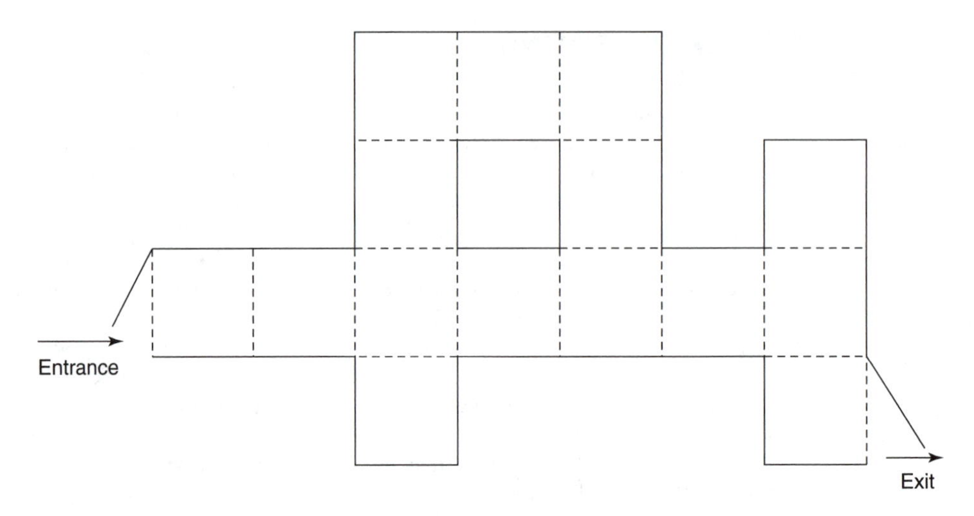

Figure 16-14 Another Maze Problem.
This maze is a little more complicated, but directions for getting through it are still simple. Can you come up with a single set of directions that would work on either maze?

17

VISION SYSTEMS

OVERVIEW

A robot working in a perfect world can do many tasks without the use of vision. As the robot's world becomes less perfect, however, vision will become more and more important to the robot, enabling it to adjust to changing or unknown circumstances. At present, vision systems are being used to help robots locate and determine the orientation of parts and to assist them in doing some assembly tasks.

How does the human vision system work? What are robot vision systems like? How can robot vision systems be improved? These are some of the questions answered in this chapter.

OBJECTIVES

When you have completed this chapter, you should be familiar with:

- The steps involved in human vision processing
- The hardware interfacing techniques that are used to connect simple vision devices to a controller
- The programming problems for vision systems

KEY TERMS

A/D converter
Array processor
Averaging
Burns
Cadmium sulfite cell
Cellular array processor
Comparator
Contrasting
Edge detection
Element
Expectational information
Histogram
Human vision
Image understanding
Information filtering
Interrupt control
Machine vision
Motion detection
Object line detection
Object shape detection
Phototransistor
Prewitt edge detector
Roberts gradient

Scatomas
Sobel edge detector
Threshold detection
Thresholding
Wizard

HOW HUMAN VISION WORKS

Human vision serves as both a short-range and a long-range sensor. Indeed, it has the longest range of any human sensor.

The sense of vision provides a method for rapid input of vast amounts of information. But this is not achieved without paying a high price. Humans may use half of their total computing power just to process visual information. This is why blind persons are able to get more information from their senses of touch, smell, and hearing. Without vision, they have freed up more processing power to use for their other senses.

Vision also takes many years to learn to use fully. A newborn baby has masses of vision input, but doesn't know what to make of all the information. Even after many years of refining their use of vision, no two humans see exactly the same thing.

To obtain a picture with the same resolution possessed by a color television screen, a human must process about 4.8 million pieces of information per eye per second. That works out to 30 pictures per second per eye of a resolution of 400 lines by 400 lines. This is analog, not digital, information. In fact, though, the resolution of the eye is much better than that of a color television screen. Each eye has 120 million rods for black-and-white detail and 6 million cones—2 million for each primary color—for a total of 252 million vision sensors that require monitoring at least 24 times per second to provide flicker-free vision. Not bad for a system whose microprocessors (neurons) run only in the millisecond range.

With such a slow processing speed, the work must be divided up and done in parallel. The area in the center of vision is processed for greatest detail, greatest color detail, and some slight motion, among other things. The outer edges of vision are processed for greatest motion detection (to warn of danger), some color, but not much detail. This arrangement requires that greater areas of processing power and memory be set aside in the brain to process the center portion of vision than to process the edges of vision.

Motion detection is the lowest level of vision processing; that is, motion detection is done first. Details detection is the highest level of processing and requires the most processing power and memory. It is only done if sufficient time is available. The details of the picture are greatly enhanced by the eye's edge detectors in the retina, which are another function of the rods and cones. These require that the eye be constantly moving in small circles. Some of the picture information gathered is compared to previously recorded information, and the new information may be stored in short-term and/or long-term memory.

The brain attempts to perform error corrections on vision information. That is, it fills in holes in the visual field such as the blind spots caused by the optic nerve connection. If the brain has not received enough information to be certain of how something should appear, it guesses at it by trying to average out the picture information or by relying on prior experiences. Vision processing therefore requires the combined efforts of both dominant and nondominant brain functions.

You can make use of the fact that edge detection takes place slightly removed from the center of vision. On a dark night, you may be able to see an otherwise invisible object by looking off to one side of it, rather than directly at it. This is because, although there is not enough light to get detailed information, there may still be enough light to allow edge detection by the most edge-sensitive part of your eye.

Figure 17–1 shows a diagram of the paths of human vision processing. An in-depth treatment of the process of vision is provided in Ernest Kent's book *The Brain of Men and Machines.*[1] The processing of vision starts at the eye. Visual information leaves each eye by two different paths: one could be called the "what was seen" path, while the other could be called the "where was it seen" path. The "where was it seen" path goes from the retina by way of the optic nerve to the

[1]Peterborough, N.H.: BYTE/McGraw-Hill, 1981.

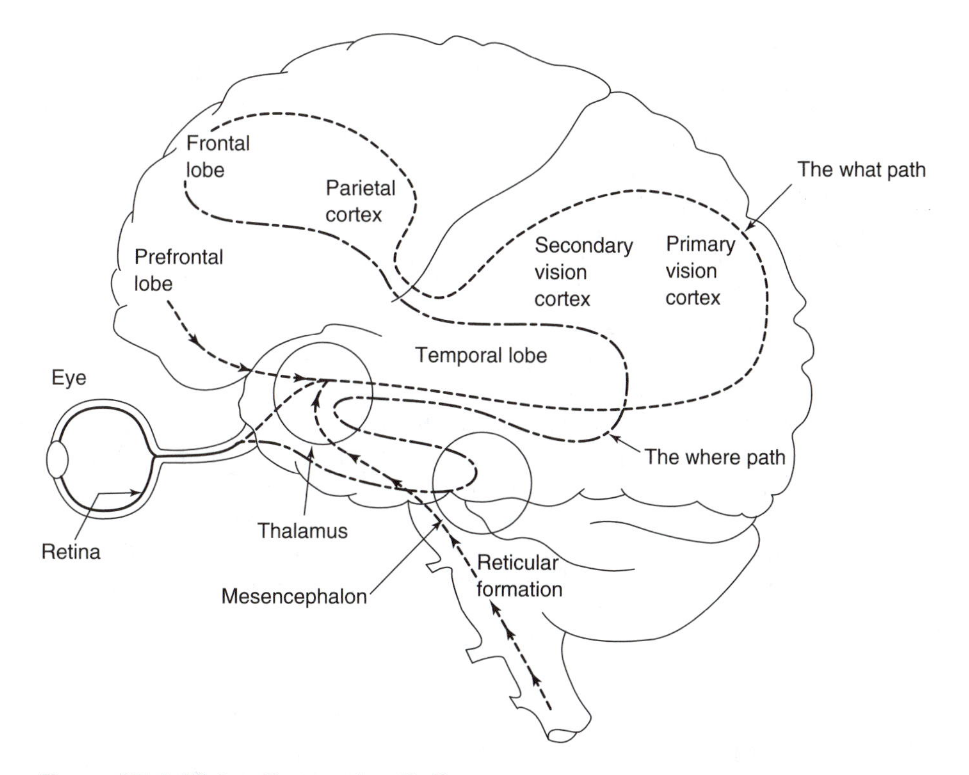

Figure 17–1 Vision Processing Paths.
Human vision uses two vision processing paths. The "where was it seen" path, which transmits data for motion detection in the mesencephalon, is also known as subconscious vision. The "what was seen" path is the conscious vision path.

ancient vision processing center in the superior colliculus of the mesencephalon, then successively through the thalamus, the secondary vision cortex, the temporal lobe, and the parietal cortex, and finally to the logic and emotion processing center of the frontal lobe. The "what was seen" path exits the retina by way of the optic nerve, then passes successively through the thalamus (where it picks up needs and goals direction from the reticular formation and the prefrontal lobe), the primary vision cortex, the secondary vision cortex, the temporal lobe, and the parietal cortex, and finally comes to the logic and emotion processing center of the frontal lobe.

The eye itself is a very intelligent input device. It performs automatic focus and automatic adjustment to varying light levels, and it is self-cleaning. The information entering the optic nerve is preprocessed by the retina such that only boundary or edge information is transmitted. Nonedge information is not sent to the brain, since the brain will fill in these areas as if the whole image were a color-by-number painting. This greatly reduces the amount of information that needs to be transmitted.

The information headed for the ancient vision center in the mesencephalon deals with motion detectors and warnings of danger. This center triggers the reflex action of getting out of the way of the object or danger. This refined information then passes through the thalamus (an interface to the cortices of the brain), through the secondary vision cortex (where the speed, direction, and perhaps size of the moving object may be deduced), through the temporal lobe (where past memory experiences add information about the movement and probable threat of the object), through the parietal cortex (where this information is related to the rest of the vision field), and finally to the frontal lobe where a final decision is made about the moving object. This path for motion detection does not use the primary vision cortex, so it can still function in a person whose primary vision cortex has been destroyed.

The "what was seen" information moves from the retina by the optic nerve to the thalamus, which again acts as an interface to the brain's cortices. Here interpretive guidance is added by the body's need center (reticular formation) and its short- and long-term goals director (prefrontal lobe) as to what objects need to be identified in order to achieve the body's needs and goals. This information serves to speed up vision processing by comparing the visual image to patterns of wanted objects. Unfortunately, it also makes the human prone to seeing what he or she expects to see, even when the object seen is really something else.

The "what was seen" information next goes to the primary vision center, where conscious vision processing takes place. Here the edges (arcs, lines, and spots) are converted into simple shapes as the start of object recognition. These simple shapes, along with their visual location, are then sent to the secondary vision cortex, where they are combined into complicated shapes and final object recognition takes place. The recognition information then passes through the infero-temporal cortex of the temporal cortex, where the objects are related first to words or symbols from past experiences and then to categories of objects. This information then passes through the parietal cortex, where relationships are established among the objects identified. The final step is to relay the information to the frontal lobe, where the final construction of the complete visual picture, tainted by emotion, puts the eyes, ears, nose, fur, etc., together to recognize the family dog or cat or the escaped lion. Emotion can alter the perceived size and distance of an object out of all proportion to the real picture.

ELEMENTS OF HUMAN VISION PROCESSING

The vision process starts with an object that is observed under some type of visible light. Some of the light energy reflects off the object and is received by the eye, the vision sensor. The eye converts the light energy into electrical signals that can be processed by the rest of the vision system. The signals scan the image for motion and edges, the edges are then analyzed for object lines, and the lines are then analyzed for object shapes. The shapes are put together to give image understanding. Color, texture, and depth may be factored in to increase the viewer's understanding of the image. Expectational information may be used to speed up processing of the vision by excluding alternative interpretations that are not anticipated to be helpful.

Illumination Source

The illumination source for humans consists of electromagnetic energy in the visible light range.

Vision Sensor

The vision sensor for humans is the eye, which acts as a smart camera that does considerably more than just convert light energy into electrical pulses that can be understood by the brain.

Edge Detection

Edge detection is the process of finding the boundaries between areas of different intensities of vision energy. Edges define the shape of an object.

The human eye—specifically, the rods in the retina of the eye—performs the system's edge detection processing.

Motion Detection

Motion detection is the process of detecting things that are moving in the field of vision. Human motion detection occurs in the ancient vision center of the mesencephalon.

Object Line Detection

Object line detection is the process of connecting detected edges into lines. Some areas of the eye's retina are best suited for finding vertical lines, some for horizontal lines, and some for diagonal lines. Since all the points of a line may not show up during edge detection, some method must be used to connect disconnected edges into continuous lines. Humans rely on past experiences and expectational information to help fill in missing information.

Object Shape Detection

Object shape detection is the process of taking detected lines and seeing if they form some known shape. If you are dealing with a single object, this may be relatively simple—especially if you know the size and orientation of the expected object. If you are dealing with overlapping objects, you have to decide which object a common line belongs to. In addition, lines may be hidden from view by objects in front of them or by shadows falling on them.

The human vision system is quite good at identifying objects from just a part of the shape. This ability comes from much experience. Figure 17–2 shows a partially hidden object that a human can identify at once. Most industrial robots equipped with a vision system work in a nearly perfect world where an object is always expected to have the same shape, size, angle, and lighting conditions. If the robot moves around, then it must learn to identify objects from different angles, under different lighting conditions, and perhaps from only a part of the whole shape.

Figure 17-2 Object Identification When Only Part of It Can Be Seen.
A human can often identify objects on the basis of partial information about the object. This ability comes from experience, and it sometimes leads to misidentifications.

Image Understanding

Image understanding is the final step of vision processing. Sometimes humans do not complete this step, perhaps because there is insufficient time to complete the task or because the scene was too unfamiliar to permit recognition to follow.

Color, Depth, and Texture

Color, depth, and texture provide additional information about a viewed scene.

Color detection is the process of separating different frequencies of visual energy. A human can use color to help pick out objects in a scene. Red and yellow tend to stand out from other colors. Brown, blue, and green are harder to pick out. However, color can actually diminish a person's ability to see small details. While it is easier to spot a yellow ball in a color scene than a whitish ball in a black-and-white scene, it may be harder to detect surface lines on the yellow ball. This is because the human eye has 6 million color sensors and 246 million black-and-white sensors. However, a single color can be used to highlight parts of a black-and-white scene to bring those areas to a human's attention. For example, red can be used to signal an unusual condition on a computer CRT display, while the rest of the information is given in black and white.

Depth perception information is not used as much by humans as might be expected. Humans depend on many other clues for depth information than just three-dimensional vision. This is why so many optical illusions involving object size can be played on humans.

The texture of an object may also be useful for helping to distinguish the object from other superficially similar objects.

Expectational Information

Expectational information, which takes into account all of a human's past experiences, is the key to the human vision system's ability to process vision so fast. By expecting to see something, a human can greatly reduce the amount of time needed to process visual information.

Using expectational information is a type of **information filtering.** If every input available to the human brain were to be processed by the conscious mind, the mind would bog down from the overload. For one thing, most of the inputs at any given instant supply redundant information about things that have not changed since the last time they were sensed. Take, for instance, the act of watching television. People typically look at most of one wall of the room where the television is located, and not just at the television set, but all the information about the rest of the scene remains unaltered and does not require updating. Only the actively changing portion of the television picture needs to be processed.

Another way to avoid overload problems is to operate sensing devices under interrupt control and threshold detection. When **interrupt control** is exerted over an input device, information is sent to the processor only when the input contains something new or unique to be processed. In addition, the processor can enable or disable a particular sensor interrupt and thereby choose to accept or ignore all inputs from that device. Interrupts can be arranged by priority, so that important inputs get serviced before less significant inputs, as when moving objects get processed before stationary objects, or when unexpected pain overrides concurrent inputs, or when a child's low voice draws a parent's attention even in a noisy environment.

Threshold detection involves allowing the input to cause an interrupt only if the new information constitutes a change from the previously received information by at least some preset amount. This is the way the thermostat on a heater or air conditioner works. The heater turns on when the air temperature reaches a preset value and turns off again when the air temperature exceeds some other preset value. Temperature readings between the two thresholds do not elicit any action and require no transfer of information. Similarly, as long as the ambient air temperature and humidity around a human are within his or her comfort zone, the human is not aware of the temperature. If they exceed the comfort zone, the human attempts to correct the situation by sweating, by taking off some clothing, by moving to a cooler place, or by otherwise changing the temperature. If the temperature and humidity fall below the comfort zone, the human responds by shivering, by putting on more clothing, by turning up the heat, or by moving to a warmer place.

Sensor filtering and threshold detection can be labor-saving operations. All of our activities that work from habits depend on these processes. Remember when you were learning to ride a bike or drive a car? How overwhelming all the sensory information was to you! But after having learned what to do under normal circumstances, you committed these operations to habit, and now your attention is trained on identifying traffic conditions some distance ahead and remaining alert for unexpected emergencies. There is not any less sensory information; you have just made better use of filtering and threshold detection.

However, people sometimes unintentionally set up sensor filtering that can block information they need to receive. It is possible to set up a person's sensor information filters externally by giving the person incorrect or excessive directions on how to interpret some sensory information that is about to be transmitted. This setting of the sensor filters is known as **scatomas** and is accomplished by the power of suggestion. Try the following example of scatomas.

Refer to Figure 17–3, and read that sentence out loud once; then return here. Now return to Figure 17–3 and count the number of Fs in the sentence; then

FINISHED FILES ARE THE RE-
SULT OF YEARS OF SCIENTIF-
IC STUDY COMBINED WITH THE
EXPERIENCE OF MANY YEARS.

Figure 17-3 Phonic Reading Sentence.
We see what we expect to see, even if our filters are set up by our own hearing.

return here once more. How many Fs did you count? If you found three Fs or six Fs, you are normal. If you found four or five Fs, you are either unduly suspicious of being tricked or inconsistent in the way you looked for the Fs. If you found three Fs, you probably learned to read by using phonics. By reading the sentence out loud first, you set up a phonic filter that did not interpret the three *ofs* as containing Fs. Then, when you went back to count the Fs in the sentence, you did not notice the Fs in the three ofs. In case you are still not sure, there are really six Fs in the sentence in Figure 17–3.

The next picture you are going to look at contains a group of five silhouette figures. One of the silhouettes looks like a comb with teeth missing from the ends; another looks like a living room with a fireplace; a third looks like an arrowhead and the remaining two look like the outlines of other rooms. Refer to Figure 17–4 and see if you can make out what the five objects are. When you have finished, return here and read on.

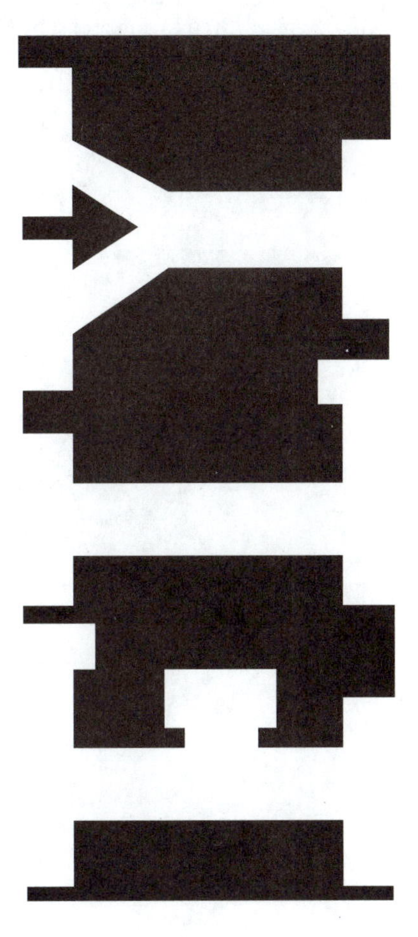

Figure 17–4 Related Objects. This time your filters are set up by the introductory instructions and by past visual experiences.

As you may have guessed by now, you were given misleading directions about Figure 17–4. The deception was aided by your being used to seeing black letters on a white background. To help overcome this scatoma, try imaging two vertical and two horizontal black lines forming a rectangle connecting the outer edges of the five objects. Do you now see the white lettering on a black background spelling out the word FLY? The objects in Figure 17–4 are really the background pieces separating the letters. With this knowledge, try looking at Figure 17–5, and see if you can find the word WIN.

Another information filtering problem involving lack of proper filtering could be described as not seeing the forest because of the trees in your line of sight. In this situation, you are too busy with the details to see the overall picture. Figure 17–6 shows an example of this. Look at it closely and then return here.

Figure 17–5 More Related Objects. By now you should have been ignoring all clues given to you and started trying to see what is actually there.

Figure 17-6 Not Seeing the Forest Because of the Trees.
We can still be fooled, even when we expect trouble.

Did you have trouble seeing anything meaningful in Figure 17–6? If so, you are normal. Now try placing Figure 17–6 about 10 to 20 feet away from you, and look at it again. Is it easier to recognize? It is a dot picture of Mickey Mouse; but with the picture up close you were too distracted by the dots to notice the picture that the dots form.

The human vision system also suffers from having two masters: the logical processor of the brain, generally the left brain; and the creative processor of the brain, generally the right brain. These two processors can disagree about what is seen, creating a conflict between the dominant and nondominant processors. Figure 17–7 shows the young lady and old lady conflict picture. Most people, when first looking at this picture, do not see anything. After a while, the dominant processor is able to recognize a figure of an old lady. It takes time and usually some explanation for people to see the young lady. For those with this problem, Figure 17–8 reproduces only the part of the picture that defines the young lady. With practice, it is possible to see both ladies at the same time. That is, you can communicate with both sides of the brain at once.

Another picture that helps show the conflict between the dominant and nondominant processors is the vase or two faces picture. It is shown in Figure 17–9.

Figure 17-7 Young Lady and Old Lady Conflict.
Vision processing is complicated by the fact that we have two different processors. These work on vision processing from different viewpoints.

Figure 17-8 The Young Lady.
This is the harder of the two ladies to find in Figure 17–7. She is identified by the creative, subconscious, or nondominant processor.

Figure 17-9 Vase or Two Faces Conflict.
This is another drawing that has different meanings to the two vision processors.

HOW MOST MACHINE VISION SYSTEMS WORK

Most **machine vision** systems start with a television camera device for collecting visual information. Additional hardware is used to convert this information into binary ones and zeros, which are then stored in a computer memory as raw image. The raw image is then fed to a series of programs designed to deal with imaging. The output of these programs is a sketch that shows the edges of the images. The sketch is then input to a program that has knowledge about applications, and the output of this program is a description of the contents of the vision field.

Among the programming techniques that can be used by a vision system are thresholding, averaging, contrasting, edge detecting, and line finding.

Thresholding can be used on a front-lighted image, with pixel values from 0 to 255, to change the image into a binary black-and-white image with pixel values of 0 and 1. To do this, the program first accepts as input some value between 0 and 255. The value is used to decide which image pixels should be black and which should be white. All pixel values below or equal to the input value are taken to be black; all pixel values above the input value are taken to be white. By adjusting the thresholding value, it is possible to emphasize or deemphasize objects in the image.

Averaging is used to reduce the noise in an image. The program scans an image and replaces the values in individual pixels with the average value contained in surrounding 3 by 3 blocks or neighborhoods of pixels. If the noise is mainly the result of isolated single pixels, averaging will greatly reduce the noise. One side effect of averaging is that the image tends to blur.

Contrasting is used to heighten the contrast of an image. The program accomplishes this by removing a portion of the image's intensity values at the top and at the bottom of a **histogram**—a graphic display of the number of occurrences of each intensity value in an image—and then stretching the rest of the image to fill the full range of intensity values. The histogram shows how well the values are distributed in the image. If the values are concentrated at one end of the scale, visual details are hard to see. If the values are mainly at the low end, the image is overexposed; if the values are mainly at the high end, the image is underexposed.

Three common programming techniques are used for doing edge detection: Roberts gradient, Prewitt edge detection, and Sobel edge detection. Edge detectors try to locate edges of objects as part of the process of identifying the whole objects. A program can detect edges by looking for differences between a pixel's value and the values of its neighbors. The greater the difference, the more likely the point is to be part of an edge.

The **Roberts gradient,** named after its developer, is the simplest and fastest edge detector, although it is not the best. Roberts uses a 2 by 2 neighborhood of pixels to determine the magnitude of a pixel. The greater the resulting magnitude, the greater the chance that the pixel is part of an edge. Four pixels have to be used for each pixel edge detection test. This is a lot of processing, but each of the calculations can be done in parallel if the computer hardware supports such an activity. Figure 17–10(a) gives the formula for the Roberts gradient.

The **Sobel edge detector** works with a 3 by 3 neighborhood of pixels to determine the likelihood that the center pixel is part of an edge. This technique uses more calculations than does the Roberts gradient. Figure 17–10(b) gives the formula for the Sobel edge detector.

Figure 17–10 Three Edge Detector Formulas.
(a) The Roberts gradient. (b) The Prewitt edge detector. (c) The Sobel edge detector.

a	b
c	d

Pixel location

$\Delta_1 = b - c$

$\Delta_2 = a - d$

Magnitude of pixel $a = \sqrt{\Delta_1^2 + \Delta_2^2}$

(a)

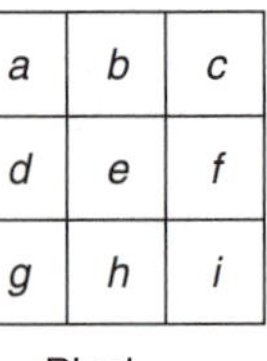

Pixel location

$\Delta_1 = \frac{1}{3}[(c - a) + (f - d) + (i - g)]$

$\Delta_2 = \frac{1}{3}[(a - g) + (b - h) + (c - i)]$

Magnitude of pixel $e = \sqrt{\Delta_1^2 + \Delta_2^2}$

(b)

a	b	c
d	e	f
g	h	i

Pixel location

$\Delta_1 = \frac{1}{4}[(c - a) + 2(f - d) + (i - g)]$

$\Delta_2 = \frac{1}{4}[(a - g) + 2(b - h) + (c - i)]$

Magnitude of pixel $e = \sqrt{\Delta_1^2 + \Delta_2^2}$

(c)

The **Prewitt edge detector** also uses a 3 by 3 neighborhood of pixels to determine the likelihood that the center pixel is part of an edge. However, extra weight is given to the horizontal and vertical neighbors of the pixel under consideration. This method is the slowest of the three edge detection techniques. Figure 17–10(c) gives the formula for the Prewitt edge detector.

Line-finding programs use regions of edge gradients as a basis for locating lines. The problem with trying to find lines is that the edges are not always complete. There may well be missing pixels in a line. Burns developed an algorithm for finding lines, but it uses a lot of memory and computational power.

While most vision systems use a single computer processor to do the calculations, some parallel-array processors are available. The Cray I computer, the first commercial supercomputer, is an **array processor.** Since the Cray I was introduced, array processors have been incorporated into smaller and smaller computers. These small parallel processors are sometimes called **cellular array processors,** a term that refers to a two-dimensional array of processors working on the same task. Cellular arrays of up to 128 by 128 processors have been built. Using cellular array hardware greatly reduces the amount of time required to do vision processing.

A machine vision system capable of matching the vision processing capacity of a human being has been estimated to require the computer processing power of 1,000 Cray II computers, each of which can execute 200 million instructions per second!

ADVANTAGES OF HARDWARE-BASED EDGE DETECTION

One area of machine vision processing that can benefit from hardware changes is that of edge detection. One way to use hardware in edge detection with a binary picture is to place EXCLUSIVE OR gates between each pair of pixels in the picture. (OR gates are discussed in Appendix B.) If one of the pixels is a one and the other pixel is a zero, the intersection of the two pixels represents an edge. Figure 17–11 shows the difference between program-based and hardware-based edge detection. The hardware can perform the edge detection procedure in a few nanoseconds, while the program may take milliseconds, seconds, or longer. Edge detection is one of the main steps in need of improvement in vision processing for a robot that is to pick up parts or navigate down a cluttered hallway.

While first-generation edge-processing hardware, using hundreds of EXCLUSIVE OR gates, is likely to be big and difficult to build, a small integrated circuit could be designed to replace the larger model's wires and gates. The integrated circuit could include the raw image memory, the EXCLUSIVE OR gates, the edge detection memory, and all the wiring for at least a 200-dot by 200-dot image. This is not impossible, given that a present-day integrated circuit can hold 4 million bits of memory.

In a gray-scale vision system, the EXCLUSIVE OR gates would have to be replaced by differential amplifiers.

Motion detection can be accomplished by saving the last edge detection information in memory and comparing it to the next results of the edge detection hardware. Figure 17–12 shows how this might be done.

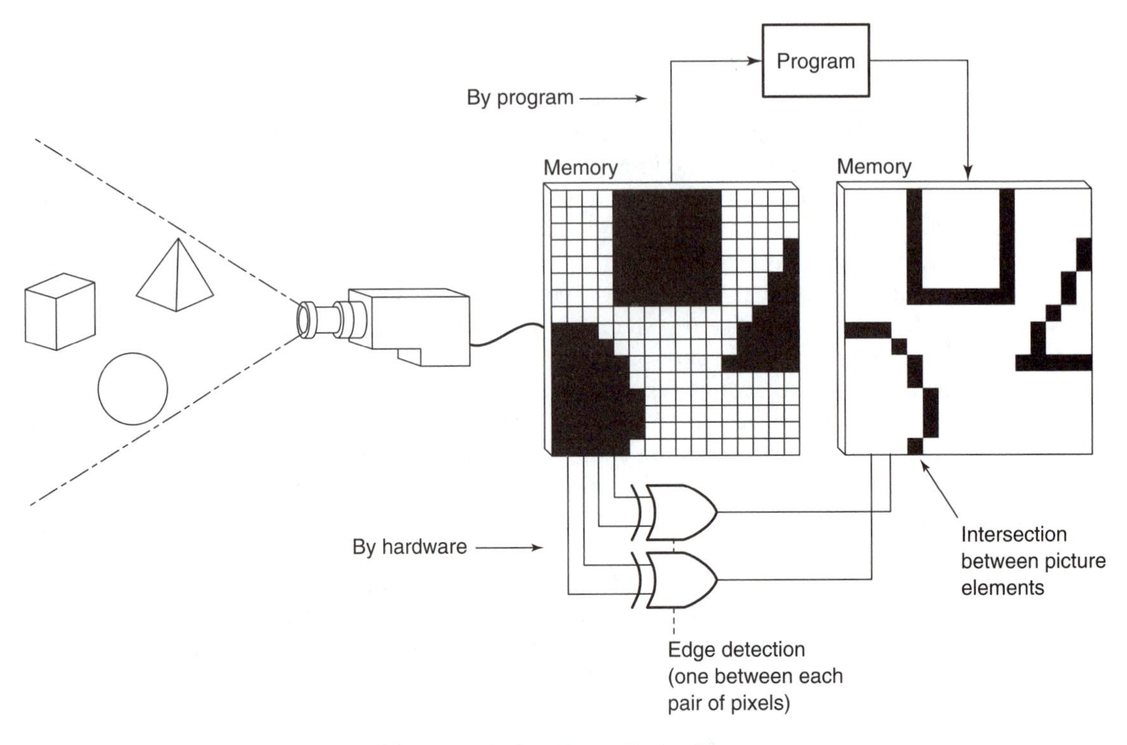

Figure 17-11 Alternatives for Machine Vision Edge Detection.
Improving the hardware used in a vision system can speed up edge detection.

Improving the Speed of Processing

Many present-day robot vision tasks require only the identification of objects by their outline. Use of edge detection hardware would greatly speed up this type of vision processing and greatly reduce the computer processing power required to identify an object. Consequently, the vision system could be used for tasks that it formerly was too slow to do.

Making New Vision Tasks Practical

Many tasks that are not now practical for vision systems to do—because the systems are too slow—would become practical with hardware advances. Robots would be able to do more complicated assembly tasks, and they would also be able to adjust to handle slightly misplaced or misaligned parts.

Reducing Overall Cost of Vision Systems

While the hardware at the front of the vision system would increase in complexity and cost, the computer processing power and programs would diminish, lowering the overall cost of the system. As edge detection and motion detection circuits become standard features of integrated circuits, their costs will be greatly reduced.

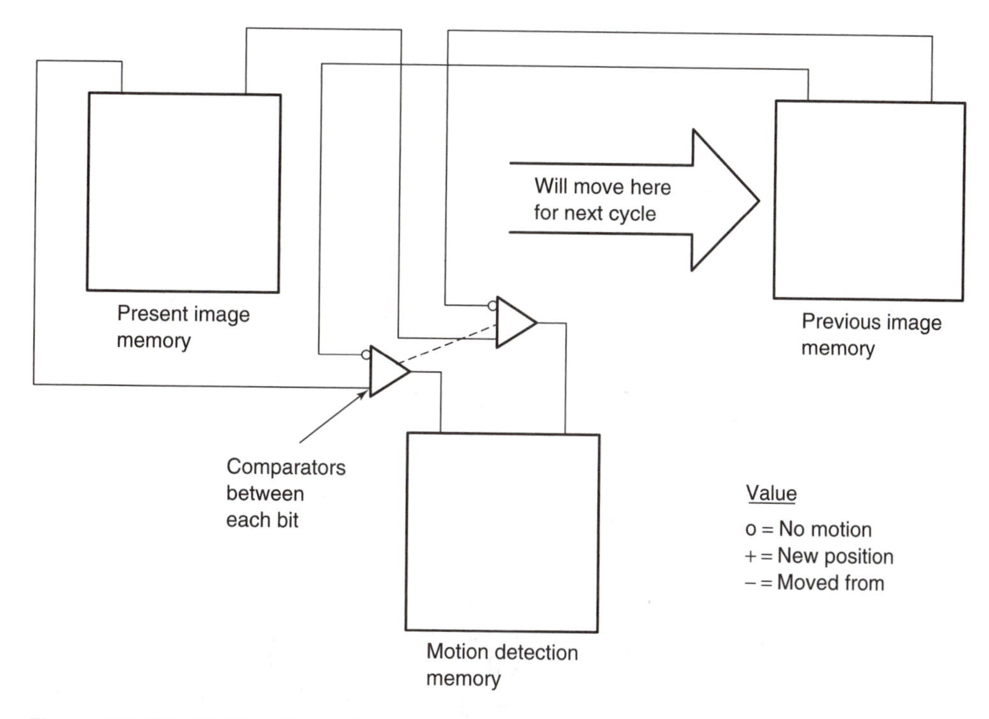

Figure 17–12 Motion Detection Hardware.
Motion detection can be speeded up by using appropriate hardware.

DOES THE ROBOT'S VISION DO WHAT WE THINK IT IS DOING?

When watching a robot use vision to do a task, do we really see how the vision is being used? Probably not. The following example shows the difference between what we first think the vision system needs to do and what the vision system is really doing. Task setup can greatly reduce the complexity of the vision task requirements.

Suppose that you go to a robotics show, and there you watch an industrial robot that is equipped with a vision system do a mock assembly of a hard disk drive. Figure 17–13 shows the task layout. The task includes the following steps:

1. The empty spindle is spun so that the holes in the top of it are at a random location.
2. The robot moves four disks with spacers between them, from storage pole to the disk drive spindle.
3. The robot places the outside bracket on the spindle.
4. The vision system locates the hole in the disk spindle.
5. The robot picks up a screwdriver attachment, moves three screws from a storage bracket, and screws them into the holes on the disk spindle.
6. The robot then disassembles the drive and starts the task all over again.

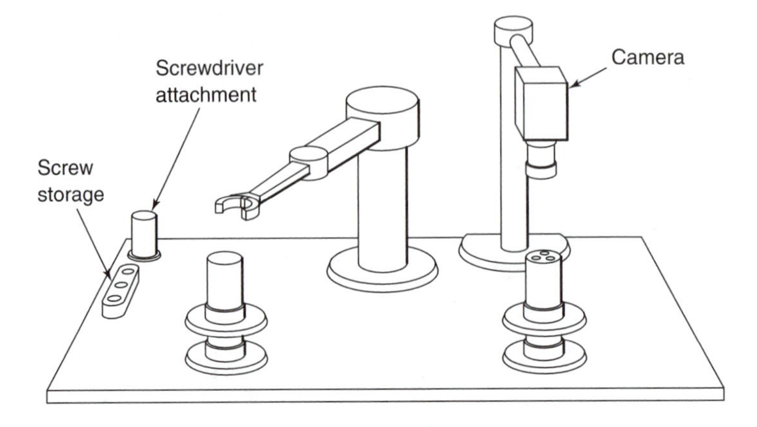

Figure 17-13 Robotic Vision Task.
Properly setting up a vision task can greatly reduce the amount of information that needs to be supplied by the vision system.

After watching the robot work for a while, you notice that the only random part of the task appears to be where to find the holes in the disk spindle, so you concentrate on the part of the task where the vision system locates the holes. First, the outline of the spindle shows up on the vision monitor almost instantly. After a while, one of the holes appears on the spindle, and suddenly the other two holes appear. You ask yourself, How did the vision system find the spindle so fast, and how did it find the last two spindle holes so fast? The answer to these questions lies in the setup of the robotic task.

When you look closely at the location of the camera for the vision system, you discover that the camera is exactly centered over the spindle. The vision system didn't have to find the spindle, because it was already trained on it. Finding the first hole took a while (a second or two), because the vision system did not know exactly where the hole would be. However, since it knew where the spindle was, it knew where to find the circle whose radius represented the distance of the holes from the center of the spindle. Once it found a hole on the circle, it knew how to calculate the locations of the other two holes. That is why it found the second and third holes so fast.

Was this demonstration cheating or showing good use of a vision system? It was showing good use of a vision system. Whenever you can predict or calculate where a vision system is to find something, you should do so. This amounts to using expectational information to make the task run faster.

THE WIZARD VISION SYSTEM

Wizard uses a neural architecture of highly interconnected parallel processing similar to the human brain. Wizard is a vision-equipped robot that creates its own internal picture of what it sees and then uses this information to make future decisions. It can be shown a group of good oranges and told to remember what they look like. It can then be shown a group of bad oranges and told to remember what a bad orange looks like. Finally it can be given a mixed group of good and bad oranges and told to pick out only the good ones. It accomplishes this by comparing the present orange in each case to past experiences. Wizard is said to have random wiring connecting its neurons, in order to resemble how humans wire up their brain connections during the early years of development.

INTERFACING A SINGLE-ELEMENT EYE

A single-element eye can be given to a robot by using a **phototransistor** or a **cadmium sulfite cell.** These two basic devices are wired as shown in Figure 17–14. The resistor values are approximate and may have to be adjusted, depending on the characteristics of the sensing device.

Both of these eyes are analog devices. That is, they can output any value over a continuous range of values. If the device is to be used just to detect the presence or absence of a part, the output needs to be converted to a digital one or zero. Figure 17–15 gives the circuits for converting either device's output to a binary one or zero using one-fourth of an LM339 quad comparator integrated circuit. The output of the comparator can be fed directly into an input line on a computer port, a robot controller, or a programmable logic controller. The circuits include an adjustment for the threshold point of the eye—the point at which the eye changes its output from a binary zero to a binary one.

The programming requirements for this simple zero/one or on/off type sensor are fairly straightforward. First, the operating system program must make this input available to the application programs. Second, the application program must check the value of this input. Third, the application program must decide what the signal value means. If the sensor is the light detector on a photoelectric sensor looking

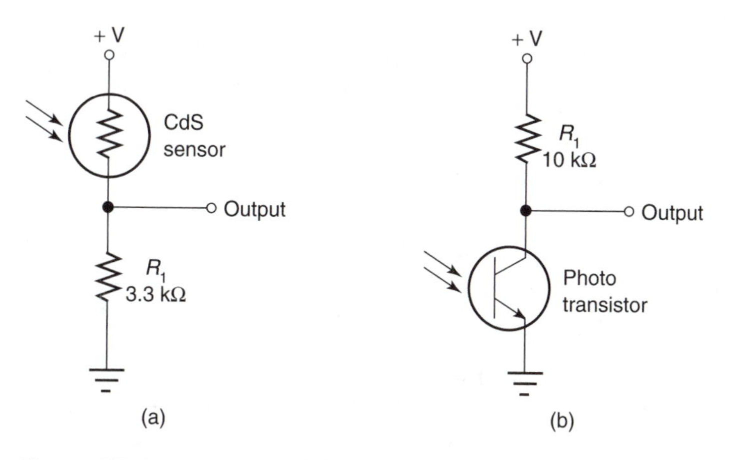

Figure 17-14 Single-element Eye Devices.
(a) A cadmium sulfite cell. (b) A phototransistor.

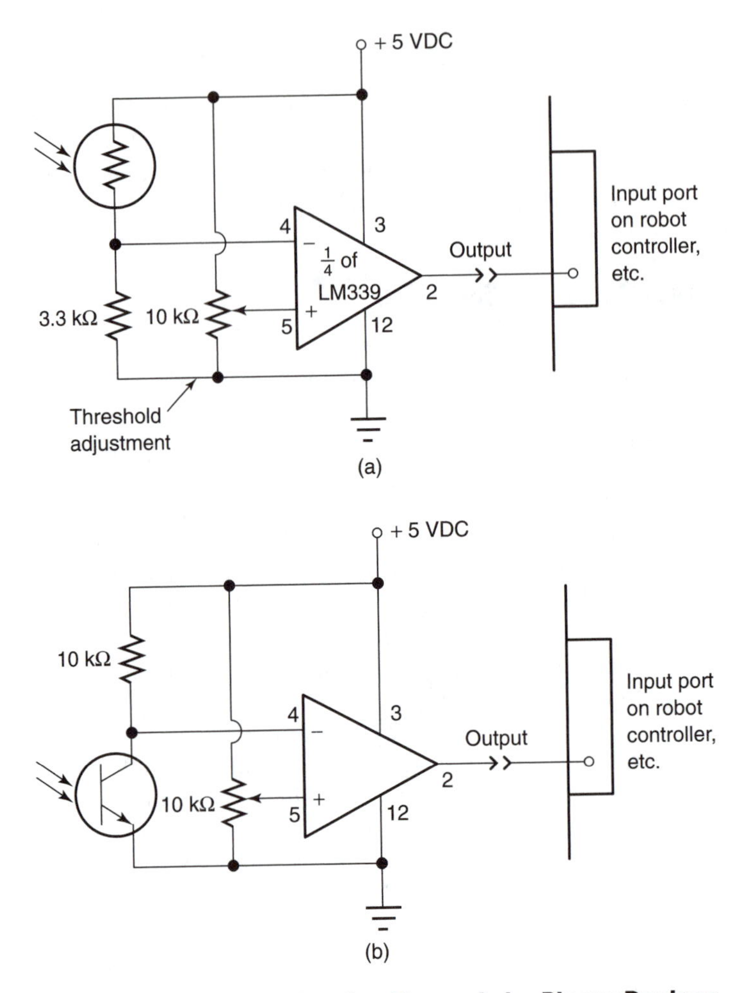

Figure 17-15 Connecting Eye Elements by Binary Devices.
Eye elements can be connected as binary devices by using comparator ICs. (a) The cadmium sulfite cell interface. (b) The phototransistor interface.

for the presence of a part, a value of zero would indicate that a part was between the photo sensor and the light source; on the other hand, if the light source is like a flashlight being shined away from the robot, with the light sensor on the robot, a value of one on the input would indicate that the light was being reflected off something in the robot's path.

If you want a greater range of values from the eye, you will need an **analog-to-digital (A/D) converter** and at least 8 input lines to a computer port or whatever. These 8 lines will allow for values from 0 to 255, which provides good visual range. Some microprocessor-based A/D converter integrated circuits suit this purpose. Actually, these devices come with 8 and 16 inputs, so you can connect 8 single-element eyes or a single 8-element eye to the device called an ADC0809 integrated circuit. The 16-input device is called an ADC0816 integrated circuit. Figure 17–16 shows a circuit for connecting the 8 sensor elements to a vision controller or whatever.

The programming requirements for a single 256-state device would be for the operating system to allow the application programs to read these 8 input lines. The

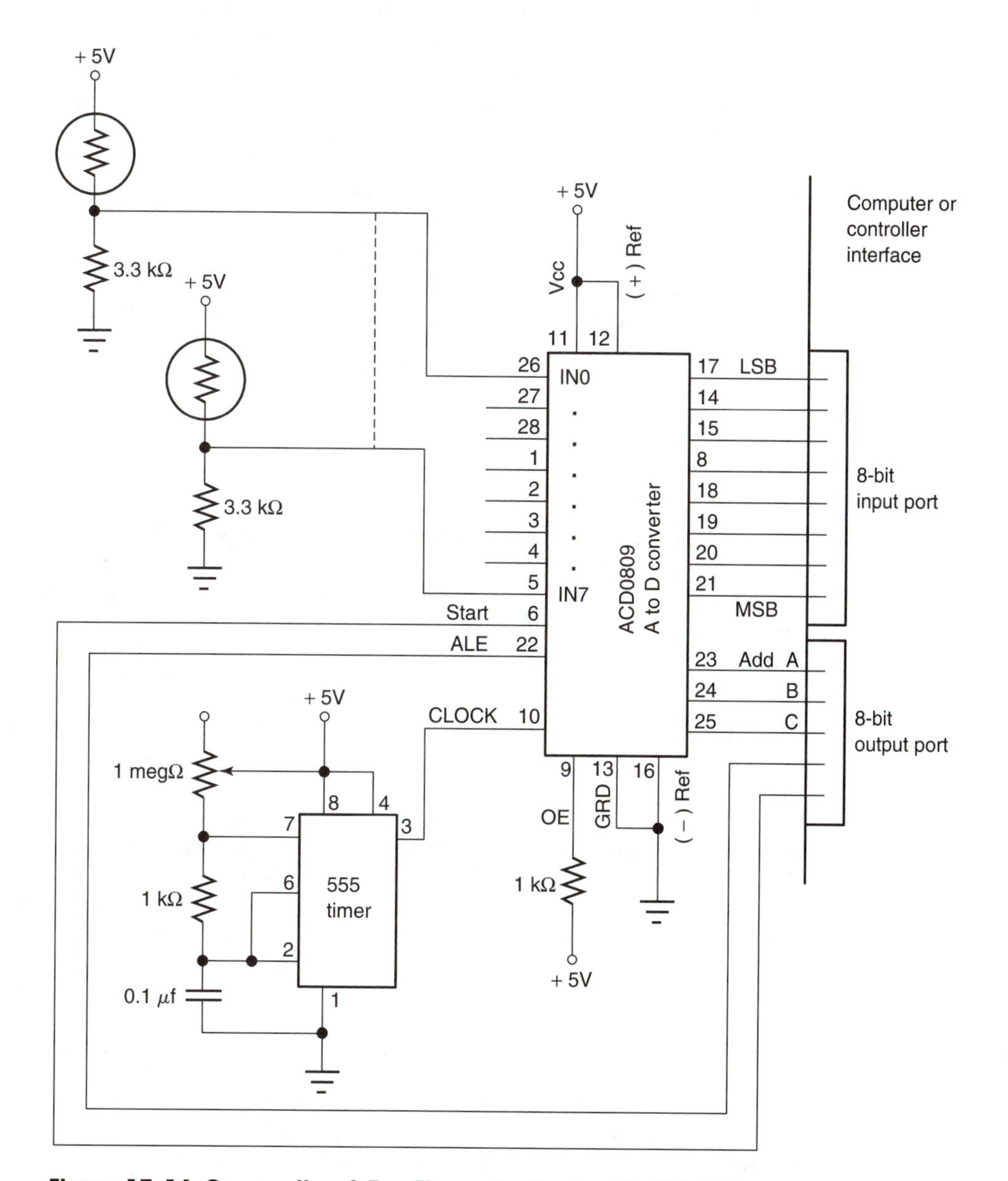

Figure 17–16 Connecting 8 Eye Elements, Each with 256 Different Values.
Up to 8 eye elements, each with 256 different possible values, can be interfaced using an ADC0809 microprocessor-based A/D converter.

application program would then have to decide what the input values mean. Could values from 0 to 63 mean that the path is probably clear? And could values above 64 mean probable danger?

The programming requirements for multiple 256-state devices are more complicated. First the application program must tell the operating system which device value it wishes to read. The operating system then uses an output port to send this information to the A/D converter. Then a time delay ensues while the converter does its work. This delay can be controlled by connecting the end-of-conversion (EOC) line on pin 7 of the ADC0809 IC to another input port line on the controller. Then the applications system must pass the resulting signal value to the applications program. Now the applications program has the value for one of several visual pixels. It must process this value in relation to all the other pixel values.

Most robot controllers are not able to handle the output of a television camera. The information has to go through some type of vision system and then the results can be sent to the robot. As robot controllers become faster and more powerful, this restriction may be removed.

SUMMARY

Both humans and machines need vision-processing systems. Most vision systems, including the one possessed by human beings, have the following parts: illumination source, vision sensor, motion detector, edge detector, line detector, object shape detector, image understanding, and expectational information. Optionally, they may use color, texture, and depth detection. The human vision system relies on parallel processing at most stages of the vision process. It uses very slow processing devices (neurons), while doing rapid vision processing. Machine vision systems tend to use a single high-speed processor, and they usually work on the information one piece at a time. If machine vision systems could use some of the hardware techniques of the human eye, it would greatly increase their speed. Experimental vision systems are now using arrays of processors to work on vision problems in parallel.

Interfacing a single-element vision sensor for just an on/off reading can be done through a single line of a computer input port, using only one-fourth of a comparator IC. To interface the same sensor and thus get value readings over the 0 to 255 range requires using an analog-to-digital (A/D) converter IC and all 8 lines of a computer input port. A microprocessor-based A/D convertor can handle up to 16 different single-element vision sensors.

The programming requirements for a simple on/off sensor are that the operating system make the signal available to application programs. The application programs must then test the on/off value and take the appropriate action. With an input signal varying from 0 to 255, the application program must decide what significance the value has and must respond by taking the appropriate action. If up to 16 different sensors are coming in on the same port, the application program must send a signal to the A/D convertor telling it which device value to place on the port. At present, most robot controllers cannot process the output of a full television camera; they require a vision system controller between themselves and the camera.

REVIEW QUESTIONS

1. What characteristics of an object can be used to help identify the object?
2. How can some of the hardware techniques of the eye be used in a machine vision system?
3. What type of expectational information might be useful to a robot using vision in a manufacturing situation?
4. Compare human vision processing, present-day industrial vision processing, and the Wizard vision system.
5. Construct a machine vision system that has four pixels. Each pixel is the base of a phototransistor or a cadmium sulfide cell. Use a bracket and tubes to point each pixel at one-quarter of a template that could fit around a human face. If the face is illuminated with a 60-watt light bulb, a pixel value for each quarter of the face can be obtained. Input the pixel values into a computer through an analog-to-digital converter. Values of 0 to 255 should be enough for each pixel. Calculate a light adjustment value based on the value of the first pixel, plus one-half of the value of the second pixel, plus one-third of the value of the third pixel, and one-fourth of the value of the fourth pixel. Adjust the raw data based on this average. As an alternative, you could use a fifth pixel to determine the light adjustment level. You could also use a door of light gray color to cover the face hole, and then you could read an adjustment average value using just the original four pixels. Store the total value of the adjusted information under the name of the person being viewed. See if the system can tell ten different persons apart. See what difference a person's smiling or frowning has on the stored value.
6. Give an example of how the placement of the camera for a vision system can make the vision task simpler.
7. How does processing vision toward some goal speed up the vision interpretation process?

18

TACTILE SYSTEMS

OVERVIEW

Some tasks using touch are mistakenly thought of as using mainly vision. **Tactile systems** attempt to give the robot the same touch sensitivity that humans gain from sensors in their skin. Of course, the robot will probably never have the need for a touch sensor that covers its entire outer surface.

How are tactile sensors interfaced to a robot? What programming problems are involved in interpreting tactile sensors? What limits are the tactile sensors trying to reach? These are some of the questions discussed in this chapter.

OBJECTIVES

When you have completed this chapter, you should be familiar with:

- Interfacing a single-point tactile sensor to a robot
- Interfacing multiple-point tactile systems
- Software considerations for tactile systems

KEY TERMS

Artificial skin
Contact
Depth sensor
Give
Interrupt line
Matrix of depth sensors
Roughness
Single-point contact
Tactile systems
Touch

UNDERSTANDING ROBOTIC TACTILE INTERFACING

Robotic tactile interfacing deals with determining what hardware and software are needed to interface tactile sensors to a robot controller. As with many sensors, it is much easier to connect the sensor information to the controller than to interpret the meaning of the information. Our discussion will start with the simplest tactile sensor, the microswitch, and it will conclude with the most complicated sensor, artificial skin.

INTERFACING THE MICROSWITCH

The microswitch outputs a binary value that can be called on/off, high/low, or one/zero. The interface can be as simple as grounding one side of the switch and running the other side to the line of an input port of a computer, programmable logic controller, or robot controller. The side of the switch connected to the port should have a pull-up resistor to 5 volts DC. This will keep the input from picking up electrical noise while the switch is open. Figure 18–1 shows the circuit for this input.

The programming requirements are also fairly simple. The operating system must allow the application programs to see this signal line. The application need only check to see whether the value is low (or on) or high (or off). If the value is high, nothing is touching the switch contact.

Suppose that the robot is to be given three microswitches as bumpers to detect when an object is touching it, and that it is not to be concerned about the bumpers until one of them is touched. In this case, the interface would need three input port lines, one three-input AND gate operating as a NOR gate, and an interrupt line connecting to the controller. The **interrupt line** is a special input to a controller (or computer) that causes the controller to stop what it is doing and process its input. Relying on the interrupt line frees the controller from having to monitor the input, because the interrupt line is certain to get the controller's attention when something happens at its input. Figure 18–2 shows this circuit.

The programming requirements now get more complicated. First, the operating system must support interrupts and have a program to handle such interrupts.

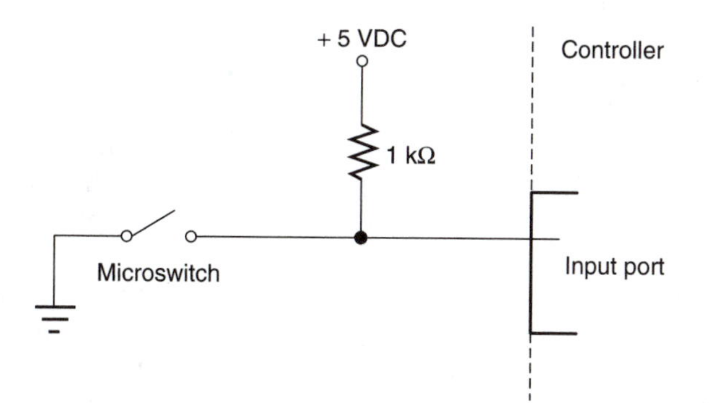

Figure 18–1 Interfacing a Microswitch to a Controller.
A microswitch is the simplest sensor to interface to a controller. It requires only a pull-up resistor and a switchable connection to ground.

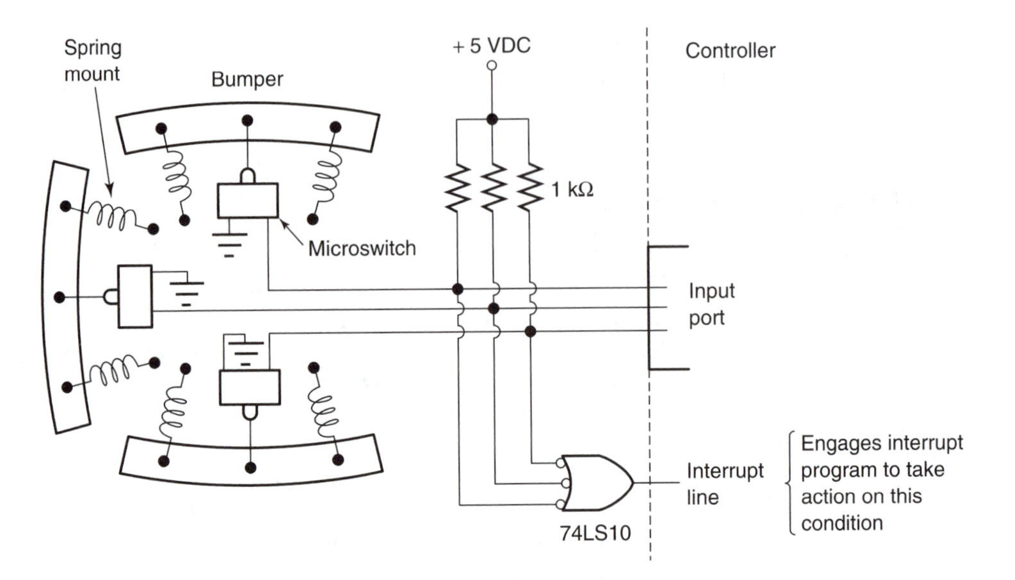

Figure 18–2 An Interrupt-driven Bumper Interface.
Bumpers can be used to activate microswitches. Several microswitches can be logically connected to a single interrupt line, which is responsible for telling the controller when to process the bumper program.

When an interrupt occurs, the operating system must pass the input line values from the interrupt to the application program, which has switched to an interrupt program called by the operating system. The interrupt program interprets the meaning of the input and responds by taking the appropriate action; then it returns control to the operating system interrupt program, which in turn transfers control back to whatever program was running before the interrupt occurred.

A matrix of 64 microswitches could be interfaced to a controller as 8 rows and 8 columns. This would be similar to the way the keys of a keyboard are connected on many computer terminals. A part pressing against this matrix would reveal the part's shape and orientation. The matrix could be read a row at a time through an 8-line input port. The row is selected by reading the decoded value of 3 lines from an output port. Figure 18–3 shows such a circuit. It is possible to handle a matrix of 256 by 256 sensors by using 8 decoder ICs off the output port and 8 encoder ICs on the input port. Of course, we are now dealing with something approaching **artificial skin.**

The programming requirements for artificial skin composed of microswitches are very similar to those for processing a high-contrast black-and-white image. It is possible to do edge detection, line detection, object detection, and so forth, on this matrix of information. But first the application program must transfer the sensor matrix of information into internal storage within the controller. To do this, the application program sends a column address to the output port lines (via the operating system). Then the column of information is read through the input port, via

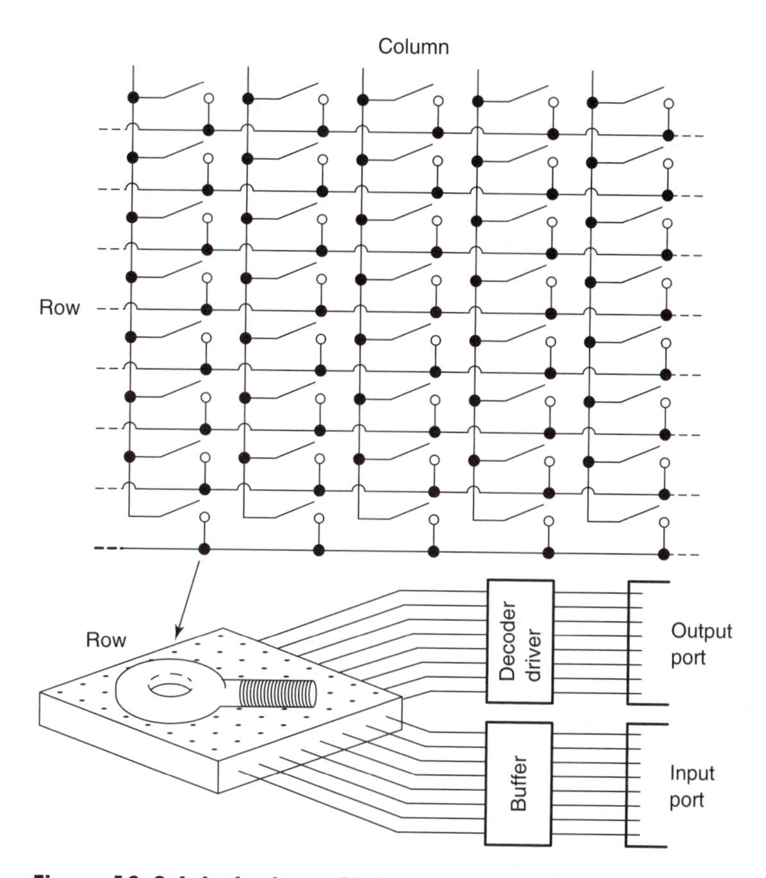

Figure 18-3 Interfacing a Matrix of Microswitches.
Interfacing a matrix of microswitches requires the use of an output port and an input port. One port connects to the rows, and the other port connects to the columns.

the operating system, to the application program, which stores it in its internal matrix or array. This procedure is repeated for each column.

INTERFACING A STRAIN GAUGE

Strain gauges can be glued or bonded on the manipulator parts to measure the acceleration acting on the parts (a form of an accelerometer). They can be glued to the gripper to measure the force being exerted on the part in the gripper. Strain gauges can be used to measure hydraulic pressure or fluid level. They can even measure the torque delivered to a joint.

A strain gauge is an analog device, whether it is used to measure the force on a joint or to measure the force on a gripper. Most strain gauges are temperature-sensitive, so their output must be corrected for temperature. This is usually done by using a second strain gauge element as a reference leg in the bridge circuit for the sensor. The reference element must not have any force acting on it. If the force is applied only in a single direction, the reference element can be on the same backing as the active element, but it should be 90° out of phase with it. Otherwise the temperature correction would have to be calculated by the controller. The output of the sensor passes through an A/D converter and then through at least an 8-line input port on the controller. Figure 18–4 shows a possible circuit.

The programming requirements for a strain gauge sensor are very similar to those for the 256-state single-vision element sensor discussed in Chapter 17.

Conductive foam, conductive rubber, and spring-loaded potentiometers can be interfaced very similarly to the way the phototransistor was interfaced in Chapter 17. The programming problems are also very similar.

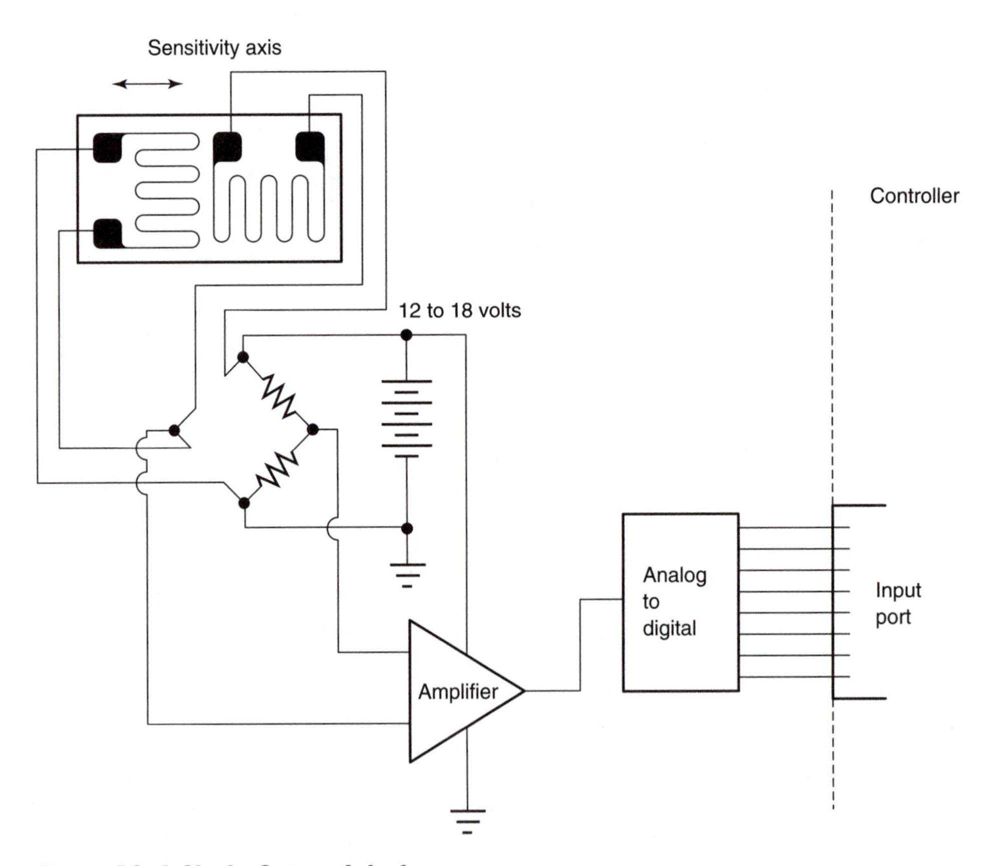

Figure 18–4 Strain Gauge Interface.
A strain gauge interface requires an active sensor for the force reading and a dummy sensor for temperature compensation. These are connected through an amplifier and an A/D converter.

THREE-DIMENSIONAL TOUCH

Humans can figure out the shape of a part by **touch.** A robot could use a **matrix of depth sensors** to determine the shape of a part. Each depth sensor could be made from a potentiometer with a dull-needle drive mechanism attached to its shaft and a spring to keep the needle fully extended until it touches something. Any movement of the needle would change the value of the potentiometer, so the signal from the potentiometer is an analog signal whose value represents a relative distance or depth reading. Figure 18–5 shows a possible depth sensor. A matrix of these sensors forms a bed of needles, and a part placed on the bed yields a three-dimensional reading of its shape.

The interface hardware shown in Figure 18–5 requires A/D converter ICs similar to those used for interfacing a 255-state gray-scale vision sensor. The operating

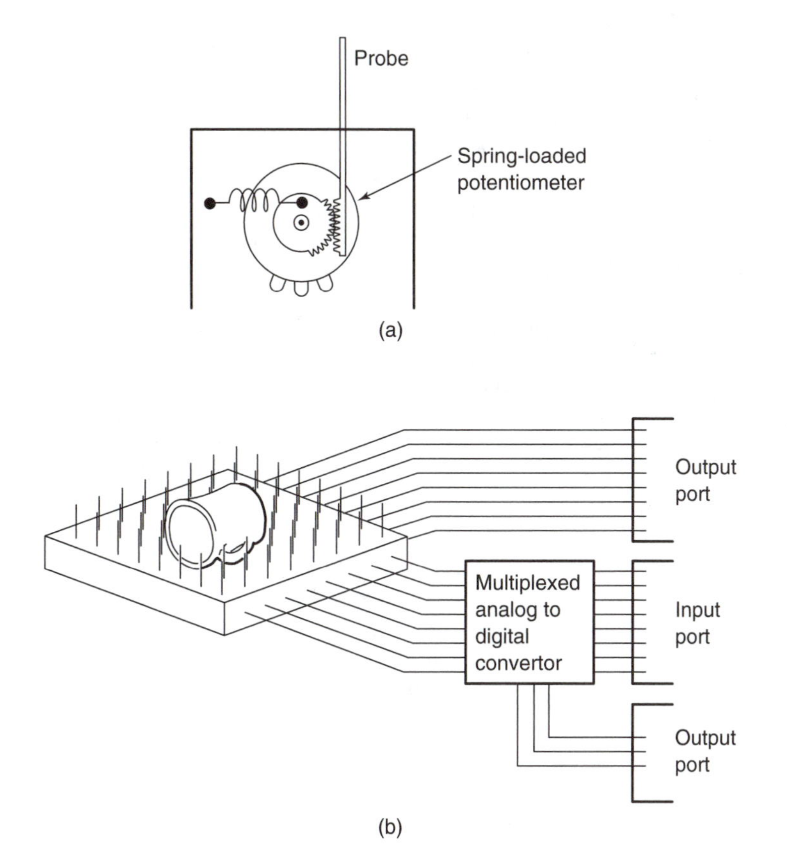

Figure 18–5 Three-dimensional Depth Sensor Interfacing.
Multiple analog sensors can be used to produce a three-dimensional touch picture. (a) A single depth sensor. (b) A bed of needles three-dimensional depth sensor.

system interface is also the same. The application program must obtain and store the information in a matrix or array and then process the information much as it would vision information.

SUMMARY

Interfacing and programming touch sensors can be as simple as connecting a microswitch to a line of an input port on a controller and then having the application program respond to the on/off value from the switch. It can also involve getting three-dimensional information from a bed of needles depth sensor matrix, inputting this information through A/D converters to the controller, and finally having the application program process the information much as it would vision information.

REVIEW QUESTIONS

1. Carefully experiment with a pin to determine how close together the touch sensors are on your fingertips. How does this compare to the spacing of pain sensors on the back of your hand?
2. Why would you bother to use a strain gauge as a sensor on a gripper to enable the robot to tell when it has a part in the gripper, when a simple microswitch could also tell this information?
3. Why is touch not the preferable way for a large mobile robot to find out that it has obstacles in its path?
4. List some tasks that you use vision to do, but that you could do using just your sense of touch.
5. What robot tasks can be helped with tactile sensors?
6. How would artificial skin be useful for finding out the orientation of a part?
7. List some applications for strain gauges.
8. Do you tie your shoelaces using a sense of touch, vision, or both? Can that be done using just one sense?

19

PROXIMITY SENSORS

OVERVIEW

Proximity sensors cover receptors from tactile through visual. This chapter investigates proximity sensors that fall somewhere between touch and vision.

How is a magnetic sensor interfaced and programmed? What about ultrasonic range finders and inductive and temperature sensors? Where can one find interfacing information on a ready-made sensor device? These are some of the questions discussed in this chapter.

OBJECTIVES

When you have completed this chapter, you should be familiar with:

- How nonvision, nontouch, nonspeech recognition proximity sensors can be interfaced and programmed for use in robotics
- How to interface and program encoders to robots
- Where to get interfacing information on ready-made devices

KEY TERMS

ASCII
BCD gray code
Binary
Binary coded decimal (BCD)
Gray code

THE REMAINING PROXIMITY DETECTORS

Many proximity sensor interfaces have already been covered in the chapters on vision and touch. All others, except for speech recognition, are covered in this chapter. Specifically they include magnetic, sound (sonar), capacitance, tuned radio circuit, temperature, radiation, and radar.

GENERAL INTERFACING INFORMATION

Industrial robots work in factories with high levels of electrical noise generated by all the electrical motors and other electrical devices. Many sensing devices produce tiny levels of electrical voltages and currents. This makes it very difficult to read the sensory information. For this reason, amplifiers need to be placed close to the sensor and the resulting sensory information needs to be converted to digital signals for better noise resistance during transmission of the information to the robot controller.

The interfacing information that accompanies a proximity sensor kit is sometimes as big as a manual on the device. It should answer a wide range of questions about the sensor. For example, does the information the sensor transmits to the outside world take the form of an analog signal (such as would be transmitted by a cadmium sulfite cell) or as a serial stream of binary ones and zeros? If the signal is serial, what are the bit rate and the protocol used? Which bit is transmitted first—the least significant or most significant bit? Will the information be transferred as straight binary or as some binary character set, such as ASCII code? Finally, what is the output voltage level for a binary one? Table 19–1 shows the ASCII character code set that is used for exchanging information between computers and terminals.

Most sensors use some type of parallel transfer of information, usually in binary number form. How many bits are used in the transfer? A standard of 8 bits of information is fairly common. If more bits are needed, the transfer is done in 8-bit groups. With an 8-bit port, information can be transferred as binary numbers, binary-coded decimal numbers, or ASCII characters. Most sensors use binary, binary-coded decimal, or one of several binary gray codes. Table 19–2 shows these codes for values from 1 to 20.

If all this is a little confusing, it is normal. Before you buy a sensor device, read its interface directions carefully. Consult the interface directions for the controller being used, too.

Encoders Encoding with a potentiometer was discussed in Chapter 18. This section deals with absolute encoders.

An absolute encoder gives its present position as a binary number. Each bit of the number must go to a separate line of a controller input port. If a value of ±359.999° is to be supported, then 20 to 25 bits of information must be transferred to the robot controller—1 bit for the ± sign, 9 to 12 bits for the 359, and 10 to 12 bits for the .999. The reason 9 to 12 bits are required for a value of 359 is that the information could be coming off the encoder as a binary or binary-coded decimal value. Either way, the value would probably take the form of a gray code and therefore would always be within 1 bit of being correct.

Table 19–2 shows decimal digits coded in binary-coded decimal (BCD), BCD gray code, and gray code. If all of these bits are run straight into input port lines, then up to 25 lines (or 3 plus input ports) would be used. Alternatively, the bits can

Table 19–1
ASCII Character Codes

Code	Hexadecimal Value	Code	Hexadecimal Value	Code	Hexadecimal Value	Code	Hexadecimal Value
NUL	00	SP	20	@	40	′	60
SOH	01	!	21	A	41	a	61
STX	02	″	22	B	42	b	62
ETX	03	#	23	C	43	c	63
EOT	04	$	24	D	44	d	64
ENQ	05	%	25	E	45	e	65
ACK	06	&	26	F	46	f	66
BEL	07	’	27	G	47	g	67
BS	08	(	28	H	48	h	68
HT	09	)	29	I	49	i	69
LF	0A	*	2A	J	4A	j	6A
VT	0B	+	2B	K	4B	k	6B
FF	0C	,	2C	L	4C	l	6C
CR	0D	–	2D	M	4D	m	6D
SO	0E	.	2E	N	4E	n	6E
SI	0F	/	2F	O	4F	o	6F
DLE	10	0	30	P	50	p	70
DC1	11	1	31	Q	51	q	71
DC2	12	2	32	R	52	r	72
DC3	13	3	33	S	53	s	73
DC4	14	4	34	T	54	t	74
NAK	15	5	35	U	55	u	75
SYN	16	6	36	V	56	v	76
ETB	17	7	37	W	57	w	77
CAN	18	8	38	X	58	x	78
EM	19	9	39	Y	59	y	79
SUB	1A	:	3A	Z	5A	z	7A
ESC	1B	;	3B	[	5B	{	7B
FS	1C	⟨	3C	\	5C	|	7C
GS	1D	=	3D	]	5D	}	7D
RS	1E	⟩	3E	^	5E	~	7E
US	1F	?	3F	—	5F	DEL	7F

be multiplexed together so that the first 8 bits are sent, then the second 8 bits, and so forth. Figure 19–1 shows a possible circuit for these connections. In addition, most encoders require a 5-volt DC power source.

The applications program would have to send the signals for transferring each part of the value. It would also have to reassemble the pieces of the value and then evaluate the value. If the value is unexpected, corrective action must be taken.

Table 19-2
Binary Codes for the Numbers 1-20

Decimal Value	Binary	Binary-coded Decimal (BCD)	BCD Gray Code	Gray Code
0	0	0000 0000	0000 0000	0000
1	1	0000 0001	0000 0001	0001
2	10	0000 0010	0000 0011	0011
3	11	0000 0011	0000 0010	0010
4	100	0000 0100	0000 0110	0110
5	101	0000 0101	0000 0111	0111
6	110	0000 0110	0000 0101	0101
7	111	0000 0111	0000 0100	0100
8	1000	0000 1000	0000 1100	1100
9	1001	0000 1001	0000 1101	1101
10	1010	0001 0000	0001 1101	1111
11	1011	0001 0001	0001 1100	1110
12	1100	0001 0010	0001 0100	1010
13	1101	0001 0011	0001 0101	1011
14	1110	0001 0100	0001 0111	1001
15	1111	0001 0101	0001 0110	1000
16	10000	0001 0110	0001 0010	11000
17	10001	0001 0111	0001 0011	11001
18	10010	0001 1000	0001 0001	11011
19	10011	0001 1001	0001 0000	11010
20	10100	0010 0000	0011 0000	11110

The incremental encoder just emits pulses—one for each increment of rotation of its shaft. This information could be transferred over a single-input port line as a series of on/off signals. Cumulatively, the pulses produce a sine wave signal and must be converted to DC pulses. If the encoder is to tell the direction in which it is turning, it must output two sets of pulses that are 90° out of phase with each other. The direction is determined by which phase signal is 90° ahead of the other signal.

The programming requirements for an encoder are fairly complex. If the encoder is used to tell the location of a joint, for example, the applications program must have some way of discovering when the joint is at its home position. This might involve its sending a command to the operating system to home a joint. In response, the operating system could do one of two things:

1. If the joint's home position is up against some mechanical stop, the operating system could send commands that would move the joint far enough to ensure that it hits the stop. Thus, the Hero I robot would issue enough step commands to the joint's stepper motor to move the joint from the farthest possible point away from home to home.

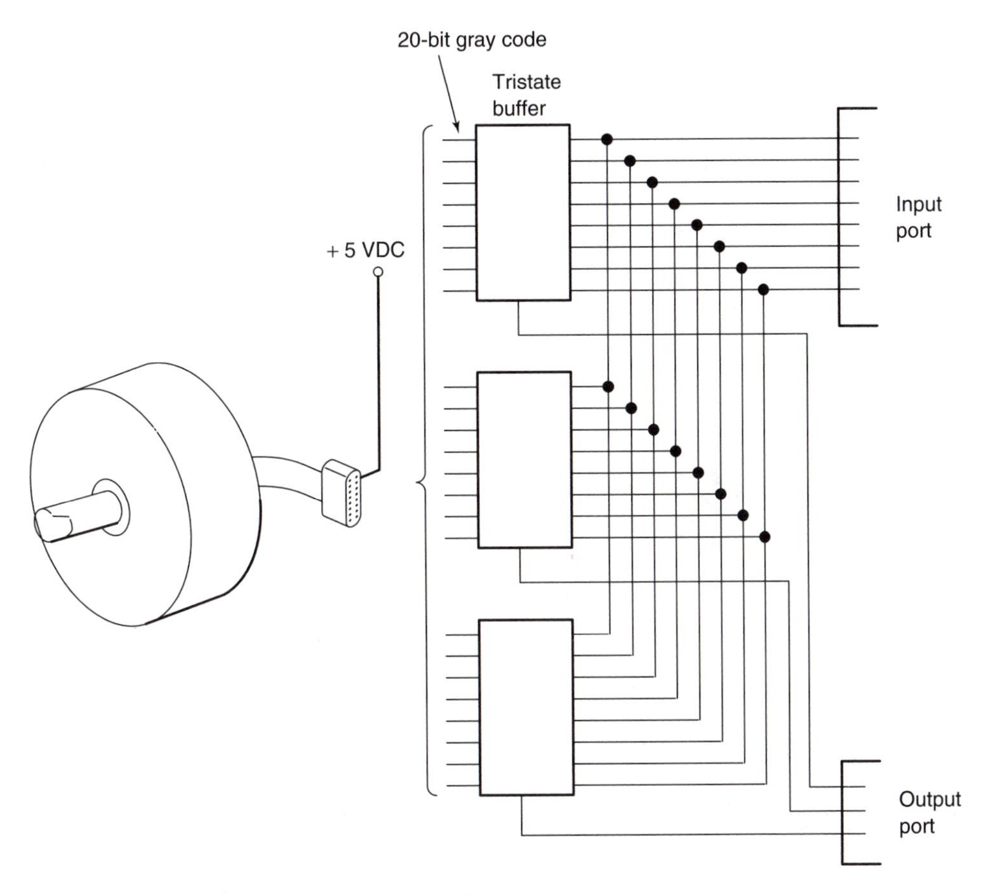

Figure 19–1 Interfacing an Absolute Encoder.
The output of an absolute encoder can be read 2 gray-coded digits (8 bits) at a time, if the output lines are multiplexed.

2. The operating system could move the joint until it trips a limit switch (microswitch) that signals that the joint is at home. This is the method the Rhino robot uses to find home.

Once the joint has reached home, the operating system program must keep count of the direction and number of pulses received thereafter. This position-defining value is sent to an application program upon request.

Magnetic Detectors A magnetic detector is used to detect the presence of magnetic materials, such as iron, cobalt, some ceramics, and some steels. It can also detect magnetic fields, such as the earth's magnetic field.

If your robot carpenter is going to use a swinging magnet stud detector, it had better have some type of photo-detector in its gripper to detect the magnet's motion.

The photo-detector interface was covered in Chapter 17. A Hall-effect device in the gripper could also detect the movement of the magnet. Once the output signal from the Hall-effect device is amplified, it can be treated like any other on/off signal. The on/off signal interface was covered in Chapter 18.

Most magnetic flux gate compasses have a digital output. If this output is accessible, handling it is just like transferring any other binary value. If it is not accessible, a vision system will be needed to read the compass.

Sound (SONAR) Range Detectors The electronic fish finder's display has its own microcomputer. Any attempt at routing its output to a robot controller would require an understanding of its internal circuits that goes beyond the scope of this book.

The Sony electronic tape measure outputs its results to a liquid crystal display as a binary value. This value could be intercepted and routed to the lines of an input port on the robot controller. The application program would then have to decide how to react to the distance value.

Polaroid's ultrasonic range-finding kit has a digital interface that gives the value in binary code. This value is then transferred to the controller, just like any other binary value. The output is expressed in feet and tenths of feet. The output of the Hero I's ultrasonic range finder is a binary value expressed in centimeters.

Capacitance Detectors Capacitance proximity detectors, such as the density meter stud finder, again produce an on/off output. The interfacing of the on/off signal is covered in Chapter 18.

Tuned Radio Circuits The simplest type of tuned radio metal detector or inductive detector uses an on/off switch action and is interfaced like a microswitch. Such circuits give no indication of the distance or size of the object being detected. They are probably the only metal detector that would be used by a robot.

Temperature Detectors If the temperature detector is a bimetallic strip, it is interfaced and programmed like the on/off signal of a microswitch.

If the temperature detector is a thermistor or a thermocouple, its output is an analog signal and its interface is similar to that of the 255-point gray-scale cadmium sulfite vision sensor. The programming problems are also similar (see Chapter 18). The only real difference involves the temperature detector's large temperature-sensing range, as a result of which a range of 0 to 255 may not provide enough accuracy.

For example, one thermistor comes with a calibration chart for –50°C to +110°C—a range of 160°. At –50°, it has a resistance of 329,200 ohms; at +110°, it has a resistance of 757.9 ohms. At the –50° end of the scale, a change of 1° is equivalent to approximately 16,340 ohms; at the +110° end of the scale, a change of 1° is equivalent to approximately 19.92 ohms. So this device is not linear. If an accuracy of 1° is wanted over the entire range of temperatures, a reading as small as 9.96 ohms needs to be detected—that is, 9.96 ohms out of 329,200 ohms, or

1 in 33,052. This level of accuracy would take 16 binary bits of information to produce, so a 16-bit A/D IC would be needed to interface the sensor to a controller.

Radiation Digital-readout radiation detectors are available on the market. If their output can drive a digital readout, the signals can be used as a digital interface to a controller. The main difficulty in establishing the interface involves multiplexing the number of display bits into an 8-bit controller input port. The application program must request the radiation reading and then evaluate and act on the results.

Radar The type of radar detector called a fuzz buster outputs a simple present/not present signal, which is interfaced and programmed just like a microswitch signal.

If the radar is used to determine the robot's speed or its distance from some object, the device probably gives a digital readout. The signals sent to the readout can also be sent to a controller through an input port and may be coded in binary-coded decimal. Binary-coded decimal uses four binary bits to represent each decimal digit.

If the radar presents a display—as radar on a modern fighter plane does—it requires a computer of its own to process the information. This processing is similar in complexity to that of a very fancy vision system, and it cannot be handled by a robot controller. Instead, the robot controller might be able to ask questions of the radar controller.

SUMMARY

Proximity detectors are used to tell a robot when it is close to an object. Almost every type of sensor can be used as a proximity detector, so the interfacing and programming problems that proximity detectors have include those that light and touch sensors have. Other types of proximity detectors are ultrasonic, tuned radio frequency, magnetic, temperature, radar, and radiation.

REVIEW QUESTIONS

1. What safety advantages can be gained by giving robots proximity detectors?
2. How do proximity detectors help robots do some tasks better?
3. How do the size, material, and distance of the object being detected influence the type of sensor that should be used?
4. List some consumer products that use proximity detectors.
5. Why is a noncontact proximity sensor preferable for a robot to use in detecting humans?
6. Why is the factory a difficult place to handle small sensor signals?
7. Why would gray code be a better way to report position than straight binary code in a shaft position encoder?

20

SPEECH

OVERVIEW

Speech deals with two areas: speech synthesis and speech understanding. The present-day industrial robot does not use speech synthesis, but this will change when the robot moves out of the noisy factory. A robot delivering mail, medicine, or food, or a robot guard could use speech synthesis.

What is speech synthesis? Of what value to a robot is speech synthesis? What methods are available for developing speech synthesis? What type of programming requirements are needed for speech synthesis? These are some of the questions explored in this chapter.

Speech understanding is an extremely useful and often undervalued skill. To be successful in home and office environments, robots must eventually master speech understanding.

How important is speech recognition or hearing to humans? What types of speech recognition systems are being developed for robots? These questions are also explored in this chapter.

OBJECTIVES

When you have completed this chapter, you should be familiar with:

- The three main methods used for speech synthesis
- The hardware used for each speech synthesis technique
- The programming problems involved with each speech synthesis technique
- The problems involved with developing a multilingual robot
- The advantages of speech synthesis for robots
- The difference between speaker-dependent and speaker-independent speech input systems
- The difference between isolated-word and continuous-speech recognition systems
- Why speech understanding may be more complicated than vision processing

KEY TERMS

- Acoustical approach
- Continuous-speech recognition
- Digitized human voice
- Homonyms
- Isolated-word recognition
- Phonemes
- Phonic integrated circuits
- Pragmatic analysis
- Prerecorded human voice
- Prosopic analysis
- Semantic analysis
- Speaker-dependent
- Speaker-independent
- Speech perception approach
- Speech production approach
- Speech reception approach
- Speech recognition
- Speech synthesis
- Syntactic analysis
- Voice input
- Voice output
- Voice prints
- Voice recognition
- Voice synthesizer
- Word understanding

SPEECH SYNTHESIS

Speech synthesis is the process by which a machine produces speech. Robots working in a noisy factory have no need for a voice, but robots in a quieter environment could benefit from one. In the office, for example, voice would make robots safer and would help them get the job done if a human gets in their way. Robots assisting the disabled need a voice for simplifying interaction and for making them more user-friendly.

Voice output or speech synthesis is considerably easier to add to a machine than vision processing or voice input. When phonic integrated circuits are used, voice output requires very little computing power. All the computer has to do is supply a series of addresses to the integrated circuit.

A machine capable of speaking could communicate with a human without requiring the human to look at the lights or other output devices on the machine. Speech is omnidirectional and is understood by most humans; thus, humans could listen to the speech output of a machine while focusing their vision on some other task. In an emergency, a machine could use voice or speech output to transmit essential information to the human faster than it could by using lights, sirens, displays, and/or printed output.

HOW IS SPEECH SYNTHESIS DONE?

The earliest recorded work on machine **speech synthesis** is credited to Alexander Graham Bell, in the mid-1800s. Through the use of bellows and other mechanical devices, Bell tried to build a reproduction of the human speech mechanism. He was successful simulating a few words, and this work contributed to his later invention of the telephone.

Three main methods are used to produce machine speech output: using prerecorded messages produced by human voices; using digitized words produced by human voices; and using phonic integrated circuits. All of these methods rely on the use of electronics, and each method has its advantages and disadvantages. No matter what method is used for speech synthesis, the speech must be understandable to humans. Figure 20–1 shows the frequency pattern of the word *ready* as spoken by a human, by Digitalker, and by the SP0256-AL2 phonetic IC.

The military has found that men respond best to male voices for routine messages and to female voices for emergency messages.

Using Prerecorded Human Voice

Early electronic speech units used tape recordings of a human voice. This method produces a machine voice that sounds like a natural human voice. The human voice is an analog form of energy composed of a complex mixture of tones. Because it is not a simple tone, it is not easy to reproduce by machine.

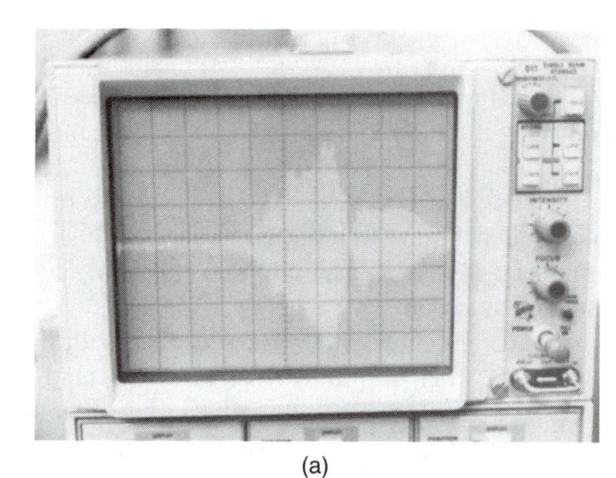

(a)

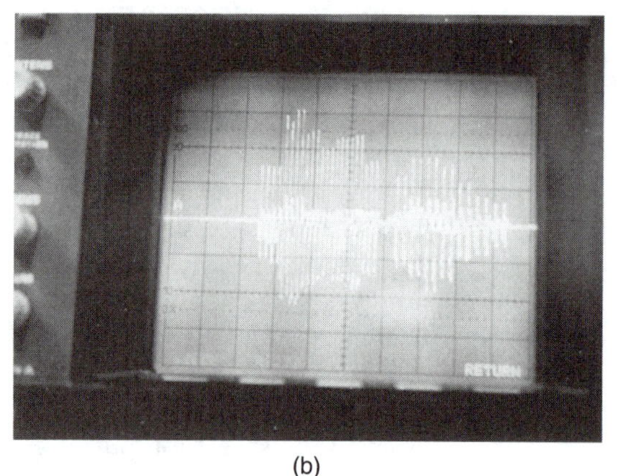

(b)

(c)

Figure 20–1 The Word *Ready* as Spoken by Three Different Voice Apparatuses.
(a) Frequency pattern of the word *ready* as spoken by a human. (b) Frequency pattern of the word *ready* as spoken by Digitalker, using digitized recorded human speech. (c) Frequency pattern of the word *ready* as spoken by BERT, using an SP0256-AL2 phonetic IC.

Of course, using prerecorded messages requires anticipating every possible message that the machine will have to speak. Then there is the problem of storing and accessing these messages. If they are stored on a single magnetic tape, the tape will have to be searched each time for the appropriate message. If each message is placed on a separate tape, you must supply a tape player for each message. Perhaps the messages can be put on short loops of tape and you can use a tape player with a method of finding, playing, and storing the different loops—something like a jukebox. However, both methods of using a single tape player will involve a time delay between when the machine realizes the need for an output message and when the message is finally spoken. Such a time delay could be a problem in an emergency.

Samples of prerecorded human voice output are still used in some toy dolls. In some cases, a miniature record is used, rather than a tape recording; some dolls make a few simple sounds, such as *mama,* using mechanical devices.

With the advent of the compact disk for audio recordings and laser disk technology, prerecorded human voice messages may make a comeback. The digital audio recordings used by the compact disk give high-quality sound, and the laser disk allows a large amount of memory to be accessed in less than 1 second.

To use prerecorded human voice messages on magnetic tape, the application program would have to search the tape for the message it wanted to play, and then play that message. If the tape contained more than a few messages, the search could take minutes.

With laser disk technology, the disk location of the message desired can be looked up in a directory. The disk can then be accessed by the listed surface and track, and the message can be played almost instantaneously.

Using Digitized Human Voice

A system that uses a digitized human voice has the advantage of being able to store the sounds in the computer's memory, like any other numeric information. Once again the sounds are originally made by a human being, but in this case the sounds are converted into numbers and stored as digital information. While present-day digitally recorded compact disk systems sound as good as the older analog recorded records, digitized speech systems have not been as successful, since they cannot record all the tones in human speech. Still, the sounds do closely resemble human speech. With the information being stored in the computer's memory, it is possible to find the proper message for each occasion in less than 1 second, so the digitized system has a fast response and still sounds good. Once again, however, all possible voice message responses have to be anticipated and prerecorded.

It is possible to record and store digitized messages one word at a time. However, humans speak not just in individual words but in whole sentences with wide variations in tone, speed, and emphasis. When individual words are recorded and later strung together into messages, the messages sound flat, mechanical, and emotionless.

National Semiconductor makes a digitized human voice system called Digitalker that uses from three to five special ICs (integrated circuits) and several support ICs for address selection and audio amplification. The main IC is the MM54104 microprocessor-based controller; then there are two or four ROM ICs such as the MM52164 SSR1, SSR2, SSR5, and SSR6. The four ROMs hold a total of 274 words. Any message made up of these words can be spoken by the Digitalker. Special ROMs can be created to suit the speech problems of a specific company. David A. Ward wrote a detailed article on a project he developed using Digitalker.[1] Figure 20–2 shows the circuit from his project. Tables 20–1 and 20–2 show the ROM words and addresses in this system.

The Digitalker can interface to a personal computer through the parallel printer port. Programming the Digitalker involves sending it a series of binary (hexadecimal) addresses of words that are to be spoken. The addresses for the proper pauses between words must also be included. One drawback of the Dig-

[1]David A. Ward, "Build This Speech Synthesizer," *Radio-Electronics* (December 1988): 80–85.

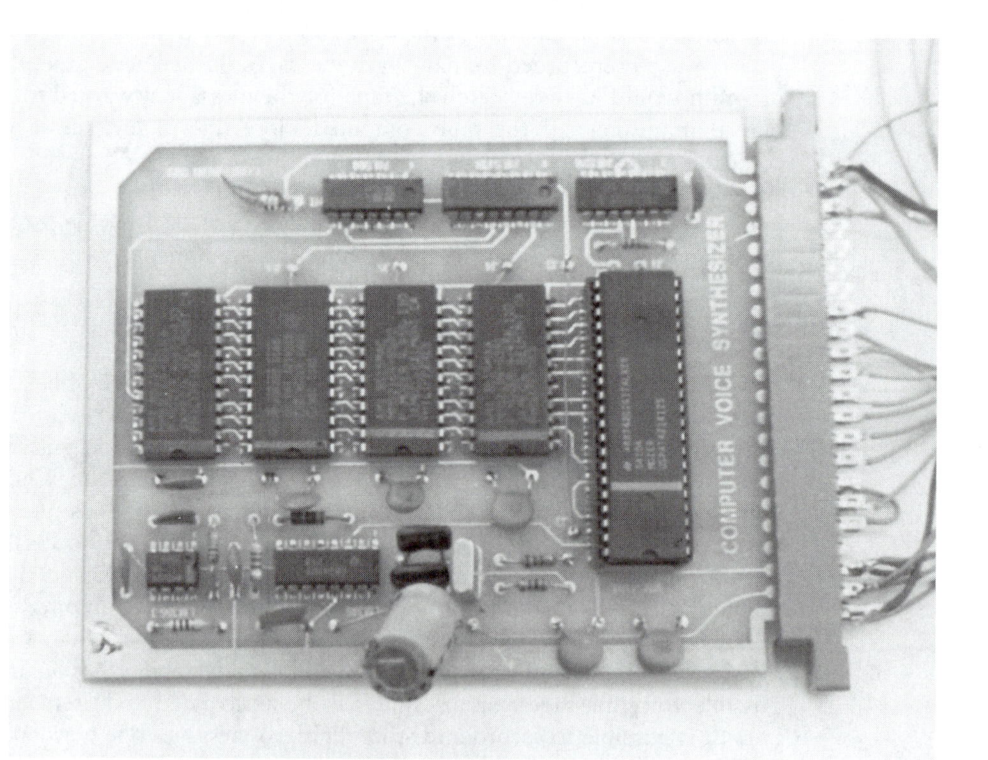

Figure 20–2 The Digitalker Project's Circuit.
This circuit can say the words listed in Tables 20–1 and 20–2. Its sound quality is very close to that of human speech. The largest IC is the Digitalker microprocessor-controlled speech controller. The four large ICs next to it are the ROMs containing the recorded words.

Table 20-1
Word List for SSR1 and SSR2 ROM

Word or Phrase	Hexadecimal Address	Word or Phrase	Hexadecimal Address	Word or Phrase	Hexadecimal Address
This is Digitalker	00	Q	30	Is	60
One	01	R	31	It	61
Two	02	S	32	Kilo	62
Three	03	T	33	Left	63
Four	04	U	34	Less	64
Five	05	V	35	Lesser	65
Six	06	W	36	Limit	66
Seven	07	X	37	Low	67
Eight	08	Y	38	Lower	68
Nine	09	Z	39	Mark	69
Ten	0A	Again	3A	Meter	6A
Eleven	0B	Ampere	3B	Mile	6B
Twelve	0C	And	3C	Milli	6C
Thirteen	0D	At	3D	Minus	6D
Fourteen	0E	Cancel	3E	Minute	6E
Fifteen	0F	Case	3F	Near	6F
Sixteen	10	Cent	40	Number	70
Seventeen	11	400 Hz tone	41	Of	71
Eighteen	12	80 Hz tone	42	Off	72
Nineteen	13	20 mS silence	43	On	73
Twenty	14	40 mS silence	44	Out	74
Thirty	15	80 mS silence	45	Over	75
Forty	16	160 mS silence	46	Parenthesis	76
Fifty	17	320 mS silence	47	Percent	77
Sixty	18	Centi	48	Please	78
Seventy	19	Check	49	Plus	79
Eighty	1A	Comma	4A	Point	7A
Ninety	1B	Control	4B	Pound	7B
Hundred	1C	Danger	4C	Pulses	7C
Thousand	1D	Degree	4D	Rate	7D
Million	1E	Dollar	4E	RE	7E
Zero	1F	Down	4F	Ready	7F
A	20	Equal	50	Right	80
B	21	Error	51	SS	81*
C	22	Feet	52	Second	82
D	23	Flow	53	Set	83
E	24	Fuel	54	Space	84
F	25	Gallon	55	Speed	85
G	26	Go	56	Star	86
H	27	Gram	57	Start	87

(continued)

*This item can be used to make singular words plural.

Table 20-1 (continued)

Word or Phrase	Hexadecimal Address	Word or Phrase	Hexadecimal Address	Word or Phrase	Hexadecimal Address
I	28	Great	58	Stop	88
J	29	Greater	59	Than	89
K	2A	Have	5A	The	8A
L	2B	High	5B	Time	8B
M	2C	Higher	5C	Try	8C
N	2D	Hour	5D	Up	8D
O	2E	In	5E	Volt	8E
P	2F	Inches	5F	Weight	8F

Table 20-2
Word List for SSR5 and SSR6 ROM

Word	Hexadecimal Address	Word	Hexadecimal Address	Word	Hexadecimal Address
Abort	00	Farad	2C	Per	58
Add	01	Fast	2D	Pico	59
Adjust	02	Faster	2E	Place	5A
Alarm	03	Fifth	2F	Press	5B
Alert	04	Fire	30	Pressure	5C
All	05	First	31	Quarter	5D
Ask	06	Floor	32	Range	5E
Assistance	07	Forward	33	Reach	5F
Attention	08	From	34	Receive	60
Brake	09	Gas	35	Record	61
Button	0A	Get	36	Replace	62
Buy	0B	Going	37	Reverse	63
Call	0C	Half	38	Room	64
Caution	0D	Hello	39	Safe	65
Change	0E	Help	3A	Secure	66
Circuit	0F	Hertz	3B	Select	67
Clear	10	Hold	3C	Send	68
Close	11	Incorrect	3D	Service	69
Complete	12	Increase	3E	Side	6A
Connect	13	Intruder	3F	Slow	6B
Continue	14	Just	40	Slower	6C
Copy	15	Key	41	Smoke	6D
Correct	16	Level	42	South	6E
Date	17	Load	43	Station	6F
Day	18	Lock	44	Switch	70
Decrease	19	Meg	45	System	71
Deposit	1A	Mega	46	Test	72

Word	Hexadecimal Address	Word	Hexadecimal Address	Word	Hexadecimal Address
Dial	1B	Micro	47	Th	73[2]
Divide	1C	More	48	Thank	74
Door	1D	Move	49	Third	75
East	1E	Nano	4A	This	76
Ed	1F[1]	Need	4B	Total	77
Ed	20[1]	Next	4C	Turn	78
Ed	21	No	4D	Use	79
Ed	22[1]	Normal	4E	Uth	7A[3]
Emergency	23	North	4F	Waiting	7B
End	24	Not	50	Warning	7C
Enter	25	Notice	51	Water	7D
Entry	26	Ohms	52	West	7E
Er	27	Onward	53	Switch	7F
Evacuate	28	Open	54	Window	80
Exit	29	Operator	55	Yes	81
Fail	2A	Or	56	Zone	82
Failure	2B	Pass	57		

[1]Eds 1F and 20 work best with words that end with *t* or *d*. Ed 22 works best with words that end with soft sounds.
[2]This item can be added to words such as *six, seven,* and *eight* to make *sixth, seventh,* and *eighth.*
[3]This item can be added to words such as *twenty, thirty,* and *forty* to make *twentieth, thirtieth,* and *fortieth.*

italker is that its ICs are very easy to damage with a static charge—even when the Digitalker is fully assembled.

Using Phonic Integrated Circuits

The least expensive and most popular method of producing speech output for machines is to use phonic integrated circuits. A phonic integrated circuit uses **phonemes**—the smallest distinct units of speech—to produce speech.

A typical phonic integrated circuit may have up to 64 phonemes, including 5 different lengths of pauses. The other 59 phonemes are such things as the "OY" in *boy* and the "AY" in *sky.* Table 20–3 lists the 64 phonemes available from the SP0256-AL2 speech processor integrated circuit made by Radio Shack. Each phoneme is stored in the speech processor integrated circuit at one of 64 addresses.

When the external parts are connected to the processor to form a working audio amplifier, all you need to do to get a sound is to give the processor the address of the desired phoneme. By giving the processor a series of addresses, you can produce words and sentences. To produce the word *may,* for example, you would simply address an "MM," as in *milk,* and then an "EY," as in *beige.* Producing the word *six* is a little more complicated. It requires first an "SS," as in *vest*; then another "SS"; then an "IH" as in *sit*; then another "IH"; then a pause

Table 20-3
Phonemes for SP0256-AL2 Integrated Circuit

Hexadecimal Address	Phonemes	Meaning or Sound	Hexademical Address	Phonemes	Meaning or Sound
00	PA1	Pause 10 MS	20	AW	out
01	PA2	Pause 30 MS	21	DD2	do
02	PA3	Pause 50 MS	22	GG3	wig
03	PA4	Pause 100 MS	23	VV	vest
04	PA5	Pause 200 MS	24	GG1	got
05	OY	boy	25	SH	ship
06	AY	sky	26	ZH	azure
07	EH	end	27	RR2	brain
08	KK3	comb	28	FF	food
09	PP	pow	29	KK2	sky
0A	JH	dodge	2A	KK1	can't
0B	NN1	thin	2B	ZZ	zoo
0C	IH	sit	2C	NG	anchor
0D	TT2	to	2D	LL	lake
0E	RR1	rural	2E	WW	wool
0F	AX	succeed	2F	XR	repair
10	MM	milk	30	WH	whig
11	TT1	part	31	YY1	yes
12	DH1	they	32	CH	church
13	IY	see	33	ER1	fir
14	EY	beige	34	ER2	fir
15	DD1	could	35	OW	beau
16	UW1	to	36	DH2	they
17	AO	aught	37	SS	vest
18	AA	hot	38	NN2	no
19	YY2	yes	39	HH2	hoe
1A	AE	hat	3A	OR	store
1B	HH1	he	3B	AR	alarm
1C	BB1	business	3C	YR	clear
1D	TH	thin	3D	GG2	quest
1E	UH	book	3E	EL	saddle
1F	UW2	food	3F	BB2	business

of 50 milliseconds called "PA3"; then a "KK2," as in *sky*; and finally another "SS." As you can see, using a phonic integrated circuit requires a little bit of work. Most manufacturers include a phonic dictionary of common words with the voice synthesizer to help you use the system.

Most phonic integrated circuits cannot remember a series of commands. They work on only one command at a time. Additional circuits (or perhaps an entire computer) are required to get the integrated circuit to produce more than just isolated sounds. The CTS256A-AL2 text-to-speech controller IC handles many of the support functions for the SP0256-AL2 speech synthesizer IC. It accepts ASCII characters as input, and outputs the phoneme codes to the SP0256-AL2. The CTS256A-AL2 furnishes a buffer for the ASCII characters, and it can forward the characters for pronunciation as single letters, single words, or groups of words separated by appropriate punctuation marks. Figure 20–3 shows CTS256A-AL2 and SP0256-AL2 ICs connected in a circuit.

The sounds in an integrated circuit are always pronounced the same way each time. And while the base frequency of the voice can be varied to sound like a man, a woman, or a child, it is still not the most pleasant of voices. The phonemes do not represent regional speech differences, and they do not handle accents well. Above all, the phonemes are made for American English; if this speech system is used to pronounce Spanish or French, it will be hard for someone who speaks that language to recognize what is being said, because each spoken language has its own combination of phonemes.

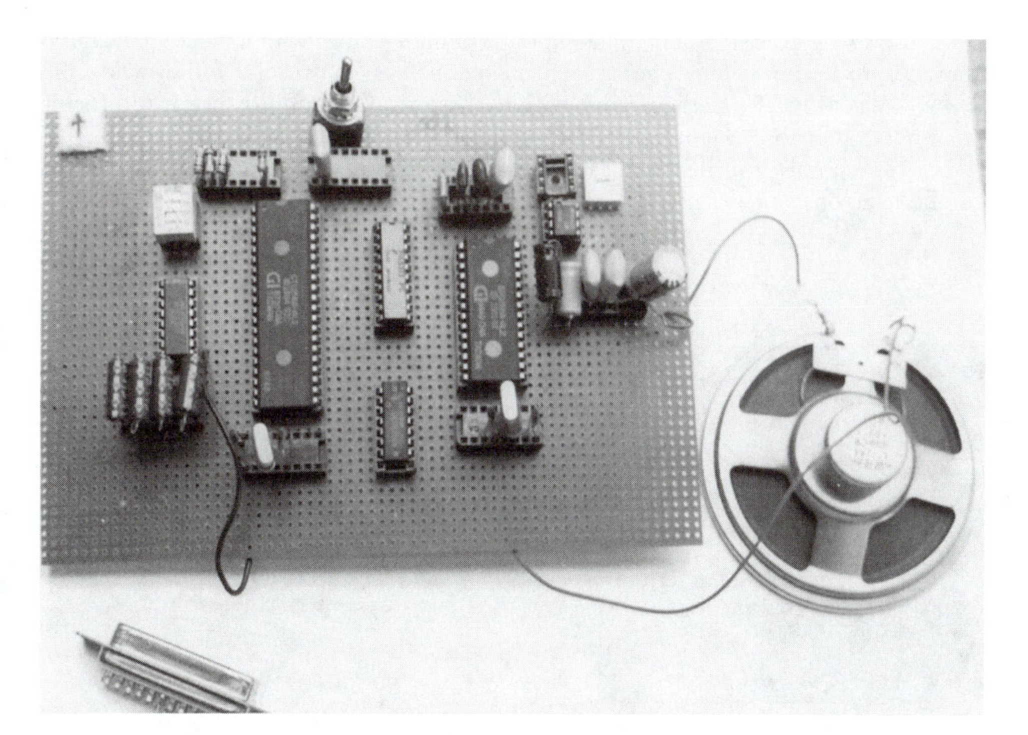

Figure 20–3 CTS256A-AL2 and SP0256-AL2 Circuit for a Speech Synthesizer. Combining the CTS256A-AL2 speech-to-text IC with the SP0256-AL2 phonetic speech synthesizer makes outputting speech as easy as sending a series of ASCII characters to the CTS256A-AL2.

Programming the SP0256-AL2 involves sending a series of addresses to the IC. These addresses must include phonemes for each word, including the pauses within each word and the pauses between words.

The Hero I robot from Heathkit can be bought with a speech synthesizer that uses a phonic integrated circuit. To get an idea of how hard it is to get good-sounding speech from a machine, you should try teaching a Hero I robot to pronounce someone's full name! Figure 20–4 shows the Hero I speech system. The Hero I contains several speeches in ROM. To get the Hero I to say speech, you simply enter the starting address of the speech and press Enter. User-defined speeches consist of machine language instruction placed around the phoneme addresses.

Speech or voice synthesizers are available for many home and personal computers. These enable the computer to tell you what is wrong, without having to wait for you to read any screen messages. Figure 20–5 shows a speech system for a Radio Shack Model IV computer. The Model IV speech system includes a program that performs the functions of a text-to-speech IC. When the speech program is active, you need only type the speech at the computer keyboard, and it will be spoken.

BERT—the Basic Educational Robot Trainer—comes with a voice that can be programmed from canned words or from phonemes. It is useful for showing the importance of pauses in the pronunciation of words. Continuous speech is difficult to render intelligible. Figure 20–6 shows BERT's speech board. To program BERT's voice, you simply give BERT a T*xx* command, where *xx* represents the address of a phoneme or canned word.

Figure 20–4 Hero I Speech System.
(a) The speaker for the Hero I speech system, with a cardboard shield around it. (b) The speech board.

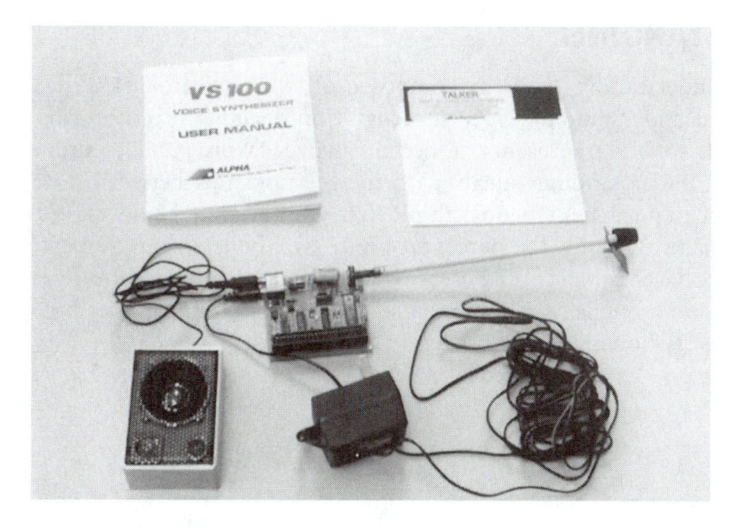

Figure 20-5 Speech System for Radio Shack Model IV Computer.
The speech system for the Radio Shack Model IV computer uses software to translate ASCII characters into phonetic commands. The circuit board contains the phonetic speech synthesizer.

Figure 20-6 BERT's Speech Board.
The little circuit board on top of BERT is its speech board. The big IC on it is an SP0256-AL2 speech synthesizer.

SPEECH UNDERSTANDING

Speech understanding is really composed of two separate tasks: speech recognition and word understanding. **Speech recognition** deals with recognizing that a certain series of sounds represents a certain word. **Word understanding**—or natural language understanding—deals with the relationships between different words. It turns out that, at best, people think that they hear and understand 80 percent of what is said. Actually, the percentage may well be less than 50 percent.

One factor that contributes to our low speech recognition percentage is the prevalence in English of **homonyms**—words that sound alike, but have different meanings. Following is a short list of homonyms:

ate	eight	
all	awl	
bare	bear	
be	bee	b
dear	deer	
dew	do	due
eye	I	
for	fore	four
hall	haul	
hear	here	
hole	whole	
knot	not	
meat	meet	
one	won	
their	there	they're
to	too	two

Homonyms may also consist of two groups of words. For example, try saying the following phrases over and over quickly:

recognize speech
wreck a nice beach

Figure 20–7 shows the frequency patterns produced by the phrases "recognize speech" and "wreck a nice beach," as spoken by the same individual. Notice how similar they look.

In addition, every person has a unique micro-accent, speaks at different frequencies and varying speeds, and tends to run words together. Unless we all learn to speak the way a phonic speech output machine does, such individual differences and complicating factors are not likely to disappear.

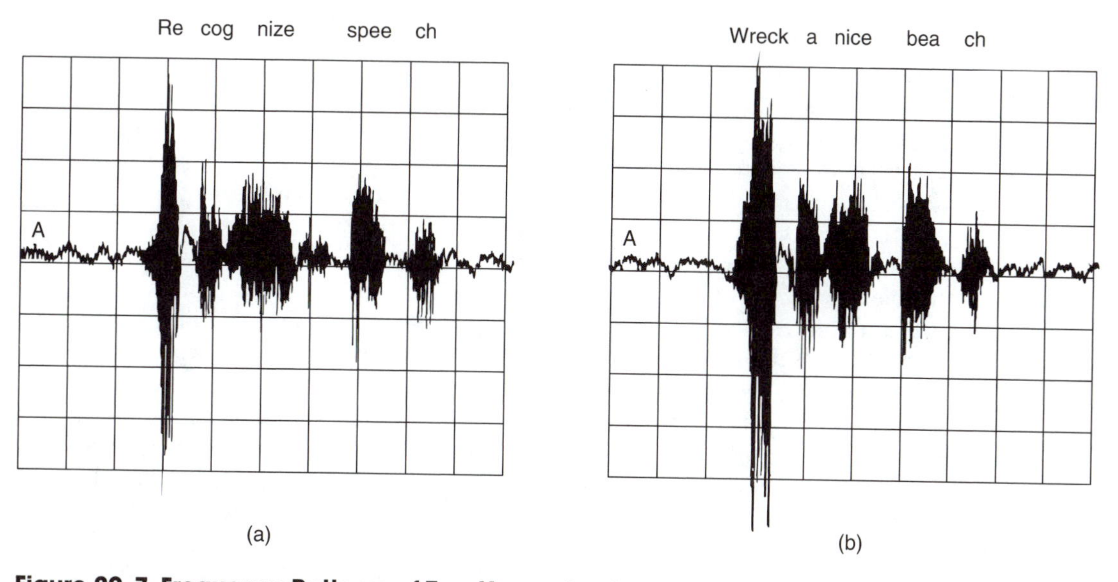

Figure 20-7 Frequency Patterns of Two Homophonic Phrases.
(a) An oscilloscope recording of the phrase "recognize speech." (b) A comparable recording of the phrase "wreck a nice beach." These phrases sound very much alike, even when spoken at a normal rate of speech.

WHY IS SPEECH UNDERSTANDING DESIRABLE?

Having a robot respond to spoken commands can make programming the robot easier. Currently, most robots are programmed through the pressing of keys or buttons on a keyboard, control panel, or teaching pendant. A person whose hands are full at a time when a command must be given to the robot cannot manage this job. In an emergency, it would be quicker to speak to a robot than to have to find and push the correct button. As more and more robots become mobile, voice commands will become more important.

Voice-controlled robots are also helpful to the disabled. A quadriplegic, for example, can use a voice-controlled educational or small industrial robot to perform tasks.

HEARING

Speech understanding starts with hearing sounds, which must then be recognized as words. Finally the words must be combined to produce idea understanding.

Hearing is the process of responding to sound waves. This generally involves having some type of specialized sound-wave receptor, or ear. For a machine, the ear frequently takes the form of some type of microphone, which is a transducer that converts sound waves to electrical energy.

The Hero I robot uses a microphone—actually, it is a loudspeaker running backward—to detect noises. The Hero I is also equipped with an ultrasonic microphone, for detecting sounds outside the range of human hearing that are produced by moving objects. However, the Hero I is incapable of genuine speech recognition; but even so, it can be made to respond to noise commands, because it can distinguish between the level and the frequency of the noise commands.

To gain some insight into the problems of speech recognition, we will look at some isolated spoken words. Figure 20–8 shows the electronic signals generated by the word *one* as spoken by three different persons speaking into a microphone connected to an oscilloscope. Figures 20–9 and 20–10 show the electronic signals generated by the words *hello* and *three,* respectively, as spoken by two different persons. Figures 20–11 and 20–12 show the signals produced by the words *up, down, left,* and *right, forward,* and *reverse,* respectively. The similarity in pattern among these words is especially a problem for a voice recognition system that looks only at the length of the words, as the VERBOT robot's system does. (The VERBOT robot is a speaker-independent, voice-commanded hobbyist robot.)

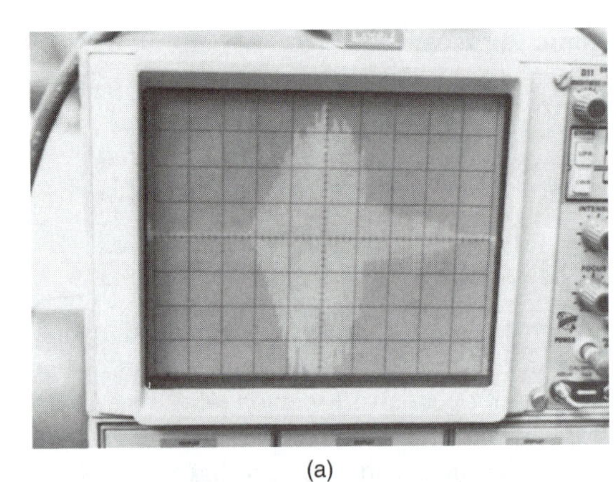

(a)

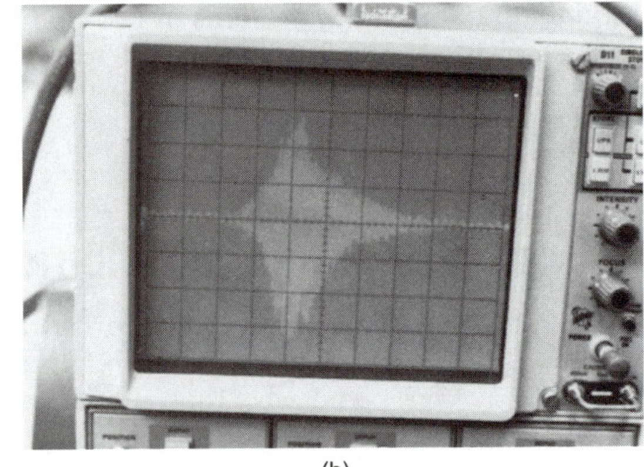

(b)

(c)

Figure 20-8 Frequency Patterns of the Word *One* as Spoken by Three Different Persons.
These pictures show oscilloscope recordings produced when three different persons spoke the word *one.* Each pattern is unique.

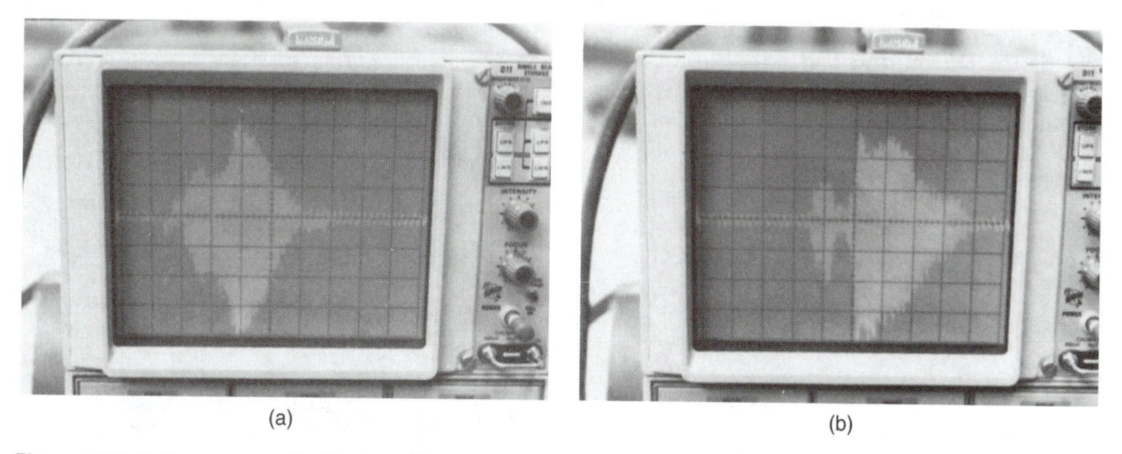

Figure 20-9 Frequency Patterns of the Word *Hello* as Spoken by Two Different Persons.
These pictures show oscilloscope recordings of the word *hello,* as spoken by two different persons. The differences here are quite noticeable.

Figure 20-10 Frequency Patterns of the Word *Three* as Spoken by Two Different Persons.
These pictures show oscilloscope recordings of the word *three,* as spoken by two different persons. The differences are noticeable.

Figure 20-11 Frequency Patterns of the Words *Up, Down,* and *Left.*
The differences in the words *up, down,* and *left* do not show up as distinctly in their oscilloscope pictures as we might expect. (a) The word *up*. (b) The word *down*. (c) The word *left*.

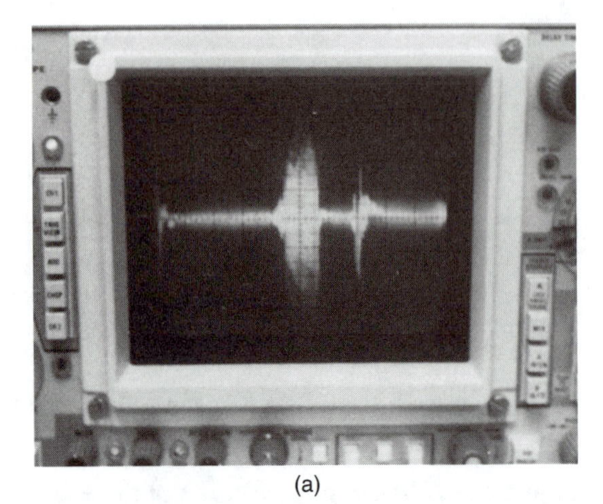

(a)

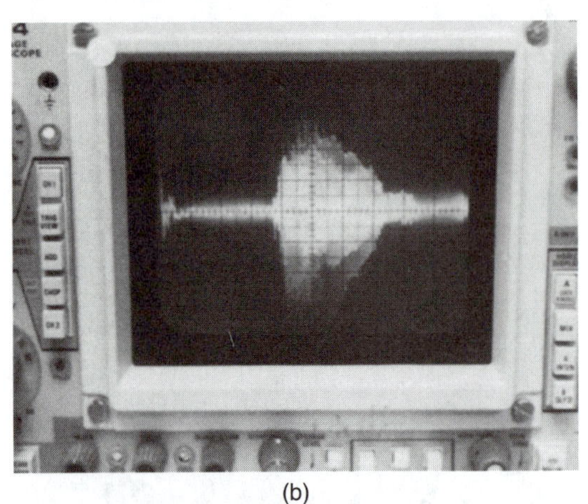

(b)

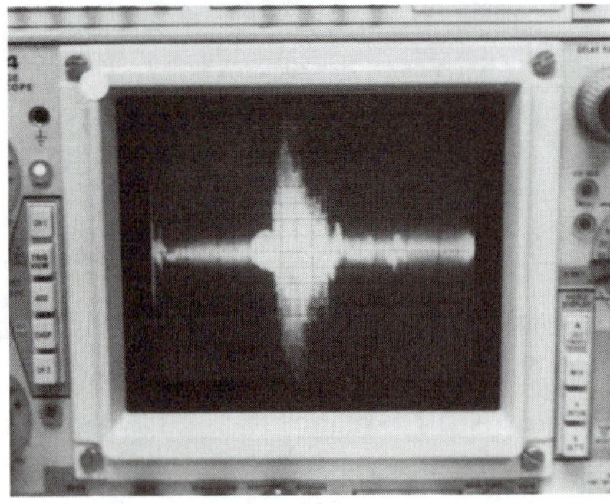

(c)

Figure 20-12 Frequency Patterns of the Words *Right, Forward,* and *Reverse.*
The words *right, forward,* and *reverse* complete a common six-word command set. (a) The word *right.* (b) The word *forward.* (c) The word *reverse.*

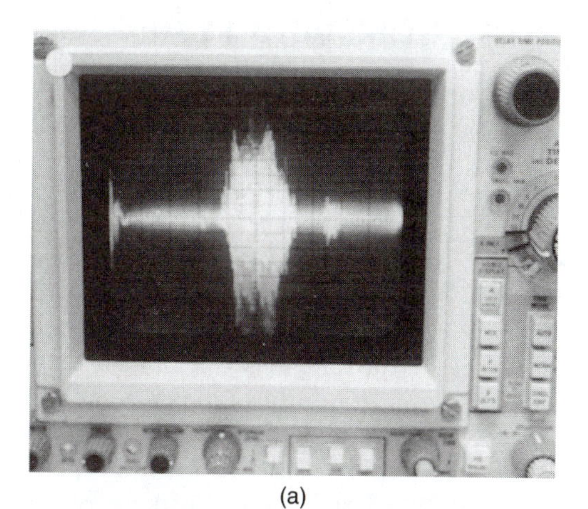

(a)

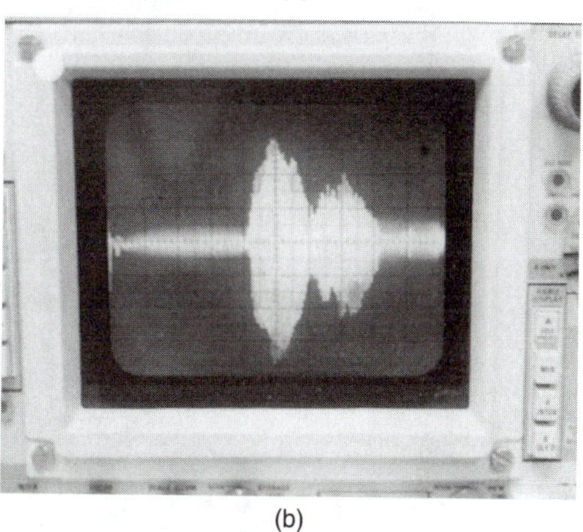

(b)

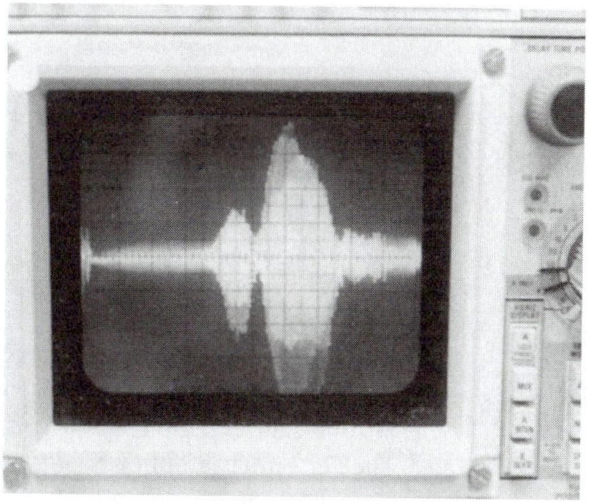

(c)

NOISE COMMAND SYSTEMS

Noise command systems use the most primitive level of hearing. The simplest of them wait for and respond to noises above some minimum volume. The first such sound might cause the device to start running forward; a second such sound might cause it to reverse directions; and a third such sound might cause it to stop. Hobbyist robots in the $20 to $50 range—including the "Supersonic Piper Mouse" and the "UFO"—have such hearing.

Even when dealing with such simple hearing, it is possible to work out a series of commands for a robot. For instance, the robot might measure the length of time between noises. Or it might count the number of noises in a given time period; thus, one noise in a given time period might be the command to start, two noise detections in one time period might be the command to stop, and three and four noises in the time period might be the commands for left and right. The Hero I robot is capable of being commanded through such simple noise command programming.

Figure 20–13 Petster Penguin, a Noise-controlled Pet Robot.
The Petster has an obey mode, where it responds to clap combinations for forward, left, right, back up, chirp, and return to ready-to-play mode. In its ready-to-play mode, it responds to clap combinations for come here, go away, and switch to obey mode. Prolonged silence will cause Petster to enter its asleep mode.

Axlon Corporation makes a series of electronic pets that respond to sound commands. Figure 20–13 shows Axlon's Petster Penguin, which has three modes—the asleep mode, the ready-to-play mode, and the obey mode. Any noise will wake the robot out of its asleep mode. Silence for 10 seconds will put it back to sleep, where it can remain for up to three weeks, before the batteries go dead. In the obey and ready-to-play modes, the robot responds to commands transmitted through series of claps and pauses. Axlon also markets a Super Petster, which has additional training and playback modes.

Such simple hearing and command sequences may be useful for the hobbyist, but they would be dangerous in an industrial setting. Industrial robots are most often used in noisy factories, where random noises could cause the robot to make unexpected (and potentially unsafe) moves. Therefore, industrial robots should not use noise control for anything but intruder detection.

SPEECH RECOGNITION SYSTEMS

Speech recognition systems are used to recognize spoken words. These may work on one word at a time or on continuous speech. The system may be tuned to a single person's voice, or it may be usable by many different persons. The least expensive systems operate with isolated words and are speaker-dependent. The most expensive systems, while still highly experimental, are being directed toward recognition of continuous speech and toward speaker independence.

Speaker-dependent Systems

A **speaker-dependent** system is a speech recognition system that recognizes only the words or commands spoken by one particular human. As you have seen, a word spoken by a human has a very complicated waveform that is as unique as a fingerprint. In fact, the electronic recordings of human words are called **voice prints.** A speaker-dependent system is easier to build than a speaker-independent system; however, it must be trained by having the intended user speak the command words while the system records them. These voice prints are later used by the system as reference forms for comparing the present input against.

An inexpensive speaker-dependent system may record only the length of each word or command. More expensive systems digitize the voice print and break it down into its different frequency components. The more information saved about the voice, the more secure the system is from unauthorized persons' successfully faking the commands. Unfortunately, a good voice print system may not always recognize the authorized user of the system. If the person is talking under stress, has a cold or laryngitis, is tired or scared, or happens to be speaking as a heavy truck drives by outside, the system may not recognize the commands given. Most speaker-dependent systems have a very limited number of words in their vocabulary.

One obvious problem with using a speaker-dependent system is that it must be retrained when you switch workers. Indeed, if the machine is used on more than one shift, it may require retraining at the beginning of each shift. Alternatively, the speech recognition system might be required to memorize voice prints of the same commands as given by several persons.

Speaker-independent Systems

A **speaker-independent** system is a voice recognition system designed to recognize commands given to it no matter who does the speaking. This presents many problems to the system, since different persons say words at different speeds, frequencies, and inflections. Consequently, a speaker-independent system must analyze the speech in several steps and then base its identification on these efforts.

First it must break the sounds it hears into **phonemes;** that is, it must find the pauses in the words. Then it must look the phonemes up in its memory and try to find the words they represent, which may require storing several pronunciations of the words. Then it must compare these words against known commands. Finally, it must carry out the command. If a command is not understood, the robot—or at least the speech recognition system—must request that the command be given again.

It may be necessary to train people in the use of the speaker-independent system. In particular, they must make an effort to speak very distinctly.

For vocabularies of only a few words, the voice print technique used by a speaker-dependent system takes less complicated hardware and less memory space. For large vocabularies, the phonic technique takes less memory space. Even the speaker-independent system has problems, however, under very noisy conditions or with persons with heavy accents.

Isolated-word Recognition Systems

The term **isolated-word recognition** is applied to a speech recognition system that hears only one word at a time. It processes the word it hears and performs the associated command. If multiple words are spoken to an isolated-word recognition system, it treats them as a single word or command. If two commands are spoken too close together, they are taken as a single unrecognized command. As a result, an isolated-word recognition system is slow and cannot follow a conversation.

Most present-day word recognition systems operate with isolated words, whether they are speaker-dependent or speaker-independent. These systems are all right for giving simple commands to a machine or robot, but they do not approach the intelligence of human speech understanding.

Radio Shack now sells an isolated-word, speaker-independent voice recognition integrated circuit (IC) called the VCP200. The IC understands the words *go, stop, left turn, turn right,* and *reverse.* Alternately, it can understand *yes/no* or *on/off.* Figure 20–14 shows the VCP200 IC mounted in a test circuit. This test circuit would not be good for controlling a free-moving robot, since the microphone works only with words spoken within a radius of a couple of feet. A speech compressor circuit to amplify all of the sounds received by the microphone to a fixed amplitude would have to be added to enable the test circuit to work over a longer range. The present circuit might work for commanding parts of a stationary robot arm.

Olivetti of Italy is working on a listening and talking machine that might function as a listening typewriter. It is envisioned as an isolated-word, speaker-independent word recognition system. It works with a vocabulary of 10,000 words. When a word is first spoken, the machine selects the best three or four candidates for the actual word and places the likeliest of these in the text. If later words give a better clue to the word just spoken, the machine goes back and changes that word in the text. Figure 20–15 provides a schematic diagram of an isolated-word recognition system.

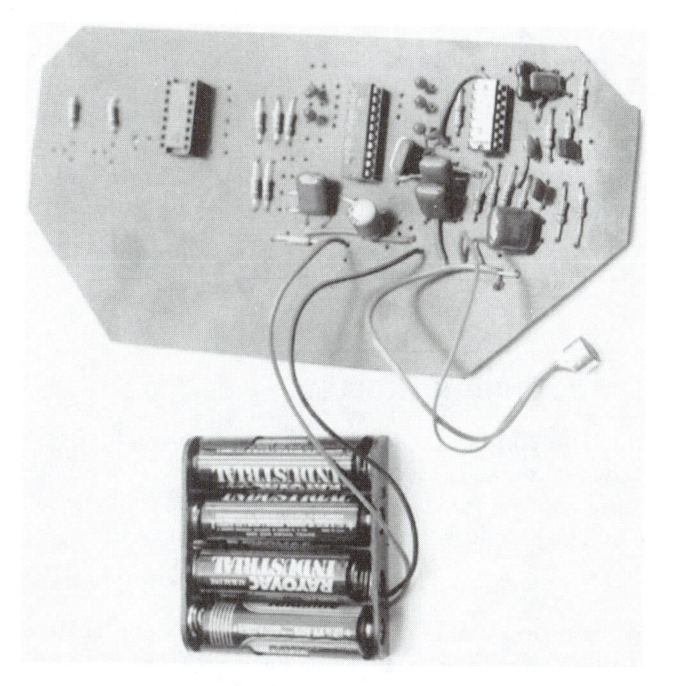

Figure 20–14 VCP200 Voice Command System.
The VCP200 voice command IC (shown here in the middle of the figure) is connected in a circuit to a small microphone (shown on the right side of the figure). The circuit responds to the spoken commands *go, stop, left turn, turn right,* and *reverse.*

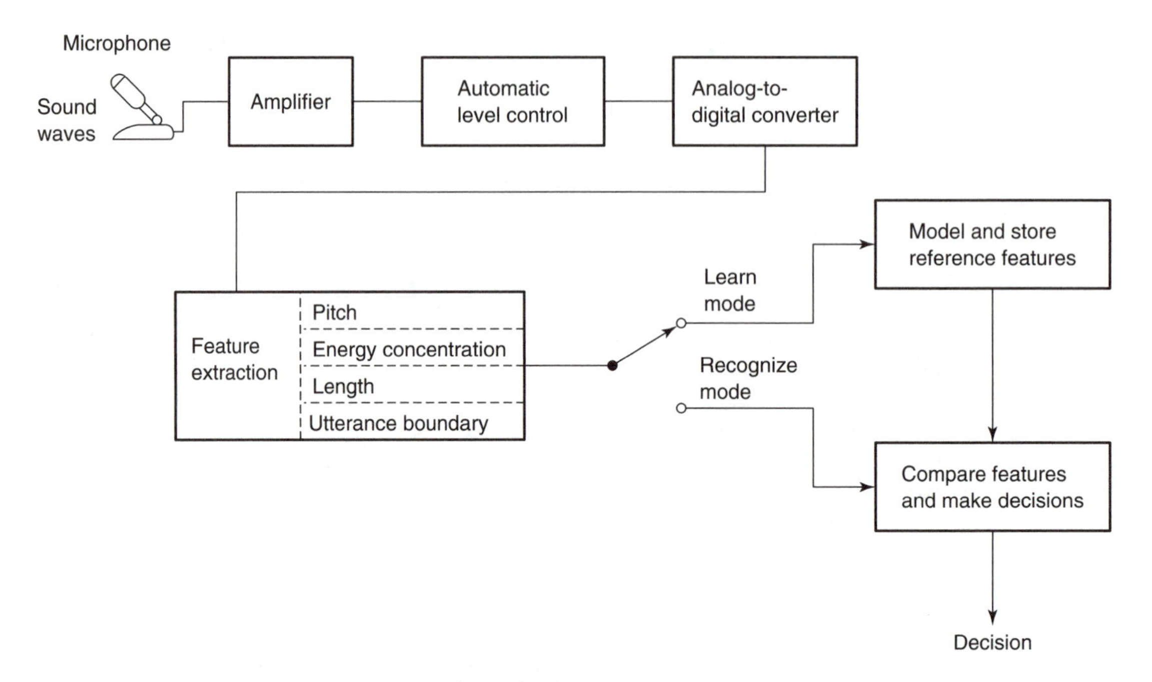

Figure 20–15 Isolated-word Recognition System.
An isolated-word recognition system must first adjust the level of all spoken commands to the same amplitude. Then the analog signal is changed to a digital signal. Next the pitch, energy concentration, length, and utterance boundary of the command are extracted. In the learn mode, these features are saved for future command recognition. In the recognize mode, the extracted features are compared with stored features in order to identify a command.

SAMPLE ISOLATED-WORD RECOGNITION PROBLEM

Suppose that you wanted to build an isolated-word recognition system capable of recognizing the two commands "forward" and "reverse." Figure 20–16 shows the wave forms for the words *forward* and *reverse,* along with their real timing in tenths of seconds. How could you differentiate between them?

Answer

Since both words take approximately 0.9 second to say, their length cannot be used to separate the two commands. Each word has two parts with one approximate zero crossing, so the number of parts will not separate the commands either. However, the length of the first part of *forward* is approximately 0.4 second, while the first part of *reverse* is approximately 0.3 second. Likewise, the second parts of the words are of different lengths. These differences reflect the fact that the emphasized syllable of *forward* is the first one, whereas the emphasized syllable of *reverse* is the

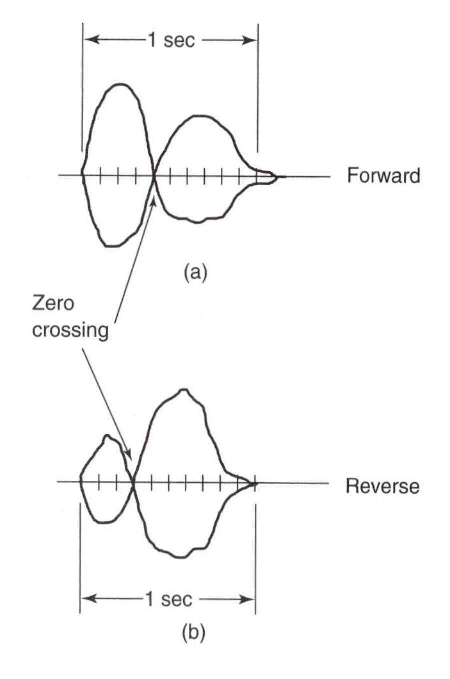

Figure 20-16 Wave Forms and Timing of the Words *Forward* and *Reverse*.
(a) Wave form and timing of the word *forward*. (b) Wave form and timing of the word *reverse*. Comparing these two forms suggests how to distinguish between them.

second one. If only these two commands are to be used in this system, either of the length differences noted can be used to distinguish between them.

Continuous-speech Recognition Systems

The ultimate in speech recognition systems will be the **continuous-speech recognition system,** capable of understanding sentence-length concepts and ideas. Although no such systems are currently in production, research is slowly heading in this direction.

A continuous-speech recognition system will have to overcome many problems that are not faced by an isolated-word recognition system. First, a continuous-speech recognition system must work in real time; that is, it must understand what is being said within a fraction of a second of hearing each word. Second, it must be able to break the speech it receives into the correct words.

Continuous-speech recognition systems can be approached as something similar to Morse Code. First, you learn to listen for the dots, dashes, and spaces. Next, you learn to hear the characters without having to count the dots and dashes. And then, with practice, you learn to recognize whole words at once such as *of, the,* and *and*. Eventually, you get to the point where you can just type out the incoming message on a typewriter, without consciously trying to understand it. To achieve fast speech recognition, the system must be able to recognize common phrases. And since speech recognition systems are used in specific application areas, the language of a particular application could be made the focus of the system's common-word vocabulary.

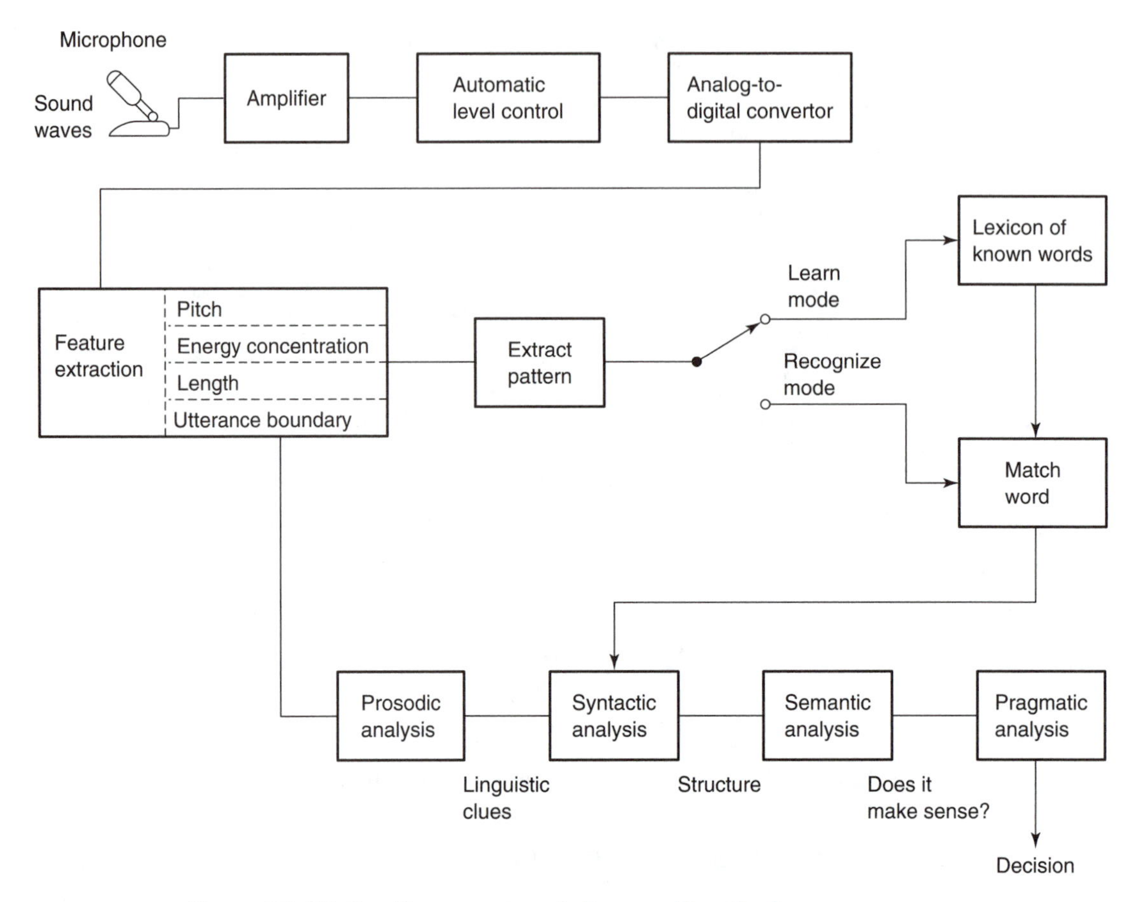

Figure 20-17 Continuous-speech Recognition System.
A continuous-speech recognition system must respond to spoken commands at the real-time rate in which they are given. They must also recognize sequences of words, not just isolated words.

Figure 20–17 shows a schematic drawing of a continuous-speech recognition system.

APPROACHES TO SPEECH RECOGNITION

There are four approaches to doing electronic speech recognition: acoustical, speech production, sensory reception, and speech perception.

The **acoustical approach** to speech recognition comes from the idea that, since speech is a wave form, it can be analyzed by means of general signal analysis techniques, including Fourier frequency analysis, principal component analysis, and statistical decision procedures.

The **speech perception approach** to speech recognition comes from the idea that speech interpretation should be based on what humans think are important parts of speech. The essential components of speech are the length of the sounds, the separation of vowel and consonant sounds, feature detection, and linguistic categories.

The **speech production approach** to speech recognition comes from the idea that, if we understand how speech is produced, we can easily understand its meaning.

The **speech reception approach** to speech recognition comes from the idea that the best way to achieve speech recognition is to duplicate the human auditory reception process in a machine. By observing how the ear receives sound and converts it into electrical impulses, and how the nerves transmit the sound information, we can attempt to replicate the process mechanically.

Whatever else is said about speech, remember that the spoken words received by a human do not yield all of the information inferred by humans. Just as humans read between the lines, humans hear the contextual meaning between spoken words. Past experiences also enter into speech interpretation, as do the physical gestures made by the speaker and seen by the hearer.

A POSSIBLE SPEECH RECOGNITION SYSTEM

All speech recognition systems have some problems in common. First, they need to be able to identify the most probable answer. The only system that can be expected to find exact matches is a voice print system used as part of a security procedure. Such systems, of course, are only able to recognize how one person says the word in question.

Second, all systems require some means of receiving the sound waves that make up speech and of converting them to electronic signals. This task requires some type of microphone and amplifier circuit, as well as some type of automatic level control to make the sounds peak at the same level.

You are now ready to do some type of signal analysis. Spectrum analysis can be done on the analog signals, as can zero crossing or utterance boundary detection. Alternatively, you can convert the analog signals into digital signals and store them in a computer memory. The computer memory can then be scanned and compared to the present input for a voice print match, assuming that it has previously been trained through recording of the voice print. Spectrum analysis, zero crossings, utterance boundary detection, and so forth, can also be done on digital information stored in memory.

If the system is performing continuous-word recognition, you have the additional tasks of prosopic analysis, syntactic analysis, semantic analysis, and pragmatic analysis. **Prosopic analysis** is used to find clues to linguistic structures, stressed words, and areas of phonetic reliability. **Syntactic analysis** finds sentence structures and grammatical relations. **Semantic analysis** checks the interpretation of the words for a meaningful sentence. **Pragmatic analysis** verifies or rules out hypothesized word combinations and produces the final machine response to the speech.

VOICE INPUT LANGUAGE FOR ROBOTS

Once you have decided to use voice control for a robot, the next step is to decide what commands the robot must be able to understand and whether the robot should have more than one master. The list of commands needed varies from one type or style of robot to another. Even if the commands are concerned only with the movement of a single arm and a single gripper, commands will be needed for moving the shoulder and elbow, for moving the wrist and gripper, and for coordinating the first two sets of commands. Of course, a robot with more than one gripper, more than one arm, or one vehicle-mounted base will require additional groups of commands for each of these features. Table 20–4 lists some of the spoken commands that a robot might be required to recognize and carry out.

The coordination commands would have to include a record or store command mode, as well as a run or execute mode for executing a sequence of commands stored in memory. In addition, it would be desirable to have grouping or mode commands, to reduce the number of words such as "lower hand" or "lower arm" or "lower left hand" in each command. With a mode command, the hand or

Table 20–4
Voice Commands for Robots

Coordination Mode	
Off	—Shut down all power
Home	—Move all joints to known starting position
Record	—Start recording commands for future use
Stop	—Stop recording commands
Body	—The following instructions are body mode commands
Arm	—The following instructions are arm mode commands (if the robot has more than one arm, there might be "right arm" and "left arm" modes)
Hand	—The following are hand mode commands (if the robot has more than one hand, there might be a right hand, a left hand, and perhaps a middle hand mode)
Run	—Carry out the prerecorded commands
Vehicle	—The following instructions are commands for moving the robot from one location to another
Body Mode	
Right	—Turn the robot's body to the right
Left	—Turn the robot's body to the left
Halt	—Stop movement
Arm Mode	
Right	—Move arm to the right
Left	—Move arm to the left
Raise	—Raise arm
Lower	—Lower the arm
Forward	—Move arm forward
Back	—Move arm backward
Halt	—Stop movement

arm words would be assumed by the fact of the command's being in hand or arm mode, and just using the word *lower* would suffice. Such commands could be compounded of a body command mode for moving the base and arm together and an arm command mode for moving the arm.

A wrist command mode for moving the wrist would be used for the final movement or orientation of the hand to do its work. Such an adjustment might involve tilting the hand from side to side, moving it up or down, or rotating it.

A hand command mode for moving the gripper would include the actual work commands: opening and closing the fingers as the hand grasps something; starting and stopping the flow of paint during spray painting; turning a spot-welding tool on and off; detaching and attaching different tools or grippers; activating and deactivating an inflatable device in the gripper.

A robot equipped with legs or a vehicle would require an additional command mode, as would a robot equipped with sensors.

As more and more commands are added for the robot to execute, it becomes more difficult for the robot to understand the spoken commands.

The vocabulary for the robot should be kept as small as possible, each command should be as short as possible, and all words used should sound as different from one another as possible.

Wrist Mode	
Rotate	—Rotate hand to the left
Turn	—Rotate hand to the right
Left	—Move hand to the left
Right	—Move hand to the right
Raise	—Raise hand
Lower	—Lower hand
Halt	—Stop movement
Hand Mode	
Halt	—Stop movement
Open	—Open hand
Close	—Close hand
Activate	—Activate hand device
Deactivate	—Deactivate hand device
Attach	—Pick up a different hand or gripper
Detach	—Drop off a hand or gripper
Vehicle Mode	
Left	—Move whole robot to left
Right	—Move whole robot to right
Forward	—Move whole robot forward
Back	—Move whole robot backward
Rotate	—Rotate whole robot to the left
Turn	—Rotate whole robot to the right
Halt	—Stop movement
Faster	—Speed up movement
Slower	—Slow down movement

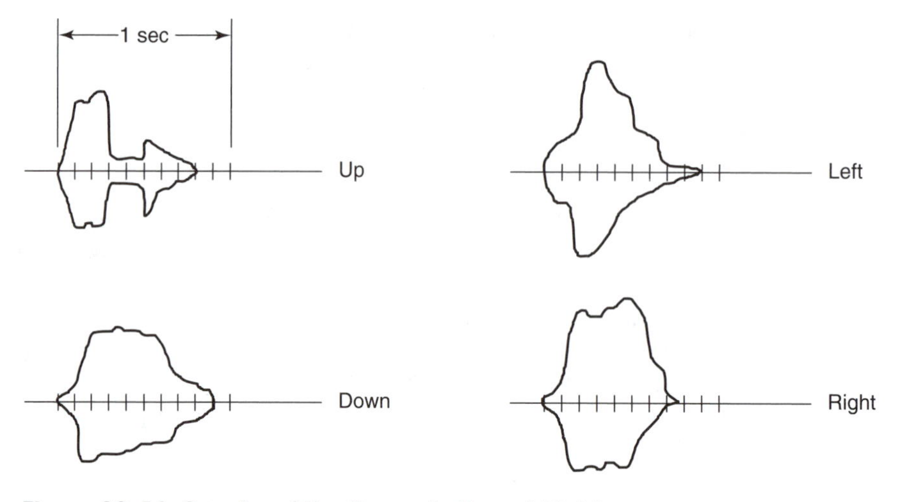

Figure 20–18 Graphs of Up, Down, Left, and Right.

It may be desirable to add numbers as part of the movement command so that the robot will know how far to move. For rotary motions, the numbers represent degrees and fractions of degrees. For linear motions, the numbers represent inches or (in some cases) hundredths of inches.

Jorg R. Jemelta adapted a Rhino XR-1 robot to voice input commands, using a Heuristics Speechplot card in an Apple II personal computer. He found that the four English words *left, right, up,* and *down* were not different enough for inexpensive voice recognition systems to distinguish among (see Figure 20–18). Consequently, he replaced these words with their French equivalents—*gauche, droit, avant,* and *arrière*—which worked well. Unfortunately, such a change would decrease (at least initially) the ease with which the average English speaker could use the robot commands.[2]

SUMMARY

The ability to speak can make a robot easier and safer to use, because you do not have to be facing the control panel display to be informed of a problem the robot is having, and because the robot can warn a human who is not facing the robot of some imminent danger. Three main methods are used to develop machine speech synthesizers or voice synthesizers: prerecording a human voice, digitizing a human voice, and using phonic integrated circuits. The prerecorded human voice sounds the most natural, but it is the slowest method of speech. Digitized human voice dictionaries sound less natural but work much faster. Phonic integrated circuits offer the least expensive and most versatile approach, but they have the poorest sound quality.

[2]Jorg R. Jemelta, "Designing a Reliable Voice-input Robot Control Language," *Robotic Age* (February 1984): 30–34.

Speech understanding involves trying to enable machines or robots to hear. Such an ability could greatly improve the ease of commanding and programming a machine. And because it is faster to give voice commands than to find and use a keyboard or locate and punch a button, a voice command system would be safer in an emergency.

Types of speech or voice recognition systems include speaker-dependent and speaker-independent systems. Speaker-dependent systems must be trained by a single user and will work only for that person, and speaker-independent systems can be used by many different persons. A speech recognition system can work with isolated words or with continuous speech. Present-day speech recognition systems are isolated-word recognition systems that can work on only one word or command at a time. Continuous-speech recognition systems must function at real-time speeds and must make sense of whole sentences.

REVIEW QUESTIONS

Speech Synthesis

1. How might equipping your personal computer with voice output make the computer easier to use?
2. What are some circumstances under which a robot with a voice would be safer than a voiceless robot?
3. Why might you use phonic integrated circuits to produce a voice even though other methods sound more natural?
4. Should a robot (or a machine) have a voice that sounds human or should it have one that is merely understandable to humans? Explain your answer.
5. How many warning messages can you make from the Digitalker word lists in Tables 20–1 and 20–2? Are there some warning messages you can't express because of missing words?
6. Would speech synthesis be useful for a robovan operating in a warehouse or factory?
7. How might delivery robots in offices and hospitals make use of speech synthesis?
8. How might a robot guard get along without speech synthesis? How might it use speech synthesis?

Speech Understanding

9. Why would it be desirable to equip a computer, robot, or other machine with word recognition abilities?
10. What robot tasks could profit from voice recognition?
11. Why is it accurate to say that speech is a complicated wave form?
12. Why is continuous speech harder to recognize than isolated words?
13. What is meant by "hearing between the words"?
14. How can background noise and emotions affect speech understanding?
15. Based on the graph information in Figure 20–18, showing the words *up*, *down*, *left*, and *right*, how would you try to tell these words apart?
16. Refer to the frequency patterns shown in Figure 20–7. How might you distinguish between the two phrases depicted there?
17. Robots working on the production line have no need of speech understanding and probably couldn't hear well in the factory. Why might a robot guard need speech understanding?
18. Would a delivery robot in a hospital or office need speech understanding?

PART 3

ROBOTICS PROJECTS

This section of the book deals with the actual steps of constructing laboratory experiments.

Not all experiments in robotics require expensive equipment. Many can be done with equipment that is already on hand for other technical courses. Some experimental materials can be bought as inexpensive parts or kits from electronics retailers. And while the more expensive educational robots can be used to teach techniques used by industrial robots, an entire lab full of these robots is rarely necessary. If your school has available an industrial robot, or if a company near you has one or more of them, you should visit and try to use the robot so that you get some of your training on the real thing.

21

CONSTRUCTION OF SIMPLE ROBOT PROJECTS

OVERVIEW

The best way to appreciate the complexities of a robotics project is to try doing one yourself. Even small robotics projects can be surprisingly complex. Electronics or computer majors cannot appreciate the engineering problems involved in getting a simple mechanical device to work properly until they try to make one. Conversely, manufacturing and mechanical majors do not appreciate the electronic and programming problems of a robotics project until they undertake one. To get the most out of this chapter, try to do projects in areas other than your major field.

OBJECTIVES

When you have completed this chapter, you should be familiar with:

- The mechanical, electronic, and programming problems associated with constructing a robot
- How to try out some of the principles you have been reading about

GETTING STARTED IN ROBOTICS PROJECTS

This chapter gives you a chance to try out some of the principles of robotics by constructing part or all of a robot. The construction projects are kept as small and as simple as possible, to reduce their cost and their difficulty. The mechanical and electronic tasks are designed to enable a nonmechanical and nonelectronic engineer to complete most of them. Since most computer technology students are not experienced at making hydraulic or pneumatic devices, the robot projects are designed to use electrical power.

While the mechanical parts of a robot are difficult to make, their materials are not terribly expensive. The robot controller, on the other hand, is often both complicated and expensive. Most schools with computer technology departments have a course in microprocessors, along with a laboratory for studying the microprocessor that includes microprocessor training stations or kits that enable the microprocessor to transmit and receive signals from input and output devices. Such a microprocessor can serve as a controller for a simple robot and using it for this purpose reduces the special equipment needed by a robotics laboratory. An 8085 microprocessor training station would suffice for a simple robot controller; a 68000 microprocessor training station would make a very smart robot controller.

Some of the projects that follow involve building robot manipulators that lack sensors. Others involve interfacing and using various types of sensor devices.

One of the simplest ways to control the positioning of the joints of the arm and hand of an electrical robot without using sensor feedback is to rely on stepper motors for positioning. This method resembles the methods used by inexpensive flexible and hard disk drives and some printers for positioning read/write heads, paper, and print heads; and indeed, stepper motors suitable for robot experiments can be salvaged from old printers and disk drives. Manufacturers of disk drives and printers may be willing to supply defective stepper motors that still work well enough to use for robotics projects.

To simplify the construction of the hand, you can design the hand to use one fixed-position finger and one movable finger, on a fixed-position wrist. To simplify its construction, the arm can be limited to one or possibly two axes of movement in the horizontal plane of a tabletop. That is, the arm moves along the surface of the table, rather than moving above the table's surface. This makes the arm easier to construct and simplifies the problems of balancing the arm as it moves with a load.

The number and complexity of the projects you can complete on robotics depends on the length of the term and your level of interest.

Using Microprocessor Trainers

Most schools that offer electronics or computer programs include a course in microprocessors. If so, they probably already have microprocessor trainers available. Any microprocessor trainer that allows connection of external devices will work for a robotics project. If you use a robotics kit, such as the Fischertechnik robot programming kit, you may well be able to use a microcomputer or personal computer in place of a microprocessor trainer. Heathkit sells several microprocessor trainers that could be used for robotics. Radio Shack sells a microcomputer trainer for under $30 that could be used for robotics. Figure 21–1 shows an MT-85 microprocessor trainer.

For $200 or less, you can build a Basic Educational Robot Trainer (BERT). This microprocessor-based system handles four simple sensors and controls up to three motors, a light, a noisemaker, and a voice synthesizer (see Figure 21–2). It is easily programmed through a serial RS-232 port.

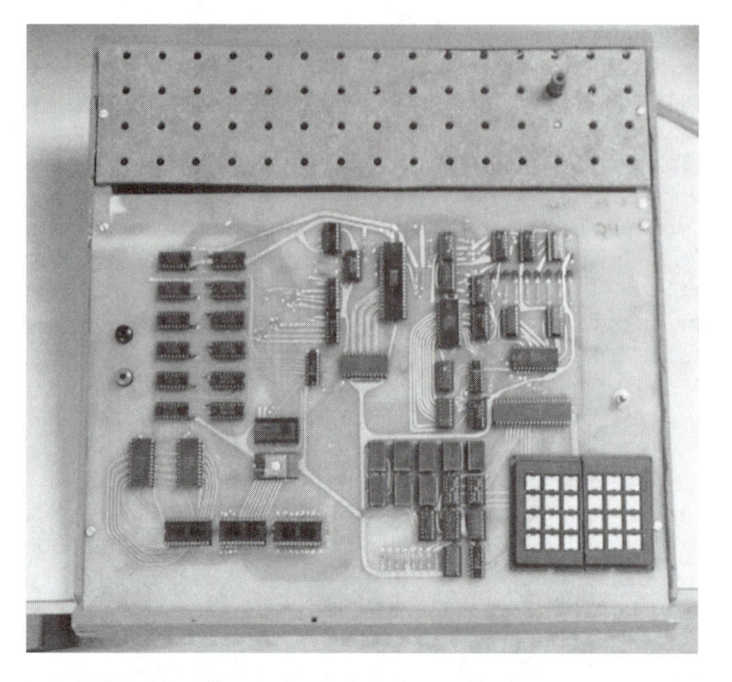

Figure 21-1 MT-85 Microprocessor Trainer.
The MT-85 microprocessor trainer uses an 8085 microprocessor, which was designed as an input/output controller. It works well for robotic sensor experiments.

Using Inexpensive Sensor Training Kits

Students interested in studying sensors might begin with the Robotics Sensor kit from Radio Shack, which costs less than $15. Figure 21–3 shows the old version of this kit. The kit comes with instructions for performing 20 experiments and includes built-in sensors that respond to light, magnetism, moisture, and sound. Some very simple sensor labs can be created using just microswitches and potentiometers. Microswitches represent digital sensors, and potentiometers represent analog sensors.

Using the Fischertechnik Robot Training Kits

Fischertechnik makes several robot training kits, including the Robot Programming Training Kit, the Robot Training Kit, and the Pneumatic Training Kit.

Robot Programming Training Kit Fischertechnik markets a robot programming kit for under $300. The kit, which uses a personal computer as the robot controller and simple DC motors for power, employs limit switches and potentiometers as sensors to control the starting and stopping of the motors. Assembling the kit gives you some insight into the mechanical problems of robot setup. The kit comes

Figure 21-2 BERT, the Basic Educational Robot Trainer.
BERT is easy to program through an RS-232 port from an external personal computer.

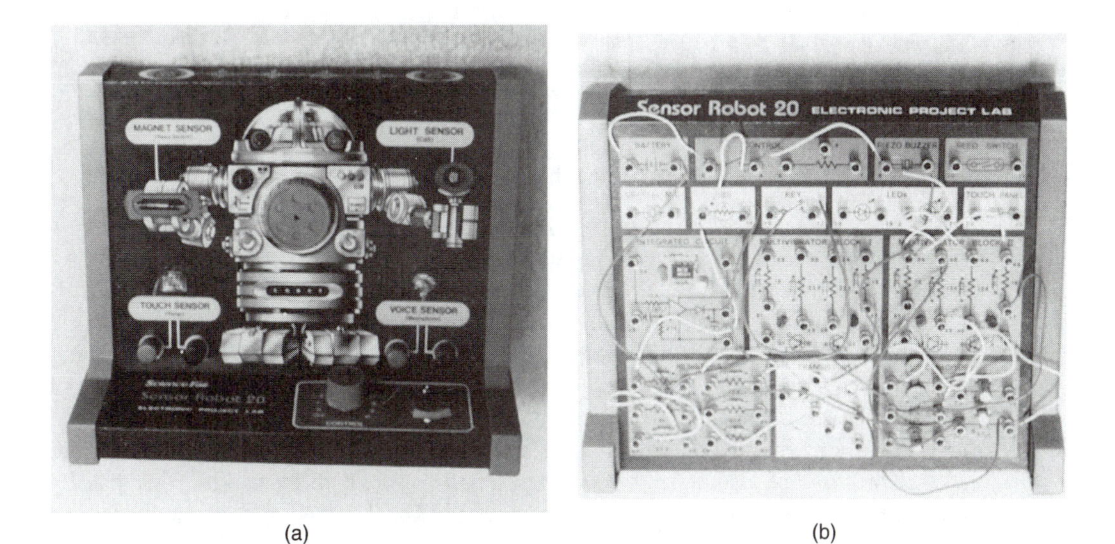

Figure 21-3 Sensor Robot 20 Kit.
This robot sensor kit from Radio Shack allows you to conduct 20 simple sensor experiments. (a) The fully assembled kit. (b) The assembled internal circuitry.

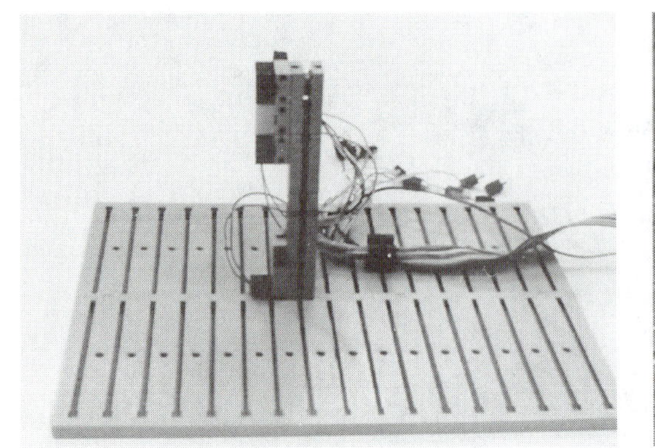
(a)

(b)

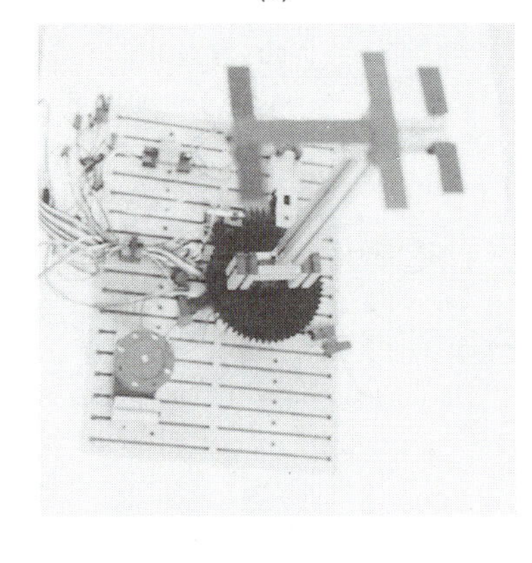
(c)

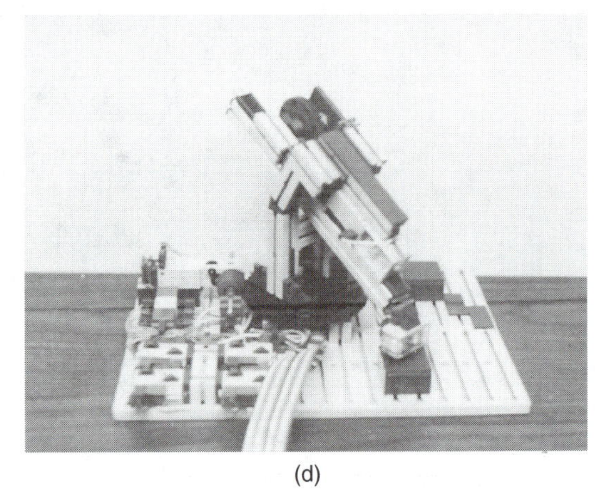
(d)

Figure 21–4 Four Experiments in the Fischertechnik Robot Programming Training Kit.
(a) The stop light, an introduction lab. (b) The machine shop. (c) The antenna rotor, a closed-loop servo system. (d) The teachable robot.

with directions for building 10 projects, which range from controlling a simple traffic light to operating a solar tracking platform; Figure 21–4 shows several of these.

Early versions of the kit came with programming instructions for programming the robot in Forth and included a diskette supporting PaRCL, an expanded version of Forth. Later versions of the kit come with BASIC programming instructions.

For about $165, you can buy an ultrasonic range-finding kit from Polaroid. Several sensors are also available on the Hero I robot from Heathkit. Sony has recently developed an electronic tape measure that can be used as a range finder.

Robot Training Kit The Fischertechnik Robot Training Kit makes a single project: a three-axis jointed-arm robot with a gripper. It uses BASIC programming and sells for just under $600 (see Figure 21–5).

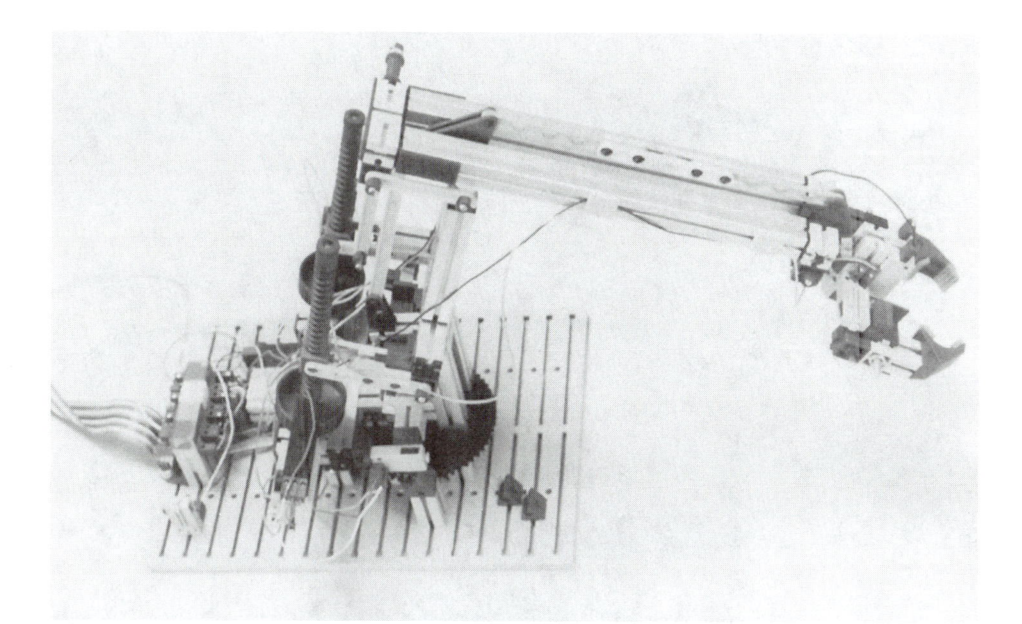

Figure 21-5 Assembled Fischertechnik Robot Trainer.
This robot trainer is equipped with a three-axis jointed-arm manipulator and optical encoders as sensors. The gripper has a range of moves.

Pneumatic Training Kit The Fischertechnik Pneumatic Training Kit provides a low-cost introduction to using pneumatic power. The compressor kit and the pneumatic control kit each cost about $80. Add-on kits are available for both of these basic kits.

Hero I Robot

Heathkit manufactures the Hero I robot, which can be purchased ready-built or as a kit. Fully assembled, with a voice and an arm, it sells for around $2,150 (see Figure 21–6). As a kit, with arm, voice, and radio remote control, it costs around $1,300. The Hero I can be controlled from a built-in keyboard, from a teaching pendant, or (optionally) from a radio remote control. It has a built-in 6808 microprocessor capable of following preprogrammed instructions.

The Hero I has sensors for detecting sound, light, motion, and obstacles. It uses an ultrasonic range finder to find obstacles. Its voice is supplied by a voice synthesizer. All of these devices can be programmed as experiments or as robotics labs. Manual operation and programming are possible by means of the teaching pendant.

Heathkit also sells a Hero 2000, for up to $4,500 (see Figure 21–7). This model has an 8088 master microprocessor that controls 11 peripheral microprocessors. It uses its own form of BASIC and may be equipped with an optional

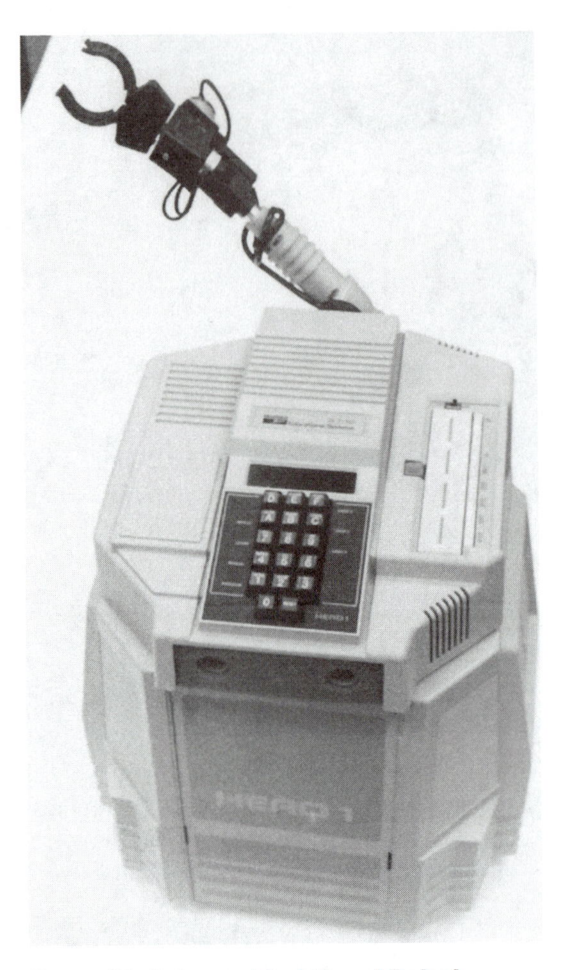

Figure 21-6 Assembled Hero I Robot.

Auto-Docking Accessory that allows it to search out, find, and connect to its charger automatically. The 2000 includes infrared vision and two ultrasonic range finders—one fixed and one movable.

The Hero 2000 arm may be bought separately, for around $1,500 (see Figure 21–8). It is a stepper-motor-driven microprocessor-controlled arm without any sensors.

Rhino Educational Robot

The Rhino educational robot can be bought with an *X*–*Y* table, a conveyor belt, and a turntable (see Figure 21–9). It can be used to simulate some simple production jobs, and it can be programmed either off-line or point to point from a teaching pendant. Unfortunately, the complete Rhino can cost from $5,000 to $10,000.

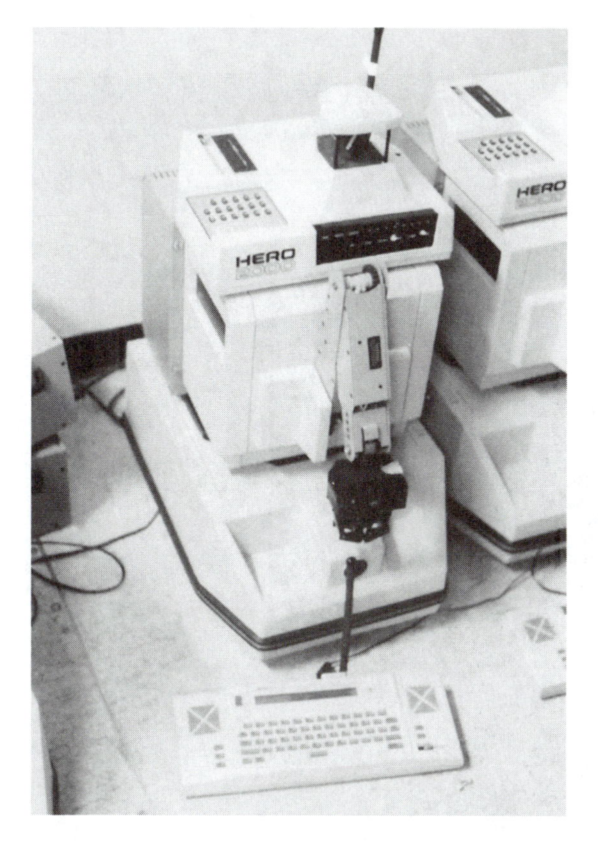

Figure 21-7 Assembled Hero 2000 Robot.

ROBOT PROJECTS

1. You are given a stepper motor that has two-phase windings and the following four states:

   State 1: Both windings with positive voltages on them.

   State 2: Winding 1 with positive voltage, and winding 2 with negative (reversed) voltage.

   State 3: Both windings with negative (reversed) voltages on them.

   State 4: Winding 1 with negative voltage, and winding 2 with positive voltage.

   Construct a test setup that will allow you to step the motor by sequencing two double-pole, double-throw switches. Count the number of switch throwings that are required to rotate the stepper motor through one complete revolution. What direction does the motor turn when going

Figure 21-8 Separate Hero 2000 Arm.

through states 1 through 4? What direction does it turn when going through these states in reverse order? Figure 21–10 depicts a circuit for controlling the four-wire, conventionally wound stepper motor, and a circuit for controlling the six-wire bifilar-wound stepper motor.

2. Using the stepper motors described in Project 1, replace the switches with 5-volt power relays. Design a counter circuit that can run the relays so that they step continuously through these four states. Make provisions to allow the states to be run in reverse order. Figure 21–11 shows a possible circuit for running the stepper motors in a single direction.
3. Using the stepper motors described in Project 1, design a circuit that accepts a direction command and a count value; then give the motor the desired number of counts.
4. Using the stepper motors described in Project 1, replace the control and driver circuits with the Motorola SAA1042 stepper motor driver integrated circuit.
5. Using the stepper motors described in Project 1, place a pulley on the motor shaft. Attach a string to the shaft, and hang a weight from it. Using the circuit in Project 4, determine the count values necessary to

Figure 21-9 Rhino Robot.

move the weight up or down by 1 foot or by any smaller distance, within the step accuracy of the motor.

6. Using the stepper motors and circuits described in Project 5, program a microprocessor trainer to control the moving of the weight up and down to the following positions:
   a. Start at 0 feet.
   b. Go to 1 foot up.
   c. Go down 0.5 foot.
   d. Return to 0 feet.

   Program the microprocessor to pause for 2 seconds at each position.
7. Use a Signetics SSA1027 stepper motor driver integrated circuit to drive a four-phase, bifilar-wound stepper motor.
8. Using a voice synthesizer integrated circuit similar to the Radio Shack SP0256-AL2, construct a voice for your microprocessor trainer. Program the microprocessor to count from 0 to 20.

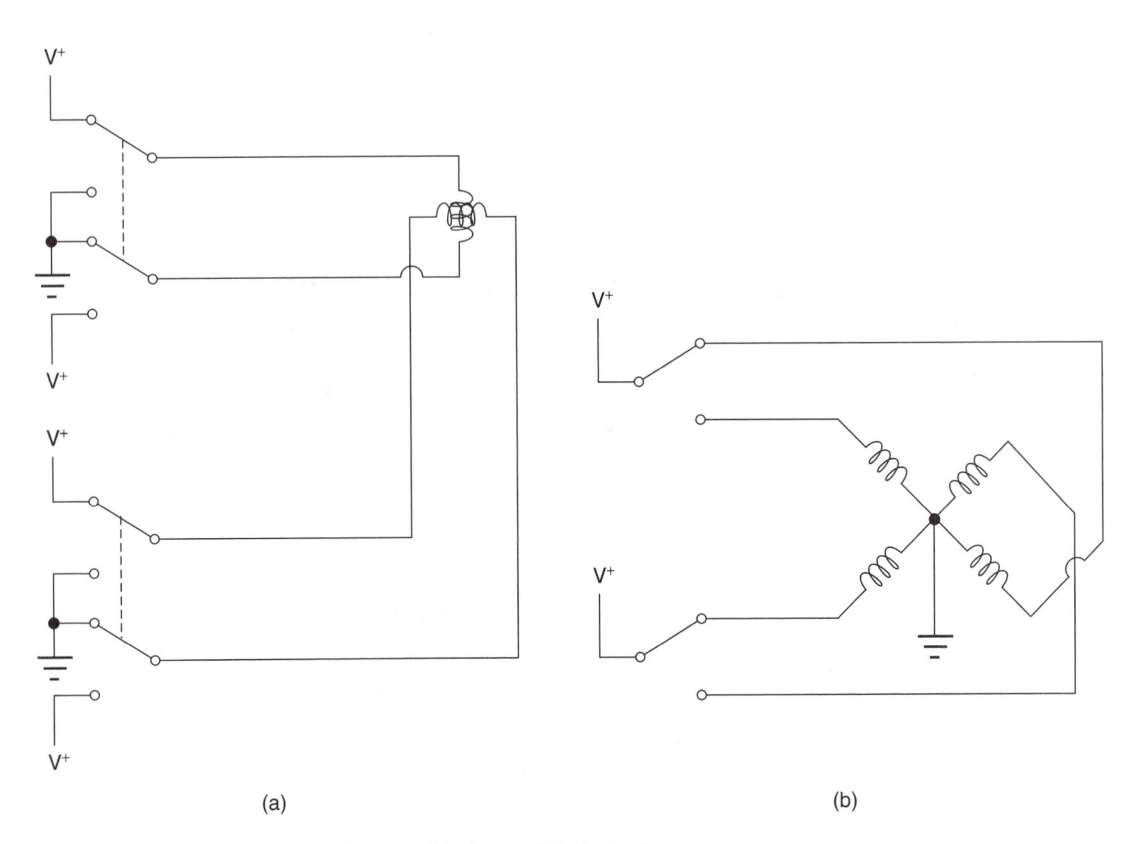

Figure 21-10 Controlling Stepper Motors with Switches.
(a) A simple drive circuit for a conventionally wound stepper motor. Here, a four-wire stepper motor has been connected to two double-pole double-throw switches. By sequencing the switches, you can cause the stepper motor to rotate. (b) A simple drive circuit for a bifilar-wound stepper motor. Here, a six-wire stepper motor has been connected to two single-pole double-throw switches.

9. Using a voice recognition integrated circuit similar to the Radio Shack VCP200, construct a voice recognition system for your microprocessor trainer. It would be simplest to use it to control a motorized base, since it understands the words *go, stop, reverse, left turn,* and *turn right.* It could also be used to position a jointed-arm robot. In this case, the go command would extend the arm (the elbow joint), and the reverse command would retract the arm. The on/off command mode would be used to open and close a simple gripper.
10. Program the microprocessor trainer to act as a controller for an automatic elevator that cycles up and down, servicing eight floors. When going up, it continues up until it has stopped at every floor where there is a request to go up. Similarly, when going down, it does not stop at floors that have no down request, even if they do have an up request. When there are no requests, the elevator stays at the last floor serviced, with the door closed.

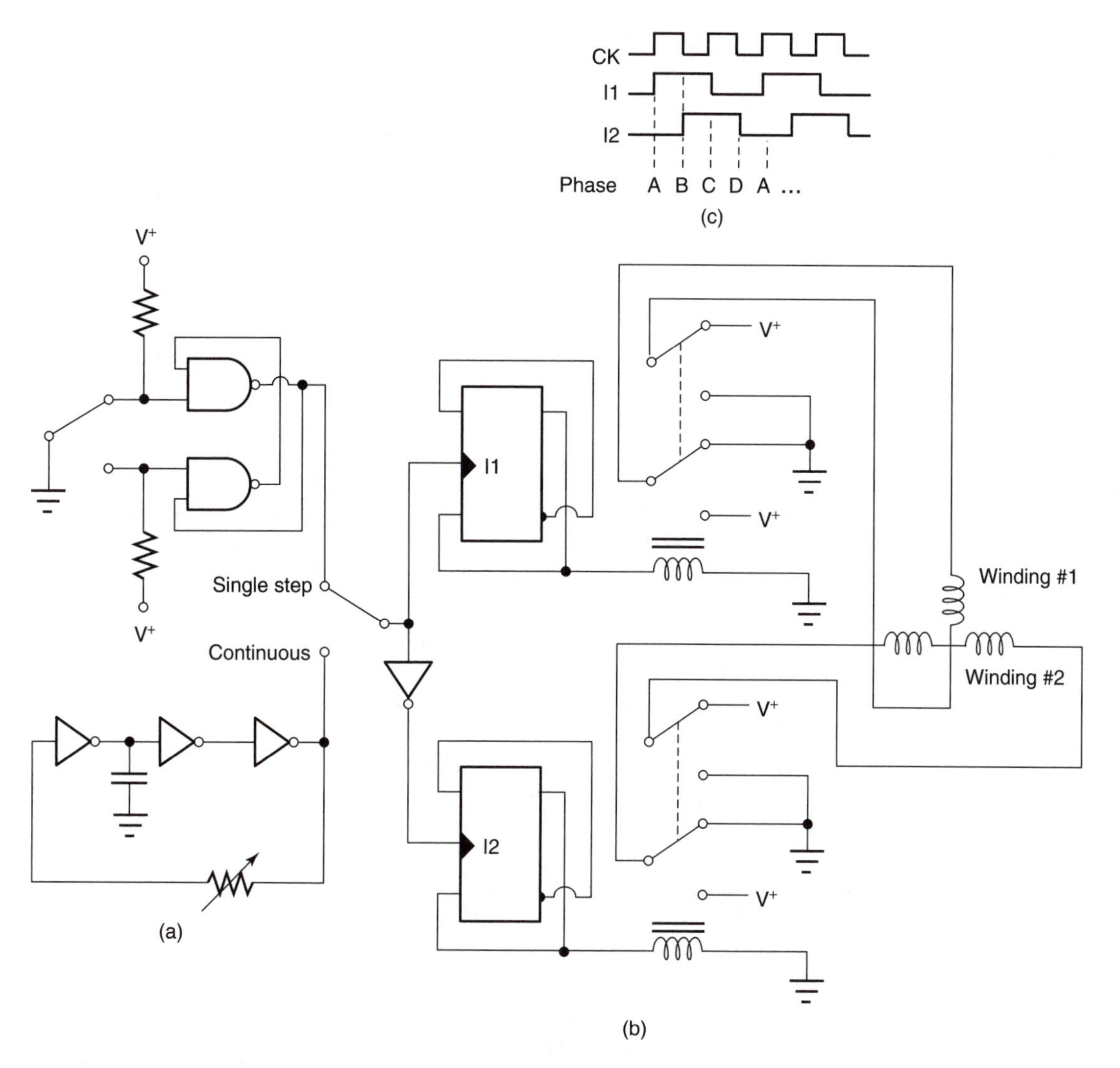

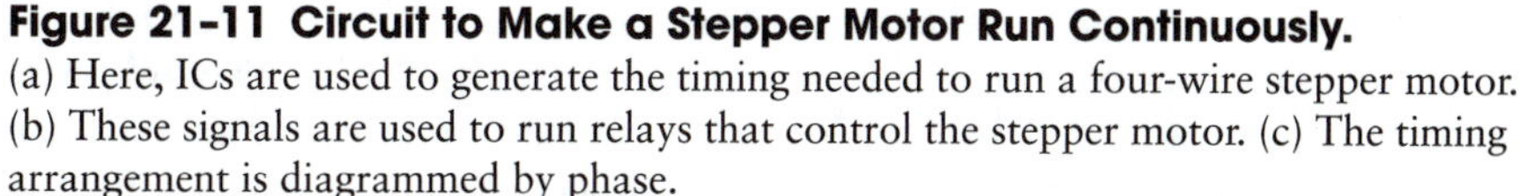
Figure 21-11 Circuit to Make a Stepper Motor Run Continuously.
(a) Here, ICs are used to generate the timing needed to run a four-wire stepper motor. (b) These signals are used to run relays that control the stepper motor. (c) The timing arrangement is diagrammed by phase.

The general layout of elevator controls and displays is as follows. There are two dip switches for each of the eight floors serviced—a separate one at each floor for up and down. There is also a light for each floor, as well as up, down, door open, pause, and door close lights. There is a separate push-button door switch. Floors are spaced 10.0 feet apart, with the ground floor at 0.0 feet. As the elevator moves, the display shows its location.

Whenever any switch is activated, the elevator begins to travel to that floor, at a rate of 2 feet per second. The location of the elevator is maintained on a display of the microprocessor trainer. Any up or down requests are noted and remembered. If the request is for travel in the direction of current motion, the elevator modifies its travel to include the new request, unless it has already passed that floor, in which case it remembers the request but waits until the next appropriate cycle to service it. The elevator can service a request made for a particular floor unless it is already within 5 feet of the floor in question, in which case it must wait until its next trip to service the request.

The elevator doesn't stop at floors where there is no relevant request. At floors where there is such a request, the elevator stops and goes through the following passenger-loading cycle. For 1 second, the elevator "opens the door" (that is, it lights the door open light). It then pauses for 5 seconds while lighting the pause light and extinguishing the door open light. If switch closures of the door switch occur during the 5-second interval, the elevator extends the time it pauses for another 3 seconds beyond the last closure. The elevator then extinguishes the pause light and begins the door close cycle. During the door close cycle, which lasts 2 seconds, the elevator lights the door close light, while continuing to monitor the door switch. If a closure of the door switch occurs, the elevator resumes its pause mode.

11. Using the microprocessor trainer program as a base, together with the program for controlling the stepper motor, make a working model of an elevator, where 1 foot of elevator travel is represented by 1 inch of travel by the weight controlled by the stepper motor.
12. Using the voice synthesizer circuit from Project 7, give the elevator of Project 10 a voice that calls out the direction of travel and the floor as it progresses.
13. Using a BERT robot or a Hero I robot, program it to say, "Here comes [your name goes here]." This may not be as easy as it sounds, since many names do not follow English rules of pronunciation. Include appropriate pauses, and end the speech with a pause.
14. Build a BERT robot.
15. Design and build your own version of BERT. Naturally, it will have more capabilities than the original BERT.
16. A *bathysphere* is an object capable of being submerged to great depths. This project involves creating a simulated research bathysphere. Its permissible depth range is from 1 fathom (6 feet) to 512 fathoms (3,072 feet). It can collect specimens for a certain minimum time, extended when necessary.

    Specifically, you are to design and construct the necessary hardware (switches, gates, and the like), and design and code a program to simulate the bathysphere. Descend at a rate of 2 fathoms per second, until the specified depth is reached. At that depth, enter the hatch open phase. Prolong the hatch open cycle whenever the extend switch is depressed momentarily. (*Note:* You must allow for multiple extensions.)

    The depth is specified by a dip switch connected to a port of your microtrainer. The 8 bits of the dip switch are used to specify an 8-bit number representing the depth to which the bathysphere is to descend (or ascend, as appropriate). The depth is displayed in a 7-segment display. When initialized, the bathysphere begins with a depth of 00 fathoms.

When the hatch open/closed cycle is completed, the program must scan the switch specifying depth; then it either rises or descends, as appropriate, at the specified rate until reaching the specified depth; then it opens the hatch again, and repeats the same steps. Continue the ascent/descent cycle until a depth of 00 is specified, at which time the bathysphere returns to the surface.

Additional features of the program might include counting the depth in decimals and adding a switch to measure the depth and quantity of specimens collected (measured by the number of times the extend switch is pressed, regardless of time during the hatch open cycle).

17. Using stepper motors, construct a two-axis manipulator that moves in a horizontal plane. The manipulator will be easier to build and balance if the joints are supported by casters on a tabletop.
18. Add a two-finger gripper to the manipulator described in Project 17. The opening and closing of the gripper may be controlled either by a solenoid that provides all-the-way-open and all-the-way-closed action or by a stepper motor that gives degrees of open and closed.
19. Simulate a stop light. The primary purpose of this project is to imitate a real traffic situation with an interrupt-driven microprocessor system. The software should be modularized as much as possible. The hardware should be on a wire-wrap board, and thin white tape should be used to outline the streets. The intersection being modeled is the nearest stoplight to your school. For example, you might model the intersection "where Campus Drive crosses the Bypass, near McDonald's." Your board should have approximately the following size and setup:

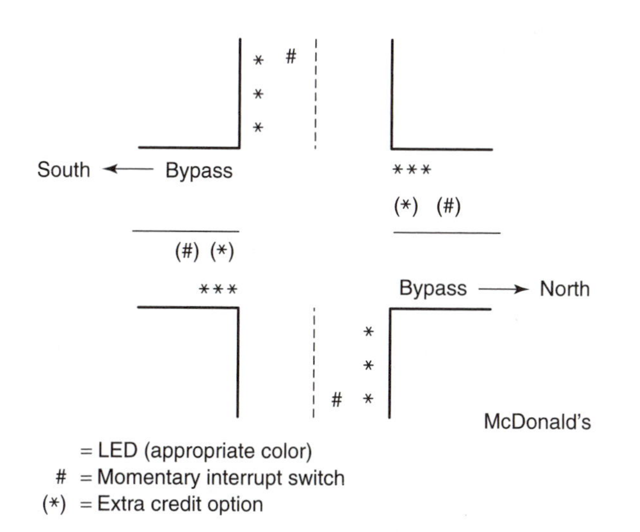

Points to Consider

a. The Bypass should be left green all of the time, until an interrupt switch is toggled.
b. Since the speed limit is higher on the Bypass, a longer yellow cycle will be necessary for the lights on it.
c. Any of the momentary interrupt switches may be toggled multiple times in a short period, but they should register only one green cycle.
d. Eastbound and westbound Campus Drive should never be green at the same time.

Extra Credit

a. Implement the left-turn sensors and arrow lights off the Bypass.
b. If there are any questions about program implementation, remember: "Reality Rules!"

20. Refer to the Kawasaki Robot Specifications in Appendix D and choose a robot for each of the following applications:
    a. Use water jet cutting to cut in half an 80-inch-wide sheet of metal.
    b. Spot weld while reaching down into a foot-high container and weld the final joints together.
    c. Paint a three-foot by six-foot panel.
    d. Weld together two halves of a pressure tank that is 16 inches in diameter.
    e. Palletize 3″ × 6″ × 12″ boxes that weigh 40 pounds each.
21. Suppose that you want to sort paint cans of different heights automatically. How could this task be done with whisker sensors? How could it be done with photoelectric sensors? How could it be done using weight sensors?
22. Do the Vision System Lab from Appendix E.

APPENDIX A: MECHANICS

No book on robots would be complete without a discussion of mechanics and motion. After all, the main difference between an ordinary computer and a robot is that the robot can move. This motion requires the use of mechanical parts and follows the laws of mechanics.

POWER TRANSMISSION

How do you get the power from the motor to the wheels? Usually the motor itself turns faster than you want the wheels to turn; the exceptions to this are some stepper motors and some servo motors. Most motors turn at rates of 1,000 to 12,000 revolutions per minute (rpm), while a joint or gripper may require a speed of 1 rpm or less. When robots have vehicles, their wheels usually run at 50 to 250 rpm. Therefore, direct driving of the wheels, joints, and grippers by the motor is usually not practical. And even if the motor does turn slowly enough to do the job directly, it may be too heavy to mount right on the gripper or other part of the arm. Consequently, some way must be found to transfer power from the motor to where the work is to be done.

Friction Drive

One of the simplest drive techniques is to have the motor shaft turn against the outside edge of a drive wheel that has a much larger diameter. Figure A–1 shows the drive mechanism for the UFO robot. Here, the weight of the robot forces the motor shaft against the drive wheel. The drive wheel then turns in the direction opposite that of the motor shaft.

If the motor shaft is $\frac{1}{16}$ inch in diameter and the drive wheel is $1\frac{1}{2}$ inches in diameter, the resulting speed reduction between the motor and the drive wheel is 24 to 1 (since $1\frac{1}{2} \div \frac{1}{16} = 24$). If the motor turns at 6,000 rpm, the drive wheel in this case will turn at 250 rpm (ignoring slippage). Using a drive wheel with a rubber surface increases the friction between the motor shaft and the drive wheel, thereby reducing slippage.

Friction drives have been used on some sewing machines.

Figure A-1 UFO Drive Mechanism.
The UFO robot uses friction drive. Its motor shaft rests against a rubber wheel attached to the drive shaft.

Pulleys and Belts

A more reliable friction drive mechanism uses pulleys and a belt. Figure A–2 shows such a drive system being used to control the spindle speed on a drill press. The speed is changed by moving the belt to pulleys of different sizes. In this case, both wheels connected by the belt turn in the same direction. The ratio of the diameters of the

Figure A-2 Spindle Power for a Drill Press.
A drill press uses a set of pulleys of different sizes to change the speed of the drill spindle.

Figure A-3 Waist-axis Power for the Rhino Robot.
The waist axis of the Rhino robot has a pulley-and-belt drive.

two wheels determines the step-down or step-up ratio of the speed. Some slippage occurs between the belt and the wheels, but this can be minimized by finding the proper tension for the belt. As the speed is stepped down, the power is increased.

Figure A–3 shows the waist-axis drive belt and pulley on the Rhino robot. Pulleys may also be used with cables for longer-distance power transfers.

Timing Belts or Ribbed Belts

Using a timing belt with ribbed pulleys is one way to eliminate slippage between the belt and the wheels. Such a system is used by some automobiles to control the timing between the crankshaft and camshaft. Besides preventing the belt from slipping on the ribbed pulleys, the ribs on the belt also reduced the tension required to make the belt and pulleys work. The ratio of the numbers of clods or of slots on the two pulleys determines the step-up or step-down ratio of the speed. You might think that you could identify the step-up or step-down ratio of the system simply by measuring the diameters of the two pulleys, but this does not give the true ratio, because of distortion of the timing belt as it bends around the small pulley.

Figure A–4 shows a timing belt, stepper motor, and pulleys used for driving the paper advance on a printer. The motor pulley has 10 slots, while the paper roller pulley has 40 slots. The slot ratio indicates that there is a 4 to 1 speed reduction on this system.

Chains

Using chains in place of drive belts allows more power to be transferred between the two drive sprockets. Chains are also more flexible than belts, and cause less friction. The ratio of the numbers of teeth on the two sprockets determines the

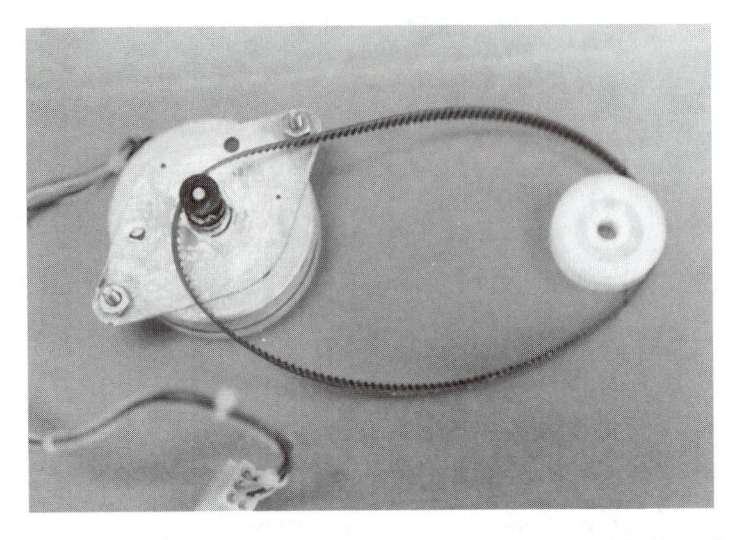

Figure A-4 Printer Timing Belt, Stepper Motor, and Pulleys for Paper Advance.

step-up or step-down ratio of the speed. Figure A–5 shows some of the chains and sprockets used on the Rhino robot.

Chain drives do have a noticeable backlash. For example, if a point is approached in a clockwise direction and recorded, it will not be the same point you would record if the approach were counterclockwise. Therefore, you should be careful to approach all points from the same direction when teaching a chain-driven robot.

Gears

Gears are useful for changing speed, power, and the direction of power transfer. Spur gears and external gears change speed, alter power ratios, and reverse the direction of rotation of the power. Two spur or external gears may be placed on a single shaft to make the gear train more compact. Bevel gears, worm gears, and helical gears make perpendicular transfers of power possible. Worm and harmonic gears allow large step-downs in gear ratios within a small space. Figure A–6 shows some of these gears. Every time two gears mesh, the direction of rotation of the second gear is the opposite of the direction of the gear driving it.

The step-up or step-down speed ratio for a set of gears is determined by the ratio of the numbers of teeth on the two gears. For a gear train, the step-up or step-down speed ratio is the product of the individual pairs of gear ratios.

A gear that is driven by a single gear and in turn drives only one gear is called an *idler gear.* It does not affect the gear ratio of the gear train, but it does reverse the direction of the remaining driven gears in the gear train.

Figure A-5 Chain Drive on Rhino Robot.
The Rhino robot uses chain drives for its shoulder, elbow, and wrist axes.

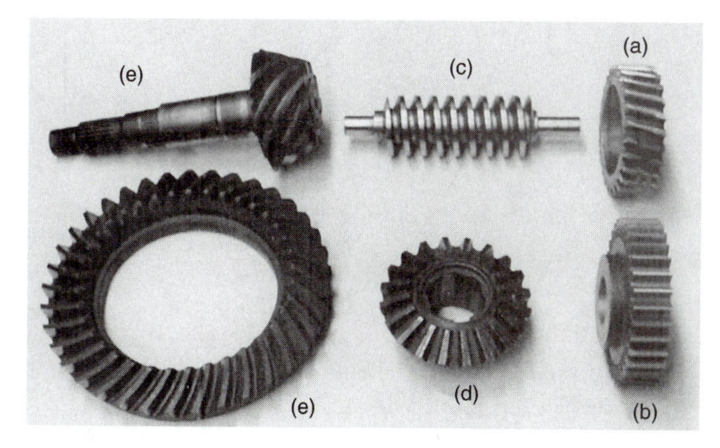

Figure A-6 Typical Gears.
(a) A spur helical gear. (b) A spur gear. (c) A worm gear. (d) A beveled gear. (e) Beveled helical gear.

SAMPLE PROBLEM

Suppose that you are given a set of gears connected as shown in Figure A–7. The motor runs at 9,000 rpm in a clockwise rotation, and it develops a torque of 0.5 oz-in. What will be the direction of rotation, the rpm, and the torque of the output shaft? (This gear train is identical to the one used in the "Big Trak" tank gearbox.)

Answer

The direction of rotation for the tire and wheel is found by tracing the gear's movement, starting with the gear on the motor and ending with the tire and wheel. Each time two gears mesh, the direction of rotation of the second gear is the reverse of that of the driver gear. Therefore, first label the axles of the gear train A through D, from left to right. Given that axle A is rotating clockwise, axle B rotates counterclockwise; then axle C rotates clockwise, and finally axle D rotates counterclockwise. Counterclockwise is thus the direction of rotation of the tire and wheel.

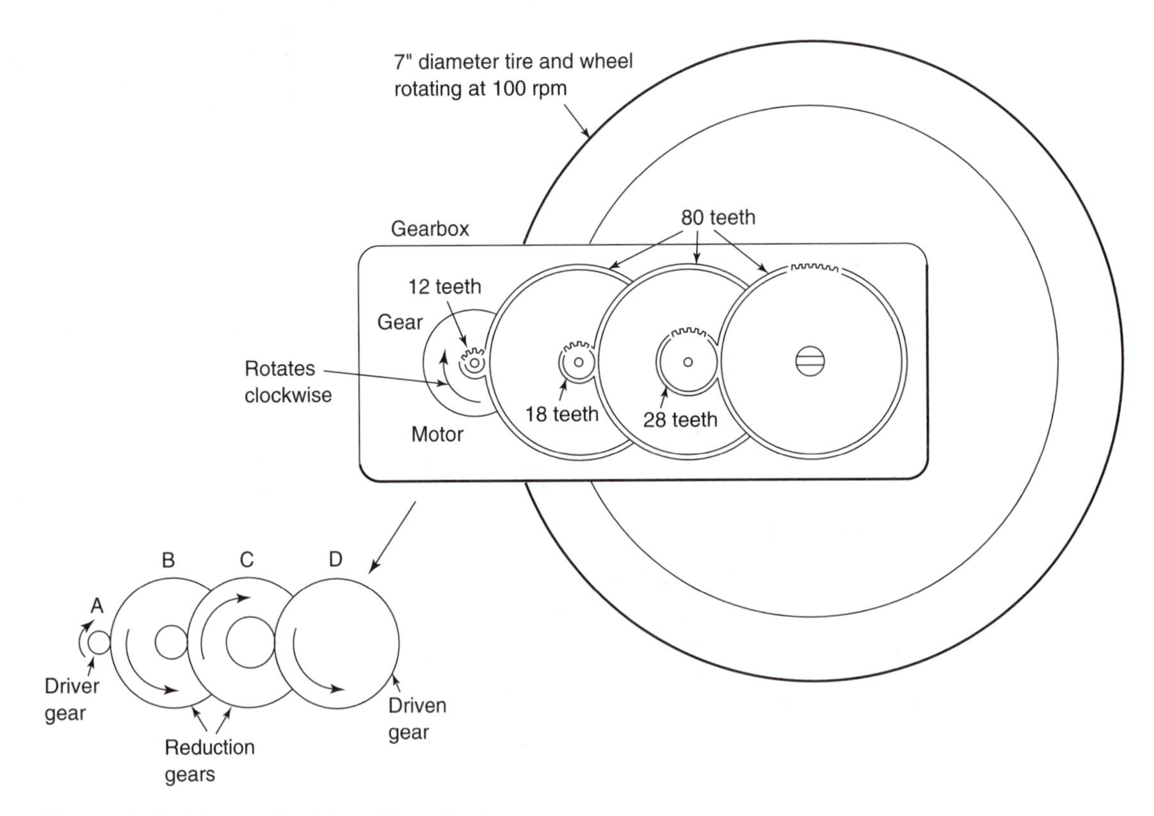

Figure A–7 Sample Problem Gear Train.

The gear reduction ratio of the gear train must be calculated as the first step in finding the rpm of the driven axle. The motor gear, with 12 teeth, drives an 80-tooth gear on the first reduction axle. This gives a ratio of 80 to 12, or 6.667 to 1. The first reduction axle gear, with 18 teeth, drives the second reduction axle gear, which has 80 teeth. This gives a ratio of 80 to 18, or 4.44 to 1. The second reduction axle gear, with 28 teeth, drives the driven axle gear, which has 80 teeth. This gives a ratio of 80 to 28, or 2.86 to 1. The total gear ratio is obtained by multiplying these three ratios together:

$$6.667 \times 4.44 \times 2.86 = 83.94$$

To get the output rpm of the gearbox, divide the input rpm by the total ratio:

$$9{,}000 \div 83.94 = 107.22 \text{ rpm}$$

If we assume that there are no losses in the gear train, the output torque is equal to the input torque times the gear reduction ratio:

$$0.5 \text{ oz-in.} \times 83.94 = 41.97 \text{ oz-in.}$$

In reality, however, the motor and gearbox have an unloaded output of 100 rpm, with 20 oz-in. of torque. Frictional losses load the motor down and use up half its power just to move the reduction gears.

Gear trains have some backlash; that is, when the direction of rotation is reversed, some slippage occurs before the gears engage in the reverse direction. This slippage is due to the gears' not fitting perfectly together. The more gears there are in a gear train, the greater the backlash. The more precise the gears are made, the less the backlash.

Harmonic Drive

Harmonic drives or harmonic gears can provide gear ratios of 1:50 or more in a very compact space. Harmonic drives consist of three basic parts: a rigid circular spline, whose position is fixed; a flexspline, which is the output device for the drive and rotates in the opposite direction to the direction of the input device; and an elliptical wave generator input, which deflects the flexspline to engage the teeth of the fixed rigid circular spline. Figure A–8 shows these three parts, plus assembled and disassembled harmonic drives. The gear ratio of the drive is determined by the number of teeth on the fixed rigid circular spline, divided by the difference between the number of teeth on the flexspline and the number of teeth on the fixed spline. For example, if the fixed spline has 100 teeth and the flexspline has 98 teeth, the gear ratio is $100 \div (100 - 98) = 50$; thus the ratio is 50 to 1. Because the harmonic drive has a large number of teeth meshed at one time, it has a high torque capacity and little or no backlash.

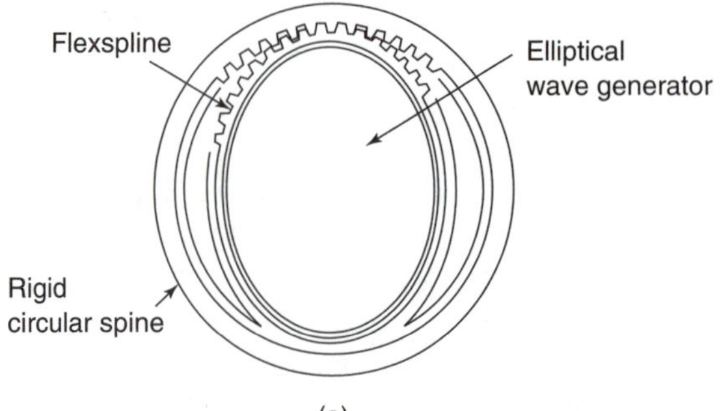

(a)

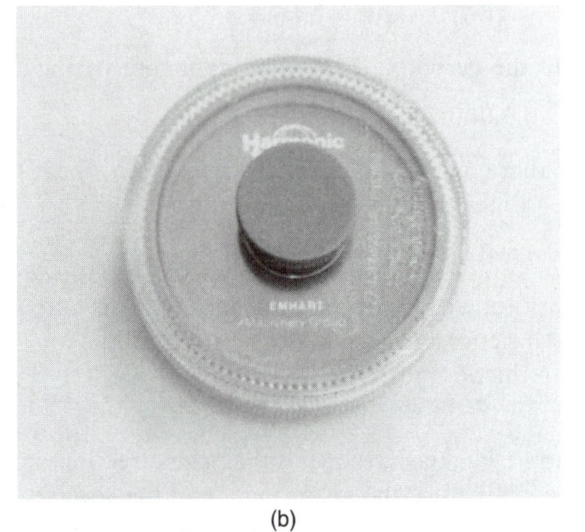

(b)

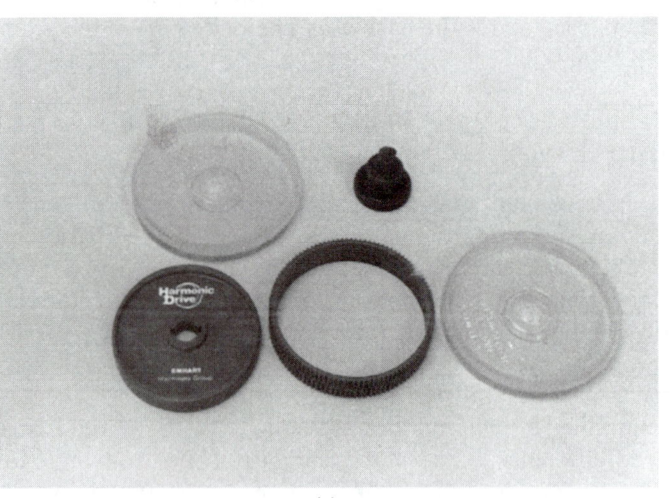

(c)

Figure A-8 Harmonic Drive.
(a) The part names of the harmonic drive. (b) An assembled harmonic drive. (c) A disassembled harmonic drive.

Linear Motion

Linear motion can be generated by rotary motion through the use of a drive screw or a rack and pinion (see Figure A–9). If the motor turning the screw of a screw drive has a stationary mounting, a block mounted on the screw will move linearly when the screw is turned by the motor. The direction of the block's movement depends on the direction of the threads on the screw and the direction of motion of the motor shaft. Some computer flexible and hard disk drive head assemblies use drive screws to position their read/write heads.

When a stationary rack is used, the pinion moves along the rack as the pinion is rotated. It is also possible to have a stationary pinion with mobile racks above and below it; in such a case, the two racks move in opposite directions when the pinion rotates. If jaws are attached to the outside ends of the racks, the racks' movement can be used to move the jaws. This is the motion required for a parallel jaw gripper. Rack-and-pinion systems are used in the steering assemblies of some automobiles.

The band drive is a variation of the rack-and-pinion drive (see Figure A–10). The band drive is used on some positioning assemblies for computer hard disk drive heads.

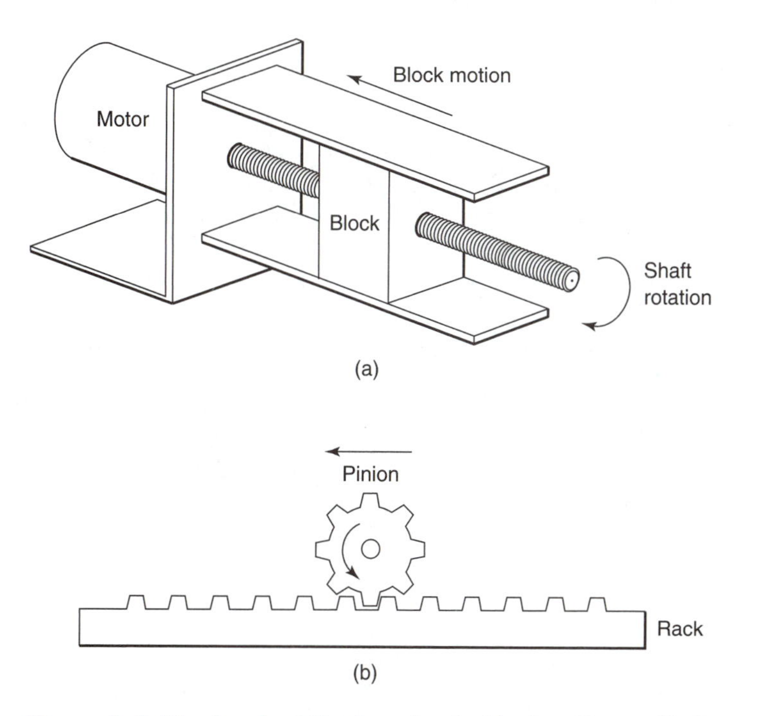

Figure A-9 Mechanical Devices for Achieving Linear Motion.
(a) A drive screw. (b) A rack and pinion.

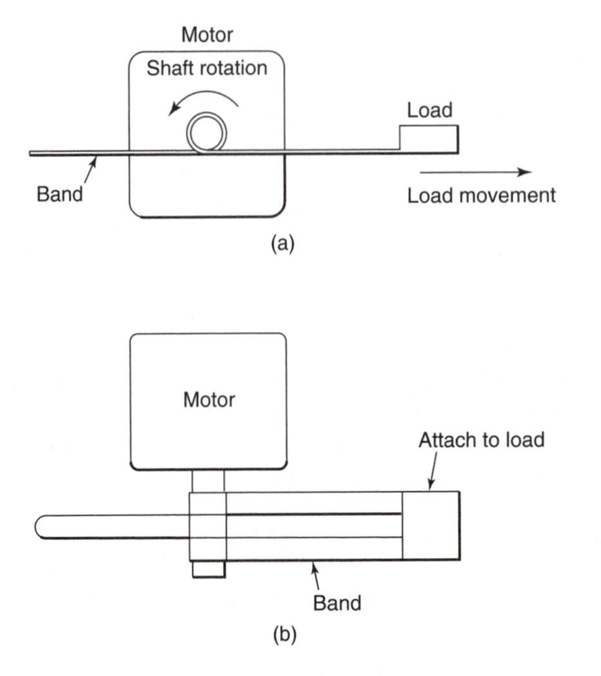

Figure A-10 Band Drive System.
(a) Front view. (b) Top view.

STATICS

Statics is the science or study of the relationships between the forces that produce equilibrium. Thus, statics deals with stationary objects. Introductory-level statics usually focuses on rigid, nondeforming bodies. More advanced statics takes into account the actual deforming of bodies under loads. This nonrigidness contributes to positioning error. While the members of a robot are not completely rigid, they are nearly so.

When a robot is not moving its joints, it should remain stationary in its position. If the structural members of a robot are too weak, the robot may collapse under its own weight. If the center of gravity of a robot falls outside the wheelbase or anchor base of the robot, gravity will tip the robot over. In other words, if a robot loses its equilibrium state, it is in danger of falling over. And should the robot start to fall, it passes from the field of statics to the field of dynamics.

DYNAMICS

Dynamics is the branch of mechanics that deals with objects in motion and the forces that produce or change such motion. The study of objects in motion is termed *kinematics*; the study of forces that produce or change motion is termed *kinetics*. The movement of a robot's manipulator follows the rules of dynamics, which include inertia, momentum, gravity, and centrifugal force. Thus, a lot of applied physics is involved in robotics.

Robot manipulators and grippers and the materials transferred by the robot all have mass. Consequently, they resist changes in motion. In many cases, the robot must undergo controlled acceleration and deceleration to prevent mechanical ringing.

REVIEW QUESTION

1. Figures A–11 and A–12 show the gear train and gear teeth used by the Movit Line Follower II robot. The motor gear turns counterclockwise at 4,500 rpm, with a torque of 0.2 oz-in. What are the direction, the rpm, and the torque of the wheel? (*Note:* The two idler gears do not contribute to gear reduction.)

Figure A-11 Movit Line Follower II Gearbox.
This view shows the drive train for one side of the Movit Lane Follower II robot.

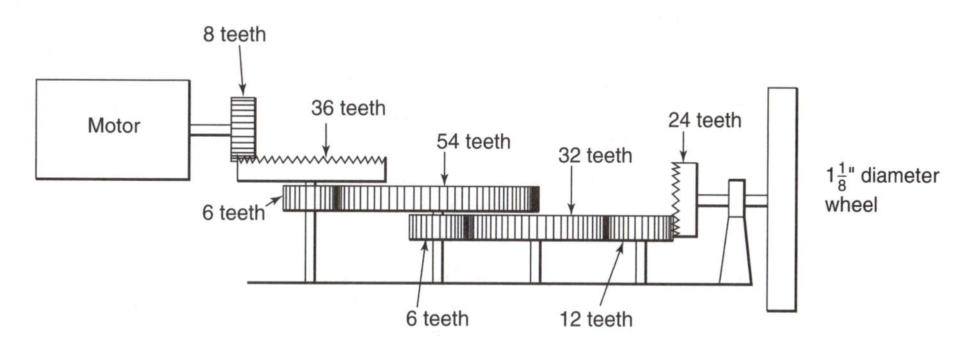

Figure A-12 Gearbox Close-up.
This view shows the number of teeth on each gear of the gear train and how they are interconnected.

APPENDIX B: ELECTRICITY, ELECTRONICS, AND COMPUTER FUNDAMENTALS

ELECTRICITY

The electrical charge that makes up electricity consists of negatively charged particles called *electrons*. Electrons normally orbit around the nuclei of atoms. Figure B–1 depicts the structure of hydrogen and helium atoms. Under some circumstances, an electron can be stripped from an atom, leaving a positively charged atom and a negatively charged electron. Materials that easily give up free electrons are called *conductors*. Materials that are very difficult to make give up free electrons are called *insulators*. Materials that fall somewhere between these two categories are called *semiconductors*.

Electrical current can be defined as the flow of electrons. In this case, the electrons flow from a negatively charged area toward a positively charged or electron-deficient area. Such charged areas can be produced chemically or mechanically. Batteries use chemical reactions to set up electric potentials and currents. An electric generator produces similar electric potentials and currents by moving a conductor through a magnetic field. Electrostatic charges can be generated by mechanically rubbing two materials together in an attempt to pull some electrons free from their atoms.

When imbalances in electrical charge exist, they tend to equalize. For example, if one body has a slight excess of electrons, and a second body has a greater excess of electrons, electrons will tend to move from the place of greater excess to the place of less excess. Again, such a movement of electrons is known as an electric current flow, and its volume is measured in amperes (or amps). The dif-

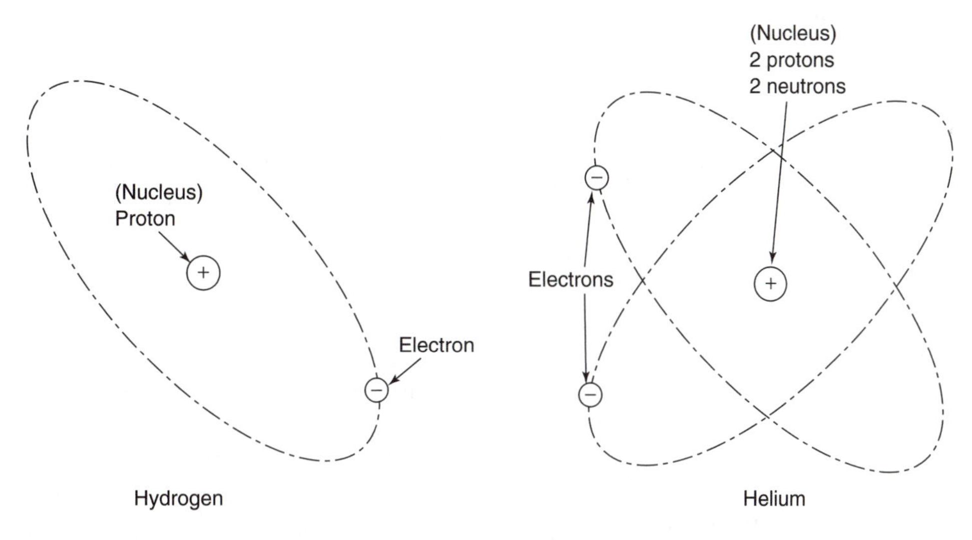

Figure B–1 Hydrogen and Helium Atoms.
Electricity comes from the electrons found in atoms. It is one of the fundamental forces of the universe.

ference in charge between two bodies is called a *potential difference;* since it is an electrical difference, this potential is known as *electromagnetic force (EMF)* and is measured in volts. EMF can be thought of as a measure of electrical pressure, analogous to water pressure in hydraulics or air pressure pneumatics. The opposition to the flow of electrical current is called *resistance* and is measured in ohms.

Ohm's Law

Georg Simon Ohm discovered the mathematical relationship of voltage, current, and resistance. The formula, called Ohm's law, states that current flow (I) is equal to voltage (E) divided by resistance (R).

$$I = E/R$$

Mathematical transpositions produce two equivalent expressions of Ohm's law that solve for E and R:

$$E = I \times R \qquad \text{and} \qquad R = E/I$$

Ohm's law can also be remembered by drawing the following magic circle, covering up the component to be ascertained, and observing the relationship of the remaining two components:

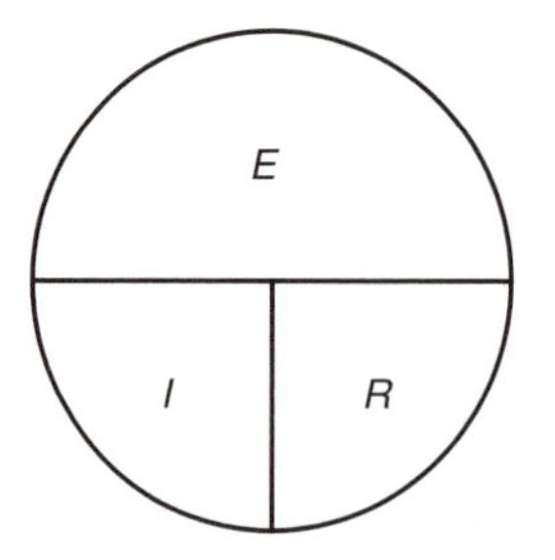

For example, if you have an electric heater whose operating voltage is 115 volts and whose resistance is 10 ohms, you can find its current by solving as follows:

$$\frac{115 \text{ volts}}{10 \text{ ohms}} = 11.5 \text{ amps}$$

Power

Electrical power, which is measured is watts, is calculated by multiplying the current by the voltage:

$$P = I \times E$$

The power of the electric heater we discussed a little while ago can therefore be calculated as follows:

$$P = 11.5 \times 115 = 1{,}322.5 \text{ watts}$$

This may sound like a lot of power, but it represents the demand of a small electrical space heater. The oven on an electric range can use 3,000 to 5,000 watts, an electric clothes dryer can use 6,000 watts, and an electric furnace for heating a home can use 50,000 watts.

Calculating power losses over a transmission line involves measuring the resistance of the line and the current through the line. This power loss is called an I^2R loss, since it is computed from the current and resistance. The actual formula comes from substituting the $I \times R$ of Ohm's law for E in the power formula $P = I \times E$. This yields the formula

$$P = I^2R$$

Generating Electricity

Electricity can be generated through chemical reaction, mechanical motion, or photovoltaic effect. If the generating method is based on chemical reactions, the device is called a *battery.* Mechanical motion devices are called *electrical generators.* Devices for generating electricity from light or other radiant energy are called *photovoltaic cells, solar cells,* or *solar batteries.*

Alessandro Volta developed the first electrochemical battery in the eighteenth century. Called *Volta's pile,* it consisted of several cells, each of which had a copper electrode and a zinc electrode that were separated by moist paper. Batteries produce an electric reservoir in which one electrode has a positive charge and the other has a negative charge. The difference (EMF) between these two charges is determined by the state of charge of the battery and the materials used for the electrodes.

Using carbon as the positive electrode and zinc as the negative electrode yields a battery cell that, when fully charged has a no-load voltage of 1.55 volts. A nickel-cadmium cell has an open-circuit voltage of 1.30 volts when fully charged. Fully charged lead–acid storage batteries have an open-circuit voltage of 2.2 volts per cell. A fully charged alkaline battery cell has an open-circuit voltage of 1.6 volts. A lithium battery yields an EMF of 3 volts per cell.

Each type of battery has characteristics that make it most suitable for particular types of applications. Carbon–zinc and alkaline batteries are dry-cell batteries that are normally not rechargeable. An alkaline battery produces approximately twice the energy of a carbon–zinc battery the same size. Alkaline batteries also have a long shelf life.

Rechargeable batteries include unsealed lead–acid batteries, their sealed cousin the gel cell, and nickel–cadmium batteries. Unsealed lead–acid batteries must be kept upright, or they can leak acid. Lead–acid and gel cell batteries can be obtained in much larger sizes than most other types of batteries. The voltage charge per cell for lead–acid and gel cell batteries over their useful discharge lives is much greater than for other types of batteries. Nickel–cadmium batteries have a relatively short shelf life, but they yield higher currents for their size than alkaline or carbon–zinc batteries do. Figure B–2 shows the discharge voltage curve for several types of batteries. Figure B–3 shows several different types of batteries.

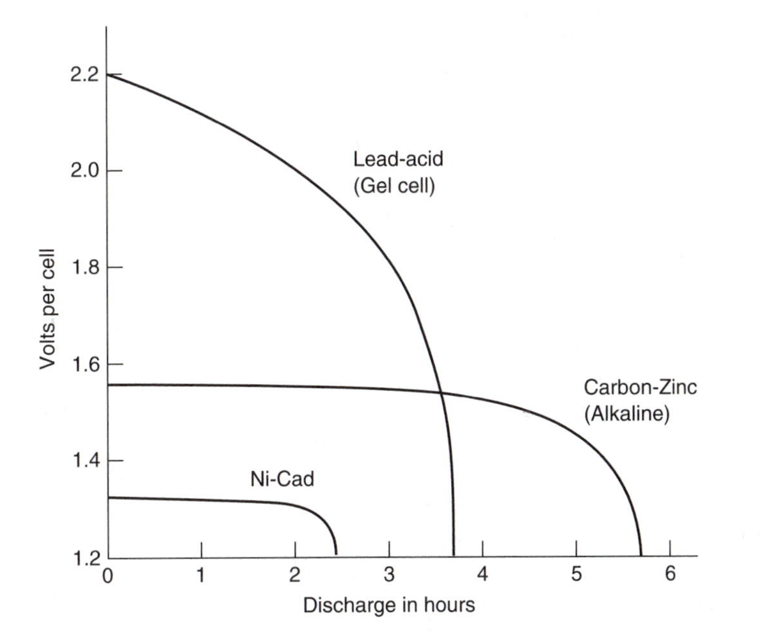

Figure B–2 Discharge Curves for Nickel-Cadmium Lead-Acid (Gel Cell), and Carbon-Zinc (Alkaline) Batteries.

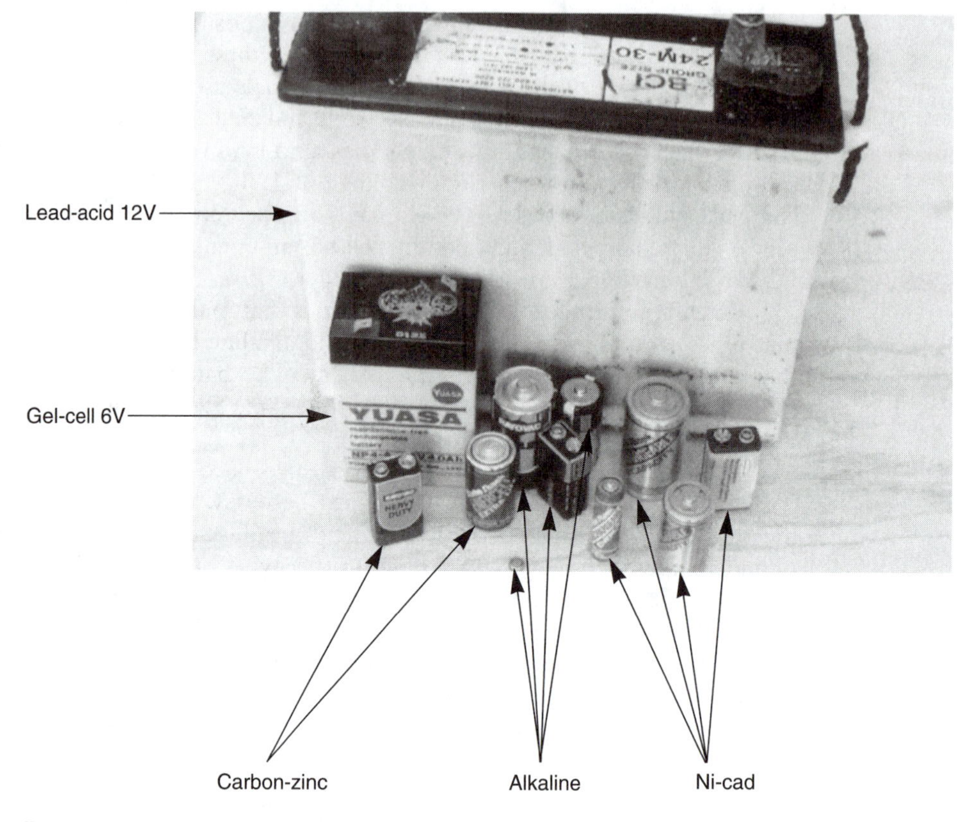

Figure B–3 Different Types of Batteries.

Nonrechargeable batteries are also called *primary cells,* while rechargeable batteries are also called *secondary cells.*

The most efficient method of mechanically generating electricity involves moving an electrical conductor through a magnetic field. The magnetic field can be generated by two bar magnets or by the two poles of a horseshoe magnet. If the conductor moves at a constant speed, so that it cuts a constant number of magnetic flux lines per second, a constant voltage is generated. It is also possible to hold the conductor stationary and move the magnet or magnetic field. Figure B–4 shows the results of moving a conductor linearly through a magnetic field and of spinning a conductor in a circular pattern through a magnetic field. It is much easier to move a conductor in a circular pattern than to move it linearly. The voltage generated in Figure B–4(a) is given by the following formula:

$$E = B \times L \times V$$

where

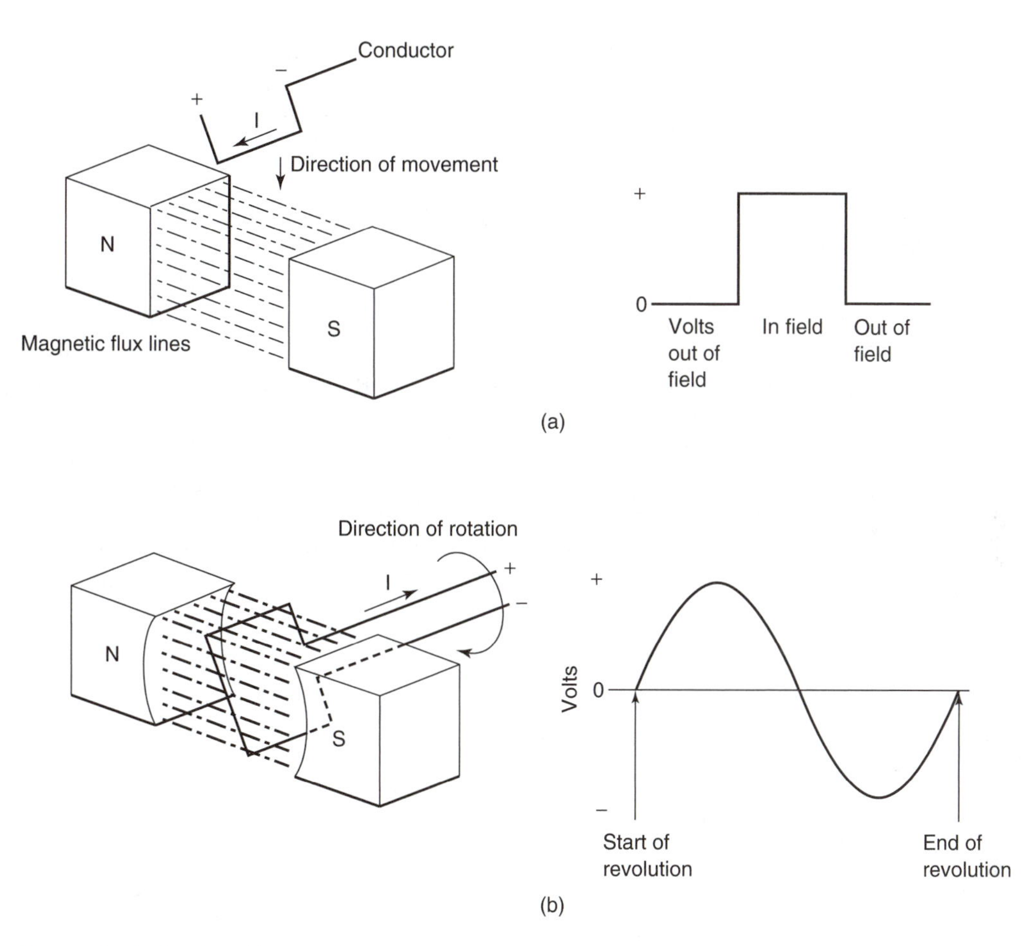

Figure B-4 Electric Generator.
(a) Linear motion of the conductor through the magnetic field. (b) Circular (rotational) motion of the conductor through the magnetic field.

E = EMF, expressed in volts

B = Magnetic flux density, expressed in webers per square meter

L = Length of the conductor interacting with the magnetic field, expressed in meters

V = Velocity of L, expressed in meters per second

The output of Figure B–4(b) is a sine wave. Here the voltage amplitude changes direction every half-revolution. Such a current, known as *alternating current (AC),* is the easiest voltage to generate with a rotating conductor. The voltage generated by batteries is called *direct current (DC).* Getting direct current from a rotating generator requires using some type of rectifier circuit, which limits the flow of electricity to one direction. The rectifier circuit works much as a check valve does in plumbing, hydraulic, or pneumatic systems. It may be mechanical—such as a split-ring commutator and brushes—or electronic—such as diodes.

Magnetism

Magnetism refers to the magnetic powers displayed by materials such as iron, nickel, cobalt, and ceramics. All magnets have magnetic lines of force around them, just as the earth itself does (see Figure B–5). If two magnets are placed next to each other, they will try to align their poles north to south; that is, each magnet's north pole is attracted to the other's south pole.

When electricity flows through a conductor, it creates magnetic lines of force around the conductor (see Figure B–6(a)). This conducting device is then called an *electromagnet*. The magnetic field can be made stronger by winding a coil onto a piece of magnetic material such as an iron rod (see Figure B–6(b)). Magnetic fields are used by mechanical electric generators, electric motors, relays, and other electromagnetic devices such as loudspeakers.

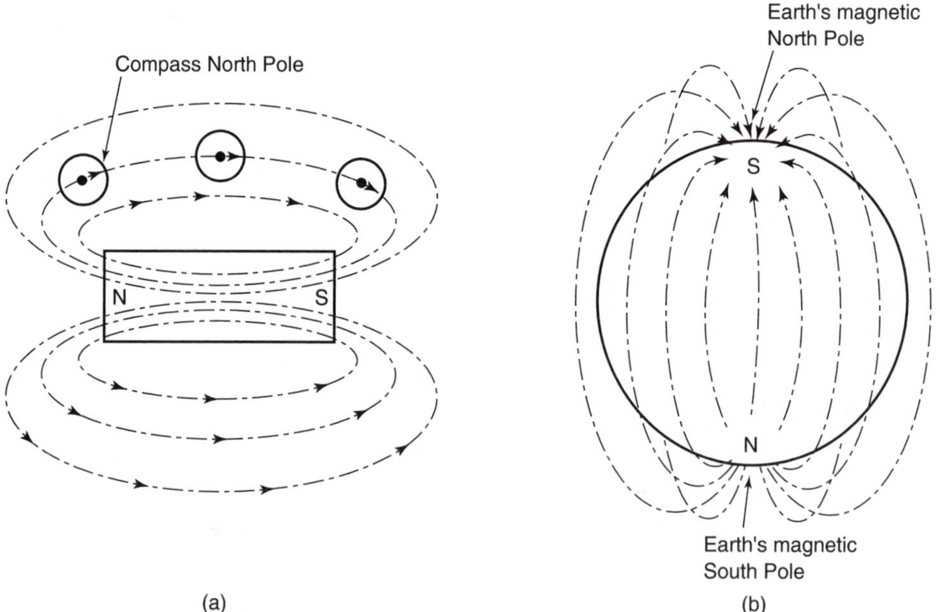

Figure B–5 Magnets.
(a) A bar magnet. (b) Earth.

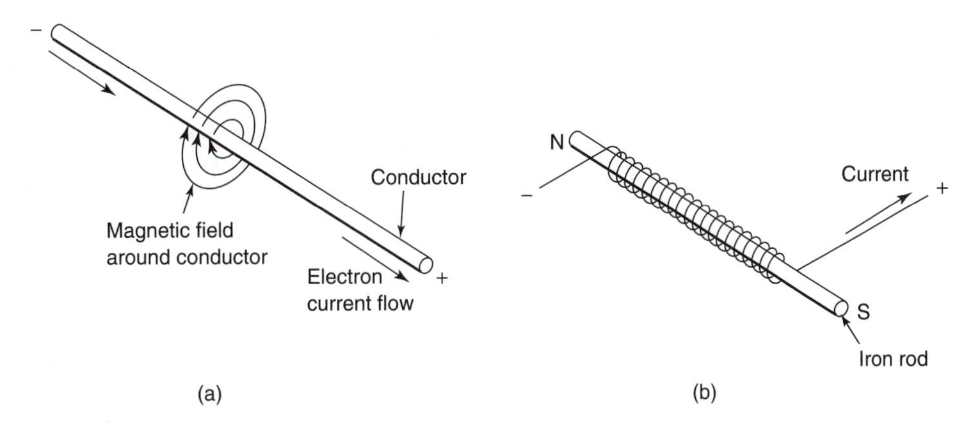

Figure B-6 Electromagnets.
(a) An electromagnet consisting of a straight conductor through which electric current passes. (b) An electromagnet consisting of a conductor coiled around a straight piece of magnetic material. Electric current passes through the conductor.

Energy Storage Devices

Electrical charge can be stored in three types of devices: batteries, capacitors, and inductors. Batteries have already been discussed.

Capacitors A capacitor or condenser is a device that can store an electrical charge, blocking the flow of direct current, while allowing alternating current to flow through it to some extent. The extent of alternating current flow permitted is directly proportional to the frequency of the alternating current and the size of the capacitor.

Capacitors are used in power supplies to help smooth out the variations in voltage levels encountered when alternating current is converted into direct current. They store energy at moments when the voltage tries to increase, and they release energy at moments when the voltage tries to decrease. Capacitors are also used in the factory environment to reduce the electrical noise from motors and switching.

Inductors Inductors or coils resist changes in the rate of flow of current and in the flow of alternating current. In both cases, the resistance of the inductor is proportional to its size and to the frequency of the alternating current. Inductors can also store energy in an electromagnetic field.

If two inductors are placed close to each other, the current flow through one of them will induce current flow in the other. This induced current is caused by the cyclical expansion and collapse of the electromagnetic field surrounding the first inductor. Remember, current can be generated in a conductor either by moving the conductor through a magnetic field or by moving the magnetic field around the conductor. The induced voltage of the second inductor is proportional to the turns ratio of the two inductors times the voltage across the first inductor. This two-inductor device is called a *transformer.* Using an iron core inside the transformer improves the efficiency of the transfer of lower-frequency AC (such as 60 Hz).

Motors

One method of putting electricity to work is to convert electrical energy into mechanical motion. Electric motors, which can be powered by AC, DC, or DC pulses, are the transducers that perform this conversion. AC motors are brushless; DC motors use brushes; DC pulse motors, such as stepper motors, do not use brushes. Very powerful electrical robots use 460-volt, three-phase AC motors. A simple two-position motor, the solenoid, can be powered by AC. A DC solenoid—or its cousin, the relay—does not require brushes. Hobbyist robots make use of small DC motors, solenoids, and relays. Figure B–7 shows some DC brush-type motors. Figure B–8 shows some DC stepper motors.

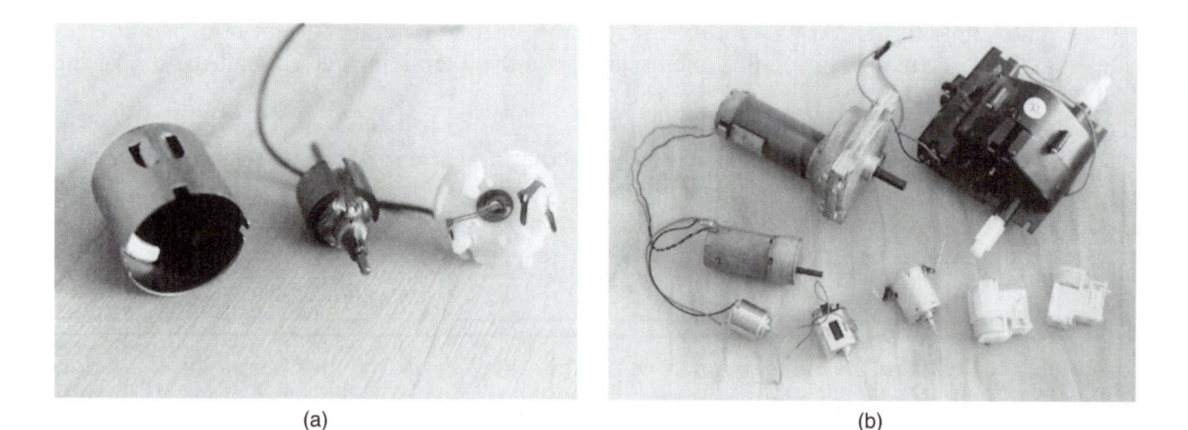

(a) (b)

Figure B–7 Some DC Brush-type Motors.
(a) From left to right, the case and a two-pole permanent magnet stator; a three-pole armature; and the two brushes (one of which is damaged). (b) An assortment of DC motors and gearboxes.

(a) (b)

Figure B–8 DC Stepper Motors.
(a) Left, the permanent magnet armature for a stepper motor, which is brushless; right, the stator windings of the stepper motor. (b) Several DC stepper motors.

Relays and solenoids are the simplest of motors. Springs hold them in their unenergized state, and an electromagnetic field moves them to their energized state. Thus, solenoids and relays are two-state or binary devices. In the case of solenoids, the two-state mechanism is used to move some attachment that is connected to the movable part of the solenoid. In the case of relays, the device is used to open and close electrical connections. Figure B–9 shows several relays.

Figure B–9 Relays.

ELECTRONICS

Electronics is the field of electricity that deals with controlling the behavior of electrons in devices to perform useful tasks. It is widely associated with the use of vacuum tubes, transistors, and other solid-state electronic devices. Electronic devices used in the computer include the integrated circuit, the microprocessor, and the computer memory chip.

Vacuum tubes and transistors are two control devices. A small amount of current at the control element of a vacuum tube or transistor can exert considerable control at the device's output. The control element of the vacuum tube is the control grid; the control element of a transistor is the base. Figure B–10 shows a vacuum tube and a transistor set up to amplify a small electronic signal. As the signal at the control element moves in the positive direction, the signal at the output moves in the negative direction. For this reason, these amplifiers are said to invert the signal as they amplify it. Circuits that can amplify a signal over a range of values are called *analog circuits.* Telephones, radios, and televisions use analog signals and need analog circuits to process these signals.

Vacuum tubes and transistors can be operated in a mode in which they are either all the way on or all the way off. In fact, this is the way a computer would use them. Running in on/off mode enables electronic circuits to represent binary values of 1 and 0. Circuits that use the two states of on and off exclusively are known as *digital circuits.*

Integrated circuits, which are made up of many different transistor circuits, can be analog or digital circuits. The digital integrated circuit is the heart of a modern computer. Some simple circuit devices used to make decisions in computers include

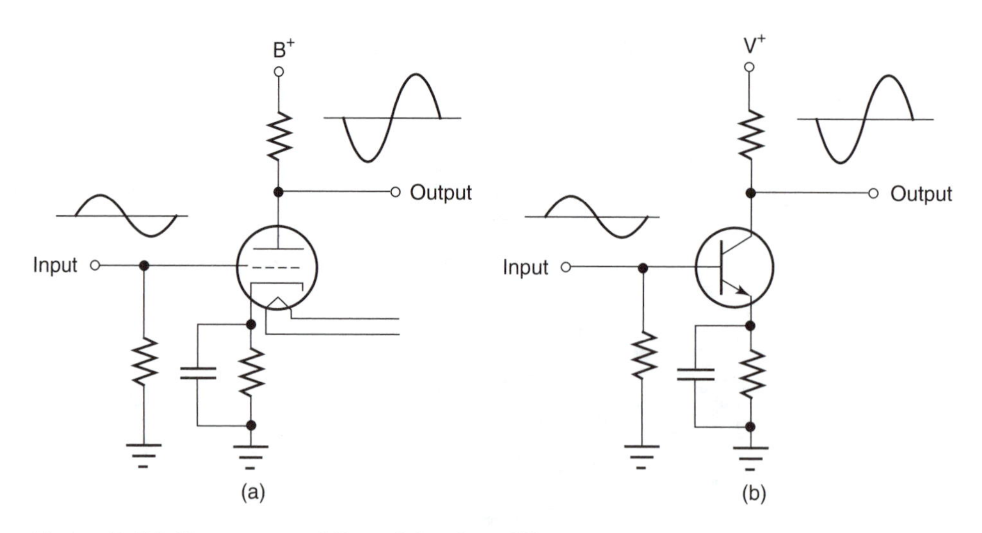

Figure B–10 Vacuum and Transistor Amplifiers.
(a) Schematic drawing of a vacuum tube amplifier. (b) Schematic drawing of a transistor amplifier.

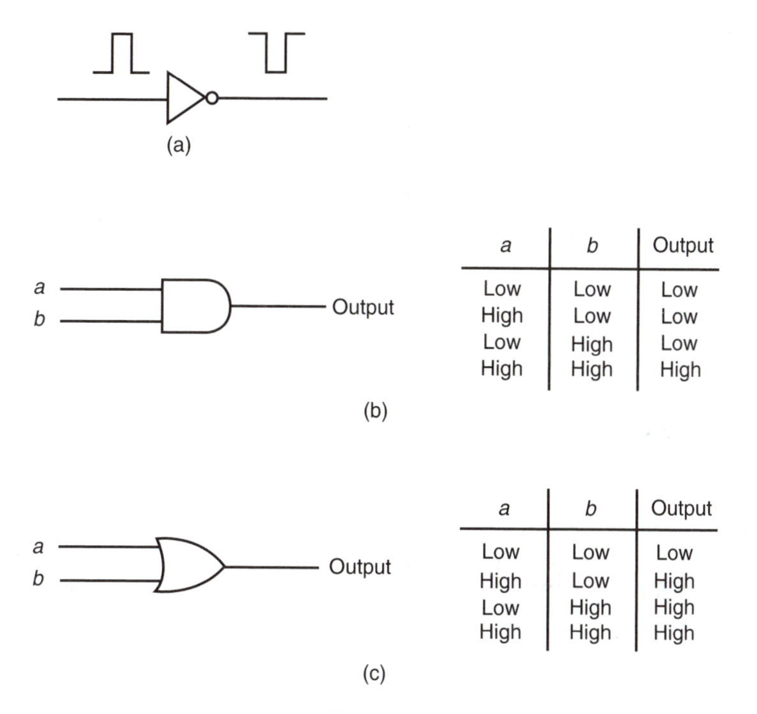

a	b	Output
Low	Low	Low
High	Low	Low
Low	High	Low
High	High	High

a	b	Output
Low	Low	Low
High	Low	High
Low	High	High
High	High	High

Figure B–11 Simple Circuit Devices.
(a) Diagram of an invertor. (b) Diagram of an AND gate, with its associated truth table. (c) Diagram of an OR gate, with its associated truth table.

the *invertor*, the *AND gate*, and the *OR gate*. The invertor changes a 1 to a 0 and a 0 to a 1. The AND gate requires that both of its inputs be 1s before the output becomes 1. The OR gate output becomes 1 if either or both of its inputs are 1. Figure B–11 shows these gates, along with the associated truth tables of their inputs and outputs.

COMPUTER FUNDAMENTALS

The present-day electronic computer is a dumb machine that has three redeeming qualities: it follows directions very faithfully; it does so very quickly; and it has built-in mathematical capabilities. The controller for a modern robot uses an electronic computer.

Numbers, Math, and Data Representation

The modern digital electronic computer, the programmable logic controller, and the robot controller represent everything internally as binary numbers. A person must understand these representations in order to design, repair, or program these controllers.

English	Uninary	Mayan	Roman	Decimal	Binary	Octal	Hexadecimal	Pears
Zero		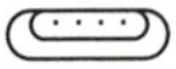 (····)		0	0000	0	0	
One	I	·	I	1	0001	1	1	
Two	II	··	II	2	0010	2	2	
Three	III	···	III	3	0011	3	3	plus
Four	IIII	····	IIII or IV	4	0100	4	4	
Five	𝍸	—	V	5	0101	5	5	
Six	𝍸 I	· —	VI	6	0110	6	6	equals
Seven	𝍸 II	·· —	VII	7	0111	7	7	
Eight	𝍸 III	··· —	VIII	8	1000	10	8	
Nine	𝍸 IIII	···· —	VIIII or IX	9	1001	11	9	
Ten	𝍸 𝍸	— —	X	10	1010	12	A	
Eleven	𝍸 𝍸 I	· — —	XI	11	1011	13	B	
Twelve	𝍸 𝍸 II	·· — —	XII	12	1100	14	C	
Thirteen	𝍸 𝍸 III	··· — —	XIII	13	1101	15	D	
Fourteen	𝍸 𝍸 IIII	···· — —	XIIII or XIV	14	1110	16	E	
Fifteen	𝍸 𝍸 𝍸	— — —	XV	15	1111	17	F	
Twenty		— — — —	XX	20	10100	24	14	
Fifty			L	50	110010			
One Hundred			C	100				
Five Hundred			D	500				
One Thousand			M	1000				

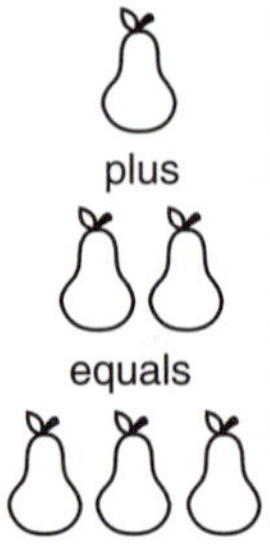

Figure B–12 Different Human Numbering Systems.

Most present-day adults are accustomed to representing numbers and doing mathematics in the base-10 (or decimal) numbering system. However, this is not the only numbering system used by human societies, nor is it the numbering system that children first use when learning about numbers (see Figure B–12). Most people are aware of the decimal numbers, Roman numbers, and written numbers in their own language. In addition, most persons have used unitary numbers or tally marks. Humans who work closely with the electronic computer have some knowledge of the binary, octal, and hexadecimal numbering systems.

Human Math Learning The present-day decimal (base-10) numbering system, the system most widely used by humans, was developed by the Arabs. In recognition of this, the symbols used to represent decimal values (0, 1, 2, 3, 4, 5, 6, 7, 8, and 9) are called Arabic numerals. However, humans are not born knowing the decimal numbering system; numbers and mathematics must be learned.

In kindergarten, children learn the rudiments of mathematics. They start by learning to recognize objects of like shapes, such as squares, circles, triangles, and rectangles. Then they learn to determine which of two groups of objects is more numerous by matching up (or pairing) objects from each group—whichever group has leftover objects is the larger group. Thus the first step in mathematics involves learning to recognize differences in the size of numbers.

The next step is to learn to recognize an Arabic numeral by relating the numeral to a group of objects. This is reinforced by learning to circle the quantity of objects equal to an Arabic numeral. Then children learn to relate a number to a numeral. Next comes the hard concept of the number zero.

The order of numbers can be reinforced by connecting dots in the proper order to draw pictures. The next step might be to learn to pick out the number of oranges that are mixed in with another fruit. From there, children might move on to learning the concepts of which number is more, which is less, and which comes next in a sequence of numbers. Next comes the idea of physical size, such as shorter and longer, and finally how to measure size in units. All of this foundation in numbers must precede the basic arithmetic operations.

The arithmetic operations of addition and subtraction are introduced through pictorial addition and subtraction, using the idea of "how many objects there are in all" and "how many objects are left." The numerals and mathematical operators are printed below each picture. By this point, the numerals 0 through 12 have been introduced, and children can practice telling what hour it is on the traditional clock face. Finally, children are introduced to counting cents, using pennies, nickels, and dimes.

In first grade, children continue working with the preceding concepts and learn to print the Arabic numerals and mathematical operators. They learn to write math problem answers by using picture examples. Next comes performing addition and subtraction entirely in Arabic numerals. Finally children learn to use numbers with more than one digit and to recognize values up to the size of 100.

Throughout the remainder of grade school, children continue to develop their mathematical ability, learning multiplication, division, fractions, and even symbolic mathematics (algebra). Simultaneously with their acquisition of mathematical skills, children develop their language abilities, their memories, and their ability to manipulate information. Learning in all of these areas continues into a person's adulthood and may never cease.

Electronic Math and Data Representation The present-day electronic computer does its mathematics in the binary numbering system, in the decimal numbering system, or in both. Computer programmers also use the octal and hexadecimal numbering systems for entering and examining information in the electronic computer.

Octal and hexadecimal numbering systems offer a means of representing binary numbers in groups of three or four digits at a time, which makes the numbers easier to remember and saves paper when the contents of an electronic computer's memory are printed out. The early electromechanical and electronic computers performed their mathematical operations using the base-10 numbering system. Relays and vacuum tubes are binary devices because they work best in one of two states such as open and on, or closed and off. If four such binary devices are used, it is possible to represent decimal values from 0 to 9. The following listing shows how a decimal value can be represented in four binary digit positions:

Decimal	Binary
0	0000
1	0001
2	0010
3	0011
4	0100
5	0101
6	0110
7	0111
8	1000
9	1001

This system of representing decimal numbers by four binary digits, called Binary Coded Decimal, is used by most electronic pocket calculators. The pocket calculator uses firmware (factory built-in programs) to do addition, subtraction, multiplication, division, and many other forms of arithmetic.

Electronic computers come with the built-in ability to do addition (in binary, decimal, or both, depending on the type of computer), and most of them also have the built-in ability to do subtraction. However, some electronic computers have to learn to perform multiplication and division through the use of software programs.

All information stored in an electronic computer is expressed in ones and zeros. That is, the machine language used in a computer consists entirely of binary numbers. Consequently, the computer cannot determine whether it is looking at mathematical numbers or at nonmathematical symbols, such as letters, label numbers, and special characters, without being told what it is looking at. Mathematical numbers (data) can be represented in many computers as either Binary Coded Decimal (BCD) or pure binary numbers (see Figure B–13). As Figure B–13 shows, storing numbers in a computer in pure binary form takes fewer binary digits than does storing them in BCD.

In the base-10 numbering system, the rightmost digit in a multidigit number represents the number of units or ones, and the digit next to it represents the number of tens; then come the hundreds, the thousands, the ten thousands, the hundred thousands, the millions, and so forth. You move from the value of one digit position to the next higher digit position by multiplying by 10, the base of the system. In exponential notation, the units digit in base 10 carries the position value of 10^0; the tens digit has the position value of 10^1; the hundreds digit has the position value of 10^2; and so on.

In the binary (base-2) numbering system, the rightmost digit still represents the units or ones, but the next digit represents the twos, the next digit represents

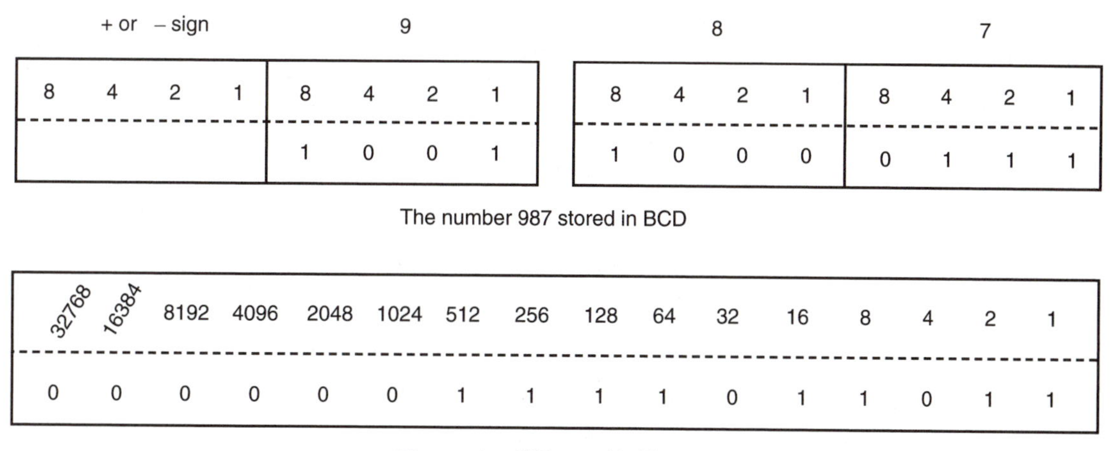

Figure B–13 Numbers Stored in BCD and in Binary.

the fours, the next digit represents the eights, and so on. You move from the value of one digit position to the next higher digit position by multiplying by 2, the base of the system. Figure B–14 shows the place value of a digit in the base-10, base-2, base-8, and base-16 numbering systems.

Within any numbering system of base 2 or higher, the largest-value digit allowed in any digit position is equal to the base number minus 1. Thus, in base 2, the largest permissible digit value is 2 – 1 = 1. Consequently, binary digits can only have values of 0 and 1. Similarly, in base 3, the largest permissible digit value is 3 – 1 = 2. In base 10, the largest permissible digit value is 10 – 1 = 9. Therefore, decimal digits can have values of from 0 to 9. In base 8, the largest permissible digit value is 8 – 1 = 7. And in base 16, the largest permissible digit value is 16 – 1 = 15; but since 15 occupies two digit positions, in base 16 it has to be represented by a single character. Therefore, it is called F. Figure B–15 shows how to count from 0 to 15 in bases 2, 8, 10, and 16.

Because computer programmers found binary numbers—ones and zeros—difficult to keep track of, they came up with the idea of grouping binary digits into sets of three and calling it an octal (base-8) numbering system. For the octal values 0 to 7, it takes exactly three binary digits to handle all eight values (in binary, 7 is 111). Printing out computer memory dumps and writing, entering, and reading binary values as octal numbers saves two-thirds of the paper and effort that would have to be expended for a pure binary system.

Another group of programmers found that they could save even more paper and effort by writing binary numbers four digits at a time. This base-16 numbering system was called hexadecimal. In it, four binary digits or bits are needed to represent all sixteen values from 0 to 15. The hexadecimal values 0 to 9 use the

10^5	10^4	10^3	10^2	10^1	10^0
100,000	10,000	1,000	100	10	1

Place values for the decimal numbering system

2^8	2^7	2^6	2^5	2^4	2^3	2^2	2^1	2^0
256	128	64	32	16	8	4	2	1

Place values for the binary numbering system

8^5	8^4	8^3	8^2	8^1	8^0
32,768	4,096	512	64	8	1

Place values for the octal numbering system

16^4	16^3	16^2	16^1	16^0
65,536	4,096	256	16	1

Place values for the hexadecimal numbering system

Figure B–14 Place Values of Digits in Different Numbering Systems.

same symbols as in the decimal system, but single-character symbols must be used for the hexadecimal equivalents of 10 through 15. Ultimately, the letters A through F were adopted for this purpose. Similarly, base 36 would use the symbols 0 through 9 and A through Z as equivalents for the decimal system numbers 0 through 35.

One problem encountered by programmers (and computers) is how to convert numbers from one numbering system into another. Converting base-10 numbers into base-2 numbers is essential if the numbers are to be stored in the computer in binary code. Therefore, programmers need to know how to convert from base 10 into base 2. One method of doing this is to divide the base-10 number repeatedly by the base into which the number is to be converted (in this case, 2), each time saving the remainder from the division, until a quotient of zero is reached. The remainders make up the new number in the desired numbering system, with the first remainder occupying the units position, the next remainder occupying the next higher digit position, and so forth. Figure B–16 shows this method of converting the decimal number 465 into the binary number 111010001. Although this

Hexadecimal (place values)	Decimal (place values)		Octal (place values)		Binary (place values)			
1	10	1	8	1	8	4	2	1
0		0	0	0	0	0	0	0
1		1	0	1	0	0	0	1
2		2	0	2	0	0	1	0
3		3	0	3	0	0	1	1
4		4	0	4	0	1	0	0
5		5	0	5	0	1	0	1
6		6	0	6	0	1	1	0
7		7	0	7	0	1	1	1
8		8	1	0	1	0	0	0
9		9	1	1	1	0	0	1
A	1	0	1	2	1	0	1	0
B	1	1	1	3	1	0	1	1
C	1	2	1	4	1	1	0	0
D	1	3	1	5	1	1	0	1
E	1	4	1	6	1	1	1	0
F	1	5	1	7	1	1	1	1

Figure B–15 Counting from 0 to 15 in Different Numbering Systems.

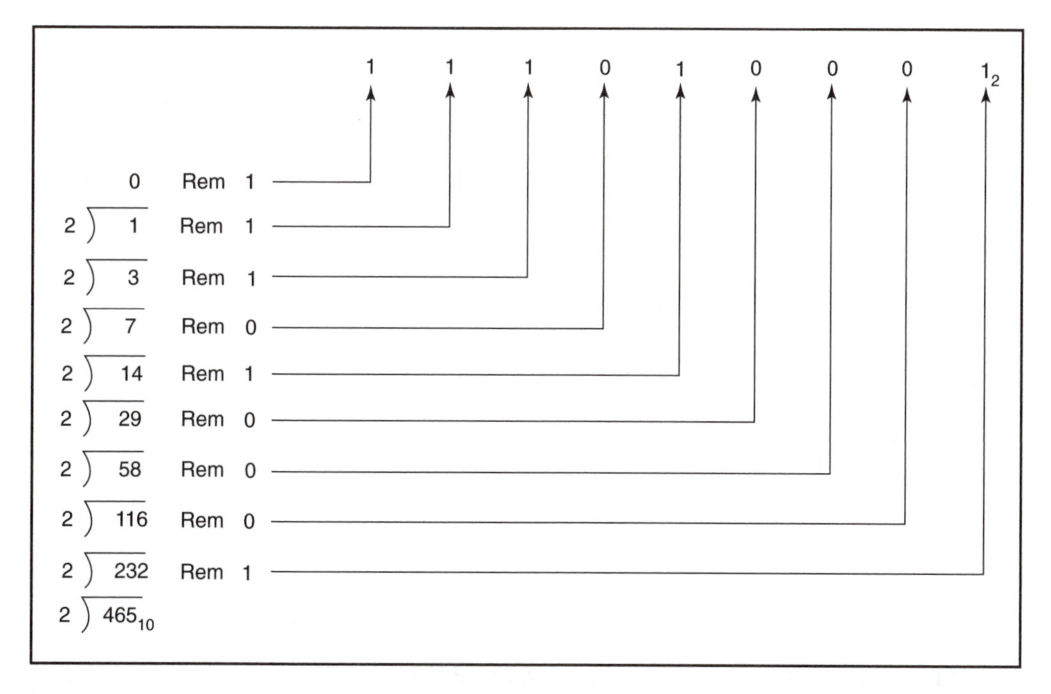

Figure B–16 Division Method of Converting Decimal Numbers into Binary Numbers.

division conversion technique is simple, you may require the help of a pocket calculator to do the division in the higher-number bases. The biggest problem with the division method is that people often forget to do the final division that gives the result of zero and some remainder.

Another way to convert decimal numbers into binary numbers is to write out the values of the binary digit positions from right to left, starting at the units position, until a value is found that is bigger than the decimal number. When this happens, place a zero in that digit position; then subtract the next lower digit position value from the decimal number and place a 1 in that binary digit position. Next, check to see if the remainder is equal to or larger than the next lower binary digit position value. If it is, subtract that digit value from the decimal, and place a 1 in that binary digit position. If the decimal remainder is less than that binary digit value, place a zero in that binary digit position and go look at the next lower binary digit position. Repeat the procedure until the decimal number has been reduced to zero. Figure B–17 shows the use of this subtraction method for converting the decimal number 465 into the binary number 111010001.

Figure B–18 shows the conversion of the decimal number 465 into the octal number 721 by the division method; similar conversions of the decimal numbers 777 and 1,000 are also provided. Figure B–19 shows the conversion of these numbers by the subtraction method. Notice that with base 8 you may have to subtract a column value as many as seven times. Thus, the subtraction method is not as sim-

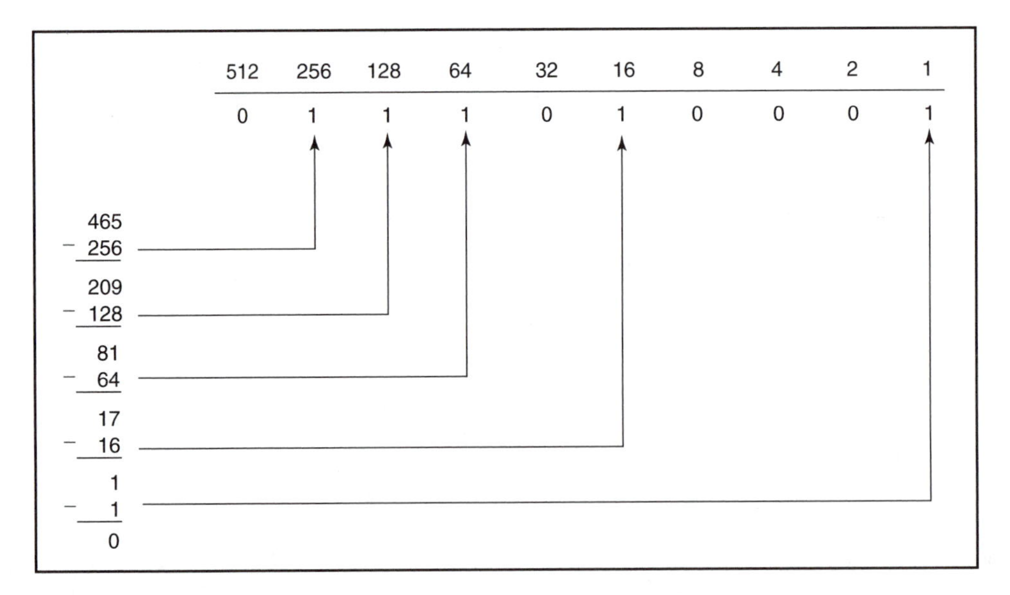

Figure B–17 Subtraction Method of Converting Decimal Numbers into Binary Numbers.

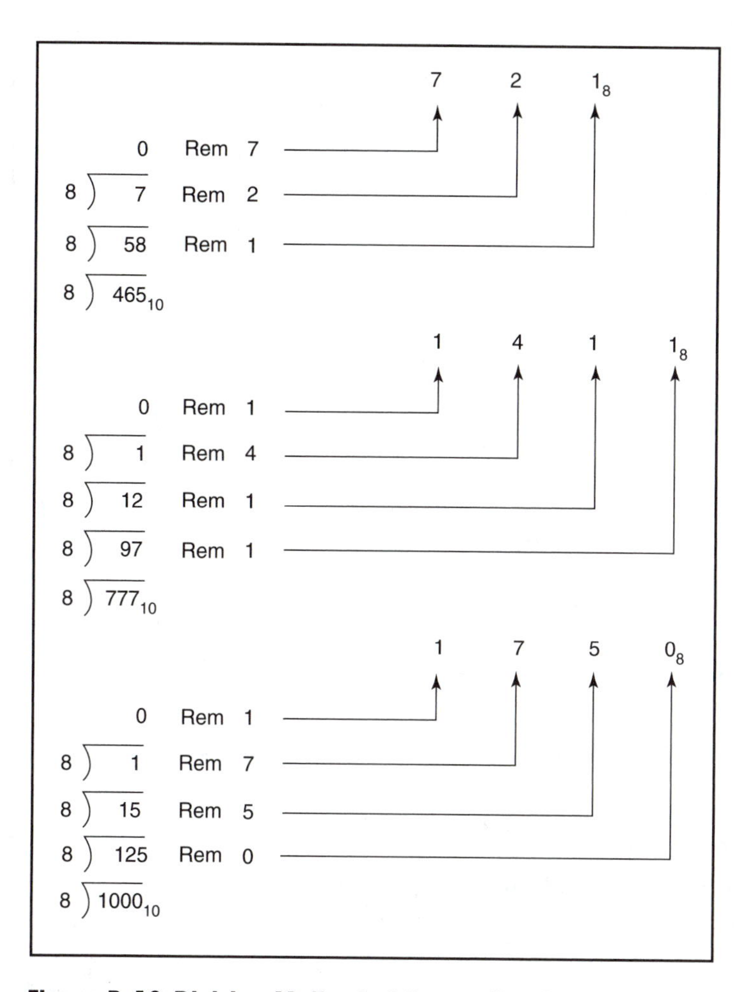

Figure B–18 Division Method of Converting Decimal Numbers into Octal Numbers.

ple as it was in the case of base-10 to base-2 conversions. Figure B–20 shows the conversion of decimal numbers into hexadecimal by the division method, and Figure B–21 shows how difficult the subtraction method becomes for conversions between these bases.

Programmers and computers need to convert binary numbers in the computer's memory into decimal numbers for use in reports and other documents that are to be read by humans. One method for doing this is to evaluate the weight of each digit position and then add the results together to get the total value. Figure B–22 shows how the total weight of the digit positions gives the total value for a decimal number. Figures B–23 through B–25 demonstrate this technique for converting from bases 2, 8, and 16 to decimal, respectively.

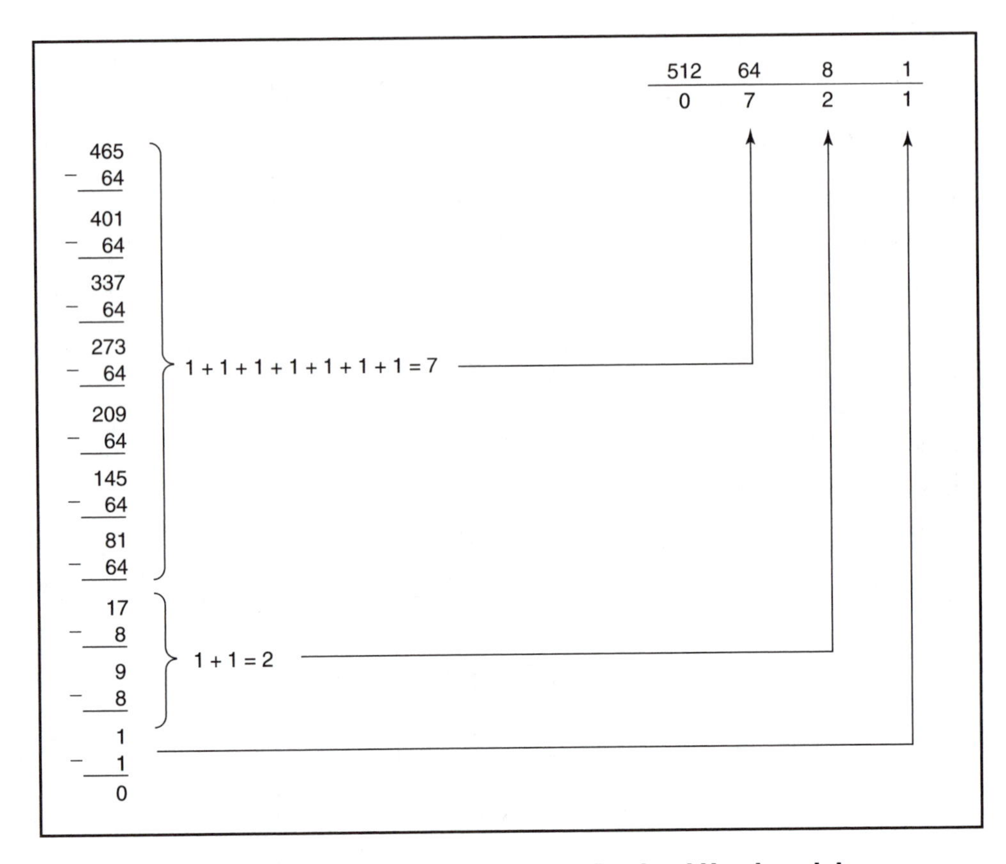

Figure B-19 Subtraction Method of Converting Decimal Numbers into Octal Numbers.

Another mechanical method of number conversion from any number base to base 10 is the X Dabble method. Basically this method uses the following steps:

1. Starting with the leftmost digit, determine the value of that digit in base 10.
2. Multiply that value by the base from which the number is being converted; place the result under the next digit of the number.
3. Add this column.
4. If this is the last column, the sum of the addition is equal to the decimal value of the number. Otherwise, go back to step 2.

Figure B–26 shows the generalized method of X Dabble. Figure B–27 shows X Dabble in the specific cases of binary conversion (Double Dabble), octal conversion (Octal Dabble), and hexadecimal conversion (Hexadecimal Dabble).

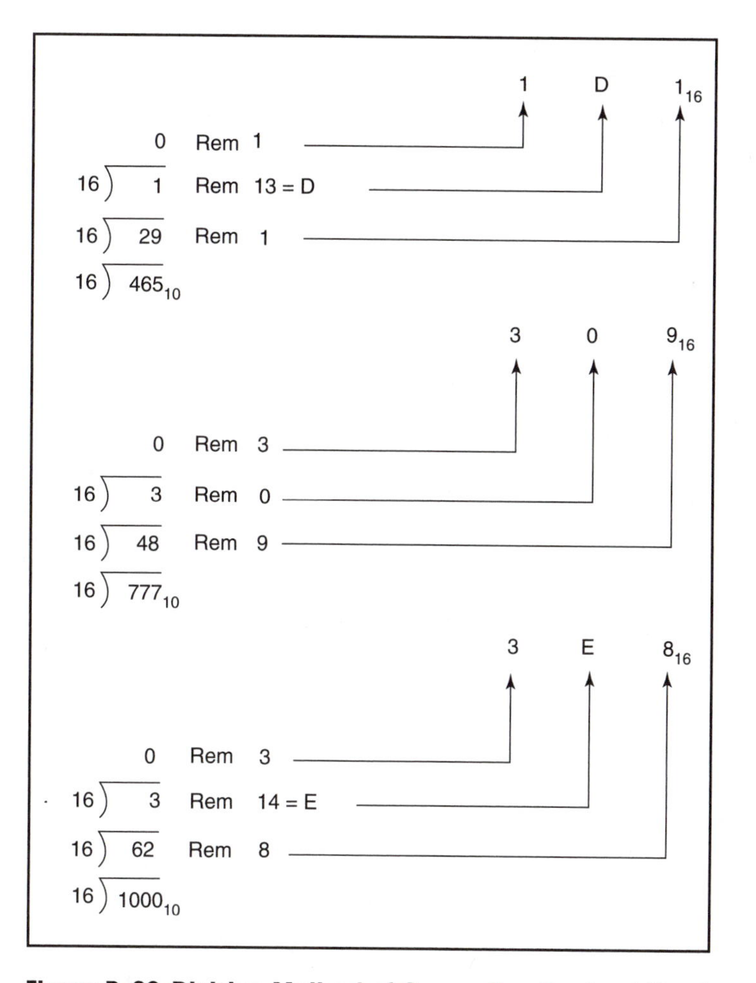

Figure B–20 Division Method of Converting Decimal Numbers into Hexadecimal Numbers.

The electronic computer stores data in various sizes and forms. Figure B–28 shows some of the storage sizes used. The smallest unit of storage in the electronic computer is the *bit* (short for binary digit), which can store a value of either 1 or 0. This information can be used to represent a numerical value of 1 or 0, a condition of true or false, a yes or no, or even a ready or not ready. When groups of four bits each are used to store a data item, they are known as *nibbles*. A nibble is large enough to hold one decimal digit value from 0 to 9. Most pocket calculators store the numbers they calculate in nibble units.

Eight bits, or two nibbles (also known as a *byte*), are generally required to store a character of information. A character (symbol) may be A through Z, a through z, 0 through 9, or some special character such as !, #, or &. The symbols

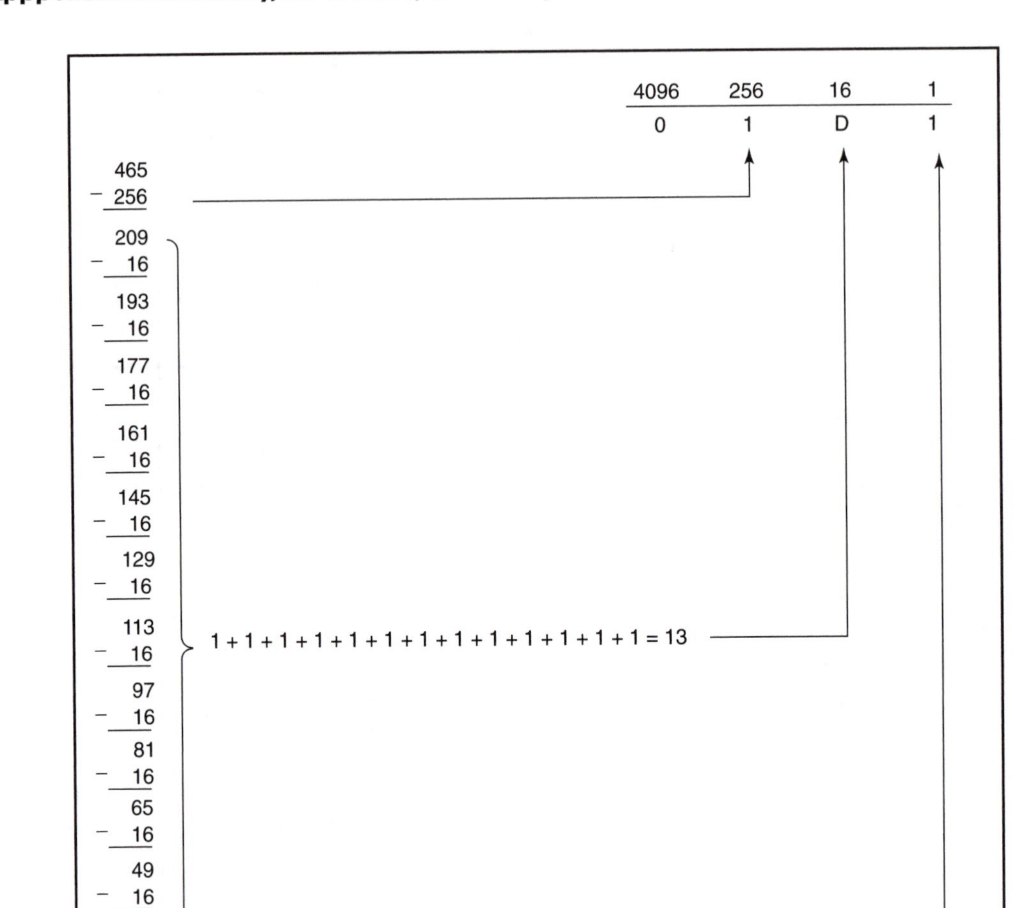

Figure B-21 Subtraction Method of Converting Decimal Numbers into Hexadecimal Numbers.

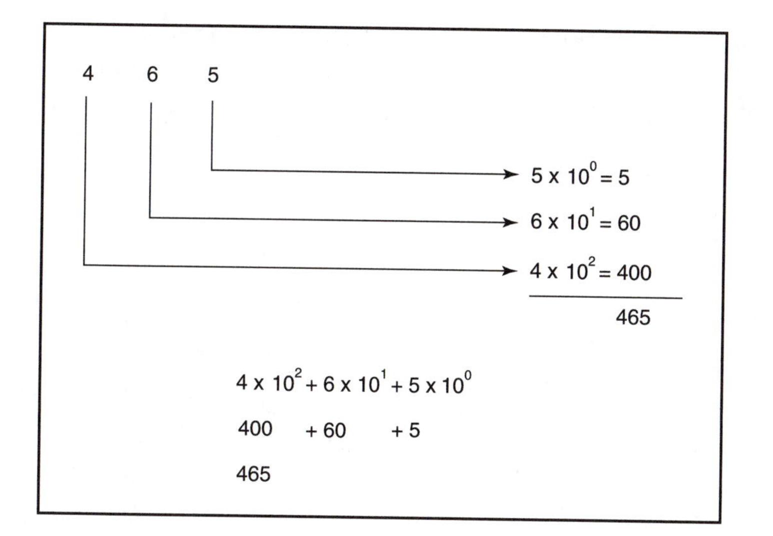

Figure B-22 Positional Value Method of Decimal to Decimal Conversion.

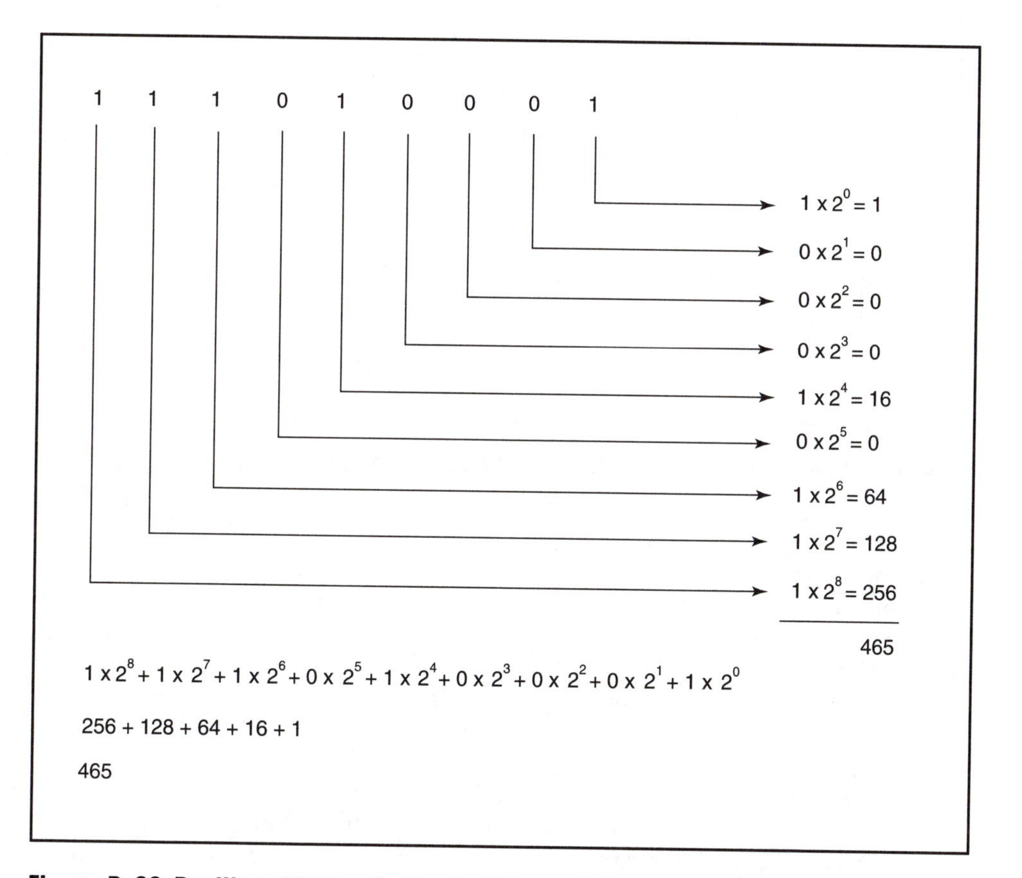

Figure B-23 Positional Value Method of Binary to Decimal Conversion.

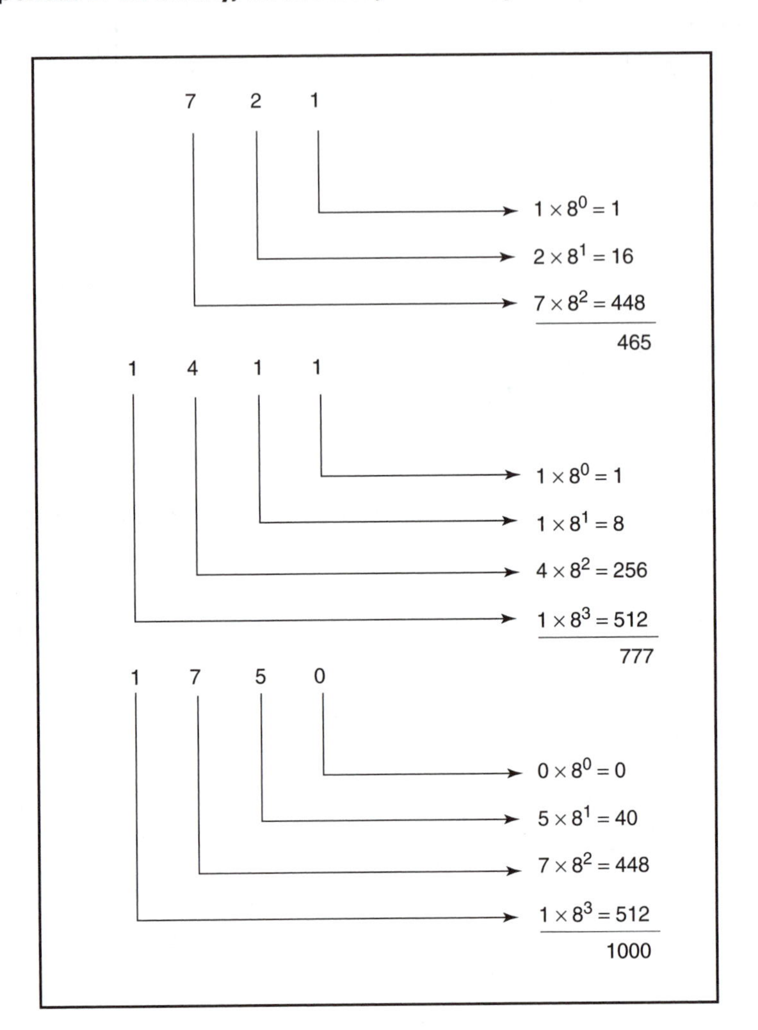

Figure B–24 Positional Value Method of Octal to Decimal Conversion.

0 through 9 should not be confused with the numeric values 0 through 9. In most computers, the byte is the smallest unit of memory that is given a unique address.

The storage of symbols or characters within the electronic computer must be done as patterns of ones and zeros. Figure B–29 shows three different methods of representing symbols in a computer. The alphanumeric Binary Coded Decimal (BCD) method uses six bits and can represent up to 64 different symbols. This was the first popular method of representing symbols in computers. The Extended Binary Coded Decimal Interchange Code (EBCDIC) method uses eight bits (one byte) per symbol and can represent up to 256 different symbols. The American Standard Code for Information Interchange (ASCII) uses seven bits per symbol, plus a check

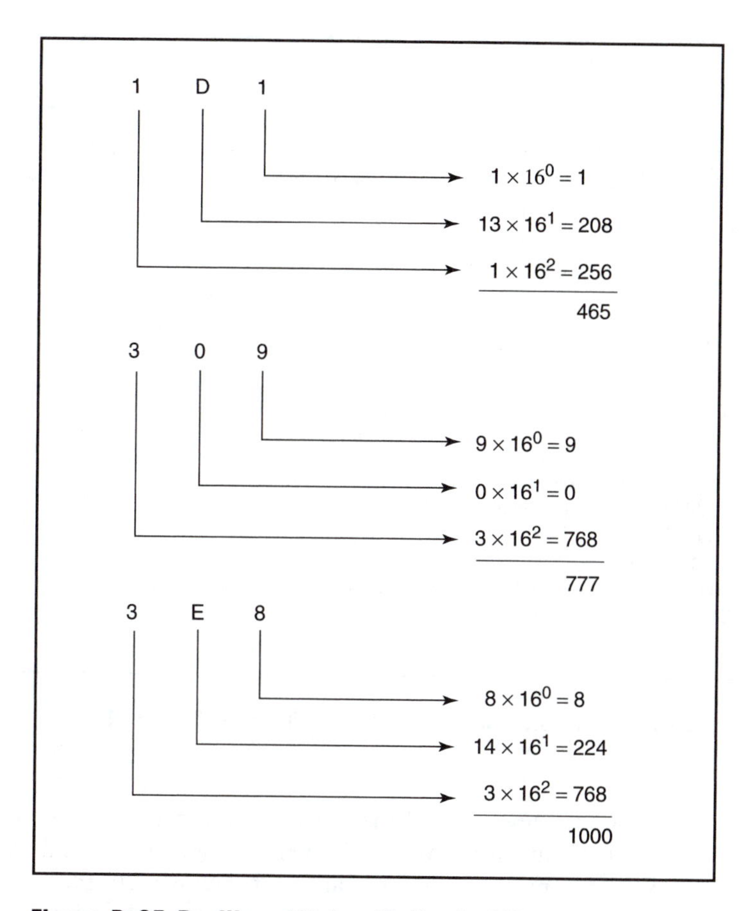

Figure B–25 Positional Value Method of Hexadecimal to Decimal Conversion.

bit (parity bit); it can represent up to 128 different symbols. ASCII is popular for sending information between a computer and a terminal or another computer.

The Electronic Computer

The electronic computer first emerged in the mid-1940s, although the principles of automatic computing were discovered by Charles Babbage in the early 1800s. The computer's design, while continuously undergoing improvements, is based on the dominant or logical processor of the human brain. Figure B–30 shows a block diagram of the electronic computer.

The central part of the computer is its main memory or primary storage. All of the information being entered into, output from, or worked on by the computer goes through main memory. Main memory also holds any instructions currently processed by the computer.

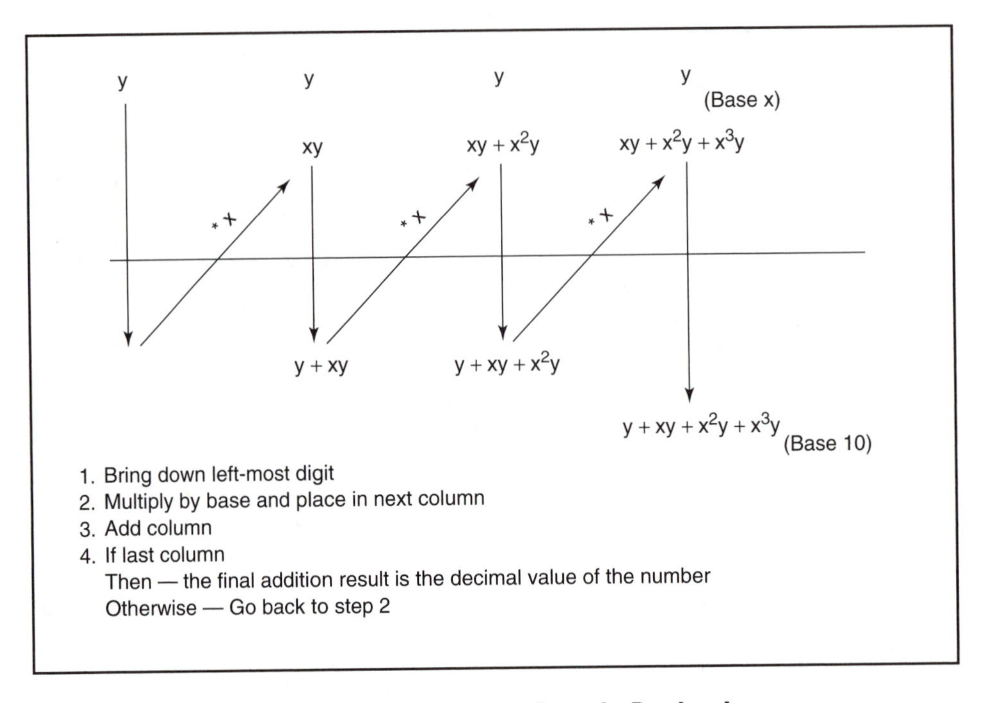

Figure B-26 Conversion from Any Number Base to Decimal.

The control unit directs all of the other parts of the computer. It receives an instruction from main memory, determines its meaning, and issues directions to the other parts of the computer on how to carry out the instruction.

The arithmetic logic unit handles the mathematics and decision-making steps for the computer. The latter includes comparing different values and performing logical operations such as and, or, and not.

The input block is needed to get information from the outside world into the computer. Common input devices for computers include keyboards, magnetic tape, magnetic disks, cards, switches, and bar code readers. Vision and voice inputs are also being developed.

The output block is needed to get information from the computer to the outside world. Output devices include displays, printers, cards, tapes, disks, voice output, lights, relays, and motors.

Files provide long-term storage for the computer. A file device must be both an input device and an output device. Common file devices include cards, tapes, and disks.

The telecommunications block of the computer enables it to communicate with other computers and with remote input and output devices.

The power supply block supplies the power requirements of all the other blocks. If the power supply is not working correctly, it can cause the other blocks to work improperly. Such malfunctions can mask a power supply problem.

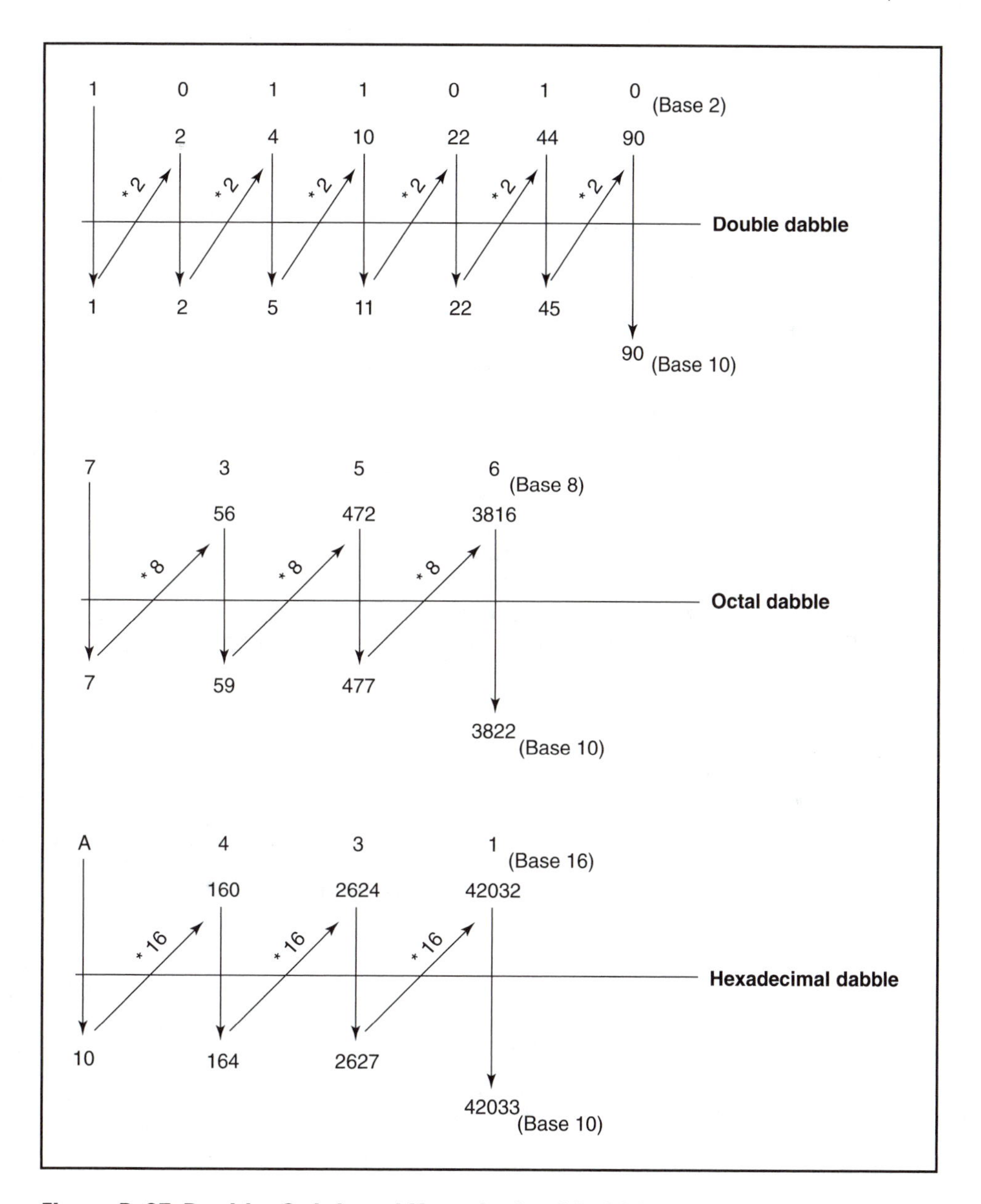

Figure B-27 Double, Octal, and Hexadecimal Dabble.

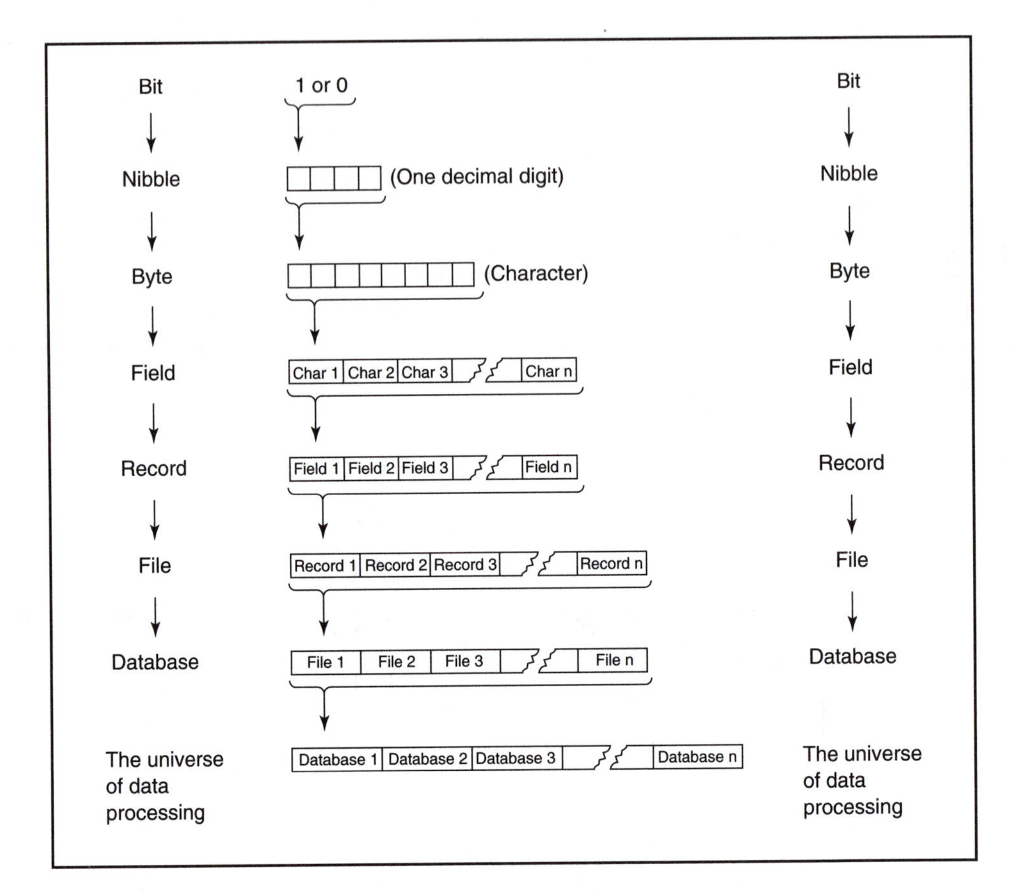

Figure B-28 Storage Units of the Electronic Computer.

Dec	Hex	ASCII (8th bit = 0)	EBCDIC	BCD
0	0	NULL	NULL	Blank
1	1	SOM		1
2	2	STX		2
3	3	ETX		3
4	4	EOT	PF	4
5	5	ENQ	HT	5
6	6	ACK	LC	6
7	7	BELL	DEL	7
8	8	BS		8
9	9	HT		9
10	A	LF		0
11	B	VT		#
12	C	FF		@
13	D	CR		:
14	E	SO		>
15	F	S1		√
16	10	DLE		b̸
17	11	DC1		/
18	12	DC2		S
19	13	DC3		T
20	14	DC4	RES	U
21	15	NAK	NL	V
22	16	SYN	BS	W
23	17	ETB	IDL	X
24	18	CAN		Y
25	19	EM		Z
26	1A	SAB		≠
27	1B	ESC		,
28	1C	FS		%
29	1D	GS		⁀
30	1E	RS		\
31	1F	US		⧻

Dec	Hex	ASCII (8th bit = 0)	EBCDIC	BCD
32	20	SP		–
33	21	!		J
34	22	"		K
35	23	#		L
36	24	$	BYP	M
37	25	%	LF	N
38	26	&	EOB	O
39	27	'	PRE	P
40	28	(		Q
41	29	)		R
42	2A	*		!
43	2B	+		$
44	2C	,		*
45	2D	–		]
46	2E	.		;
47	2F	/		Δ
48	30	0		&
49	31	1		A
50	32	2		B
51	33	3		C
52	34	4	PN	D
53	35	5	RS	E
54	36	6	UC	F
55	37	7	EOT	G
56	38	8		H
57	39	9		I
58	3A	:		?
59	3B	;		.
60	3C	<		□
61	3D	=		[
62	3E	>		<
63	3F	?		‡

Dec	Hex	ASCII (8th bit = 0)	EBCDIC
64	40	@	Blank
65	41	A	
66	42	B	
67	43	C	
68	44	D	
69	45	E	
70	46	F	
71	47	G	
72	48	H	
73	49	I	
74	4A	J	¢
75	4B	K	
76	4C	L	<
77	4D	M	(
78	4E	N	+
79	4F	O	|
80	50	P	%
81	51	Q	
82	52	R	
83	53	S	
84	54	T	
85	55	U	
85	56	V	
87	57	W	
88	58	X	
89	59	Y	
90	5A	Z	!
91	5B	[	$
92	5C	\	*
93	5D	]	)
94	5E	^	.
95	5F	–	,

Dec	Hex	ASCII (8th bit = 0)	EBCDIC
96	60	\	–
97	61	a	/
98	62	b	
99	63	c	
100	64	d	
101	65	e	
102	66	f	
103	67	g	
104	68	h	
105	69	i	
106	6A	j	
107	6B	k	.
108	6C	l	%
109	6D	m	–
110	6E	n	>
111	6F	o	?
112	70	p	
113	71	q	
114	72	r	
115	73	s	
116	74	t	
117	75	u	
118	76	v	
119	77	w	
120	78	x	
121	79	y	
122	7A	z	:
123	7B	{	#
124	7C	¦	@
125	7D	}	'
126	7E	~	=
127	7F	DEL	"

Figure B-29 Symbol Representation in Electronic Computers.

Dec	Hex	ASCII (8th bit = 1)	EBCDIC	Dec	Hex	ASCII (8th bit = 1)	EBCDIC	Dec	Hex	ASCII (8th bit = 1)	EBCDIC	Dec	Hex	ASCII (8th bit = 1)	EBCDIC
128	80	NULL		160	A0	SP		192	C0	@		224	E0	'	
129	81	SOH	a	161	A1	!		193	C1	A	A	225	E1	a	
130	82	STX	b	162	A2	"	s	194	C2	B	B	226	E2	b	S
131	83	ETX	c	163	A3	#	t	195	C3	C	C	227	E3	c	T
132	84	EOT	d	164	A4	$	u	196	C4	D	D	228	E4	d	U
133	85	ENQ	e	165	A5	%	v	197	C5	E	E	229	E5	e	V
134	86	ACK	f	166	A6	&	w	198	C6	F	F	230	E6	f	W
135	87	BELL	g	167	A7	'	x	199	C7	G	G	231	E7	g	X
136	88	BS	h	168	A8	(	y	200	C8	H	H	232	E8	h	Y
137	89	HT	i	169	A9	)	z	201	C9	I	I	233	E9	i	Z
138	8A	LF		170	AA	*		202	CA	J		234	EA	j	
139	8B	VT		171	AB	+		203	CB	K		235	EB	k	
140	8C	FF		172	AC	,		204	CC	L		236	EC	l	
141	8D	CR		173	AD	–		205	CD	M		237	ED	m	
142	8E	SO		174	AE	.		206	CE	N		238	EE	n	
143	8F	S1		175	AF	/		207	CF	O		239	EF	o	
144	90	DLE		176	B0	0		208	D0	P		240	F0	p	0
145	91	DC1	j	177	B1	1		209	D1	Q	J	241	F1	q	1
146	92	DC2	k	178	B2	2		210	D2	R	K	242	F2	r	2
147	93	DC3	l	179	B3	3		211	D3	S	L	243	F3	s	3
148	94	DC4	m	180	B4	4		212	D4	T	M	244	F4	t	4
149	95	NAK	n	181	B5	5		213	D5	U	N	245	F5	u	5
150	96	SYN	o	182	B6	6		214	D6	V	O	246	F6	v	6
151	97	ETB	p	183	B7	7		215	D7	W	P	247	F7	w	7
152	98	CAN	q	184	B8	8		216	D8	X	Q	248	F8	x	8
153	99	EM	r	185	B9	9		217	D9	Y	R	249	F9	y	9
154	9A	SUB		186	BA	:		218	DA	Z		250	FA	z	
155	9B	ESC		187	BB	;		219	DB	[		251	FB	{	
156	9C	FS		188	BC	<		220	DC	\		252	FC	¦	
157	9D	GS		189	BD	=		221	DD	]		253	FD	}	
158	9E	RS		190	BE	>		222	DE	^		254	FE	~	
159	9F	US		191	BF	?		223	DF	–		255	FF	DEL	

Figure B-29 (continued)

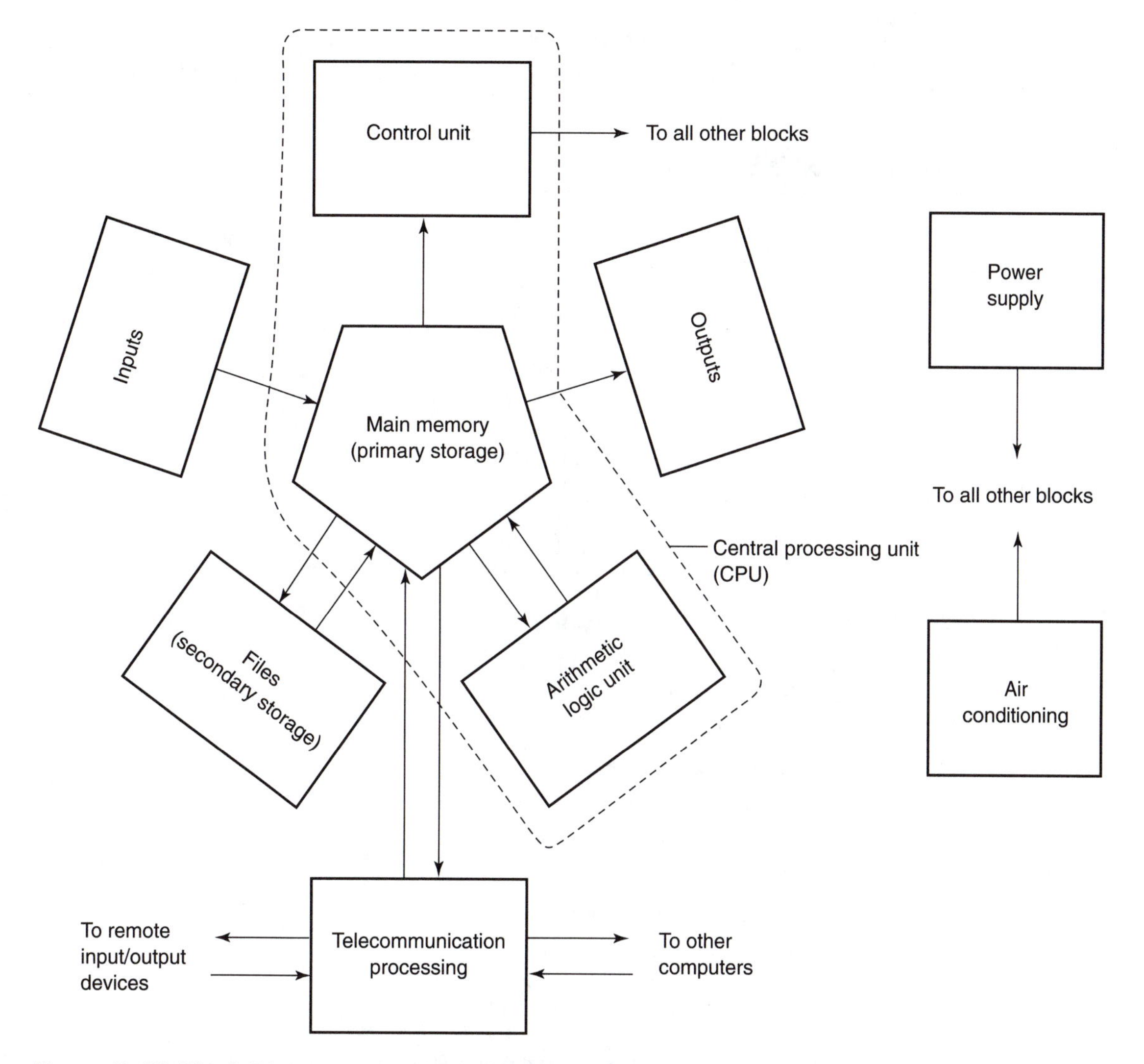

Figure B-30 Block Diagram of an Electronic Computer.

The air-conditioning block is responsible for removing heat from the computer. A computer that gets too hot or too cold will not function correctly.

REVIEW QUESTIONS

1. Given the following binary numbers, write out the binary numbers for the decimal numbers 11 through 128.

0	00000000
1	00000001
2	00000010
3	00000011
4	00000100
5	00000101
6	00000110
7	00000111
8	00001000
9	00001001
10	00001010

2. Add the following base-2 numbers:

10111	10101	11111
+ 1001	111	100
	+ 1110	+ 10

3. Add the following base-8 numbers:

	1075	1001
635	616	404
+ 25	+ 46	+ 77

4. Add the following base-16 numbers:

E862	A40	ABC
+ 78E	BF	DE
	+ F0	+ F1

5. Convert the following numbers into the indicated base:

125 (base 10) into . . . .(base 2)
10111011 (base 2) into . . . .(base 8)
10111011 (base 2) into . . . .(base 16)
3512 (base 8) into . . . .(base 16)
3512 (base 8) into . . . .(base 10)
A4C (base 16) into . . . .(base 2)

6. Change the following ASCII message from binary into English:

01001000 01101001 00100001 00100000 01000111 01110010
01100101 01100101 01100100 01100011 00100000 01100110
01110010 01101111 01101101 00100000 01000100 01110101
01100110 01100110 01111001 00100000 01100001 00100000

01010100 01010010 01010011 00101101 00111000 00110000
00100000 01100011 01101111 01101101 01110000 01110101
01110100 01100101 01110010 00101110 00100000 01001101
01100001 01111001 00100000 01100001 01101100 01101100
00100000 01111001 01101111 01110101 01110010 00100000
01100011 01101111 01101101 01110000 01110101 01110100
01100101 01110010 01110011 00100000 01100010 01100101
00100000 01110101 01110011 01100101 01110010 00100000
01100110 01110010 01101001 01100101 01101110 01100100
01101100 01111001 00101110

APPENDIX C: ROBOTICS ORGANIZATIONS

A robotics book written in the mid-1970s listed six U.S. robotics manufacturers and four European robotics manufacturers. It was a complete list for the time. By 1987, there were 400 to 600 robotics manufacturers in the U.S. alone. These can be found in the December 1987 issue of *Robotics Today.* By 1998, there were over 400 robotics-related companies listed on the Internet from all over the world and the list grows in length each day. The Robotics Industries Association, at www.robotics.org lists 118 companies in the U.S. Robot kit sources are found at www.robotstore.com. The Robotics Internet Resources Page at http://piglet.cc.umass.edu:4321/robotics.html lists many of the robotics resources on the Internet.

ROBOTICS ORGANIZATIONS

The Connecticut Robotics Society—contact: Jake Mendelssohn, 190 Mohegan Dr., W. Hartford, CT 06117, 203/233-2379.

The Dallas Personal Robotics Club—contact: Walter Bryant, 814 Mockingbird Circle, Lewisville, TX 75067.

Instrument Society of America (ISA), 67 Alexander Drive, P.O. Box 12277, Research Triangle Park, NC 27709.

National Service Robot Association—contact: Jeff Burnstein, NSRA, 900 Victors Way, P.O. Box 3724, Ann Arbor, MI 48106, 313/994-6088.

O.I.T. Robotics Club–Oregon Institute of Technology, 3201 Campus Drive, Klamath Falls, OR 97601-8801.

Robotics Society of America, 36 Newell Street, San Francisco, CA 94133. Corporate Membership, $100; Personal Membership, $25; Student Membership, $15.

Robotics Society of America, P.O. Box 54-H, Scarsdale, New York 10583, 914/633-8427. Individual Membership, $50.

Society of Manufacturing Engineers (SME), One SME Drive, P.O. Box 930, Dearborn, MI 48121-0930.

The Vancouver Robot Club, c/o Seaport Pacific Services Ltd., Suite 611, 470 Granville St., Vancouver, B.C., Canada, V6B 1V5.

APPENDIX D: ROBOT SPECIFICATIONS

The following information in this appendix is brought to you courtesy of
Advance Research & Robotics Inc.
341 Christian Street
Oxford, CT 06478
1-800-82-ROBOT

Table D-1
Kawasaki Robot Specs

Model	Axes	Payload	Horizontal Reach	Vertical Reach	Repeatability
Js-5	6 (7)*	5 (11)	1020 (40)	1250 (50)	±0.05 (±.002)
Js-10	6 (7)*	10 (22)	1475 (58)	1775 (70)	±0.1 (±.004)
Js-30	6 (7)*	30 (66)	2215 (87)	2865 (113)	±0.15 (±.006)
Js-40	6 (7)*	40 (88)	1915 (75)	2565 (101)	±0.15 (±.006)
ArcJs	6	6 (13)	1475 (58)	1775 (70)	±0.1 (±.004)
UD100	4	100 (220)	2801 (110)	2490 (98)	±0.5 (±.02)
UD150	4	150 (330)	2801 (110)	2490 (98)	±0.5 (±.02)
UT100	6 (7)*	100 (220)	3298 (130)	3733 (147)	±0.5 (±.02)
UT120	6 (7)*	120 (264)	3298 (130)	3733 (147)	±0.5 (±.02)
UT150	6 (7)*	150 (330)	3298 (130)	3733 (147)	±0.5 (±.02)
UX70	6 (7)*	70 (154)	3036 (120)	3725 (147)	±0.5 (±.02)
UX100	6 (7)*	100 (220)	2736 (108)	3425 (135)	±0.5 (±.02)
UX120	6 (7)*	120 (264)	2736 (108)	3425 (135)	±0.5 (±.02)
UX150	6 (7)*	150 (330)	2736 (108)	3425 (135)	±0.5 (±.02)
UX200	6 (7)*	200 (440)	2551 (100)	3240 (128)	±0.5 (±.02)
UZ100	6 (7)*	100 (220)	2205 (86.8)	3175 (125)	±0.3 (±.012)
UZ120	6 (7)*	120 (264)	2205 (86.8)	3175 (125)	±0.3 (±.012)
UZ150	6 (7)*	150 (330)	2205 (86.8)	3175 (125)	±0.3 (±.012)
EE-10	6 (7)*	10 (22)	2750 (108.3)	3450 (136)	±0.5 (±.02)
EH120	6	120 (264)	2396 (104)	2943 (116)	±0.5 (±.02)
*with Traversing Unit		kg (lbs)	mm (inches)	mm (inches)	mm (inches)

Kawasaki Js Series Robots

The Kawasaki Js Series robots are intended for small to medium duty, high speed, high accuracy applications within a flexible workplace. These robots handle an impressive payload with capabilities allowing them to bend over backwards. This feature enables them to work in a very large, spherical area. The Js series robot utilizes sealed construction allowing it to work under the most difficult and undesirable conditions. All wiring and piping for motors, encoders, and tooling are built into the arm and hidden inside the robot structure to eliminate application interference. Js Series robots are available for floor, ceiling or wall mounting installation.

Js 5/10 Series
Js 30/40 Series
ArcJs Series

Js 5/10 specifications
Js 30/40 specifications
ArcJs specifications

Kawasaki Home

ArcJs Series

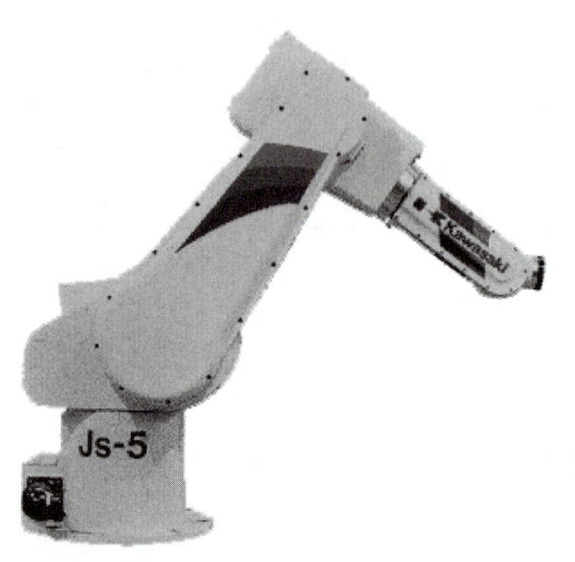

Axes: 6 standard (7 optional)

Robot Type arm: Articulated

Payload: 5-10 kg (11-22 lbs.)

Applications: Material handling, Palletizing, Machine Loading, Packaging, Dispensing, Assembly, Water Jet Cutting

Features:
- Common AD Series Controller
- Ability to Bend over Backwards
- Flexible Working Envelope

Light Payload
- High wrist torque and inertia capacity
- Maximum payload at entire work envelope at full speed

Light Payload
- Small Footprint: 400 mm sq (16 in sq)
- Slim arm construction
- No dead zone

Production Advantages
- High speed capabilities: up to 5 m/sec (16.4 ft/sec)
- Increased acceleration and deceleration
- High speed servo loop and all digital servo control

Reduced Maintenance Costs
- 5000 hour maintenance intervals
- Modular design for quick repairs
- Brushless AC servo motor
- Common spare parts
- Fittings for simple lubrication

[Js - 5/10 Specifications] [Kawasaki Home]

Specifications JS-5/10

Specifications	Type	JS-5 Base Machine
Payload		**5 kg (11)**
Axes		**6 (7)***
	JT1	320 degrees
	JT2	260 degrees
Ranges of Motion	JT3	560 degrees
	JT4	540 degrees
optional	JT5	290 degrees
	JT6	720 degrees
	JT7*	custom lengths
	JT1	220 degrees/sec. (180 degrees)**
	JT2	180 degrees/sec
Max Speed	JT3	240 degrees
	JT4	430 degrees/sec
optional	JT5	430 degrees/sec
	JT6	720 degrees/sec
	JT7*	1200 mm/sec
Wrist Rated Torque	JT4	.8km-m
	JT5	.8 kg-m
	JT6	.3 kg-m
Wrist Rated Moment of Inertia	JT4	.018 kg-m-s2
	JT5	.018 kg-m-s2
	JT6	.006 kg-m-s2
Vertical Reach (in)		1280 mm (50)
Horizontal Reach (in)		120 mm (40)
Repeatability		+/-0.05mm (+/-.002)
Weight (lbs)		85 kg (187)
Positioning feeback Drive Motors		Absolute Encoder Brushless AV servo motor
*with traversing unit **Wall mount	[Kawasaki Home]	

ArcJs Series

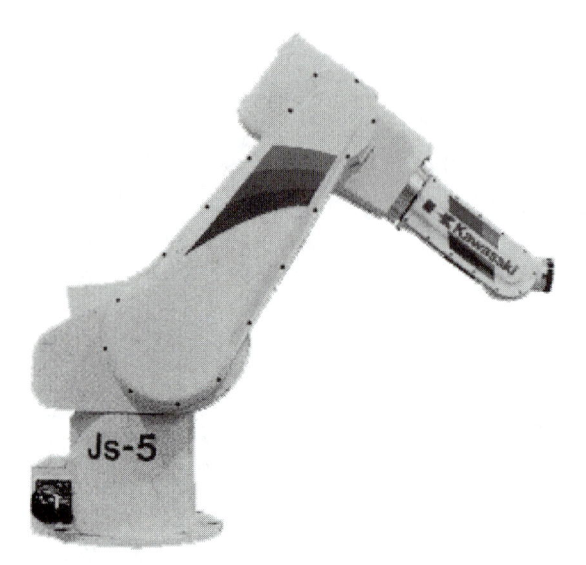

Axes: 6 standard (7 optional)

Robot Type arm: Articulated

Payload: 30-40 kg (66-88 lbs.)

Applications: Material handling, Palletizing, Dispensing, Assembly, Racking

Features:

-Common AD Series Controller

-Ability to Bend over Backwards

-Flexible Working Envelope
Medium Payload

High wrist torque and inertia capacity
Maximum payload at entire work envelope at full speed

Space Saving Design

Small Footprint: 550 mm sq (21.6 in sq)
Slim arm construction
No dead zone

Production Advantages

High speed capabilities: up to 5 m/sec (16.4 ft/sec)
Increased acceleration and deceleration rate
High speed servo loop and all digital servo control

Reduced Maintenance Costs

5000 hour maintenance intervals
Modular design for quick repairs
Brushless AC servo motor
Common spare parts
Fittings for simple lubrication

[Js - 30/40 Specifications] [Kawasaki Home]

Specifications JS-30/40

Specifications	Type	JS-5 Base Machine
Payload		**3 kg (66)**
Axes		**6 (7)***
Ranges of Motion	JT1	320 degrees
	JT2	260 degrees
	JT3	560 degrees
	JT4	540 degrees
optional	JT5	260 degrees
	JT6	720 degrees
	JT7*	custom lengths
Max Speed	JT1	160 degrees/sec.
	JT2	120 degrees/sec
	JT3	150 degrees
	JT4	240 degrees/sec
optional	JT5	340 degrees/sec
	JT6	340 degrees/sec
	JT7*	1000 mm/sec
Wrist Rated Torque	JT4	18km-m
	JT5	18 kg-m
	JT6	10 kg-m
Wrist Rated Moment of Inertia	JT4	.062 kg-m-s2
	JT5	.062 kg-m-s2
	JT6	.020 kg-m-s2
Vertical Reach (in)		2865 mm (50)
Horizontal Reach (in)		2215 mm (40)
Repeatability		+/-0.15mm (+/-.006)
Weight (lbs)		700 kg (1540)
Positioning feeback Drive Motors		Absolute Encoder Brushless AV servo motor
*with traversing unit **Wall mount	[Kawasaki Home]	

ArcJs Series

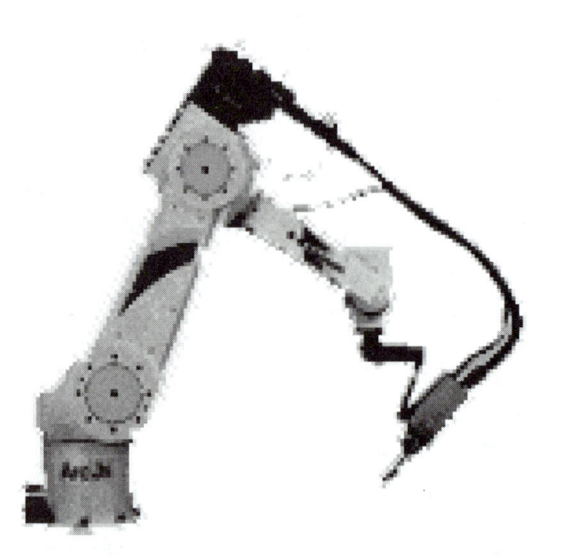

The ArcJs robot combines the JS series, with additional features designed for arc welding. The ArcJs' fast processing speed considerably reduces air-cut time, resulting in high productivity. Workers unfamiliar with robot automation, will easily be able to teach and operate the robot. It's large working envelope also provides easy continuous-circumferential welding.

Features:
Common AD Series Controller
Built-in Arc Welding Features
Flexible Working Envelope

Space Saving Design
- Small Footprint: 400 mm sq (16 in sq)
- Slim arm construction
- No dead zone

Production Advantages
- High speed capabilities: up to 1500 m/sec (in air-cut motion) up to 100 mm/sec in welding motion (4 in/sec)
- Increased acceleration and deceleration rate
- High speed servo loop and all digital servo control

Reduced Maintenance Costs
- 5000 hour maintenance intervals
- Modular design for quick repairs
- Brushless AC servo motor
- Common spare parts
- Fittings for simple lubrication

[ArcJs Specifications] [Kawasaki Home]

ArcJs Series Specifications

Specifications	Type	JS-5 Base Machine
Payload		6 kg (13)
Axes		6
Ranges of Motion	JT1	320 degrees
	JT2	260 degrees
	JT3	560 degrees
	JT4	540 degrees
optional	JT5	360 degrees
	JT6	720 degrees
	JT7*	custom lengths
Max Speed	JT1	150 degrees/sec.
	JT2	120 degrees/sec
	JT3	180 degrees/sec
	JT4	420 degrees/sec
optional	JT5	420 degrees/sec
	JT6	600 degrees/sec
	JT7*	1200 mm/sec
Wrist Rated Torque	JT4	1.2 km-m
	JT5	1.2 kg-m
	JT6	.6 kg-m
Wrist Rated Moment of Inertia	JT4	.024 kg-m-s2
	JT5	.024 kg-m-s2
	JT6	.006 kg-m-s2
Vertical Reach (in)		1775 mm (70)
Horizontal Reach (in)		1475 mm (58)
Repeatability		+/-0.1mm (+/-.004)
Weight (lbs)		140 kg (308)
Positioning feeback		Absolute Encoder
Drive Motors		Brushless AV servo motor
*with traversing unit **Wall mount	[Kawasaki Home]	

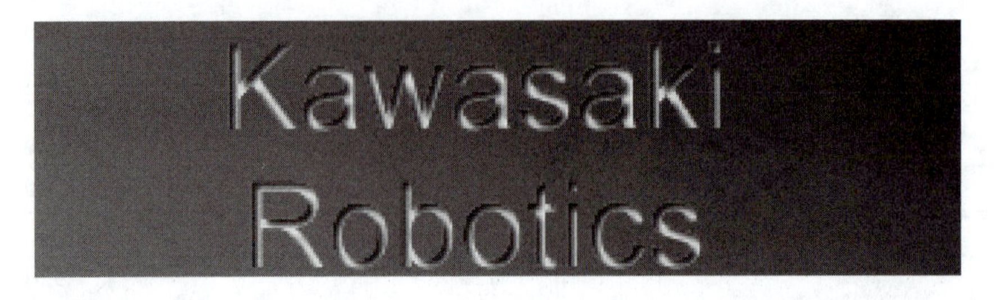

U Series Overview

The Kawasaki U Series robots are designed for a wide range of industrial applications involving high speed and high accuracy. These robots can handle a heavy payload with an exceptionally large range of motion, and in the case of the UT, even work underneath itself.

Each robot in the series has it's own set of unique capabilities including:

- The ability to bend over backwards
- A wide wrist range of motion
- The ability to work overhead or underneath itself
- A small footprint.

1

Each of the robots in the U series can be mounted on a 7th axis traversing base for even greater flexability.

U Series Applications

- Assembly
- Dispensing
- Machine Tending
- Material Handling
- Packaging
- Palletizing
- Part Transfer
- Press Tending
- Racking
- Spot Welding
- Stud Welding

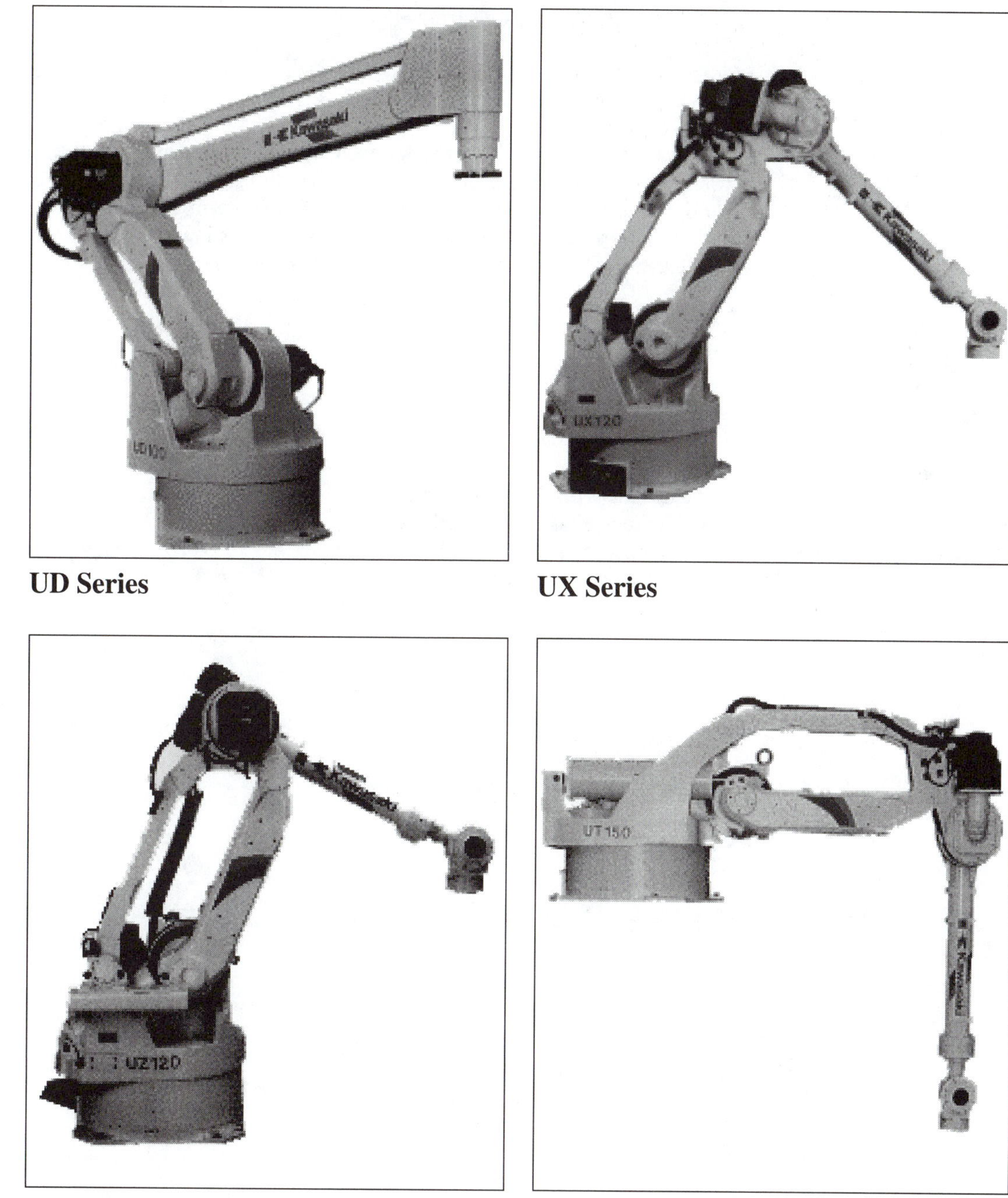

UD Series

UX Series

UZ Series

UT Series

[Kawasaki Home]

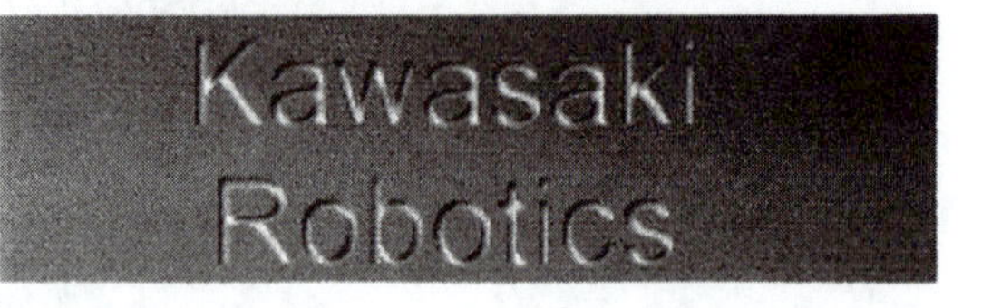

EE-10 Robot
The Kawasaki EE-10 Robot is designed for high-performance painting and sealing applications. It's lighter, stronger arm allows the EE Robot to operate at high speeds along a continuous, smooth path. In addition, the EE Robot is intrinsically safe and explosion proof and can be configured in either right or left hand versions dependent on your application.

EE-120 Robot
The Kawasaki EH Robot has a unique patent pending designenabling the arm to rotate at the top of the shoulder axis. This design eliminates rear projection, reduces side to side interference, and requires little operating space. The compact mechanical profile allows for much closer spacing of installations resulting in shorter production lines and a greater number of spot welds per station.

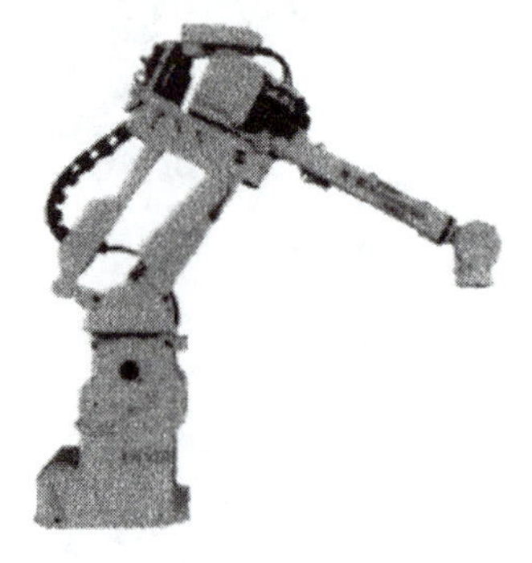

Kawasaki Robot Specifications

[AD Controller] [Js Series] [U Series] [E Series]

Home

Unimate Kawasaki Puma Contact Us Sales Service

Unimate Puma Kawasaki Sales Service Contact Us

Puma

PUMA ROBOTS GENERAL INFORMATION

- **Puma 260**
  - 6 axes
  - 16 inch reach
  - 2.2 lb payload
  - Application: assembly, material handling, machine loading, clean room (class 10) wafer handling

- **Puma 560**
  - 6 axes
  - 34 inch reach
  - 5 lb payload
  - Application: assembly, material handling, machine loading, palletizing, automated testing, clean room (class 10) applications
- **Puma 562**
  - 6 axes
  - 34 inch reach
  - 8.8 lb payload
  - Application: assembly, material handling, machine loading, palletizing, automated testing, clean room (class 10) applications
- **Puma 761**
  - 6 axes
  - 59 inch reach
  - 22 lb payload
  - Application: machine loading, palletizing, die casting, welding, forging, material handling, clean room (class 10) wafer handling, furnace loading
- **Puma 762**
  - 6 axes
  - 49 inch reach
  - 44 lb payload
  - Application: machine loading, palletizing, die casting, welding, forging, material handling, clean room (class 10) wafer handling, furnace loading

Puma Home

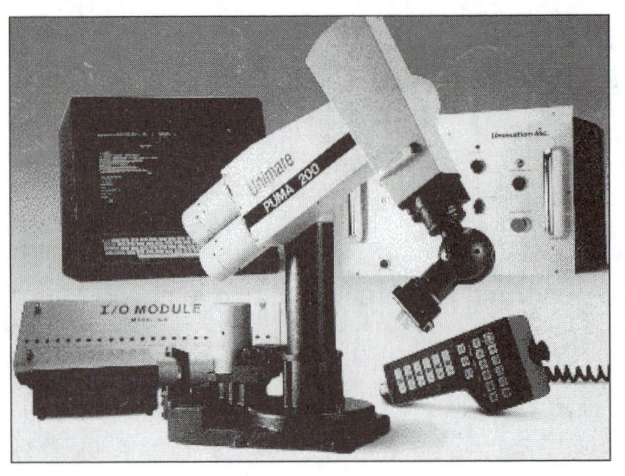

The Puma 200

A compact, computer-controlled robot for high speed, close-toleranceassembly, light materials handling, and inspection applications.

Features

The Series 200 is the most compact model in the UNIMATE PUMA line Of electricallydriven robots. With an 18 inch reach, and a 2.2 pound payload capacity,the PUMA Series 200 robot is designed for medium to high speed assemblyand materials handling. It's capabilities are particularly suited to therequirements of electronics and other industries where lightweight partshandling is highly repetitive, fast, and precise

Ease Of Use

VALtm , a revolutionary advance inrobot control systems, is used to control and program PUMA robots. The systemuses an LSI-11 as a central processing unit and communicates with individualjoint processors for servo control of robot arm motions. The results areease in set-up, high-tolerance repeatability, and greater application versatility.

VAL™ combines a sophisticated, easy-to-userobot programming capability with advanced servo control methods. IntuitiveEnglish-language instruction provides fast, efficient program generationand editing capabilities. All servo-path computations are performed in real-time, which makes it possible to interface with sensory-based systems.

Applications

With it's high speed, repeatability, and flexibility, the PUMA 200robot is suited to a wide range of small parts-handling applications, andVALtm control makes it easy to design applicationprograms to carry out the most difficult robotic tasks.

Current assembly applications include automotive instrument panels,small electric motors, printed circuit boards, subassemblies for radios,television sets, appliances and more. Other applications include packagingfunctions in the pharmaceutical, personal care, and food industries. Palletizingof small parts, inspection, and electronic-parts handling in the computer,aero-space and defense industries round out the present installed base.

Click here to see technical specificationsfor the PUMA 200.

Puma Home

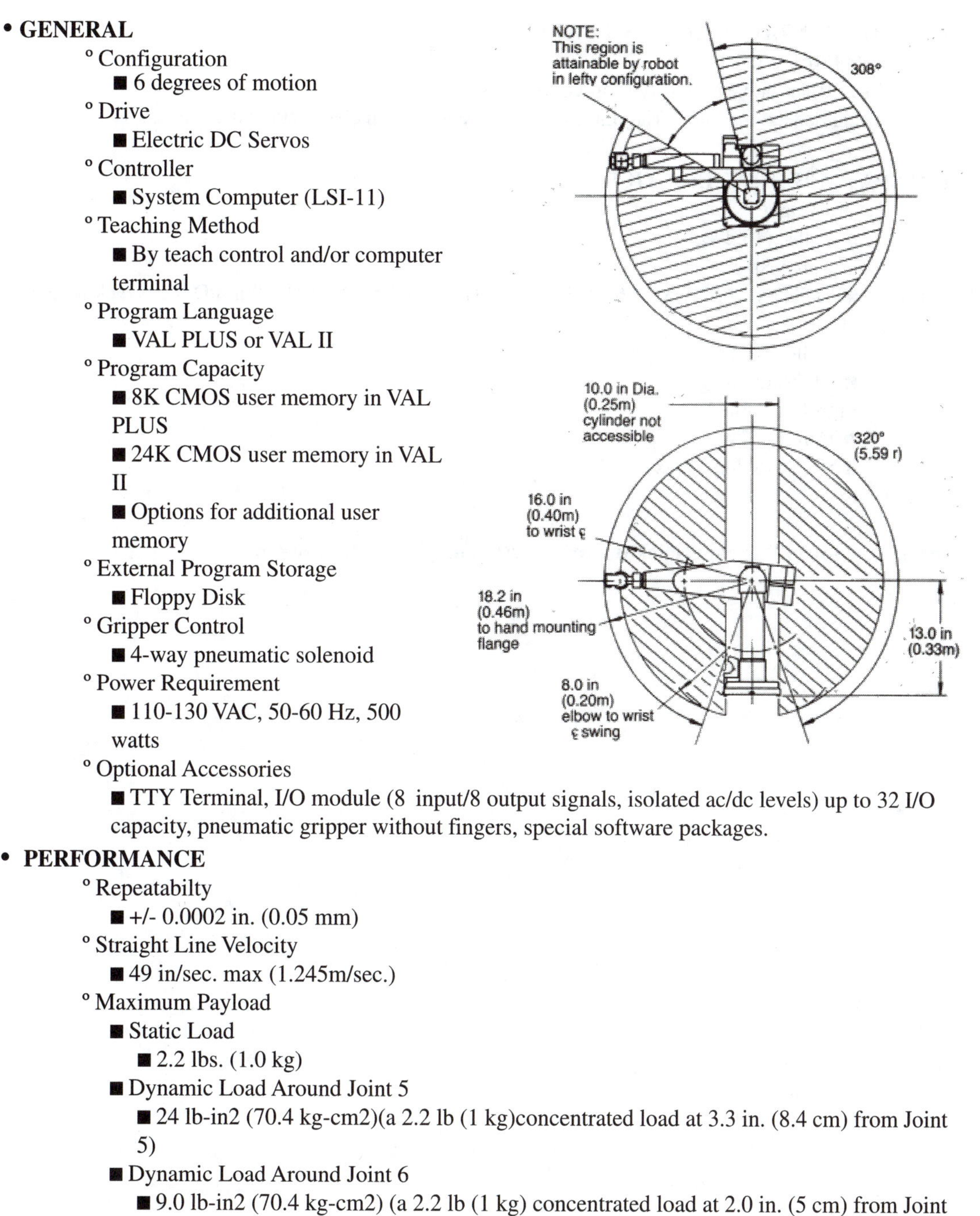

- **GENERAL**
  - ° Configuration
    - ■ 6 degrees of motion
  - ° Drive
    - ■ Electric DC Servos
  - ° Controller
    - ■ System Computer (LSI-11)
  - ° Teaching Method
    - ■ By teach control and/or computer terminal
  - ° Program Language
    - ■ VAL PLUS or VAL II
  - ° Program Capacity
    - ■ 8K CMOS user memory in VAL PLUS
    - ■ 24K CMOS user memory in VAL II
    - ■ Options for additional user memory
  - ° External Program Storage
    - ■ Floppy Disk
  - ° Gripper Control
    - ■ 4-way pneumatic solenoid
  - ° Power Requirement
    - ■ 110-130 VAC, 50-60 Hz, 500 watts
  - ° Optional Accessories
    - ■ TTY Terminal, I/O module (8 input/8 output signals, isolated ac/dc levels) up to 32 I/O capacity, pneumatic gripper without fingers, special software packages.
- **PERFORMANCE**
  - ° Repeatabilty
    - ■ +/- 0.0002 in. (0.05 mm)
  - ° Straight Line Velocity
    - ■ 49 in/sec. max (1.245m/sec.)
  - ° Maximum Payload
    - ■ Static Load
      - ■ 2.2 lbs. (1.0 kg)
    - ■ Dynamic Load Around Joint 5
      - ■ 24 lb-in2 (70.4 kg-cm2)(a 2.2 lb (1 kg)concentrated load at 3.3 in. (8.4 cm) from Joint 5)
    - ■ Dynamic Load Around Joint 6
      - ■ 9.0 lb-in2 (70.4 kg-cm2) (a 2.2 lb (1 kg) concentrated load at 2.0 in. (5 cm) from Joint 6)

- **ENVIRONMENTAL OPERATING RANGE**
  - 50-120°F (10-50°C)
  - 10-80% relative humidity (non-condensing)
  - Shielded against industrial line fluctuations and human electrostatic discharge

- **PHYSICAL CHARACTERISTICS**
  - Arm Weight
    - 29 lbs. (13.2 kg)
  - Controller Size
    - 12.5"H x 17.5"W x 19.6"D (317.5mm H x 444.5mm W x 500.0mm D) (19" rack mount able)
  - Controller Weight
    - 80 lbs. (36.4 kg)
  - Controller Cable Length
    - 15 ft. (4.57m) standard
    - 50 ft. (15.24m) max.

For a hard copy of this page, click the print button on your web browser.

Puma Home

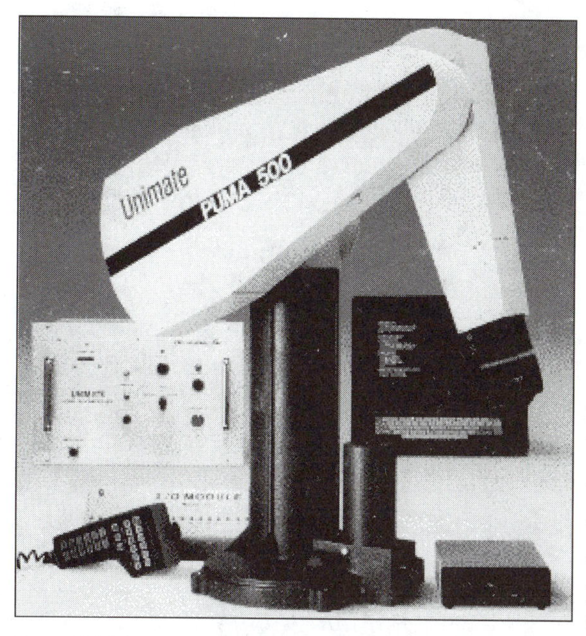

The Puma 500

A compact, computer-controlled robot for medium-to-lightweight assembly, welding, materials handling, packaging, and inspection applications.

Features

The Series 500 is the most widely used model in the UNIMATE PUMA line of electrically driven robots. With a 36 inch reach, and 5 pound payload capacity, the PUMA Series 500 robot is designed for assembly and applications requiring high degrees of flexibility and reliability.

With thousands of units in the field, its capabilities are particularly suited to the electronics and other industries where light-to-medium weight parts handling or processed functions are carried out.

Ease Of Use

VALtm, a revolutionary advance in robot control systems, is used to control and program PUMA robots. The system uses an LSI-11 as a central processing unit and communicates with individual joint processors for servo control of robot arm motions. The results are ease in set-up, high-tolerance repeatability, and greater application versatility.

VALtm combines a sophisticated, easy-to-use robot programming capability with advanced servo control methods. Intuitive English-language instruction provides fast, efficient program generation and editing capabilities. All servo-path computations are performed in real time, which makes it possible to interface with sensory-based systems.

Applications

With it's high speed, repeatability, and flexibility, the PUMA 500 robot is suited to a wide range of small parts-handling applications, and VALtm control makes it easy to design application programs to carry out the most difficult robotic tasks.

Current assembly applications include automotive instrument panels, small electric motors, printed circuit boards, subassemblies for radios, television sets, appliances and more. Other applications include packaging functions in the pharmaceutical, personal care, and food industries. Palletizing of small parts, inspection, and electronic-parts handling in the computer, aero-space and defense industries round out the present installed base.

Click here to see technical specifications for the PUMA 500.

Puma Home

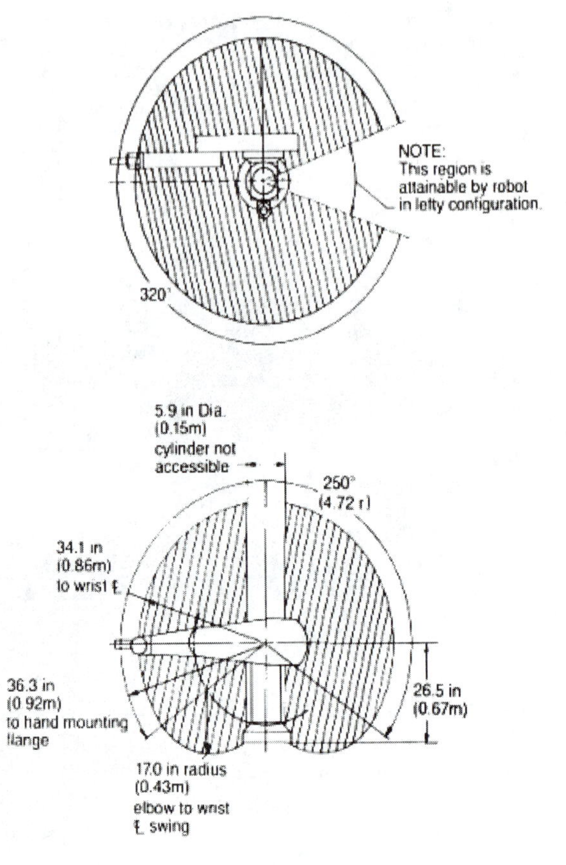

- **GENERAL**
  - Configuration
    - Up to 6 degrees of motion
  - Drive
    - Electric DC Servos
  - Controller
    - System Computer (LSI-11)
  - Teaching Method
    - By teach control and/or computer terminal
  - Program Language
    - VAL PLUS or VAL II
  - Program Capacity
    - 8K CMOS user memory in VAL PLUS
    - 24K CMOS user memory in VAL II
    - Options for additional user memory
  - External Program Storage
    - Floppy Disk
  - Gripper Control
    - 4-way pneumatic solenoid
  - Power Requirement
    - 110-130 VAC, 50-60 Hz, 500 watts
  - Optional Accessories
    - CRT or TTY Terminal, I/O module (8 input/8 output signals, isolated ac/dc levels) up to 32 I/O capacity pneumatic gripper without fingers, special software packages.

- **PERFORMANCE**
  - Repeatabilty
    - +/- 0.004 in. (0.1 mm)
  - Straight Line Velocity
    - 49 in/sec. max (1.245m/sec.)
  - Maximum Payload
    - Static Load
      - 2.2 lbs. (1.0 kg)
    - Dynamic Load Around Joint 5
      - 137.5 lb-in2 (403.2 kg-cm2)(a 5.5 lb (2.5 kg)concentrated load at 5 in. (12.7 cm) from Joint 5)
    - Dynamic Load Around Joint 6
      - 12.4 lb-in2 (35.3 kg-cm2) (a 5.5 lb (2.5 kg) concentrated load at 1.55 in. (3.76 cm) from Joint 6)

- **ENVIRONMENTAL OPERATING RANGE**
  - 50-120°F (10-50°C)
  - 10-80% relative humidity (non-condensing)
    - Shielded against industrial line fluctuations and human electrostatic discharge

- **PHYSICAL CHARACTERISTICS**
  - Arm Weight
    - 29 lbs. (13.2 kg)
  - Controller Size
    - 12.5"H x 17.5"W x 19.6"D (317.5mm H x 444.5mm W x 500.0mm D) (19" rack mountable)
  - Controller Weight
    - 80 lbs. (36.4 kg)
  - Controller Cable Length
    - 15 ft. (4.57m) standard
  - 50 ft. (15.24m) max.

For a hard copy of this page, click the print button on your web browser

Puma Home

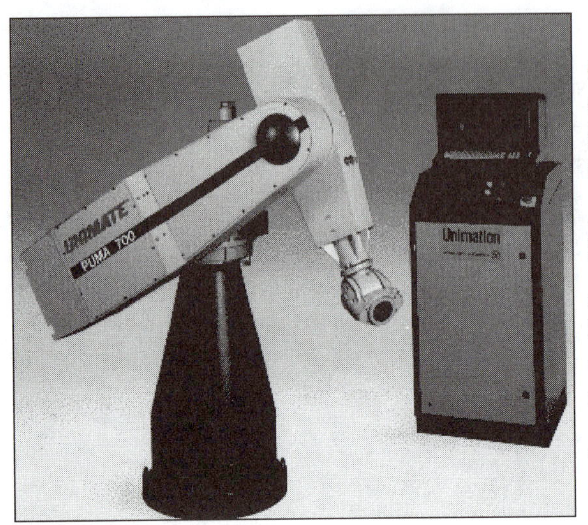

The Puma 700

Compact, long reach, medium payload capacity electric robots, for arc welding, sealant dispensing, material handling, machine loading, inspection, testing, joining and assembly.

Features

The UNIMATE Series 700 are the largest and most powerful electric robots in the PUMA line. Extensive application engineering, design experience and improved mnufacturing have resulted in a powerful new robot line.

The advanced design and versatility of the robots offers new opportunities for productivity and quality improvement in a wide range of manufacturing operations.

PUMA Series 700 are six-axis, revolute arm industrial robots.

Features include:

- A compact, ergonomically designed control cabinet which permits easy unstrained access to all system controls and status indicators.
- Industrial grade CRT with full travel keys and floppy disk drive protected to minimize exposure to environmental contamination.
- Single-piece cast inner link, outer link and base provide precise, stiff motion.
- Flexible lip seals on all rotating joints and cover plates are designed to seal the arm to IP54/NEMA 12 specifications and maximize protection for longer robot life.
- Adjustable hardware and software stops on Joints 1 and 2 can be used to selectively restrict robot arm motion.

Click here to see technical specifications for the PUMA 700.

Puma Home

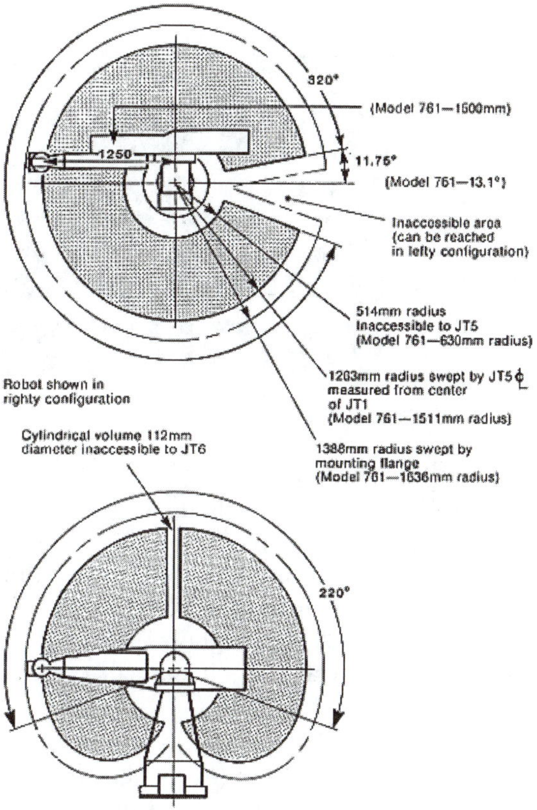

GENERAL
Configuration
Up to 6 degrees of motion
Drive
Electric DC Servos
Controller
System Computer (LSI-11), with CRT
Teaching Method
By teach control and/or computer terminal
Program Language
VAL II
Program Capacity
24KW CMOS user memory in VAL II
External Program Storage
Floppy Disk
I/O Module
32 input/32 output AC/DC user selectable voltage range, plus control I/O's
Gripper Control
4-way pneumatic solenoid
Power Requirement
240-380/415/480 Volts, 3-phase, 4500 watts (peak)
ENVIRONMENTAL OPERATING RANGE
50-120°F (10-50°C)
95% relative humidity (non-condensing)
Shielded against momentary industrial line fluctuations of -15%, +200% of nominal voltage and up to 10KV electrostatic discharge. All covers and rotary seals have gaskets to protect against water spray and dust. Designed to comply with IP54 and NEMA 12 International Electrical Packaging Specifications.

PHYSICAL CHARACTERISTICS

Model 761 Model 762 --
Arm Weight 1276 lbs (580 kg) 1298 lbs (590 kg) Base Diameter 23.6 in. (0.6m)
23.6 in. (0.6m) Control Cabinet Size (both) 45.6 in.H x 23.6 in.W x 31.5 in.D (1160mm x 600mm x 800mm) Weight 440 lbs. (200kg)

Performance

	Model 761	Model 762
Repeatability (both)	+/- 0.008 in. (0.2mm)	
Maximum Payload		
Static Load	22 lbs. (10 kg)	44 lbs (20 kg)

APPENDIX E: PC BOARD INSPECTION USING A STAND ALONE VISION SYSTEM

This lab uses the Intelledex–IntelleVue 200 Vision System as a stand alone vision system. That is, the vision system will be run without a robot or robot controller. It will show some of the capabilities and limitations of this vision system.

The IntelleVue 200 Vision System is a front-lighted vision system with up to 255 shades of gray. The brighter the light, the greater the number of gray shades that can be distinguished.

Objective

1. Learn to connect the vision system.
2. Learn the simple use of the vision commands: VACTUAL, VDIG, VSIMAGE, VSNAP, and VSUBTRACT.
3. Learn about some of the capabilities and limitations of the vision systems, through a simulated inspection task.

Equipment required

1–IBM PC or comparable computer with serial port (host)

1–Vision Computer (Intelledex Vision processor)

1–Camera (Hitachi)

1–Tripod or other means for mounting the camera in a stationary manner

1–Monitor (Hitachi)

1–Monitor cable

1–RS-232 25-pin cable

1–Camera cable—9-pin to 19-pin

1–Computer disk with IBMHOST Intelledex program on it

1–Set of 4 PC boards labeled 1-a, 1-b, 1-c, and 1-d

1–Set of 2 PC boards labeled 2-a and 2-b

Procedure

1. Find a well-lighted, large place to set up the vision system according to Figure E–1. The front edge of the camera body should be mounted about 3 feet 2 inches away from the surface holding the target. The target holder is a 13⁄16-inch-thick board with two plastic straps to register the PC boards against. Use the camera without any extension tubes, since this will get the widest angle of view.
2. Connect the vision system according to Figure E–2. The camera cable connects to the camera 0 connector on the back of the vision computer. The serial cable connects to the SI01 connector on the back of the vision computer. The monitor cable connects to the monitor connector on the back of the vision system. Switch S3 on the back of the vision computer to *on,* in order to use the vision system without the robot controller. Switches S1, S2, and S3 should be off.
3. a. Turn on the monitor and let it warm up.
   b. Turn on the IBM PC and let it initialize. When it is ready, load and execute the IBMHOST software by switching to the disk drive and directory that it is in and typing IBMHOST; then press the enter key. You will get the following message on the screen of the PC:

```
Host Program V7.04 12-08-84
```

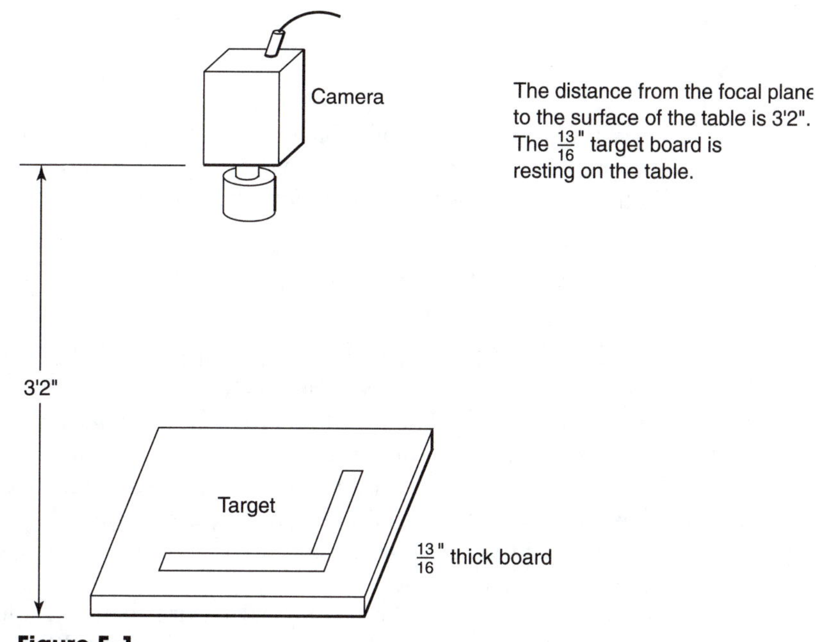

Figure E–1

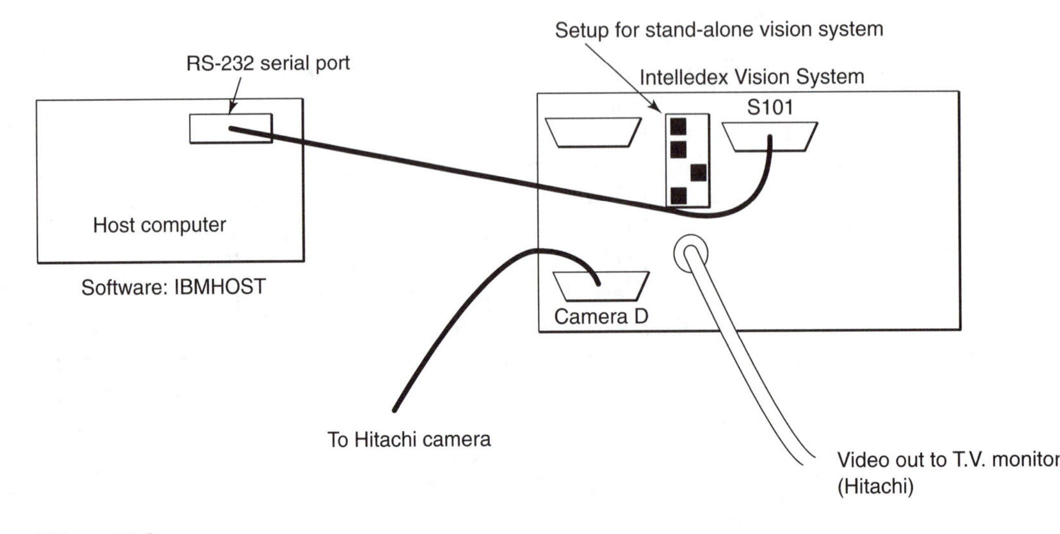

Figure E-2

and the computer will hang up.

c. Turn on the Vision computer. You will get the following message on the PC:

```
<Microsoft BASIC Version 5.2
[MS-DOS Version]
Copyright 1977-1983 (C) by Microsoft
62209 Bytes free
OK
```

and there should be a blinking cursor waiting for you to enter a command. Remove the lense cap from the camera. The monitor should show a live picture. If not, enter the following commands on the IBM PC.

```
V200 <CR>
VCAMERA 0 <CR>
VACTUAL <CR>
```

Note: <CR> stands for pressing the return or enter key. The vision system should now be in a real-time mode displaying what the camera is currently viewing. Focus on the target board, without a PC board on it. Adjust the focus (distance) and brightness settings of the camera to 3.6 feet and f-stop 4 (2⁄3 full), respectively. Adjust the monitor for the best picture of the wood grain in the target board. Now readjust the camera for the best (highest contrast) picture of the wood grain. This may take some practice.

4. Place PC board 1-a on the target board, with the protruding connector (next to U8 label), fitting into the slot in target jig. PC 1-a has all the

parts mounted on it. It will become the standard that the system will do its testing against. Type the following command:

```
VACTUAL <CR>
```

This should display the PC board on the monitor in a live mode. You should adjust the camera for best focus and high contrast. Then type the following commands:

```
VSNAP <CR>
VDIG <CR>
VSIMAGE <CR>
```

The VSNAP command saves a real-time image in display RAM. The VDIG command or VDIGITIZED writes the display RAM to the monitor screen. The VSIMAGE writes the display RAM to a selected image buffer. This saves the image for later comparing. That is, you have now saved your standard image, which you will be testing against later.

5. Place the PC board 1-b on the target jig, oriented the same as the 1-a board. This board is missing the 7805 voltage regulator. Type the following commands:

```
VSNAP <CR>
VSUBTRACT 32 <CR>
```

The VSNAP command moves the new image in to display RAM. The VSUBTRACT 32 command subtracts the image buffer (where we stored our standard image) from the display RAM. The 32 is a base value for the subtraction that is in the middle of the allowed range of 0 to 63.

The monitor should now show the differences between the saved standard image and the present image. The differences show up as white pixels. Slight misalignment of parts on the two boards are shown as white lines. The flat black body of the 7805 voltage regulator IC becomes a white rectangle when the part is missing on the second PC board. Do you see this white rectangle? ______

6. Now replace the PC board with board 1-c. This board is missing the piezo speaker. Type the following commands in order to compare this PC board to the standard:

```
VSNAP <CR>
VSUBTRACT 32 <CR>
```

The monitor should now show the difference between the saved standard image and the present image. You will probably see a white ring showing the outline of where the piezo speaker should be, along with the circuit pads under the piezo speaker. To us humans, the missing piezo speaker is quite obvious. We see high contrast between the blank spot on the board and the shiny piezo speaker. It doesn't show up that well to the vision processor. This is a possible problem.

7. Now replace the PC board with board 1-d. This board is missing both the voltage regulator and the piezo speaker. Type the following commands in order to compare this PC board to the standard.

```
VSNAP <CR>
VSUBTRACT 32 <CR>
```

The monitor should now show the white rectangle for the 7805 and the white ring for the piezo speaker. Once again, the flat black body of the 7805 shows up better in the comparison.

Can you think of some ideas for making the piezo speaker easier for the vision system to see? ________________________________

____

____

8. Now replace the PC board with board 2-a. This board has a piezo speaker that is painted flat black. Then type the following commands:

```
VSNAP <CR>
VDIG <CR>
VSIMAGE <CR>
```

This will save this board image as the new standard image.

9. Now replace the PC board with board 2-b. This board is missing the piezo speaker. This should show a solid white circle (minus the printed circuit pads) where the missing piezo speaker should have been located. The vision system finds the flat black piezo speaker much more noticeable than the shiny piezo speaker. This is an example of making a change to a part so that an automatic system can better handle it.
10. Write a summary of what you did in this experiment.

11. Write a conclusion about what you thought of this experiment, what you learned, what you would have liked to do that wasn't included in this experiment, and so forth.

GLOSSARY

ABSOLUTE SHAFT ENCODER. A sensor that divides the rotation angle of a shaft into fractions of a degree and then reads out the present angle in some binary code.

ACCELERATION. The rate of change in the velocity of a part or device.

ACCURACY. Proper positioning, as measured by the difference between the position a robot is commanded to reach and the position it actually reaches.

ACTUATOR. A motor or other device for converting electrical, hydraulic, pneumatic, or other energy into motion for a robot.

AGV. *See* AUTOMATED GUIDED VEHICLE.

AI. *See* ARTIFICIAL INTELLIGENCE.

AL. Arm Language—a robotic research language developed at Stanford University that can coordinate vision and hand movements.

ALPHA-LEVEL INTELLIGENCE. *See* BUILT-IN INTELLIGENCE.

AML. A high-level programming language developed by IBM for robot applications.

ANALOG SIGNAL. A signal that can have any value over a continuous range.

ANALOG TO DIGITAL. The process of converting an analog signal to a digital signal.

ANDROID. A robot that has the approximate appearance of a human being.

ANIMAL. A computer game that builds on previous knowledge and thus involves beta-level intelligence.

ANTHROPOMORPHIC ROBOT. *See* JOINTED-ARM ROBOT.

ARM. *See* MANIPULATOR.

ARTICULATE ROBOT. *See* JOINTED-ARM ROBOT.

ARTIFICIAL INTELLIGENCE. The operational basis for a machine, computer, or mechanism to perform some function normally associated with human intelligence.

ASCII. American Standard Code for Information Interchange—a 7-bit code used to represent 128 characters for communications control of computers.

ASRS. *See* AUTOMATED STORAGE AND RETRIEVAL SYSTEM.

ASSEMBLY LANGUAGE. A medium-level computer language that acts as an intermediary between a human language (such as English) and a machine language.

AUTOMATED FACTORY. A factory that produces goods without the help of human labor.

AUTOMATED GUIDED VEHICLE. A computer-controlled vehicle that operates without human help.

AUTOMATED STORAGE AND RETRIEVAL SYSTEM. A system for storing and retrieving parts from a warehouse under computer control.

AUTOMATION. The replacing of human labor by machine labor.

AUTOMATON. A working or moving scale model of a creature.

AXIS. A degree of freedom or basic motion about a point, along a straight line, or through an angle of rotation.

BANG-BANG ROBOT. A pneumatic-powered pick-and-place robot that makes noise as it runs into the mechanical stops that determine its extreme positions.

BASE. The foundation for a stationary robot.

BATCH PROCESSING. Saving up tasks, processing, or calculations and then working on them in a group at one's leisure.

BCD. Binary Coded Decimal—a system that represents decimal numbers in a computer by assigning 4 binary bits for each decimal digit.

BETA-LEVEL INTELLIGENCE. Intelligence that can learn from its experiences or build on previous knowledge.

BUILT-IN INTELLIGENCE. Nonlearning and probably nonlearned intelligence of the type placed in the read only memory of a computer.

CAD. Computer Aided Design—a method that uses the electronic computer to simplify and speed up the design process in various areas of research and development.

CAE. Computer Aided Engineering—a method that is basically the same as CAD.

CAM. Computer Aided Manufacturing—a method that uses the electronic computer at certain manufacturing steps to improve quality and lower costs.

CAR. Computer Aided Robotics—a method that uses simulations of robots and robot systems in designing robots, training robot operators, and programming robots off-line.

CARTESIAN COORDINATES. The system of representing a point in space by its three linear coordinate values on the *x*-axis, *y*-axis, and *z*-axis.

CARTESIAN-COORDINATES ROBOT. A robot whose manipulator moves along all three of its axes only in straight lines. This robot has a rectangular work cell, which accounts for its secondary name of rectangular-coordinates robot.

CELL. The work area of a robot; also called a work cell or work envelope. It includes all the space that the robot can reach.

CENTER OF GRAVITY. The point at which a robot's (or other object's) mass is centered.

CLOSED-LOOP SERVO-CONTROLLED SYSTEM. A control system that uses the output of the system as part of the input to the control system, in order to correct any errors in carrying out the commands.

COEFFICIENT OF FRICTION. A measure of how efficiently a gripper holds a part.

COMPILER. A program that takes statements written in a high-level computer language and translates them into the machine language used by the computer or robot.

COMPLIANCE SURFACE. A padded or soft surface used to adjust for minor mispositioning of a robot's gripper.

CONTROLLER. The brain of the robot.

CONTROL UNIT. *See* CONTROLLER.

CYBERPHOBIA. The unreasonable fear of computers. As recently as 1983, 50 percent of the adults of the United States suffered from cyberphobia.

CYBORG. Short for *cybernetic organism*—a human being with artificial limbs or organs.

CYLINDRICAL-COORDINATES ROBOT. A robot whose manipulator can rotate about its base and can move linearly in horizontal and vertical planes. The robot's work cell is cylindrical.

DATA HIGHWAYS. A local area network for exchanging electronic information or data with other devices.

DEGREES OF FREEDOM. The number of joints or axes of motion an industrial robot possesses.

DIFFERENTIAL DRIVER. A data transmission technique that sends a signal over two wires; the value of the signal is the difference in the values of the signal over the two wires.

DIGITAL SIGNAL. A signal that can be varied only in steps and not over a continuous range.

DIRECT READING ENCODER. A device that gives the present degrees of rotation directly; also called an absolute shaft encoder.

DOMESTIC ROBOT. A robot that performs tasks in the home.

EDUCATIONAL ROBOT. Any device that can be used to teach the principles of robotics.

ELBOW. The first joint out from the shoulder on a robot's (or a human's) arm.

ELECTRICAL NOISE. The noise generated by the opening and closing of electrical contacts and by electronic circuits.

ELECTRICAL POWER. Power supplied by the movement of electrons over a wire.

ELIZA. A program written by Joseph Weizenbaun to meet Turing's definition of artificial intelligence; it uses alpha-level intelligence.

EMERGENCY MAINTENANCE. Maintenance that must be performed at once, when a machine breaks down during normal production time and must be fixed before production can continue.

END EFFECTOR. A device, tool, gripper, or hand located at the end of a manipulator and used to perform work or movement on some object.

END-OF-ARM TOOLING. *See* END EFFECTOR.

ENVELOPE. *See* CELL.

EXPERT SYSTEM. Condensed human intelligence on some subject area. A human is led through the system by answering a series of questions asked by the system.

EXTERNAL SENSOR. A device that furnishes a robot with information about the outside world.

FEEDBACK. Sensory information about how well a task is being done.

FIBER OPTICS. The process of sending information down a special circuit line as light waves. Fiber optic circuits are not affected by electrical noise.

FINGER. The jaws or digits of a gripper that are used to handle a tool or part.

FIRST-GENERATION INDUSTRIAL ROBOT. A robot from the design period when robots were of a fixed-sequence type.

FLEXIBLE AUTOMATION. The use of reprogrammable, general-purpose machines, such as the industrial robot, to facilitate introducing changes in the manufacturing process.

FOUR DS OF ROBOTICS APPLICATIONS. Dirty, dangerous, dull, and difficult. A job that consists largely of work that involves one or more of these characteristics may be appropriate for a robot to do in place of a human.

GAMMA-LEVEL INTELLIGENCE. The ability to make generalizations from past experiences.

GENERATION OF ROBOTS. A group of robots that are classified together because they were produced during a particular design period.

GOVERNOR. A mechanical or electronic device for controlling the speed of another device.

GRAY CODE. A binary numbering system that changes only 1 bit as the value is increased or decreased by 1. It is used in absolute position encoders to ensure that the position can be read out with an accuracy of ± 1 bit.

GRIPPER. A specialized attachment for handling parts, tools, materials, or special instruments.

HALL EFFECT. An effect caused by rotating a magnetic field, producing a current change around a semiconductor device carrying an electric current.

HAND. *See* END EFFECTOR.

HAPTIC PERCEPTION. The ability to sense information from the joints and muscles; also called *kinesthesia.*

HARD AUTOMATION. The use of specialized tools and machinery (*not* robots) to perform specific operations very efficiently.

HELP. A high-level programming language developed for use with General Electric's Allegro assembly robots.

HER. Hexapawn Educable Robot—a robot that demonstrates beta-level intelligence in learning from its mistakes at the game of Hexapawn.

HEURISTICS. The use of generalizations and rules of thumb.

HIGH-LEVEL LANGUAGE. A computer programming language that is closer to the normal language of the person programming the computer than it is to machine language.

HIGH-TECHNOLOGY ROBOT. A continuous-path control unit robot.

HOBBYIST ROBOT. Any device that is programmed to produce automatic motion. Many hobbyist robots are remote-controlled devices.

HOME. A reference point from which all the robot's movements are measured.

HUMANOID. *See* ANDROID.

HYDRAULIC POWER. Power produced through the use of a noncompressible fluid such as hydraulic oil.

INCREMENTAL SHAFT ENCODER. A sensor that emits a pulse for each increment of the shaft rotation.

INDUSTRIAL ROBOT. A reprogrammable, multifunctional manipulator designed to move materials, parts, tools, or specialized devices through variable programmed motions, in order to perform various tasks.

INHERITED INTELLIGENCE. *See* BUILT-IN INTELLIGENCE.

INPUT DEVICE. A teaching pendant, control panel, or sensor device that can be used to gather or enter information for the robot to use in performing its duties.

INPUT SIGNAL. Any information received by a robot.

INTELLIGENCE. The ability to learn from experience and to adapt to a changing environment.

INTELLIGENT TUTORING SYSTEM. A system that can tell a person not only when an answer is wrong, but how the person's reasoning went astray.

INTERNAL SENSOR. A device that furnishes a robot with information about itself.

JOINT. *See* AXIS.

JOINTED-ARM ROBOT. A robot whose manipulator has a bend or joint in it so that it resembles the human arm. This robot has the largest work cell of any robot type for a given amount of floor space occupied.

KINESTHESIA. The sensing of information from muscles, tendons, and joints, in order to know their present positions. *See also* HAPTIC PERCEPTION.

LAN. Local Area Network—a system of techniques used to interconnect local computers and terminals.

LAWS OF ROBOTICS. Principles formulated by Isaac Asimov to govern relations between humans and robots.

LEAD SCREW. A device used to detect the position of a shaft by rotating on precision antibacklash gears.

LERT. Linear, Extensional, Rotational, and Twist—a robot classification system based on the type of motion produced by each robot axis.

LEVEL CONVERTER. A device used to convert RS-232 voltages to and from integrated circuit voltages of from 0 to +5 volts.

LIMIT SWITCH. A device that tells the robot that it is at the end of travel for some axis.

LOW-TECHNOLOGY ROBOT. A pick-and-place robot that receives no positional feedback information.

LUBRICATION REQUIREMENTS. A factory-supplied list of how frequently it is necessary to lubricate various parts and what types of lubricants should be used.

MACHINE LANGUAGE. The native language of a computer; that is, the language in which the computer actually executes instructions. An electronic computer's machine language consists of a series of ones and zeros known as binary.

MAINTENANCE SCHEDULE. The manufacturer-supplied list of when to do preventive and other maintenance.

MANDREL LIFTING DEVICE. A pneumatic gripping device that grips a part by inflating inside it.

MANIPULATOR. A device for producing motion, generally in order to move or turn something.

MAP. Manufacturing Automation Protocol—a communications standard developed for General Motors.

MCL. Manufacturing Control Language—a high-level robot application programming language developed by McDonnell Douglas.

MECHANICAL STOP. A device that physically stops the motion of an axis of a robot.

MEDICAL ROBOT. Any robotlike devices that either help with giving medical aid or substitute for or restore functions for a disabled person.

MEDIUM-TECHNOLOGY ROBOT. A point-to-point control unit robot.

MILITARY ROBOT. Any automatic system that produces motion, such as remote-controlled tanks, ships, planes, and rockets.

MODEM. A device used to convert voltage levels into tones that can be transmitted over a telephone line; short for modulator/demodulator.

NATURAL LANGUAGE UNDERSTANDING. An area of artificial intelligence research dedicated to teaching computers to understand human languages.

NONSERVO-CONTROLLED SYSTEMS. *See* OPEN-LOOP CONTROL SYSTEM.

NUMERICAL CONTROL. Control over an automated machine provided by a digital computer.

1.5-GENERATION INDUSTRIAL ROBOT. A robot produced during the design period when robots had some rudimentary sensory-controlled actions.

OPEN-LOOP CONTROL SYSTEM. A control system that does not use the results of its output as part of its input information.

OUTPUT DEVICE. A device for executing, reproducing, or storing input instructions or information. The main output device for an industrial robot is the actuator that opens or closes the gripper.

PARITY CHECK BITS. Extra information added to transmitted data to make detecting signal errors easier.

PAYBACK PERIOD. The length of time in years required for a robot to pay for itself through the savings it provides. For a robot, the payback period generally should be 2½ years or less.

PERSONAL ROBOT. *See* DOMESTIC ROBOT.

PHONEMES. The smallest units of speech that distinguish one utterance from another.

PIEZOELECTRIC. A crystalline substance that produces electricity as it is distorted.

PIN LIFTING DEVICE. A pneumatic gripper that grips a part by inflating around the outside of it.

PITCH. The up-and-down movement of a human hand or robot gripper; one of the wrist's axes of motion.

PLAY ROBOT. *See* SHOW ROBOT.

PLC. Programmable Logic Controller—a device used to control many modern production lines.

PNEUMATIC FINGER. A device that grips a part by bending when compressed air is forced into it.

PNEUMATIC POWER. Power provided by a compressible fluid such as air.

POINT-TO-POINT WIRING. Running a single wire from one point in a system to another.

POLAR-COORDINATES ROBOT. A robot whose manipulator can rotate about its head and about its base, as well as moving in and out. The work cell for the robot takes the shape of the space between two hemispheres.

POTENTIOMETER. A resistance strip with a movable contact.

PREVENTIVE MAINTENANCE. Performing periodic inspections, adjustments, and minor repairs on a machine before it malfunctions.

PROMOTIONAL ROBOT. *See* SHOW ROBOT.

PROXIMITY DETECTOR. A device that detects the presence of objects, usually without contacting them.

PULSE READING SHAFT ENCODER. *See* INCREMENTAL SHAFT ENCODER.

RAIL. A high-level programming language developed by Automatix for use with robots and vision systems.

RAM. Random Access [read and write] Memory—computer memory where nonpermanent instructions for doing a task are loaded and where sensor information is stored.

RANGE FINDING. A device used to measure the distance from it to some other object.

REAL TIME. The speed at which events unfold in the real world.

RECOMMENDED SPARE PARTS LIST. The manufacturer-supplied list of spare parts for a robot.

RECTANGULAR COORDINATES ROBOT. *See* CARTESIAN-COORDINATES ROBOT.

REMOTE CENTER COMPLIANCE DEVICE. A mechanical device incorporated into the hand or gripper of a robot to aid in adjusting for small misalignments. The device has a multiaxis float mechanism that can adjust sideways or up and down against the tapered edge of a guide hole in a part to be assembled or drilled.

REMOTE CONTROL. Control over a device exercised at a distance, by mechanical, wire, or radio means.

REPEATABILITY. *See* REPEAT ACCURACY.

REPEAT ACCURACY. Proper repetition of a task, as measured by the difference between the actual point reached and the point given during training for the robot to reach from cycle to cycle in performing a task.

RESOLVER. A device that produces an output voltage proportional to the product of an input voltage and the sine of the rotor angle.

REVOLUTE COORDINATES ROBOT. *See* JOINTED-ARM ROBOT.

RIGID FLAT COIL. A type of strain gauge.

ROBOT. An automatic, general-purpose device whose primary function is to produce motion in order to accomplish some task. The six main types of robots are educational, hobbyist, industrial, medical, military, and show.

ROBOT CHOREOGRAPHY. The process of several robots working together.

ROBOTEER. A person who helps manipulate or otherwise control a show or promotional robot.

ROBOTICS. The study of robots.

ROLL. The twisting or rotational movement of a hand or gripper; one of the wrist's axes of motion.

ROM. Read Only Memory—permanent random-access memory of information placed in the computer during the manufacturing process.

RPL. Robot Programming Language—a high-level robot application programming language developed by SRI.

RS-232. A data transmission technique that uses from ±3 to ±27 volts.

SAFETY FACTOR. A fudge factor used to correct for unaccountable errors or unforeseen factors.

SCAMP. Self-Contained Autonomous Mobile with Personality—a pet robot.

SCARA. Selective Compliance Articulated Robot Arm—a robot arm configuration. The work cell of the SCARA robot looks like that of a cylindrical robot, but instead of having a linear reach axis it has a rotational one.

SCHEDULED MAINTENANCE. Periodic maintenance done to a robot to prevent major problems.

SECOND-GENERATION INDUSTRIAL ROBOT. A robot from the design period when robots possess hand-to-eye coordination control.

SELF-REPLICATING FACTORY. A factory that is able to reproduce itself, as well as producing other goods.

SELF-REPRODUCING MACHINES. Machines that can reproduce themselves as well as performing other functions.

SENSOR. A device that furnishes a robot with information about itself or about the outside world.

SENSORY INPUT INTERPRETATION. The process of understanding sensory information.

SERVO-CONTROL SYSTEM. *See* CLOSED-LOOP SERVO-CONTROLLED SYSTEM.

SHAFT ENCODER. A device that converts rotational movement into pulses or direct degree readings.

SHAPE MEMORY ALLOY. A metal alloy wire that contracts when it is heated and stretches back to its normal length when cooled.

SHIELDED WIRE. A wiring technique that runs a ground shield around each signal wire.

SHOULDER. The point at which the arm is attached to the base of a robot or human. This is the first joint of the arm.

SHOW ROBOT. A remote-controlled device used to attract people's attention; also known as a promotional or play robot.

SIGNAL ERROR DETECTION. The process of detecting bad information due to electrical noise or other environmental causes.

SPEAKER-DEPENDENT SYSTEM. A voice recognition system that must be trained by the person who will be using it.

SPEAKER-INDEPENDENT SYSTEM. A voice recognition system that works regardless of the particular speaker who is using the system.

SPEECH RECOGNITION. The ability of a robot to receive meaningful speech input.

SPEECH SYNTHESIS. The ability of a robot to produce meaningful speech output.

SPHERICAL-COORDINATES ROBOT. *See* POLAR-COORDINATES ROBOT.

STEPWISE REFINEMENT. The process of defining and expanding each subtask of a problem until that part of the problem is fully explained or solved.

STRAIN GAUGE. A device that changes its electrical characteristics as a force is applied to it.

SUBSTITUTION OF ROBOTS. The practice of replacing a broken robot with another robot, in order to keep production going.

SYNCHRO. A motorlike device with a multiwinding stator and a rotor that can transform an electrical input into an angular output or an angular position into an electrical output.

TEACHING. The process of giving a robot a series of instructions for doing a task.

TEACHING PENDANT. A device used to position a robot manually at each point in its task and cause the robot to record these points.

TELEOPERATOR. A mechanical remote-control device or (more recently) a radio-controlled device.

THIRD-GENERATION INDUSTRIAL ROBOT. A robot from the design period when robots will be able to handle discrete parts assembly.

THREE RS OF ROBOTIC SAFETY. Robots Require Respect! This is the first step in attaining safety in robotics work.

TOOL CENTER POINT (TCP). The point at the center of the flange at the end of the manipulator where the tool attaches to the manipulator.

TOP. Technical and Office Protocol—a program developed for use in office automation by Boeing Computer Services.

TOP-DOWN PROGRAMMING. A programming design technique that describes a high-level problem in terms of more elementary subtasks. This technique describes control structures first and leaves the details of how to do the tasks until later.

TRAINING. *See* TEACHING.

TROUBLESHOOTING PROCEDURES. A manufacturer-supplied list of things to do when the robot does not work correctly.

TWISTED PAIR. A wiring technique that involves twisting two signal lines together to cancel out any electrical noise pickup.

2.5-GENERATION INDUSTRIAL ROBOT. A robot from the design period when robots will have perceptual motor functions.

VAL. A high-level robot application programming language developed for Unimation's Unimate and Puma lines of robots.

VEHICLE. A mobile base that enables a robot to reach a particular location at which it is to use its manipulator.

VISION INTERPRETATION. The process of making sense out of information from a vision sensor.

VOICE PRINTS. A computer recording of how a certain person says some word.

VOICE RECOGNITION. The ability of a machine to hear and understand spoken words.

VOICE SYNTHESIS. The process by which a machine produces and speaks words.

WAVE. A high-level programming language developed by Stanford Artificial Laboratory in 1973. This robot research language was the first robot programming language.

WORK CELL. *See* CELL.

WRIST. The multi-axis joint between a forearm and a hand or end effector.

YAW. The sideways or side-to-side movement of a hand, one of the wrist's axes of motion.

BIBLIOGRAPHY

Ahl, David H. *BASIC Computer Games*. Morristown, N.J.: Creative Computing Press, 1979.

Andrew, Alan. *ABC's of Synchros & Servos*. Indianapolis: Howard W. Sams, 1962.

APT Long Range Program Staff, IIT Research Institute. *APT Part Programming*. New York: McGraw-Hill, 1967.

Asimov, Isaac. *I, Robot*. Greenwich, Conn.: Fawcett Crest Book, 1950.

Asimov, Isaac, ed. *Robot Review*, 1(3) (Spring 1988).

Augarten, Stan. *BIT by BIT: An Illustrated History of Computers*. New York: Ticknor & Fields, 1984.

Bristow, Geoff. *Electronic Speech Recognition*. New York: McGraw-Hill, 1986.

Caporali, M., and Shahinpoor, M. "Design and Construction of a Five-Fingered Robotic Hand," *Robotics Age* (February 1984): 14–20.

Chesson, Frederick. "Robot Rover, or Build Your Own Pavlovian Pooch," *Creative Computing* (June 1979): 62–64.

CNN. "Norman Mailer Robot Mail Cart." CNN News Channel (December 1987).

Craig, John J. "Anatomy of an Off-Line Programming System," *Robotics Today* (February 1985): 45–47.

Edelson, Edward. "Robots—The Next Generation," *Popular Science* (February 1985): 39–46.

Engelberger, Joseph F. *Robotics in Practice*. American Management Associations, 1980.

Fjermedal, Grant. *The Tomorrow Makers: A Brave New World of Living-brain Machines*, New York: Macmillan, 1986.

Freitas, Robert A., Jr. "Building Athens Without the Slaves," *Technology Illustrated* (August 1983): 16–20.

Fuller, John G. "Death By Robot," *Omni* (March 1984): 44–46.

Gardner, Martin. "Mathematical Games," *Scientific American* (March 1962): 138–153.

Geduld and Gottesman, eds. *Robots, Robots, Robots*. Boston: New York Graphic Society, 1978.

Heiserman, David L. *How to Build Your Own Self-Programming Robot*. Blue Ridge Summit, Penn.: TAB Books, 1979.

Jemelta, Jorg R. "Designing a Reliable Voice-Input Robot Control Language," *Robotic Age* (February 1984): 30–34.

Kent, Ernest. *The Brains of Men and Machines*. Peterborough, N.H.: BYTE/McGraw Hill, 1981.

Krutch, John. *DOCTOR—Experiments in Artificial Intelligence for Small Computers.* Indianapolis: Howard W. Sams, 1981.

Lea, Wayne A. "The Elements of Speech Recognition." In *Electronic Speech Recognition: Techniques, Technology & Applications,* ed. by Geoff Bristow. New York: McGraw-Hill, 1986.

Michie, Donald. *Trial and Error.* Penguin Science Survey, Vol. 2, 1961.

Parker, Richard. "An Expert for Every Office," *Computer Design* (Fall 1983): 37–46.

PM Magazine. "Six-year-old with Bionic Hand," *PM Magazine* (July 6, 1982).

Popoff, David. "The Robot Game: What's Your Robot's I.Q.?" *Psychology Today* (April 1969): 33–36B.

Rehg, James. *Introduction to Robotics: A System Approach.* Upper Saddle River, N.J.: Prentice-Hall, 1985.

Safford, Edward L., Jr. *The Complete Handbook of Robotics.* Blue Ridge Summit, Penn.: TAB Books, 1978.

______. *Handbook of Advanced Robotics.* Blue Ridge Summit, Penn.: TAB Books, 1982.

Schiavone, John J., and Brandeberry, James E. "Super Armatron, an Inexpensive, Microprocessor-controlled Robot," *Robotic Age* (January 1984): 20–28.

Simons, G. L. *Robot in Industry.* National Computing Centre Limited, 1980.

Skinner, B. F. "The Machine That Is Man," *Psychology Today* (April 1969): 20–25.

Sleeman and Brown. *Intelligent Tutoring Systems.* New York: Academic Press, 1982.

Solem, Johndale C. "ANDROTEXT: A High-Level Language for Personal Robots," *Robotic Age* (May 1984): 16–26.

Spaulding, Carl P. *How to Use Shaft Encoders.* Monrovia, Calif.: Datex Corporation, 1965.

Stauffer, Robert N. "Artificial Intelligence Moves into Robotics," *Robotic Age* (1987): 11–13.

Susnjara, Ken. *A Manager's Guide to Industrial Robots,* Upper Saddle River, N.J.: Prentice-Hall, 1982.

Von Blois, J. F. "Robotic Justification Considerations." In *Robots 6 Conference Proceedings,* March 2–4, 1982, Detroit, Mich., Robotics International of SME, Dearborn, Michigan, 1982.

Ward, David A. "Build This Speech Synthesizer," *Radio-Electronics* (December 1988): 80–85.

Webb, John W. *Programmable Controllers: Principles and Applications.* Columbus, Ohio: Merrill Publishing, 1988.

Winston, Patrick H., and Horn, Berthold K. P. *LISP.* Menlo Park, Calif.: Addison-Wesley, 1984.

INDEX